LIVER REGENERATION

LIVER REGENERATION

Basic Mechanisms, Relevant Models and Clinical Applications

Edited by

UDAYAN APTE, PhD DABT

Associate Professor,
Department of Pharmacology, Toxicology and Therapeutics,
The University of Kansas Medical Center,
Kansas City, KS, USA

ELSEVIER

AMSTERDAM • BOSTON • HEIDELBERG • LONDON
NEW YORK • OXFORD • PARIS • SAN DIEGO
SAN FRANCISCO • SINGAPORE • SYDNEY • TOKYO
Academic Press is an imprint of Elsevier

AP

Academic Press is an imprint of Elsevier
125 London Wall, London, EC2Y 5AS, UK
525 B Street, Suite 1800, San Diego, CA 92101–4495, USA
225 Wyman Street,Waltham,MA 02451, USA
The Boulevard, Langford Lane, Kidlington, Oxford OX5 1GB, UK

Library of Congress Cataloging-in-Publication Data
A catalog record for this book is available from the Library of Congress

British Library Cataloguing in Publication Data
A catalogue record for this book is available from the British Library

ISBN: 978-0-12-420128-6

For information on all Academic Press publications
visit our website at http://store.elsevier.com/

Publisher: Mica Haley
Acquisition Editor: Stacy Masucci
Editorial Project Manager: Shannon Stanton
Production Project Manager: Julia Haynes
Designer: Mark Rogers

Typeset by SPi Global, India
Printed in USA

Dedication

This book is dedicated to

My loving parents,
Dr. Madhu and (late) Mrs. Snigdha Apte

and

My parents-in-law,
Mr. Bhagwan and Mrs. Medha Limaye

Without their unconditional love, support, and constant encouragement this would not have been possible.

Contents

5. Extracellular Signals Involved in Liver Regeneration: Direct and Auxiliary Mitogens

UDAYAN APTE, PALLAVI B. LIMAYE, GEORGE K. MICHALOPOULOS

6. Developmental Pathways in Liver Regeneration-I

KARI NICHOLE NEJAK-BOWEN, SATDARSHAN PAL SINGH MONGA

7. Mechanisms of Termination of Liver Regeneration

CHAD WALESKY, UDAYAN APTE

8. Role of CXC Chemokines in Liver Repair and Regeneration

CHRISTOPHER M. FREEMAN, GREGORY C. WILSON, HIROYUKI NOJIMA, ALEX B. LENTSCH

9. Bile Acid Receptors and Liver Regeneration

MINGJIE FAN, LILI DING, WENDONG HUANG

10. Role of Developmental Morphogens in Liver Regeneration

MARIANA VERDELHO MACHADO, ANNA MAE DIEHL

11. Regulation of Cell Cycle During Liver Regeneration

YULIA A. NEVZOROVA, CHRISTIAN TRAUTWEIN

Contributors

Gianfranco Alpini Department of Internal Medicine, College of Medicine, Texas A&M University Health Science Center; Division of Research, Central Texas Veterans Health Care System, and BaylorScott & White Digestive Disease Research Center, BaylorScott & White, Temple, Texas, USA

Udayan Apte Department of Pharmacology, Toxicology and Therapeutics, University of Kansas Medical Center, Kansas City, Kansas, USA

Vishakha Bhave Philadelphia College of Osteopathic Medicine, School of Pharmacy, Suwanee, Georgia, USA

Alicia R. Brown Pediatric Stem Cell Research and Hepatic Disorders, Child Health Research Institute, Department of Pediatrics, University of Florida, Gainesville, Florida, USA

Jean S. Campbell Northwest Liver Research Program, and Department of Pathology, University of Washington, Seattle, Washington, USA

Daniel Cook Daniel Baugh Institute for Functional Genomics and Computational Biology, Department of Pathology, Anatomy and Cell Biology, Sidney Kimmel Medical College, Thomas Jefferson University, Philadelphia, Pennsylvania, and Department of Chemical and Biomolecular Engineering, University of Delaware, Newark, Delaware, USA

Anna Mae Diehl Division of Gastroenterology, Duke University, Durham, North Carolina, USA

Lili Ding Department of Diabetes and Metabolic Diseases Research, Beckman Research Institute, City of Hope National Medical Center, Duarte, California, USA, and Shanghai Key Laboratory of Complex Prescriptions and MOE Key Laboratory for Standardization of Chinese Medicines, Institute of Chinese Materia Medica, Shanghai University of Traditional Chinese Medicine, Shanghai, China

Andrew W. Duncan Department of Pathology, McGowan Institute for Regenerative Medicine, University of Pittsburgh, Pittsburgh, Pennsylvania, USA

Mingjie Fan Institute of Life Science, Taishan Medical University, Taian, China

Christopher M. Freeman Department of Surgery, University of Cincinnati Medical Center, Cincinnati, Ohio, USA

Shannon Glaser Department of Internal Medicine, College of Medicine, Texas A&M University Health Science Center; Division of Research, Central Texas Veterans Health Care System; BaylorScott & White Digestive Disease Research Center, BaylorScott & White, Temple, Texas, USA

Wolfram Goessling Brigham and Women's Hospital/Dana-Farber Cancer Institute, Harvard Medical School, Boston, and Harvard Stem Cell Institute and Broad Institute of MIT and Harvard, Cambridge, Massachusetts, USA

Yuyan Han Department of Internal Medicine, College of Medicine, Texas A&M University Health Science Center, Temple, Texas, USA

Jiansheng Huang Department of Pediatrics, Washington University School of Medicine, St. Louis, Missouri, USA

Wendong Huang Department of Diabetes and Metabolic Diseases Research, Beckman Research Institute, City of Hope National Medical Center, Duarte, California, USA

Alex B. Lentsch Department of Surgery, University of Cincinnati Medical Center, Cincinnati, Ohio, USA

Yanfeng Li Department of Medicine, Division of Gastroenterology and Hepatology, and Marion Bessin Liver Research Center, Albert Einstein College of Medicine, Bronx, New York, USA

Pallavi B. Limaye Xenometrics LLC, Stilwell, Kansas, USA

Joseph Locker Department of Pathology, University of Pittsburgh, Pittsburgh, Pennsylvania, USA

Mariana Verdelho Machado Division of Gastroenterology, Duke University, Durham, North Carolina, USA

Luca Maroni Clinic of Gastroenterology, Università Politecnica delle Marche, "Ospedali Riuniti" University Hospital, Ancona, Italy

Marco Marzioni Clinic of Gastroenterology, Università Politecnica delle Marche, "Ospedali Riuniti" University Hospital, Ancona, Italy

Fanyin Meng Division of Research, Central Texas Veterans Health Care System, Temple, and BaylorScott & White Digestive Disease Research Center, BaylorScott & White, Texas, USA

George K. Michalopoulos Department of Pathology, University of Pittsburgh School of Medicine, Pittsburgh, Pennsylvania, USA

Satdarshan Pal Singh Monga Department of Pathology, University of Pittsburgh, School of Medicine, Pittsburgh, Pennsylvania, USA

Kari Nichole Nejak-Bowen Department of Pathology, University of Pittsburgh, School of Medicine, Pittsburgh, Pennsylvania, USA

Yulia A. Nevzorova Department of Medicine III, University Hospital, Aachen, Germany

Hiroyuki Nojima Department of Surgery, University of Cincinnati Medical Center, Cincinnati, Ohio, USA

Bryon E. Petersen Pediatric Stem Cell Research and Hepatic Disorders, Child Health Research Institute, Department of Pediatrics, University of Florida, Gainesville, Florida, USA

Liya Pi Pediatric Stem Cell Research and Hepatic Disorders, Child Health Research Institute, Department of Pediatrics, University of Florida, Gainesville, Florida, USA

Zsuzsanna Polgar Department of Medicine, Division of Gastroenterology and Hepatology, and Marion Bessin Liver Research Center, Albert Einstein College of Medicine, Bronx, New York, USA

Michele T. Pritchard Department of Pharmacology, Toxicology and Therapeutics, University of Kansas Medical Center, Kansas City, Kansas, USA

Debolina Ray Department of Internal Medicine, College of Medicine, Texas A&M University Health Science Center, Temple, Texas USA

Kimberly J. Riehle Northwest Liver Research Program; Department of Surgery, and Department of Pathology, University of Washington, Seattle, Washington, USA

Kevin Riggle Department of Surgery, University of Washington, Seattle, Washington, USA

Paulette M. Robinson Pediatric Stem Cell Research and Hepatic Disorders, Child Health Research Institute, Department of Pediatrics, University of Florida, Gainesville, Florida, USA

Jayanta Roy-Chowdhury Department of Medicine, Division of Gastroenterology and Hepatology; Marion Bessin Liver Research Center, and Department of Genetics, Albert Einstein College of Medicine, Bronx, New York, USA

Namita Roy-Chowdhury Department of Medicine, Division of Gastroenterology and Hepatology; Marion Bessin Liver Research Center, and Department of Genetics, Albert Einstein College of Medicine, Bronx, New York, USA

David A. Rudnick Department of Pediatrics, and Department of Developmental Biology, Washington University School of Medicine, St. Louis, Missouri, USA

David A. Shafritz Marion Bessin Liver Research Center, Division of Gastroenterology and Liver Diseases, Albert Einstein College of Medicine, Bronx, NY, USA

David C. Sullivan Pediatric Stem Cell Research and Hepatic Disorders, Child Health Research Institute, Department of Pediatrics, University of Florida, Gainesville, Florida, USA

Christian Trautwein Department of Medicine III, University Hospital, Aachen, Germany

Rajanikanth Vadigepalli Daniel Baugh Institute for Functional Genomics and Computational Biology, Department of Pathology, Anatomy and Cell Biology, Sidney Kimmel Medical College, Thomas Jefferson University, Philadelphia,

Pennsylvania, and Department of Chemical and Biomolecular Engineering, University of Delaware, Newark, Delaware, USA

Chad Walesky Department of Medicine—Genetics Division, Brigham and Women's Hospital, Harvard Medical School, Boston, Massachusetts, USA

Gregory C. Wilson Department of Surgery, University of Cincinnati Medical Center, Cincinnati, Ohio, USA

Preface

Liver regeneration has always been a fascinating subject and has remained at the center of liver biology research for almost a century. Research on liver regeneration has grown exponentially in last 50 years and thousand of original research articles have been published. However, a comprehensive volume that provides a state-of-the-art understanding of this vast original research was lacking. I hope that this book fill that void. The effort here is combining historical research with some of the most cutting edge and exciting ideas in liver regeneration field in one book. My hope is that this book will be equally valuable for the beginners, the veterans, and everybody in between who are fascinated by livers amazing ability to regenerate.

The book begins by providing historical perspective on research in liver regeneration and then outlines the major methods and models used. Several chapters provide detailed analysis of various aspects of mechanisms of liver regeneration including role of cytokines, growth factors, chemokines, developmental signals, and bile acids. Additional chapters review important role of cell cycle regulators and signals provided by changes in nutrition. An interesting new aspect covered in the book is the use of computational modeling to study liver regeneration, which is sure to be centerpiece of biomedical research in the future. Role of hepatic progenitor cells and their contribution to liver regeneration is a highly debated and exciting area of research and the chapters in the final section of the book outline advances in these fields. Overall, the book provides a balanced and comprehensive view of what is an extremely vast body of original work.

I would like to sincerely thank all the authors for their contribution. This book would not have been possible without their efforts. Every one of them have made time from their busy research, clinical and teaching schedules. A big thank you to Mark Weinstein, the creator of the amazingly funny "Miserable World of Prometheus" cartoon series, for letting us use his cartoons for the book. I want to thank Stacy Masucci and Shannon Stanton of Elsevier for all their help and support for this book. Stacy came up with the idea for this book and helped me get on my feet as an editor. Shannon has been outstanding in her support for every minute detail of the book, dealing with copyright permissions and accommodating the moving deadlines. Special thanks to my family, my loving wife Pallavi and my great kids, Mihir and Mukta, who have been a source of constant encouragement and energy in this endeavor. Finally, none of this would have been possible without the tremendous original research work conducted by hundreds of outstanding scientists around the world who are as excited about liver regeneration as I am. This book is a culmination of hard work, persistence, and genius of those unknown soldiers of liver regeneration science.

INTRODUCTION

EVERY DAY THIS EAGLE EATS MY LIVER...
...AND ALL BECAUSE ZEUS IS ANGRY WITH ME FOR GIVING **FIRE** TO MANKIND.
OH, WELL, MY SUFFERING IS WORTH IT—I'M SURE MANKIND HAS MADE GOOD USE OF MY GIFT!
© Mark Weinstein
MEANWHILE...

CHAPTER

1

Liver Regeneration: An Introduction

Udayan Apte

Department of Pharmacology, Toxicology and Therapeutics, University of Kansas Medical Center, Kansas City, Kansas, USA

One of the most interesting, unique, and intriguing capabilities of liver is the ability to truly regenerate and regulate its own size. Liver can regenerate after either surgical resection or injury-induced cell death initiated either by a chemical or viral infection. Liver regeneration involves replication of all cell types in the liver and involves rebuilding blood vessels, biliary tree, and extracellular matrix replacement. It is an extremely complicated process that is super-orchestrated and fine-tuned by a number of signaling pathways. Extensive studies have revealed underlying mechanisms including the kinetics of cell proliferation, species and strain differences, role of a number of growth factors, cytokines, chemokines, and other regulatory ligands, and involvement of plethora of intracellular signal transduction proteins such as kinases, transcriptional coactivators, transcription factors, and nuclear receptors. Despite extensive research over last eight decades, liver regeneration remains at the forefront of basic research in liver biology. The scientific literature on liver regeneration is vast and diverse with thousands of original papers and numerous reviews. However, a concise volume that gathers proven ideas as well as cutting-edge research in the area of liver regeneration is missing. This book is an effort to present "state-of-the-field" in the words of experts who have contributed to the field of regeneration over several decades.

1.1 HISTORY

No discussion of history of liver regeneration is complete without mentioning the story of Prometheus from the Greek mythology [1,2]. Prometheus, a titan, stole fire from the gods and gave it to the humans. He was condemned to "eternal torture" by Zeus. Prometheus was chained to mount Atlas where an eagle would eat his liver every day only to regenerate back later in the night, which would be eaten again by the eagle and the eternal torture continued. For most liver regeneration aficionados, the myth of Prometheus is indication that ancient Greeks new about livers amazing capability to regenerate. Whereas there is no clear-cut evidence of such knowledge, this remains a popular story shared among and by the regeneration researchers [3].

Liver Regeneration
http://dx.doi.org/10.1016/B978-0-12-420128-6.00001-4

Scientific reports on liver regeneration first appeared in the late nineteenth century and became frequent in early twentieth century [4]. While there were sporadic reports of liver regeneration in 1920s and 1930s, the first detailed description of study of liver regeneration using a rodent model was given by Higgins and Anderson in 1931 [5]. In this seminal paper, the authors described in great detail the technique to perform partial hepatectomy (PHX) surgeries, which has become the most commonly used model to study liver regeneration. The paper also describes the kinetics of cell division and changes in gross anatomy of the liver over a period of 21 days following PHX. The Higgins and Anderson paper soon become the guiding light for liver regeneration researchers (cited 5716 times so far according to Web of Science) who used the PHX model to study mechanisms of liver regeneration. Whereas the Higgins and Anderson paper has become a highly cited paper, some studies before that had looked at liver regeneration after chemical injury to the liver using model chemicals such as carbon tetrachloride (CCl_4) and chloroform [4]. Research in liver regeneration increased during mid-1950s and has since continued to catch the imagination of liver researchers (Figure 1.1). A Pubmed search with the phrase "liver regeneration" indicates that over 15,000 papers have been published since 1920s, with an average of 200 papers in 1950s and 1960s increasing to an average of 350 in 1980s and peaking to an average of 450 papers per year in the last decade. These data indicate that liver regeneration remains an interesting and active area of liver research in liver field.

1.2 MODELS OF LIVER REGENERATION

The models used to study liver regeneration can be grouped into two main categories—surgical resection or PHX and chemical injury models [6–8]. Traditionally, the PHX model remains popular since Higgins and Anderson elegantly described it in 1930s. In the PHX model, approximately 60-70% liver is surgically removed, and the remaining liver is allowed to regenerate. The process of liver regeneration takes approximately 14 days after PHX in rodents [5]. PHX proves to be a reliable model to study regeneration with minimal liver injury, synchronized cell cycle, and very high reproducibility. The chemical injury models, though traditionally used, remain relatively less common but have gained some popularity in last decade [9]. Variety of chemicals that are known to induce cell death in the liver resulting in subsequent compensatory liver regeneration have been used including CCl_4, thioacetamide (TA), acetaminophen (APAP), chloroform, galactosamine (GalN), allyl alcohol (AA), diethylnitrosamine (DEN), and bromobenzene [9]. Except AA, most other chemicals are centrilobular hepatotoxicants and substrates for CYP2E1. Out of these chemicals, CCl_4, TA, and APAP are probably the most commonly used chemicals to study the liver injury-regeneration cycle. Most researchers study liver regeneration after administration of a single toxic dose of a chemical. Studies on liver regeneration with chronic treatment of chemicals such as CCl_4 and TA have been done, but the focus of those studies remains on liver fibrosis and injury processes in pathogenesis of fibrosis [10]. Chemical injury models are more complicated than PHX because of inherent issues arising in separation of mechanisms of injury from mechanisms of regeneration, relatively higher animal-to-animal variation, significant inflammation and dose dependency of the injury and regeneration responses [6]. Apart from PHX and acute chemical injury models, some researchers have studied liver regeneration using unconventional models such as ischemia/reperfusion injury [11] and a dietary model of chronic liver injury such as DDC diet [12]. However, the majority of reliable mechanistic data on liver regeneration

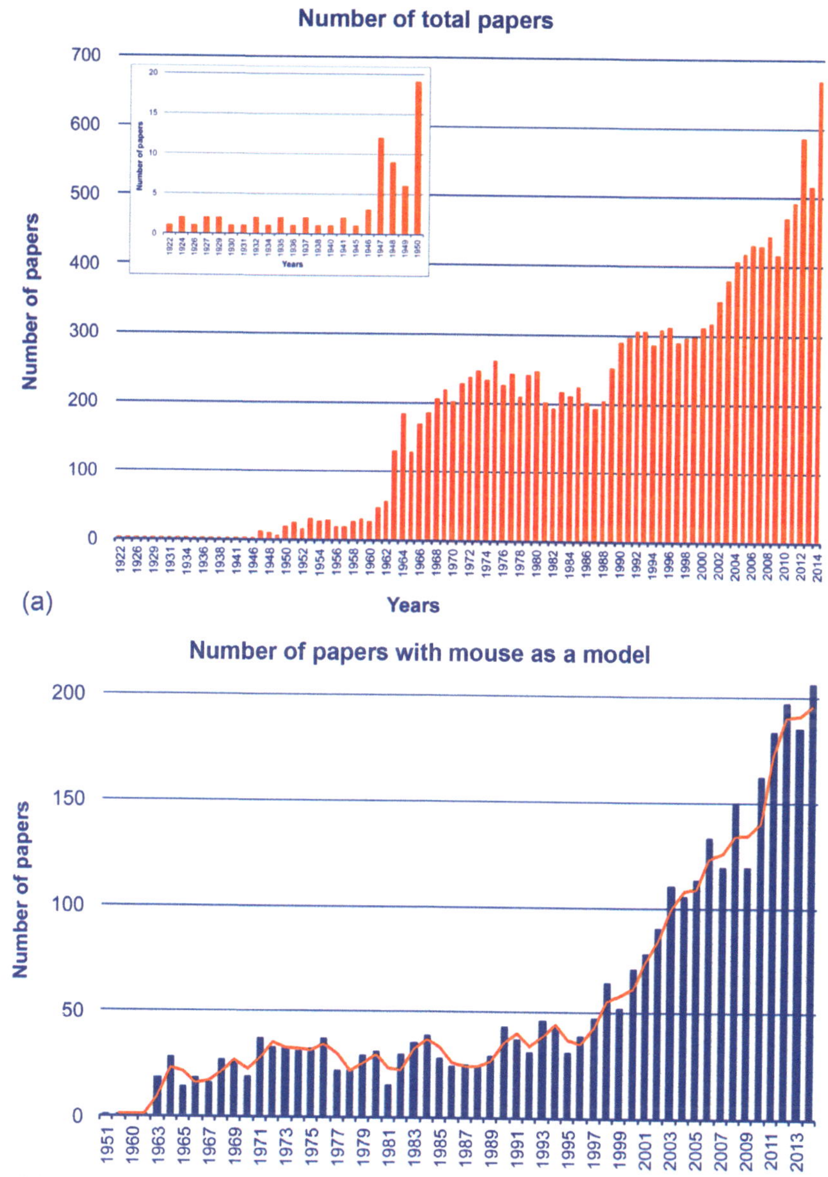

FIGURE 1.1 Bar graphs showing the number of publications on PubMed retrieved using the search term "liver regeneration" starting from 1920 to present. (a) Bar graph shows the total number of publications, inset shows number of papers from 1920 to 1950 and (b) bar graph shows number of publications with mouse as a model.

has been obtained from PHX and acute chemical injury (especially CCl_4, TA, and APAP) models.

Rodents remain the species of choice to study liver regeneration using both PHX and chemical injury models (Figure 1.1b and c). Initially, the rat was the most popular species used from 1950 till early 1990s. With the advent of knockout and transgenic technology, mouse became the species of choice in the late 1990s and has now replaced the rat because of the obvious advantages provided by progress in genetic engineering. Apart from rats and mice, other species such as gerbils [13], rabbits [14], Guinea pigs [15], dogs [16], and pigs [17,18] have also been used but much less frequently. Studies on human patients are also not uncommon [19] but remain relatively less informative because of obvious dependence on serum markers, historic histopathology samples, and radiological studies.

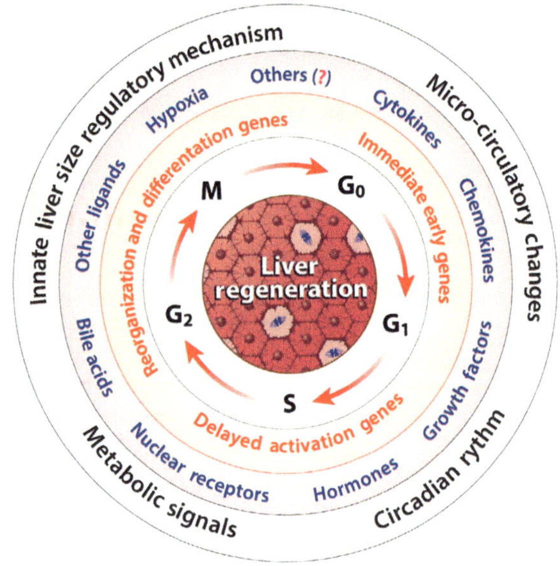

FIGURE 1.2 Schematic representation of mechanisms involved in liver regeneration. At its core, liver regeneration involves proliferation of hepatocytes and other cells in the liver. A number of factors, both intracellular and extracellular, are involved in stimulation of cell proliferation during liver regeneration.

1.3 MECHANISMS OF LIVER REGENERATION

Understanding the mechanisms that regulate the liver regeneration process has been at the center of liver biology research for over seven decades. The intense research has revealed a number of details about liver regeneration including the kinetics of cell cycle, role of a number of signaling pathways in stimulating cell proliferation, and the humoral control of liver regeneration [7,20]. Nevertheless, the mechanisms of liver regeneration are far from clear and many outstanding questions remain. It is impossible to summarize all the studies done on liver regeneration, but some highlights are worth mentioning. The extensive studies have clearly shown that liver regeneration is an extremely complicated process that is precisely orchestrated by a variety of autocrine, paracrine, and endocrine signaling, which are highly redundant (Figure 1.2). Loss-of-function studies conducted using knockout mice show that deletion or lack of any single gene does not completely stop the liver regeneration process. In most cases, lack of deletion of a particular gene results in "delayed" regeneration with slower progression of cell cycle till some sort of compensatory signaling takes over the regulation process [21]. Cases where animals die after PHX or chemical injury due to inhibition of liver regeneration, inevitably involve very high cell death and injury resulting in delayed progression of cell proliferation [22].

The kinetics of liver regeneration indicate that the process can be divided into three phases including initiation, progression, and termination. These phases are highly interlinked, and the boundaries between them are not clearly demarked. The majority of the data obtained from the PHX model have focused on the mechanisms of initiation and progression of liver regeneration. There are two schools of thought on regulation of liver regeneration [1,23]. Fausto and colleagues have put forth the "priming and progression" theory [24], which states that a cytokine-mediated priming process is essential for the quiescent hepatocytes to enter cell cycle during initiation of regeneration. Once primed, these hepatocytes respond to other mitogenic stimuli including those from growth factors. Recent studies on extremely rapid changes in activation of promitogenic regulators, such as β-catenin that initiates intracellular signaling within minutes after PHX and APAP overdose without requirement of cytokine-mediated priming process, indicate that hepatocyte proliferation may be a stepwise process as suggested in this theory [7,25]. Michalopoulos and colleagues have developed the alternative "primary and secondary mitogen theory" [7], which states that hepatocytes do not require priming but are selective to the mitogenic stimuli to which they respond to. This theory is based on the data obtained from ability of isolated hepatocytes cultured in serum-free conditioned media to enter cell cycle upon treatment with a specific factor [26]. Some growth factors such as HGF, EGF, and TGFα can stimulate hepatocyte proliferation without any other factor required and are termed the primary mitogens. Most other humoral factors, termed secondary mitogens, require the presence of primary mitogens to induce their mitogenic response on hepatocytes, which is generally additive or synergistic.

Once initiated, hepatocytes and other cells in the liver undergo several rounds of cell cycle during the progression phase till the pre-PHX liver size is reached. One of the most interesting mysteries in liver regeneration is the mechanism of termination of liver regeneration. Extensive experiments in rodents and some in higher mammals indicate that following surgical resection liver regeneration continues till it reaches approximately to the size it started with and then regeneration is stopped. It is known that a wave of apoptosis follows the wave of cell proliferation after PHX, but the mechanisms remain unknown [27]. At cellular level, cell proliferation stops, cells exit cell cycle and start to differentiate. This interesting ability of liver to regulate its own size has given rise to the concept of "hepatostat," molecular nature of which is still under investigation [7]. Initially, TGFβ was considered a prime suspect in termination of regeneration based on its ability to inhibit hepatocyte DNA synthesis. Detailed studies indicated that TGFβ alone may not be involved in termination but may require additional signaling from molecules such as Activin A [28]. Recent studies indicate that matricellular signaling via integrin-linked kinase may be involved in termination of liver regeneration [29].

Liver regeneration as a whole involves proliferation of cells in the liver followed by angiogenesis and reorganization of extracellular matrix. However, at its heart liver regeneration involves proliferation of various liver cells such as hepatocytes and cholangiocytes. In a normal liver, most cells are quiescence and in the G0 phase of cell cycle [1]. Upon PHX or chemical injury, cells enter cell cycle and go through various phases including G1, S (DNA replication occurs here), G2, and mitosis. In liver regeneration after PHX, the cell cycle is highly synchronized and cells enter S phase and mitosis together. This discovery led to the use of PHX model as an *in vivo* model of synchronized cell cycle [30]. Interestingly, the cell cycle is not well synchronized after chemical injury and differs fundamentally in various aspects including which cells enter cell cycle, the duration and timing of cell proliferation, and the signals required for stimulation [6,31].

A number of studies have demonstrated that liver regeneration is accompanied by extensive changes in gene expression [1,7,29,32,33]. The changes in gene expression are rapid and start within minutes after PHX or chemical injury and continue throughout the time course of regeneration [7,30,32]. The genes changing during liver regeneration have been loosely grouped into two groups based on their timing after the regeneration stimulus. The so-called "immediate early genes" change within minutes to hours after PHX, while the "delayed activation genes" change their expression 1 h and beyond after the regeneration stimulus is applied. The regulation of these genes has been an intense topic of research. At molecular level, these changes are driven by transcription factors and nuclear receptors such as NFκB, STAT3, AP1, TCF4, FXR, peroxisome proliferator activated receptor-alpha (PPARα), constitutive androgen receptor (CAR) aided by transcriptional coactivators including β-catenin, and Yap [7]. The regulators are in turn stimulated by a plethora of signaling pathways that are activated by circulating ligands that bind to their cognate receptors. A variety of cytokines (TNFα, IL-6) [23], chemokines (CXCL2) [11], growth factors (HGF, EGF, TGFα) [7], hormones (estrogen) [34], other ligands such as Hedgehogs and Wnts are known to stimulate liver regeneration [35,36]. Additionally, broader mechanisms such as circadian changes [37] and hemodynamic changes [38] are known to provide trigger for cell cycle movements. Finally, studies indicate that extracellular matrix-related signaling including integrin-linked kinase and the newly discovered pathways such as the Hippo Kinase pathway, which regulate organ size, may also play a role in termination of liver regeneration [29]. Whereas the number of signaling pathways involved in liver regeneration remains increasing, most studies indicate that regeneration of liver is critical for life and the mechanisms that regulate this process are highly redundant.

1.4 REGENERATION USING PROGENITOR CELLS

One of the most contested questions in the liver biology is the presence of stem cells in the liver and their role in liver regeneration. There is a general consensus that liver has hepatic progenitor cells, also called the oval cells due to their appearance in histological sections, residing in the terminal bile ducts called the Canals of Herring [39–41]. However, majority of the evidence suggests that they do not participate in regeneration after PHX [7,12]. Activation (expansion) of progenitor cells in liver regeneration after chemical injury has been noted, but their exact role in completion of liver regeneration process remains to be studied [42,43]. Similarly, role of bone marrow stem cells in hepatic regeneration has also been proposed [44,45]. Hepatocytes, which form almost 65-70% of the liver, are unique in the sense that they are highly differentiated cells that can still undergo cell proliferation when the need arises. Thus, majority of liver regeneration in normal liver is by proliferation of hepatocytes. However, extensive studies have shown in situations where hepatocytes are unable to divide and liver regeneration is taken over by hepatic progenitor cells. First noted by Farber and colleagues [39], the progenitor cells undergo proliferations and expansion followed by differentiation to give rise to functional hepatocytes [46]. The molecular markers that label the progenitors and the signals that stimulate their expansion and differentiation remain some of the most researched areas in regeneration field [47].

Similar to other aspects of liver regeneration research, the majority of initial research on progenitor cells was conducted using rat as a model. The most widely used model was the so-called "Salt-Farber Model" named after its inventors. In this model, rats were treated with 2-acetylaminofluorene, which intercalated hepatocyte DNA and induced replication defect, followed by PHX, which induced a need for regeneration. Subsequently, appearance of hepatic progenitor cells was observed which proliferated, expanded, and then differentiated into hepatocytes. In recent years, with advent of knockout and transgenic technology, new models using mice have become common. The most popular of these is mice-fed diet containing 3,5-diethoxycarbonyl-1,4-dihydrocollidine (DDC) for several weeks, which induces significant porphyria and hepatocyte injury resulting in emergence of hepatic progenitor cells. With arrival of lineage tracing technology to the study of regeneration [48], the mouse models to study hepatic progenitor cell-mediated regeneration have become even more popular.

1.5 MITOGEN-INDUCED HEPATOCYTE PROLIFERATION

In the liver, any cell injury, especially to the hepatocytes, results in compensatory cell proliferation. Without any injury or surgical removal, most hepatocytes remain in quiescence and there is no proliferation. However, some agents, both endogenous and xenobiotic in nature, can stimulate hepatocyte proliferation without any liver injury or cell death [49]. These include endogenous agents such as thyroid hormone (T3 and T4) and xenobiotics such as phenobarbital, TCPOBOP, phthalate plasticizer, and lead nitrite. These agents are direct mitogens to the hepatocytes and their mode of action generally involves nuclear receptors such as CAR and PPARα. Whereas studies have shown that PPARα-induced cell proliferation is mainly a rodent phenotype and less relevant in humans because of significantly less PPARα in human hepatocytes [50], the relevance of other mitogens to liver health and disease remains relatively unstudied and is under investigation.

1.6 FRONTIERS

Despite extensive research over 80 years and over 15,000 publications, liver regeneration remains an enigma for liver biologists. These studies have undoubtedly enriched our understanding of how liver regenerates. Nevertheless, there are many outstanding questions. The nature and role of metabolic signals in regulation of liver regeneration is not completely clear. We know that common nutrients regulate regeneration, but the precise mechanisms are unclear. With metabolic disease presenting itself as one of the major hepatic disorders, understanding metabolic regulation of liver regeneration has become more important than ever. The mechanisms of termination of liver regeneration, and the biochemical as well as molecular nature of the elusive "hepatostat" remain another major unanswered questions. Similarly interesting is the unexplored apparent connection between termination signals and pathogenesis of liver cancer. It is known that liver experiences significant microcirculatory changes immediately after PHX and also after massive hepatic injury, and it has been hypothesized that these pressure changes are some of the earliest signals for regeneration. However, precise nature of the molecular circuitry that connects blood pressure changes to cell proliferation is not understood.

The ability of liver to regenerate has caught the imagination of generations of liver researchers and has led to extensive basic science research. Liver regeneration is an essential component of liver homeostasis and is required for maintaining liver health. However, the power of liver regeneration has not been harnessed in clinical settings as much as it should have been. Research and development of regenerative therapies for liver diseases remain in infancy, and much work is needed to translate the basic science work into actual treatments. Successfully adapting the basic science understanding of regeneration into regenerative therapies seems to be the final unexplored frontier for research in liver regeneration.

References

[1] Michalopoulos GK, DeFrances MC. Liver regeneration. Science 1997;276(5309):60–6.

[2] Reuben A. Prometheus and Pandora—together again. Hepatology 2004;39(5):1460–3.

[3] Power C, Rasko JE. Whither prometheus' liver? Greek myth and the science of regeneration. Ann Intern Med 2008;149(6):421–6.

[4] Fishback FC. A morpholigical study of regeneration of the liver after partial removal. Arch Pathol 1929;7:955–77.
[5] Higgins GM, Anderson RM. Experimental pathology of liver: restoration of liver of the white rat following partial surgical removal. Arch Pathol 1931;12:186–202.
[6] Apte U, Mehendale HM. Hepatic defenses against toxicology: liver regeneration and tissue repair. In: McQueen CA, editor. Comprehensive toxicology. 2nd ed. Philadelphia: Elsevier; 2009. p. 339–67.
[7] Michalopoulos GK. Principles of liver regeneration and growth homeostasis. Compr Physiol 2013;3(1):485–513.
[8] Bucher NL. Regeneration of mammalian liver. Int Rev Cytol 1963;15:245–300.
[9] Mehendale HM. Tissue repair: an important determinant of final outcome of toxicant-induced injury. Toxicol Pathol 2005;33(1):41–51.
[10] Bissell DM. Therapy for hepatic fibrosis: revisiting the preclinical models. Clin Res Hepatol Gastroenterol 2011;35(8-9):521–5.
[11] Van Sweringen HL, Sakai N, Tevar AD, Burns JM, Edwards MJ, Lentsch AB. CXC chemokine signaling in the liver: impact on repair and regeneration. Hepatology 2011;54(4):1445–53.
[12] Shin S, Kaestner KH. The origin, biology, and therapeutic potential of facultative adult hepatic progenitor cells. Curr Top Dev Biol 2014;107:269–92.
[13] Cai ZW, Mehendale HM. Protection from CCl_4 toxicity by prestimulation of hepatocellular regeneration in partially hepatectomized gerbils. Biochem Pharmacol 1991;42(3):633–44.
[14] Gerdes RG, Rienits KG, Williams JF. Partial hepatectomy and liver regeneration in the rabbit. Cytobios 1971;4(15):193–207.
[15] Grabowska-Bochenek R, Birkner E, Birkner J, Hebrowska B. Effect of prolonged hydrocortisone administration on nucleic proteins of regenerating guinea-pig liver after partial hepatectomy. Acta Med Pol 1986;27(1-2):9–17.
[16] Schotanus BA, Penning LC, Spee B. Potential of regenerative medicine techniques in canine hepatology. Vet Q 2013;33(4):207–16.
[17] Mortensen KE, Revhaug A. Liver regeneration in surgical animal models - a historical perspective and clinical implications. Eur Surg Res 2011;46(1):1–18.
[18] Chen YL, Chen WB, Wan YY, Li WG, Huang ZQ, Wu XT, et al. Effects of partial portal vein arterialization on liver regeneration after hepatectomy in minipigs with obstructive jaundice. Chin Med J (Engl) 2012;125 (13):2302–5.
[19] Starlinger P, Assinger A, Haegele S, Wanek D, Zikeli S, Schauer D, et al. Evidence for serotonin as a relevant inducer of liver regeneration after liver resection in humans. Hepatology 2014;60(1):257–66.
[20] Moolten FL, Bucher NL. Regeneration of rat liver: transfer of humoral agent by cross circulation. Science 1967;158(3798):272–4.
[21] Borude P, Edwards G, Walesky C, Li F, Ma X, Kong B, et al. Hepatocyte-specific deletion of farnesoid X receptor delays but does not inhibit liver regeneration after partial hepatectomy in mice. Hepatology 2012;56 (6):2344–52.
[22] Huang W, Ma K, Zhang J, Qatanani M, Cuvillier J, Liu J, et al. Nuclear receptor-dependent bile acid signaling is required for normal liver regeneration. Science 2006;312 (5771):233–6.
[23] Fausto N, Campbell JS, Riehle KJ. Liver regeneration. Hepatology 2006;43(2 Suppl. 1):S45–53.
[24] Fausto N, Laird AD, Webber EM. Liver regeneration. 2. Role of growth factors and cytokines in hepatic regeneration. FASEB J 1995;9(15):1527–36.
[25] Monga SP, Pediaditakis P, Mule K, Stolz DB, Michalopoulos GK. Changes in WNT/beta-catenin pathway during regulated growth in rat liver regeneration. Hepatology 2001;33(5):1098–109.
[26] Block GD, Locker J, Bowen WC, Petersen BE, Katyal S, Strom SC, et al. Population expansion, clonal growth, and specific differentiation patterns in primary cultures of hepatocytes induced by HGF/SF, EGF and TGF alpha in a chemically defined (HGM) medium. J Cell Biol 1996;132(6):1133–49.
[27] Sakamoto T, Liu Z, Murase N, Ezure T, Yokomuro S, Poli V, et al. Mitosis and apoptosis in the liver of interleukin-6-deficient mice after partial hepatectomy. Hepatology 1999;29(2):403–11.
[28] Oe S, Lemmer ER, Conner EA, Factor VM, Leveen P, Larsson J, et al. Intact signaling by transforming growth factor beta is not required for termination of liver regeneration in mice. Hepatology 2004;40 (5):1098–105.
[29] Apte U, Gkretsi V, Bowen WC, Mars WM, Luo JH, Donthamsetty S, et al. Enhanced liver regeneration following changes induced by hepatocyte-specific genetic ablation of integrin-linked kinase. Hepatology 2009;50 (3):844–51.
[30] Taub R. Liver regeneration: from myth to mechanism. Nat Rev Mol Cell Biol 2004;5(10):836–47.
[31] Bhushan B, Walesky C, Manley M, Gallagher T, Borude P, Edwards G, et al. Pro-regenerative signaling after acetaminophen-induced acute liver injury in mice identified using a novel incremental dose model. Am J Pathol 2014;184(11):3013–25.
[32] Taub R. Liver regeneration 4: transcriptional control of liver regeneration. FASEB J 1996;10(4):413–27.

[33] Fausto N. Messenger RNA, in regenerating liver: implications for the understanding of regulated growth. Mol Cell Biochem 1984;59(1-2):131–47.
[34] Takada T, Yamamoto T, Koike K, Kanda Y, Miyake A, Sugihara A, et al. Effects of prolactin and estrogen on cell proliferation of the mouse liver induced by partial hepatectomy. In Vivo 1997;11(5):409–13.
[35] Monga SP. Role and regulation of beta-catenin signaling during physiological liver growth. Gene Expr 2014;16 (2):51–62.
[36] Omenetti A, Choi S, Michelotti G, Diehl AM. Hedgehog signaling in the liver. J Hepatol 2011;54(2):366–73.
[37] Matsuo T, Yamaguchi S, Mitsui S, Emi A, Shimoda F, Okamura H. Control mechanism of the circadian clock for timing of cell division in vivo. Science 2003;302 (5643):255–9.
[38] Abshagen K, Eipel C, Vollmar B. A critical appraisal of the hemodynamic signal driving liver regeneration. Langenbecks Arch Surg 2012;397(4):579–90.
[39] Solt DB, Medline A, Farber E. Rapid emergence of carcinogen-induced hyperplastic lesions in a new model for the sequential analysis of liver carcinogenesis. Am J Pathol 1977;88(3):595–618.
[40] Tatematsu M, Ho RH, Kaku T, Ekem JK, Farber E. Studies on the proliferation and fate of oval cells in the liver of rats treated with 2-acetylaminofluorene and partial hepatectomy. Am J Pathol 1984;114 (3):418–30.
[41] Thorgeirsson SS. Hepatic stem cells in liver regeneration. FASEB J 1996;10(11):1249–56.
[42] Kofman AV, Morgan G, Kirschenbaum A, Osbeck J, Hussain M, Swenson S, et al. Dose- and time-dependent oval cell reaction in acetaminophen-induced murine liver injury. Hepatology 2005;41(6):1252–61.
[43] Pritchard MT, Nagy LE. Hepatic fibrosis is enhanced and accompanied by robust oval cell activation after chronic carbon tetrachloride administration to Egr-1-deficient mice. Am J Pathol 2010;176(6):2743–52.
[44] Grompe M. The role of bone marrow stem cells in liver regeneration. Semin Liver Dis 2003;23(4):363–72.
[45] Petersen BE. Hepatic "stem" cells: coming full circle. Blood Cells Mol Dis 2001;27(3):590–600.
[46] Wang X, Foster M, Al-Dhalimy M, Lagasse E, Finegold M, Grompe M. The origin and liver repopulating capacity of murine oval cells. Proc Natl Acad Sci USA 2003;100(Suppl. 1):11881–8.
[47] Dorrell C, Erker L, Lanxon-Cookson KM, Abraham SL, Victoroff T, Ro S, et al. Surface markers for the murine oval cell response. Hepatology 2008;48(4):1282–91.
[48] Yanger K, Knigin D, Zong Y, Maggs L, Gu G, Akiyama H, et al. Adult hepatocytes are generated by self-duplication rather than stem cell differentiation. Cell Stem Cell 2014;15(3):340–9.
[49] Columbano A, Shinozuka H. Liver regeneration versus direct hyperplasia. FASEB J 1996;10(10):1118–28.
[50] Corton JC, Cunningham ML, Hummer BT, Lau C, Meek B, Peters JM, et al. Mode of action framework analysis for receptor-mediated toxicity: the peroxisome proliferator-activated receptor alpha (PPARalpha) as a case study. Crit Rev Toxicol 2014;44(1):1–49.

METHODS TO ASSESS LIVER REGENERATION

CHAPTER

2

Models to Study Liver Regeneration

Michele T. Pritchard, Udayan Apte

Department of Pharmacology, Toxicology and Therapeutics, University of Kansas Medical Center, Kansas City, Kansas, USA

2.1 INTRODUCTION

Complete tissue regeneration after injury is a unique property of few animals and tissues. In lower vertebrates, complete tissue regeneration is achieved after injury. For example, after limb amputation in salamanders, an exact copy of the lost tissue, including digits, is reconstructed [1]. Higher vertebrates, including rodents and humans, also exhibit complete regeneration. Specifically, fetal dermal wounds as well as oral mucosal wounds regenerate completely without scar formation [2,3]. Another tissue which regenerates across genera is the liver [4,5]. Indeed, in cases of noniterative injury, the liver is able to completely regenerate and recoup lost liver mass and hepatic function after toxic injury, traumatic or surgical loss of parenchyma. This is achieved, in large part, by inducing the proliferation of remaining hepatocytes [4–6]. Failure of liver regeneration only occurs in the most extreme of cases such as those which involve more than 90% loss of parenchyma [7]; this clearly demonstrates the power of the hepatic regenerative response.

The liver's unique ability to regenerate likely evolved due to its predominant role in xenobiotic and nutrient metabolism. In addition, the liver's anatomic location, in direct contact with the external environment via the portal circulation from the gut, places the liver in a precarious position exposed to agents, which could cause considerable liver injury. Regardless of precipitating agent, complete physical and functional liver regeneration is achieved in an astonishing number of cases. Liver regeneration is a complex and, importantly, redundant process whereby lost liver mass is recovered and nothing more [8,9]. The mechanisms behind this tight regulation in liver size relative to organism size are the subject of several investigations as it provides insights into mechanisms which contribute to normal regeneration as well as uncontrolled cell growth found in hepatocellular carcinoma. Defining the signals that turn off the hepatic regenerative response is as critical as those that induce this response.

In this chapter, we will discuss the animal models that researchers employ to explore the mechanisms of liver regeneration. This includes surgical and chemical models, in which hepatocyte proliferation is responsible for regeneration, and those that interrogate the contribution of the liver's bipotent progenitor cell population (hepatic progenitor cells, HPCs, also known as "oval cells") to this process.

Liver Regeneration
http://dx.doi.org/10.1016/B978-0-12-420128-6.00002-6

In addition, animal models of postnatal and pregnancy-induced liver growth will be presented. Finally, we will cover animal models of liver regeneration after ischemia/reflow-induced liver injury and end the chapter with a discussion of various assays employed to assess liver regeneration.

2.2 MAIN MODELS

2.2.1 Partial Hepatectomy

The most widely used model to study liver regeneration in rodents is the surgical removal procedure or partial hepatectomy (PH). The pivotal 1931 paper by Higgins and Anderson gave the first detailed account of the PH procedure; this procedure has remained essentially unchanged since then and is still widely used [10]. Higgins and Anderson's study used the laboratory rat as a model animal but the procedure has been successfully adapted to mice. The procedure in mice has been modified to account for the major anatomical difference in the two rodent species, namely, presence of gall bladder in mice. However, other parts of the procedure and the method to calculate the change in liver-to-body weight ratio remain same.

2.2.1.1 PH Procedure

PH is a major surgery and should be performed with strict adherence to standard aseptic surgical techniques required for small animal surgeries. PH is performed under anesthesia and while the choice of anesthetic used varies with the investigator, most laboratories use machine-administered isoflurane. It is a laparotomy procedure performed on the dorsal abdomen. When using rats, the abdomen is shaved but in mice this may not be necessary. The abdomen is wiped down with Betadine. An upper midline incision is made from the xyphoid down toward the tail. Anatomically, the parts of liver removed involve both arms of the median lobe and the left lateral lobe, which constitute approximately 65-70% PH [10,11].

There are two major ways to perform the surgery after this point. In one format, a bigger incision of approximately 1.5 cm is made, and the skin folds are held out using hemostatic forceps visualizing the entire liver. This is followed by manipulation of liver lobes using moistened, sterile cotton swabs. In rats, all three lobes are tied together using a silk suture, placed as proximal to the origin of the lobe as possible. This is followed by removal of the lobes using sharp scissors. In mice, care needs to be taken to avoid tying the gall bladder, which is not removed. Therefore, in mice, three separate ties are applied, one to each lobe, and then the lobes are cut one after the other.

In a variation of the basic surgical procedure called the "flowering technique," a smaller incision is made to visualize the xyphoid, which is cut providing access to the liver [11]. Then, pressure is applied to the abdomen such that the three liver lobes to be cut are pushed out through the incision. The lobes are tied (together in rats or separately in mice) and cut. Care should be taken not to apply too much pressure in order to avoid any internal organ damage. This version of the process results in much less blood loss, fewer chances of infections, and post-operative complications. It also results in faster recovery mainly due to the smaller incision and also due to the fact that other visceral organs, as well as the lobes that are not resected, are never exposed. However, this procedure requires substantial practice and experience to exteriorize enough liver to do a 65-70% PH and not damage any other organs.

Following resection of the lobes, the muscle layer is closed and secured with dissolvable sutures, followed by closure of the abdominal skin. Some laboratories use wound clips to close the skin while other investigators, including the authors, prefer dissolvable sutures. The animals should be placed in clean cages following surgery and given access to gel-based rodent diet to aid in nutrient and hydration recovery after surgery. Whether or not to use analgesia to reduce pain is an important issue and should be carefully considered. Many pain medications, including buprenorphine, interfere with cell proliferation and therefore may adversely affect the liver regenerative response. A special IACUC exemption by placing the animals in USDA category E along with veterinarian-assisted extensive monitoring and a clear euthanasia plan is recommended to avoid this problem.

2.2.1.2 *Establishing Baseline, Time Course, and Proper Controls*

Because the PH is complicated procedure, requires substantial practice, and the results obtained can vary significantly from investigator to investigator, it is critical to establish a baseline of liver weight to body weight ratio in the animals [10]. This can be done by performing PH on the animals followed by immediate sacrifice to get an estimate of the actual percentage of liver being removed. Similarly, a baseline liver-to-body weight ratio PH should also be established for each unique rat or mouse strain. These baseline values are critical in measuring regrowth of the liver after PH. The data obtained from the PH studies is generally expressed as percent of liver regrowth compared to the pre-PH liver mass.

Traditionally, liver regeneration after PH is studied over a time course of 14 days [11]. A variety of time courses have been used by different laboratories but a typical time course involves, apart from 0 h, 1, 2, 3, 5, 7, and 14 days after PH. Additional time points ranging from minutes after PH to various times in first 24 h have also been used. The time course used also depends on the species used. In rats, the peak proliferation is 24 h after PH, but in mice the peak proliferation is between 42 and 48 h [11,12]. With advent of knockout and transgenic technology, mice have become the species of choice and, typically, 0, 24, 36, 48, 72, 120, and 168 h after PH has become the most studied time course (Figure 2.1a shows liver growth after PH).

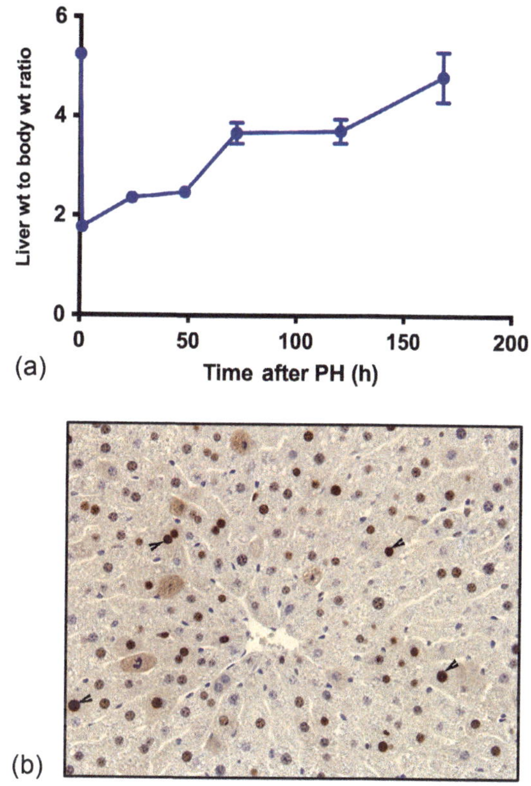

FIGURE 2.1 Liver regeneration after PH. Mice were subjected to PH and euthanized at 0, 24, 48, 72, 120, and 168 h later. (a) Liver weight to body weight ratio demonstrating liver regrowth. (b) Representative photomicrographs of PCNA immunohistochemistry on paraffin-embedded mouse liver section 48 h after PH. Arrowheads point to cells in S-phase.

Using proper controls is essential to evaluate effects of surgical procedure on the liver regeneration process. This led to use of sham surgeries where laparotomy is performed without excision of the liver. In recent reports using knockout or transgenic animals, the majority of laboratories have moved away from the sham surgeries and are using wild-type animals subjected to PH as the controls [9,11,13].

2.2.2 Models of Liver Regeneration After Chemical-Induced Liver Injury

Apart from regenerating after surgical resection, the liver is also capable of regenerating following toxic injury. In fact, regeneration after toxic chemical-induced injury is likely to be more prevalent as it is the major site of drug metabolism and is involved in detoxification of a variety of agents including pharmaceutical agents, environmental pollutants, and endogenous toxins. There are fundamental differences in the way liver regenerates after PH and liver regeneration after chemical injury. These differences arise due to inherent differences in mechanisms of injury induction in these models [14]. Some of the major differences include a role of inflammatory signaling and synchronized (PH) versus unsynchronized (chemical injury) nature of cell cycle. Here, we discuss some of the most commonly used models of chemical-induced liver injury and regeneration.

2.2.2.1 D-Galactosamine: Mechanism of Hepatotoxicity

D-Galactosamine (GalN) is a potent hepatotoxin often used in conjunction with lipopolysaccharide (LPS) or tumor necrosis factor (TNF) α in a lethal, hepatitis animal model [15–17]. In this context, GalN sensitizes hepatocytes to LPS or TNFα resulting in striking apoptotic cell death of nearly all hepatocytes; this sensitization leads to animal death within 5-9 h [17]. While the GalN/LPS or TNFα model is useful for explorations into mechanisms of lethal hepatitis, it is not ideal for investigating mechanisms of liver regeneration. Treatment of mice, rats, rabbits, and guinea pigs with GalN alone, however, at nonlethal doses (e.g., <1.4 g/kg as a single dose in rats [18]), is suitable for this purpose [19,20]. The mechanisms by which GalN induces hepatotoxicity involve a series of metabolic derangements leading to the complete consumption of uridine nucleotides (UTP) and UDP-glucose with concomitant accumulation of UDP-galactosamine derivatives [21]. UTP depletion inhibits RNA synthesis followed by inhibition of protein synthesis [22]. Synthesis of additional macromolecules is reduced after GalN exposure, including glycoproteins and glycogen, and also contributes to cell death [20]. Therefore, while coadministration of GalN and LPS or TNFα will lead to rampant apoptotic cell death as described above, when GalN is used alone, cell death is limited and allows for investigations into the hepatic regenerative response.

In addition to metabolic derangements, other factors contribute to GalN-induced hepatocyte cell death. For example, endotoxin is clearly involved in the hepatotoxic effect of GalN. Indeed, colectomy, and therefore removal of a large source of endotoxin, renders rats resistant to GalN-induced liver cell injury and death [23]. However, colectomy does not affect depletion of UTP- or UDP-glucose [23]. In intact animals, complement is activated after GalN exposure, and this is likely due to increased hepatic endotoxin release from the gut. Complement activation is manifested in complement fixation on hepatocyte cell membranes, effectively labeling those cells for complement-mediated cell lysis [23]. In colectomized, endotoxin-deficient rats, complement activation is attenuated, consistent with reduced GalN-mediated cell death in the same animals. Inflammation mediated by increased hepatic endotoxin and complement activation could also be exacerbated by mast cells. Indeed, GalN exposure induces robust mast cell degranulation in the intestine which can promote impairments in the intestinal barrier function, facilitating release of endotoxin into the portal circulation [23]. Therefore, in conjunction with metabolic issues, endotoxin exposure, complement activation, and mast cell degranulation responses all contribute to liver cell death after GalN exposure.

2.2.2.1.1 D-GALACTOSAMINE: HEPATIC REGENERATIVE RESPONSE

Nonlethal GalN doses must be employed to study liver regeneration. While one study demonstrated absence of lethality up to 1 g/kg GalN in rats, additional studies use far less GalN in the range of 200-400 mg/kg as single doses or multiple doses up 1 g/kg or more to explore liver regenerative response. At these doses, liver injury, assessed by histology and biochemical measures, increases 24 h after GalN exposure. Necrotic cells are scattered throughout the parenchyma but more prevalent in the pericentral area [20]. Indications of liver cell proliferation begin at 48 h but extend 72-96 h post-GalN. Hepatocyte proliferation contributes to liver regeneration after GalN, but contribution of HPC (i.e., oval cells) cannot be ignored, particularly when higher doses or multiple doses [24] of GalN are utilized. LacZ hepatocyte labeling studies revealed hepatocyte proliferation occurs as soon as 24 h after GalN exposure. However, in high-dose GalN studies, HPC numbers increase 48-96 h after 750 mg/kg GalN and appear to play a predominant role in liver regeneration [24].

It is interesting to note that actively proliferating hepatocytes, such as those found in neonatal rats (5-days old), are refractory to GalN-mediated hepatocyte death while adult rats (5-months old) are not [25]. After intraperitoneal injection of 400 mg/kg GalN, adult rats exhibited depletion of uracil nucleotides and glycogen. Peak plasma alanine and aspartate aminotransferase (ALT and AST) is observed 24 h after GalN exposure and by 48 h, the liver regenerative response (measured by ^{3}H-thymidine incorporation) peaks 48 h after GalN and is reduced thereafter. By contrast, neonatal rats are protected from the metabolic and hepatotoxic perturbations of GalN exposure [25]. So, while this study certainly demonstrates hepatocyte-mediated liver regeneration is possible in adult rats, a beneficial effect of ongoing hepatocyte proliferation is clearly a benefit to the outcome of hepatotoxin exposure. Similar findings regarding the hepatoprotective effect of hepatocyte proliferation to subsequent toxin exposure are observed in additional models of toxic liver injury [26–29].

2.2.2.2 *Carbon Tetrachloride: Mechanism of Hepatotoxicity*

Carbon tetrachloride (CCl_4), a volatile inorganic compound, is a very well-described hepatotoxicant. CCl_4 was once widely used as a dry cleaning solvent, precursor to refrigerant synthesis, propellant for aerosol cans, a fire extinguishing liquid, as a solvent for oils, fats, and lacquers, as a grain fumigant and curiously enough, in lava lamps [30,31]. CCl_4 exhibits ozone-depleting properties, and most of its use was discontinued by the mid-1960s [30]. Its hepatotoxic properties were revealed predominantly by those who were directly involved with the manufacture or use of CCl_4. Acute and chronic exposure to CCl_4 results in hepatic and renal toxicity. In these organs, expression of Cytochrome P450 2E1 (CYP2E1) is, in large part, responsible for CCl_4's toxicity while CYP2B1, 2B2, and 3A are also involved [32]. CCl_4 is metabolized by CYP2E1, creating highly reactive metabolites including the trichloromethyl (CCl_3^*) and trichloromethylperoxy (CCl_3OO^*) radicals [32]. In the liver, the highest concentration of CYP2E1 is in the pericentral area (Zone 3) [33] which is where liver injury is localized after CCl_4 exposure. After CYP-mediated bioactivation, the reactive metabolites modify proteins, lipids, and DNA in hepatocytes in which CCl_4 was metabolized, as well as neighboring cells by way of a usually limited bystander effect, eventually leading to necrotic cell death [28,29]. Robust hepatocyte necrosis leads to increased plasma ALT activity within 1 h to greater than 10,000 U/L within 36 h after CCl_4 exposure in mice (Figure 2.2a). By 72 h after CCl_4, plasma ALT activity is reduced >90% relative to the 36 h level (Figure 2.2a). Independent of necrosis, evidence exists that apoptosis also contributes to hepatocyte death after CCl_4 exposure [36], but this mechanism is a minor contributor to CCl_4-mediated cell death.

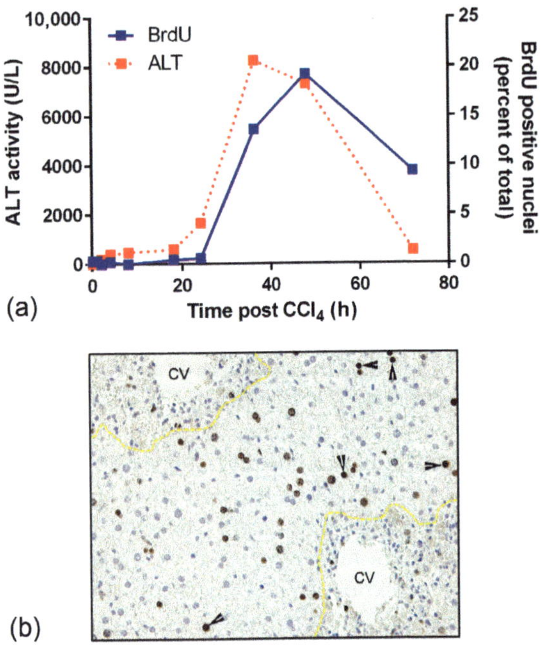

FIGURE 2.2 Liver injury and regeneration after acute CCl_4 exposure. Mice were exposed to CCl_4 and euthanized 0, 1, 2, 4, 8, 18, 24, 36, 48, and 72 h later. (a) Plasma ALT activity (red, dotted line, left *y*-axis) and BrdU incorporation (blue solid line, right *y*-axis) are plotted as a function of time post-CCl_4 exposure. The data suggest that once the liver regenerative response is induced, further liver injury is prevented as shown by others [25–29,34,35]. (b) BrdU immunolocalization in mouse liver 48 h after acute CCl_4 exposure. The cells with dark brown nuclei were in S-phase at the time this liver was excised and fixed in formalin. CV = central vein. Yellow dotted line denotes areas of necrosis. Arrowheads point to cells in S-phase.

2.2.2.2.1 CCl_4: HEPATIC REGENERATIVE RESPONSE

As mentioned above, CCl_4 can be used in acute or chronic exposure settings [37]. In the acute setting, a single dose of CCl_4, usually diluted 1:3 in olive or corn oil, is administered, intraperitoneally, at a dose of 1 μL/g body weight (0.4 mg/kg); this dosing regimen is not lethal to wild-type mice or rats. For precise dosing, many investigators use 100 μL glass Hamilton syringes, fitted with 26G 5/8″ needles, which will allow administration of CCl_4 to the microliter in mice 25 g and under. In mice over 25 g, dosing is managed to the nearest 2.5 μL using 250-μL Hamilton syringes. CCl_4 and vehicle syringes are kept separate and cleaned after each use with a strong solvent. CCl_4-containing solutions are prepared in glass tubes.

The acute CCl_4 treatment paradigm is used to evaluate the full course of the liver injury and repair, including the liver regenerative response. The chronic CCl_4 exposure paradigm involves repeated dosing of experimental animals two to three times per week over 4-12 weeks; fibrosis gets worse with increasing exposure to CCl_4, approaching the histological appearance of cirrhosis [37]. Using this paradigm, mechanisms of hepatic fibrosis and fibrosis resolution can be appreciated and explored; however, it is not usually utilized to study regeneration.

There are many parallels between liver regeneration after CCl_4 and PH with respect to mediators and molecules involved in the response. One major difference between the two is profound liver injury after CCl_4, but limited liver injury after PH. Mediators which participate in regeneration in both models include activation of complement and release of endotoxin from the gut, which are among the earliest signals required for liver regeneration. These early molecules promote the activation of latent cytoplasmic transcription factors, including NFκB, signal transducer and activator of transcription 3 (STAT3), C/EBPβ and induce expression of immediate early genes such as early growth response 1 (*Egr1*) transcription factor [38,39]. These molecules then promote the expression of cytokines including TNFα and IL6, downstream signaling by which induces hepatocyte transition from G0 to G1 of the cell cycle [6]. Once hepatocytes enter the cell cycle, a growth factor-driven progression phase pushes those cells through the rest of the cell cycle [6]. Various molecules and signaling pathways are involved in cessation of liver regeneration, including integrin-linked kinase [6,9]. Interestingly, the peak plasma ALT and peak DNA synthesis occur simultaneously. This is consistent with the dramatic reduction in plasma ALT activity between 48 and 72 h after CCl_4 as shown in Figure 2.2b. Although not synchronous, it is still possible to evaluate cell-cycle stage during hepatocyte proliferation using many of the standard methods. As long as hepatocyte cell proliferation is not inhibited or liver injury is not too severe, remaining hepatocytes are responsible for repopulating the injured liver and replacing lost liver mass within 7-10 days after CCl_4 exposure in rodents.

Even though it requires little skill to utilize CCl_4 in the lab compared to PH, there are some caveats which one should consider before utilizing "chemical hepatectomy" over surgical/PH. First, there is no human disease which CCl_4 recapitulates [37]. Second, CCl_4 is an ozone-depleting substance and must be used with a firm understanding of the rules regulating is use. Third, it is clearly very toxic, so persons using CCl_4 must be educated regarding this toxicity and utilize all proper precautions when preparing, administering, or disposing of CCl_4. Fourth, CCl_4 causes pain and distress when administered, and in recent years, IACUCs are ensuring that appropriate analgesia is provided to mitigate that pain and suffering. Use of analgesia can introduce a variable into experimental design, which researchers must take into consideration when interpreting their data. Finally, because CYP2E1 is involved with CCl_4's hepatotoxicity, every assurance must be made that investigators evaluate CYP2E1 content and activity in their different mouse strains and after various treatments to ensure their generated data are not simply artifacts due to differences in CYP2E1. Despite these caveats, CCl_4 remains one of the most well-studied and thoroughly understood models of liver injury and of the hepatic wound-healing response.

2.2.2.3 *Thioacetamide*

Thioacetamide (TA) was introduced in 1945 as a fungicide to control orange decay but was quickly realized to be a hepatotoxicant and a chemical carcinogen [40]. TA was banned in late 1960s and since then has served as a model hepatotoxicant to study liver injury, hepatic fibrosis, liver cancer, and liver regeneration. It is a popular model toxicant because of its outstanding solubility in the water, a prolonged injury and recovery pattern (up to 120 h) giving significant time to study mechanisms and a broad range of use from a single dose-induced acute liver injury, subchronic exposure leading to liver fibrosis, and chronic exposure resulting in liver cancer [26,41–43].

Like CCl_4, TA is a classic centrilobular (Zone 3) hepatotoxicant, which requires CYP450-mediated bioactivation to cause injury [26]. TA is metabolized by CYP2E1 first to TA-sulfoxide and further to TA-sulfone, which is the ultimate reactive metabolite [27]. TA-sulfone binds to cellular macromolecules and causes necrotic cell death. Unlike CCl_4 and acetaminophen (see next section), TA toxicity does not seem to depend on lipid peroxidation or GSH depletion, respectively [27]. Interestingly, TA may also be metabolized by flavin-containing monooxygenase enzyme system to unknown metabolite(s), which are less toxic [44]. However, because of its limited use as a model toxicant resulting in only a handful of studies, the exact mechanism of TA-induced necrotic cell death is not clear. These studies indicate that treatments or genetic alterations that induce or repress hepatic CYP2E1 will affect TA toxicity and subsequent liver regeneration.

TA can be administered in drinking water (0.03%, w/v) for 4-6 weeks or as a daily injection (300 μg/kg) three times a week for 8-12 weeks to reliably induce hepatic fibrosis. However, acute toxicity studies are more amenable to study liver regeneration due to the significantly prolonged time (up to 96 h) required for recovery. TA can be used over a dose range of 250-550 mg/kg dose range in rats [45]. Dose beyond 600 mg/kg are known to be lethal in Sprague-Dawley (SD) rats [26]. Interestingly, mice are more resistant to TA due to mechanisms currently unknown and can tolerate doses up to 800 mg/kg [46]. A typical dose range that allows studying both liver injury and regeneration in mice is from 400 to 800 mg/kg in various strains of mice including Swiss Webster, CD-1, and C57BL/6. Doses beyond 900 mg/kg are lethal in most mouse strains [42,46]. TA is highly soluble in aqueous solvents and thus saline is the choice of vehicle for treatment.

Like other toxic chemicals used for liver regeneration studies, TA-induced injury and subsequent liver regeneration are dose dependent. In SD rats, doses up to 450 mg/kg stimulate liver regeneration proportionate to liver injury [45]. A significant inhibition of liver regeneration due to decreased promitogenic signaling is observed at 600 mg/kg TA dose in rats. Furthermore, a variety of factors including diseases such as type I and type II diabetes, glucose feeding, moderate fatty acid supplementation, and caloric restriction are known to affect TA-induced liver injury and regeneration [28].

2.2.2.4 *Acetaminophen*

Acetaminophen (APAP, also known as paracetamol in Europe, Asia, and Africa) is a very popular analgesic and antipyretic agent used worldwide [47]. Whereas APAP is safe at therapeutic doses when used as directed, overdose of APAP results in acute liver failure [48,49]. In fact, APAP overdose is the most common cause of acute liver failure in the Western world [49,50]. Thus, studying liver regeneration after APAP overdose has a strong clinical and translational relevance. APAP overdose can be reproduced in mice and has served as an excellent rodent model to study not only mechanisms of liver injury but in recent years as a model to study mechanism of liver regeneration after toxic injury [51].

Treatment of supra-pharmacological doses of APAP results in centrilobular necrosis, and massive liver injury, which is followed by rapid liver regeneration and recovery up to a threshold dose [52]. Studies in the last three decades have revealed significant information on the mechanisms of APAP-induced liver injury [53,54]. APAP at therapeutic doses is eliminated by conjugation reactions including glucuronidation and sulfation. After administration of toxic doses of APAP, these Phase II reactions are quickly overwhelmed, and APAP is then metabolized by CYP2E1, the main enzyme involved in APAP bioactivation to its reactive metabolite *N*-acetylbenzoquinone imine (NAPQI). Hepatic glutathione (GSH) is the primary defense against NAPQI, which is conjugated to GSH and eliminated. However, at toxic doses hepatic GSH quickly depletes resulting in excess NAPQI, which triggers a cascade of cellular events resulting in necrotic cell death. These events, still under investigation, include sequential activation of kinases such as RIP and JNK, production of peroxynitrite and other reactive oxygen species, and mitochondrial permeability transition resulting in significant mitochondrial damage. Several other events including autophagy inhibition and calpain-mediated progression of liver injury are also involved in APAP-induced liver injury [55,56]. Whereas the mechanisms of APAP-induced liver injury are well studied, the mechanisms of subsequent liver regeneration response are relatively less understood. Recent studies indicate that liver regeneration after APAP overdose is dose dependent and involves Wnt/β-catenin and NFκB-mediated signaling [52].

Because of the dose dependency of the liver regeneration response, it is critical to perform preliminary studies to determine a dose that will provide a good time course to study liver regeneration. Further, APAP toxicity in mice is highly strain dependent, and therefore, it is important to consider the background of the mice, especially when using transgenic or knockout mice as models [57]. In a typical study using C57BL/6 mice, individuals are fasted between 12 and 16 h prior to APAP administration [52]. This provides a more uniform response to APAP following treatment by neutralizing any mouse-to-mouse variation in hepatic glutathione levels, which critically affect APAP toxicity. Interestingly, even short-term overnight fasting is known to induce hepatic CYP2E1 levels, which may have an effect on the liver injury as well. The typical time course to study liver regeneration extends up to 96 h in mice with 12, 24, 48, and 72 h time points. APAP is relatively less soluble in water than TA and is administered either in warm (37 °C) saline or in a half normal (0.45%) basic (pH 9) saline. It is important to keep the APAP solution warm while dosing multiple animals in order to avoid precipitation of the drug.

Another important caveat of APAP studies is knowing the fact that dimethylsulfoxide (DMSO), a popular solvent used for chemical inhibitors of kinases and other proteins, interferes with APAP toxicity both by inhibiting CYP2E1 activity and by reducing oxidative stress [58,59]. Therefore, using DMSO-soluble chemicals often leads to false results. Similarly, APAP toxicity (and by extension subsequent liver regeneration as it depends on injury) is highly dependent of hepatic CYP2E1 activity, and loss of CYP2E1 activity results in protection from APAP. This is an important consideration for using chemical inhibitors post-APAP treatment, which may inhibit hepatic P450 activity. For the same reason, it is critical to determine changes in CYP2E1 activity levels in any transgenic or knockout mice being used as before starting the experiments. These considerations are also applicable to hepatic GSH levels, changes in which may lead to differences in hepatic injury and later regeneration [51].

2.2.2.5 Regeneration After Ischemia/Reflow

Despite the incredible advances in our understanding of pathological processes leading to liver disease, transplantation remains the only cure for advanced liver disease. Though "curative," liver transplant is not without potentially serious clinical complications and caveats. For example, following liver transplantation, the patients must take immunosuppressive drugs for the rest of their lives, predisposing him or her to increased incidence of infections and cancer. In addition to necessary immune suppression, other issues exist which complicates liver transplant, namely, ischemia/reperfusion (I/R) injury and the impact that has on liver regeneration, particularly in the context of small size for liver transplantation. Understanding the mechanisms of I/R-induced liver injury and repair go beyond transplantation. Indeed, these same considerations are also critical determinants of hepatic trauma and partial liver resection. To this end, animal models exist which have explored the mechanisms of I/R hepatic injury and the impact this phenomenon has on subsequent liver regeneration in an effort to identify points of intervention to mitigate posttransplant, trauma, or resection-related complications and impact on the liver regenerative response.

I/R injury results from a period of low oxygen delivery to tissue, followed by normalization of oxygen perfusion, which leads to significant inflammation precipitating organ damage and dysfunction. Jaeschke and colleagues have defined two phases to I/R injury. In the first phase, liver-resident macrophages (i.e., Kupffer cells) are activated; complement activation contributes to this process [60,61]. Oxidants released from activated Kupffer cells induce hepatocyte stress and limited injury. Despite increased plasma ALT activity, hepatic architecture is maintained in this early phase of liver injury. Subsequently, hepatic inflammation is increased due to damage- or danger-associated molecular patterns release from dead and dying hepatocytes, which, in conjunction with complement and Kupffer cell activation, leads to the production and secretion of cytokines such as IL12, TNFα, and IL1β. This hepatic inflammation sets the stage for the second phase of the I/R response: in response to inflammation, adhesion molecule expression is induced on liver sinusoidal endothelial cells, and CXC chemokines are liberated [62]. These two changes provide the signals and microenvironment necessary to recruit neutrophils to the liver. Neutrophil sequestration in the sinusoids and contact with hepatocytes lead to further parenchymal destruction mediated by oxidant and protease release from neutrophils [63]. The neutrophil-mediated second phase of I/R injury is exacerbated by the further recruitment of neutrophils due, in large part, to chemotactic factors released by recruited neutrophils, including leukotriene B4 [64]. The second phase of I/R injury is manifested in very large increases in plasma ALT activity as well as significant alteration of hepatic architecture through predominantly necrotic hepatocyte death.

Hepatic I/R consists of a time period in which the blood supply to the entire liver or a portion of the liver is prevented, followed by reinstatement of blood flow. There are several variations of this protocol, but generally speaking, if the blood supply to the entire liver is prevented, both the portal vein and hepatic artery are occluded using microaneurysm clamps while in partial I/R, usually blood supply to the left later and median lobes is prevented using atraumatic microvascular clamps. The duration of ischemia is variable, with most reports utilizing between 30 and 90 min. The reperfusion period is also variable, and studies have evaluated liver injury and repair during this phase for 15 min to 168 h (1 week), depending upon the various mechanistic parameters in which the investigators were interested. Peak injury is observed usually within 12 h, followed by hepatocyte proliferation at 48 h which contributes to near complete hepatic recovery by 96 h [65]

2.2.2.5.1 MECHANISMS OF I/R INJURY AND REPAIR

As briefly mentioned above, several mechanisms contribute to hepatic I/R injury and regeneration, including important roles for IL6, complement pathway activation, and chemokines and chemokine signaling. In the next section, we will briefly explore those mechanisms.

IL6 is a well-known cytokine which signals via a heterohexameric receptor; two IL6Rα, the ligand-binding chain, and two gp130 (γ common chain), the signaling chain, are required for IL6 signaling [66]. Downstream of IL6 ligation, Janus kinases and STAT transcription factors induce cellular responses to IL6 which induce inflammation, cell growth or arrest, cell differentiation, and cell survival [67]. IL6 is a critical mediator of liver regeneration after PH [68,69], as well as after hepatic I/R when combined with PH. Specifically, 30 min of ischemia after 70% hepatectomy reduces the hepatic regenerative capacity relative to animals who underwent hepatectomy alone; this effect is attenuated by pretreatment with recombinant IL6 [70]. Likewise, IL6-deficient mice subjected to PH and 30 min ischemia also exhibit reduced proliferative capacity when compared to wild-type mice, again, corrected by pretreatment with recombinant IL6 [70]. These data clearly demonstrate the importance of IL6 to the hepatic regenerative response after ischemia.

The complement system is an ancient component of innate immunity and consist of approximately 30 soluble and membrane proteins which form three pathways, the classical, lectin, and alternative pathways [71,72]. These pathways differ based on how they are activated as well as on the proteins utilized in each activation cascade [73]. Once activated, the complement system performs three main tasks: antigen opsonization to facilitate phagocytosis, creation of a membrane attack complex (MAC) on the surface of targeted cells leading to their complement-mediated lysis, and recruitment of phagocytic cells including neutrophils and macrophages through production of the anaphylatoxins, C3a and C5a. Complement activation can occur due to pathogen invasion or sterile inflammation after tissue injury [73].

Complement activation is critical for timely liver regeneration after PH. Complement activation leads to the production of C3a and C5a, potent inflammatory mediators which are critical to liver regeneration [74]. Specifically, signaling mediated by C3a and C5a after binding their cognate receptors, C3aR and C5aR, respectively, are critical for NFκB and STAT3 translocation to the nucleus after PH [74,75]. In addition, TNFα and IL6 production is attenuated if mice are treated with a C5aR antagonist or in C3-deficient mice, suggesting the importance of this pathway to cytokine production. These perturbations lead to reduced liver regeneration and reduced survival after PH [74]. Although not yet explored in the liver, activation of complement and formation of sublytic amounts of the MAC on the surface of cells stimulates mitogenesis, *in vitro* [76]. Therefore, complement activation is a critical determinant of liver regeneration.

The complement system is also an important contributor to liver diseases [77]. As one might suspect knowing the dual role that complement activation plays in inflammatory liver injury and in liver regeneration as discussed above, complement activation in I/R injury and repair also has a dual role. For example, complement activation early in the time course after I/R promotes liver injury, for when complement is inhibited using complement receptor 2 (CR2), an inhibitor of C3 activation, or if I/R is performed in $C3^{-/-}$ mice, liver injury and inflammation is reduced relative to untreated or wild-type mice [78]. Conversely, pharmacologic deficiency of C3 by CR2 administration impairs liver regeneration after I/R. However, the effect of C3 blockade by CR2 administration differs based on dose: low-dose CR2 administration promotes, while high-dose CR2 inhibits liver regeneration [78], suggesting that a threshold level of complement activation is crucial for liver repair after I/R. A mechanism contributing to complement-mediated liver repair after I/R involves the C3a degradation product, acylation-stimulating protein (ASP), also known as C3a des Arg, and the putative ASP receptor, C5L2 [78].

Complement-mediated injury is not solely regulated by C3 or C3 degradation products. For example, interventions which block signaling by C5a also prevent liver injury and inflammation after I/R [79]; the effect of this treatment on liver regeneration has not been investigated. Likewise, mice deficient in CD59, a negative regulator of MAC formation, exhibit increased liver injury and MAC formation on hepatocytes after I/R; these effects were attenuated by cobra venom factor-mediated complement depletion [80].

An additional study by Saidi et al. took an alternative approach to understand the potential role of complement in I/R-mediated liver injury and repair [81]. In this study, C1 esterase inhibitor (C1-INH), a serine protease inhibitor, was used to inhibit complement activation. C1-INH can inhibit complement activation by preventing enzymatic activity of the classical pathway (C1r and C1s inhibition) [82–85], the lectin pathway (MASP2 inhibition) [86], and the alternative pathway (reversible interaction with C3b) [87]. C1-INH also inhibits other plasma cascade systems during I/R including the kinin, coagulation, and fibrinolytic systems [88]. C1-INH treatment reduces liver injury after I/R alone and enhances animal survival and liver regeneration in a combined model of I/R plus PH or CCl_4-induced liver injury models in mice [81]. The increased liver regenerative response found in this study contrasts with the studies discussed above in which single complement proteins or complement protein receptors are missing in genetically modified mice. The apparent discrepancy between this study and others may relate to the incredibly broad inhibitory function of C1-INH. Additional studies are required to more clearly understand these contrasting observations. Taken together, these data support the idea that a balance exists between complement-mediated injury and regeneration after hepatic I/R [78].

Chemokines play opposing roles in liver injury and regeneration after hepatic I/R. In the injury phase, neutrophils are recruited to the liver [63]. TNFα produced during the early phase of I/R injury are largely responsible for the production of the CXC chemokines, MIP2 and KC (CXCL2 and CXCL1, respectively) [62,89]. TNFα also contributes to adhesion molecule upregulation on liver sinusoidal endothelial cells [90]. These two events facilitate the recruitment and activation of neutrophils to the liver and subsequent neutrophil-mediated hepatocyte death. In the repair phase after I/R, chemokines inhibit hepatocyte proliferation; this contrasts with the role of chemokines in liver regeneration after PH [91,92]. Specifically, genetic deletion or pharmacologic inhibition of CXCR2, one of the receptors to which MIP2 and KC bind, is associated with enhanced hepatocyte proliferation and recovery from I/R-mediated injury, and this is associated with higher levels of CXC chemokines when compared to controls [93]. An additional study by the same group determined in which cell type CXCR2 expression was most important to I/R injury and repair through creation of bone marrow chimeras. The result of that study suggests that myeloid expression of CXCR2, most likely on neutrophils as Kupffer cells are CXCR2-deficient throughout I/R injury and repair, is critical for hepatocyte proliferation [94].

The different roles of CXC chemokines in PH versus I/R remains unclear, but some studies suggest that it is related to the amount of chemokine produced in each model of liver regeneration. Specifically, chemokine production is more than 10-fold greater after I/R than after PH [65,93].

Additional chemokine receptors are also implicated in liver recovery after I/R-induced injury. For example and in contrast to the role for CXCR2 in liver repair, CXCR1, the second receptor to which MIP2 and KC bind, is associated with a delay in liver repair. Consistently, hepatic CXCR1 expression is robustly induced 24-96 h after reperfusion [65], time points associated with liver repair after I/R-induced injury. This was demonstrated through studies, which performed I/R in mice with a genetic deletion of CXCR1 or in *CXCR2*$^{-/-}$ mice treated with the CXCR1/CXCR2 antagonist, repertaxin. Using these strategies, Clarke et al. discovered reduced hepatocyte proliferation and late-phase expansion of liver injury when compared to injury and repair in wild-type or untreated mice. These findings were supported by *in vitro* studies clearly demonstrating a proliferative effect of IL8, a CXCL1 ligand, on primary mouse hepatocyte [65]. Finally, another chemokine receptor, CXCR4, the receptor for stromal cell-derived factor-1 (SDF1), functions similar to CXCR1 as it attenuates hepatocyte proliferation after hepatic I/R. Through pharmacologic manipulation of SDF-1 signaling, Wilson et al. discovered increased hepatocyte proliferation when SDF-1 signaling was antagonized or reduced hepatocyte proliferation when SDF-1 signaling was agonized [95].

2.3 ALTERNATE MODELS OF LIVER GROWTH

2.3.1 Postnatal Liver Growth

Liver size is regulated such that, relative to body weight, liver weight is ~3-5% of total body mass. The mechanisms by which this proportion of liver to body mass is attained are incompletely understood. At birth, the liver undergoes robust growth during which it attains its adult mass relative to the rest of the body; this liver growth is unique in that it occurs at the same time as differentiated hepatic function develops. In mice, postnatal liver growth occurs from postnatal day (PND) 10 to PND25; adult liver-to-body weight ratios are attained between PND25 and 30 [96]. Liver growth is dependent on hepatocyte proliferation and independent of HPC [97]. Due to rapid liver growth and differentiation, availability of numerous transgenic models and reagents, the laboratory mouse is an attractive animal to study postnatal liver development.

Using postnatal liver growth, investigators have uncovered novel mechanisms responsible for liver cell proliferation, which improve our understanding of liver growth processes. For example, the Wnt/β-catenin pathway, a critical regulator of fetal liver development, is also reactivated after birth and is required for proper postnatal liver growth [98]. Indeed, total β-catenin levels as well as the amount of active β-catenin, reappear at PND5, peak around PND20 and are reduced to baseline by 3 months after birth. The importance of β-catenin reemergence in postnatal liver growth is revealed in hepatocyte-specific β-catenin null mice. Liver-to-body weight ratios in male mice deficient in β-catenin are reduced 28% relative to wild-type mice. Female β-catenin null mice also exhibit reduced liver growth, 17.5% less than female wild-type mice. Regardless of gender, reduced liver growth is associated with reduced markers of hepatocyte proliferation (PCNA) suggesting a direct relationship between β-catenin and liver growth. One potential mechanism by which β-catenin regulates postnatal growth involves loss of the interaction between β-catenin and c-met, the hepatocyte growth factor receptor, facilitating translocation of β-catenin to the nucleus where it can regulate transcription of genes involved in postnatal liver development. Further support for a role of β-catenin in postnatal liver growth is revealed in mice with a stable-mutant or full-length β-catenin displaying a three to fourfold or 15% increase in liver size attributed to hepatocyte proliferation, respectively [99,100].

Another molecule which plays an important role in postnatal liver growth is Yes-associated protein (Yap). Yap is a transcriptional co-activator and critical target of Hippo kinase pathway. This pathway is activated by physical cues such as cell shape, adhesion, and density and is inactivated in several cancers, enhancing cell proliferation and survival [101–103]. When the Hippo kinase pathway is activated, Yap-mediated transcription is inhibited [104–106]. TEAD, a transcription factor to which Yap binds, is critical for the expression of Yap target genes [104]. These genes include survival and growth-promoting molecules such as survivin/birc5, connective tissue growth factor, and cyclin D1. In mice, nuclear Yap localization peaks between PND15 and 20 [96]. Coincident with peak Yap nuclear localization is expression of cyclin D1 and liver cell proliferation [96]. The importance of Yap to postnatal liver growth in mice was revealed in $Yap^{+/-}$ mice. Liver-to-body weight ratios are significantly reduced in Yap-deficient mice at PND15 and 30, and this was associated with reduced hepatocyte proliferation [96]. Interestingly, Yap plays a dual role in postnatal liver development. In addition to its growth-promoting qualities, Yap is important for regulation of genes involved in hepatocyte differentiation including those involved in bile acid and retinoic acid metabolism [96].

The molecules involved in postnatal liver growth are under tight regulation. Once the liver reaches the required size, these molecules and signaling pathways are disengaged preventing further liver growth. It is perhaps not surprising that the same pathways important for postnatal liver growth are also implicated in hepatocellular carcinoma development [107,108] (and references therein). Clearly, understanding the mechanisms which lead to liver growth are important not only for promoting liver regeneration when needed but also important for development of anticancer therapies.

2.3.2 Pregnancy-Induced Liver Growth

There are several maternal adaptations to pregnancy. One of those adaptations required to sustain the incredible metabolic demand of the developing fetus is liver enlargement and this is due to hyperplasia and hypertrophy by mechanisms not completely defined [109–113]. It is interesting to note that liver growth precedes maternal weight gain, is proportional to number of fetuses [110], begins with implantation, and ceases with parturition [114]. Gestational hepatomegaly is due to largely two phenomena: hepatocyte proliferation which leads to an increase in cell number at the expense of cell size and a subsequent increase in the volume of those hepatocytes thereafter. These two processes occur independent of one another. Maternal liver growth in mice is a relatively fast process taking between 19 and 21 days. An interesting feature of maternal compensatory liver growth is the observation that waves of hepatocyte proliferation occur. Specifically, using Ki67 to label proliferating hepatocyte nuclei, Dai et al. observed peaks in hepatocyte proliferation on gestational days 8, 13, and 18 while at intervening time points (gestational days 10, 11, and 15) Ki67-positive hepatocyte numbers are significantly reduced [114]. These waves of hepatocyte proliferation are most closely paralleled by waves of cyclin A2, with more subtle differences in cycle E over time [114].

There are similarities and differences between compensatory liver regeneration after PH/chemical injury and pregnancy-associated liver growth. Specifically, both mechanisms are associated with IL6, TNFα, cJun, and IL1β, while maternal liver growth does not require HGF FGF1 or PXR [114]. Finally, hepatocyte ploidy does not change during pregnancy; there are equivalent percentages of 2N, 4N, and 8N hepatocytes in nonpregnant and pregnant rats [109], which is in contrast with liver regeneration after PH where cell ploidy increases during liver regrowth [115].

To support hepatic adaptation of pregnancy, many gene expression changes occur. In one study, hepatic mRNA accumulation of nephroblastoma overexpressed gene (*Nov/Ccn3*), transcription factor mammalian achaete-scute homolog-1 (*Ascl1*), patatin-like phospholipase domain-containing protein 3 (*Pnpla3*), neurotrophic tyrosine kinase receptor type 1 (*Ntrk1*), transcription factor *Egr1*, and a cell-surface protein prominin 1 (*Prom1*) increased as pregnancy progressed, especially during the second half of pregnancy [109]. These changes occur in parallel with increases in maternal liver growth. Another protein, α2 macroglobulin (α2M) is robustly increased in livers from pregnant rats, but only just before parturition when hepatocyte proliferation returns to baseline; it is not present before or after parturition [109]. Interestingly, a role of α2M in inhibiting the G1-S transition in hepatocytes is appreciated [116]. The expression of this gene, therefore, may provide signals to prevent further proliferation and instead prepare the liver for postpartum involution. The mechanisms of liver size reduction or involution postpartum are not clear but presumably, atrophy or apoptosis of existing hepatocytes is involved. This is an area in which further study is warranted.

Little is known regarding the role of hormones such as estrogen or prolactin in the maternal adaptive liver growth response during pregnancy. It is known that estrogen is a comitogen for hepatocytes [117–119]. Specifically, estrogen enhances the mitogenic effects of epidermal growth factor [118,120]. Conversely, estrogen can attenuate the antiproliferative effects of transforming growth factor β. These findings suggest estrogenic control of hepatocyte hyperplasia or hypertrophy may also be involved in maternal adaptive liver growth during pregnancy. Even less is understood regarding a role for prolactin in maternal liver growth adaptation during pregnancy. However, evidence does suggest that prolactin as well as placental lactogens mediate maternal pancreatic growth during pregnancy [121,122]. Clearly, future work is needed to understand the role of these hormones in hepatocyte hyperplasia and hypertrophy during pregnancy.

Nuclear factor erythroid 2-related factor 2 (Nrf2) is a transcription factor which regulates genes involved in protection from oxidative stress and inflammation [123,124]. Nrf2 also regulates liver regeneration. In its absence, liver regeneration after PH is profoundly delayed due to a transient resistance to insulin and insulin-like growth factor 1 induced by oxidative stress in hepatocytes [125,126]. Subsequent studies added to our understanding of the role of Nrf2 in liver regeneration by demonstrating the importance of Notch1 signaling [127]. In addition to the clear defects in liver regeneration after PH, Nrf2-deficient mice have smaller liver-to-body weight ratios than wild-type mice suggesting that Nrf2 is critically important for normal liver size control. Therefore, it is not surprising that Nrf2 is also a critical regulator of pregnancy-induced liver growth. Specifically, during the first half of gestation, Nrf2-deficient mice cannot reduce the size of their hepatocytes and exhibit a delay in maternal hepatocyte proliferation when compared to wild-type mice [128]. However, hepatocytes from Nrf2-deficient mice do increase in size during the second half of gestation [128], suggesting proper compensation by or engagement of at least part of the maternal liver growth program during pregnancy. Mechanistically, dysregulation of pregnancy-dependent activation of the AKT/mTOR/p70S6K, a pathway which participates in hepatocyte size control, is associated with reduced maternal liver growth during pregnancy [128].

In addition to Nrf2, the bile acid sensor, FXR is important for gestational hepatomegaly. During pregnancy, bile acids are increased in humans [129] and rodents [110] and are required for normal liver regeneration after PH [13]. The role of FXR in gestational liver growth was revealed in $FXR^{-/-}$ mice. Similar to livers in wild-type mice, livers in $FXR^{-/-}$ mice enlarge during pregnancy. However, the mechanism by which those livers grow differs between genotypes. Specifically, instead of an increase in individual hepatocyte size (hypertrophy), livers in pregnant $FXR^{-/-}$ mice enlarge due to an increase in hepatocyte proliferation and only limited individual cell growth [110]. The authors of this study found that hepatocytes from wild-type mice did not progress through mitosis after entering the cell cycle and, instead, increased in size and ploidy (2N) [110]. The reader will note the differences between this study and those mentioned earlier in which hepatocyte ploidy was not affected by pregnancy [109]. The propensity for FXR-deficient hepatocytes to proliferate more than normal hepatocytes is consistent with an increase in HCC incidence in $FXR^{-/-}$ mice [130]. In summary, mice lacking FXR have reduced hepatocyte hypertrophy, compensated for by increased hepatocyte hyperplasia, independent of bile acid content.

2.4 MODELS TO STUDY HPCs

In normal situations, liver regeneration is carried out by hepatocyte proliferation followed by proliferation of other cells including biliary epithelial cells. However, in situations where hepatocytes are injured beyond a certain limit that they cannot enter cell cycle or are incapable of cell division due to innate problems with DNA synthesis, the liver can regenerate using HPCs [131–133]. First noted in liver cancer pathogenesis studies in rats [134,135], these cells were termed "oval cells" due to their appearance on histological sections and in recent years, have been increasingly referred to as HPC. Studies indicate that the periportal areas, specifically the terminal bile ductules, termed Canals of Herring, are the niches for HPC [136]. Whereas most investigators agree on the existence, niche, and role of HPC, a significant disagreement remains on the markers expressed by HPC. It is agreed that HPC cannot be located in normal liver using any of the known markers and are only seen in chronically or severely injured livers where normal liver regeneration process has been compromised. A number of different models have been used to study HPC and are discussed below.

2.4.1 2-Acetylaminofluorene Combined with Partial Hepatectomy

2-AAF + PH (2-acetylaminofluorene combined with partial hepatectomy) is the classic rat model to study HPC biology. 2-AAF is a mutagen that causes DNA mutations in hepatocytes resulting in development of liver cancers. Whereas 2-AAF was associated with liver cancer since the mid-1950s, Emmanuel Farber demonstrated the first use of 2-AAF along with PH to study HPC in late 1970s [137,138]. Since then a slightly modified version of 2-AAF + PH model (also called the "Solt-Farber model") has been a gold-standard for studying HPC in rats. This model is based on the fact that a prolonged treatment with 2-AAF induces hepatic DNA intercalations inhibiting DNA synthesis. When PH is performed in these rats to stimulate liver regeneration, HPC compartment is stimulated because normal hepatocytes cannot support regeneration. The HPC undergo an expansion phase followed by differentiation resulting in regeneration of the liver. Subsequent studies have shown that chemical injury instead of PH can also induce an HPC response and modified versions of the model with 2-AAF + CCl_4 and 2-AAF + allyl alcohol have been utilized [139]. Further studies indicate that HPC reside in the terminal bile duct when the 2-AAF + PH model was combined with another chemical, methylene dianiline, that specifically injures and kills biliary epithelial cells; no HPC are found in this model [136].

The 2-AAF + PH model works only in rats and not in mice for reasons which are not completely clear but may be related to 2-AAF metabolism. In a typical protocol, male, Fisher-344 rats are used [139,140]. The first stage is subchronic administration of 2-AAF, which is achieved by either i.p. injection, oval gavage dosing or, most easily, by implanting a slow-release 2-AAF palette in the peritoneum or under the skin. The 2-AAF treatment is started for at least 1 week before performing PH or CCl_4 injections and continued throughout the entire duration of the experiment. A time course up to 3 weeks following PH is studied with sample collection typically on days 5, 10, 15, and 21 after PH (Figure 2.3i-iii). This 3-week time frame provides a window to study all phases of the HPC response including expansion (proliferation) and subsequent differentiation.

Because HPC are only a small percentage of the entire liver even at the peak of expansion, detection of HPC and studying cell signaling associated with them using techniques such as Western blotting, which require homogenizing the liver, are not very useful. Investigators have depended upon immunohistochemical analysis for detection of these cells and a number of markers that label the HPC have been proposed. These include CK-19, A6, OV6, Thy1, and EpCAM that work pretty well in rat tissues [132,140]. Whereas these markers are widely used to detect HPC, significant difference of opinion exists about the choice of the markers used.

2.4.2 The DDC Diet Model

With the advent of transgenic and knockout mouse technology, a need for a mouse model to study HPC developed mainly because the 2-AFF + PH model could not be established in mice. The most commonly used mouse model to study HPC is the DDC diet model where mice are fed a diet containing 0.01% DDC (3,5-diethoxycarbonyl-1,4-dihydrocollidine), which is also called DDTPDC (diethyl 1,4-dihydro-2,4,6-trimethyl-3,5- pyridine dicarboxylate) [140,141]. DDC inhibits heme biosynthesis and causes significant porphyria and liver injury resulting in activation of HPC compartment [142]. Interestingly, many investigators have noted that hepatocyte proliferation may not be completely inhibited in DDC diet model. Thus, the mechanisms of HPC activation in DDC model are most likely different than in 2-AAF + PH model. Nevertheless, DDC diet model remains the most popular model today to study HPC due to low cost, relative ease to use and its

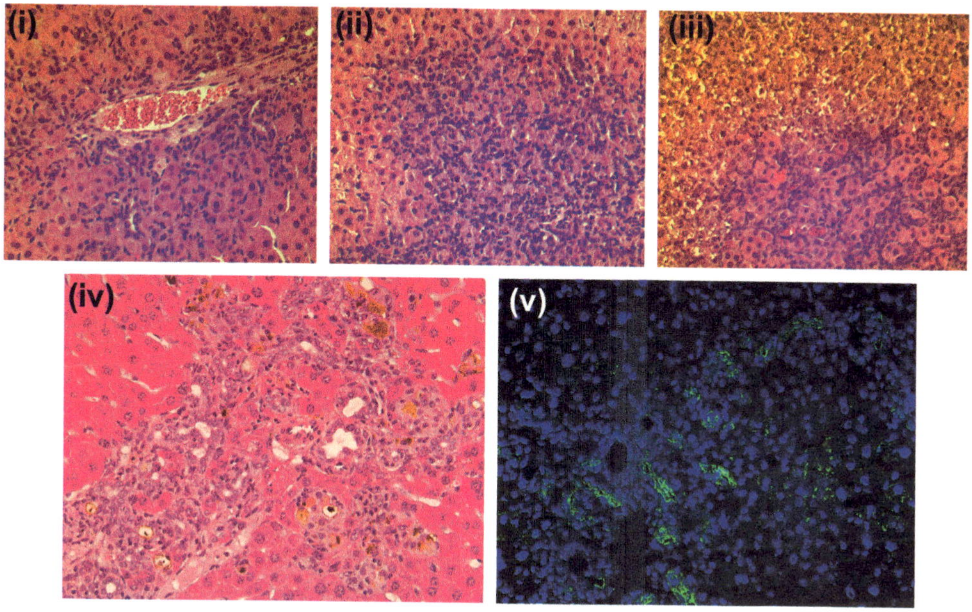

FIGURE 2.3 Hepatic progenitor cell activation in rat and mouse models. Representative photomicrographs showing HPC activation in rat 2-AAF+PH model at (i) 5 days, (ii) 10 days, and (iii) 20 days post-PH. (iv) H&E stained liver section of mouse fed DDC diet for 2 weeks showing activation of HPCs. (v) Immunofluorescence staining for CK19 (green) and DAPI (blue) on frozen liver section of mouse fed DDC diet for 2 weeks.

ability to be adapted in a number of genetically modified mice including those used for lineage-tracing experiments.

In a typical DDC study, mice are fed with a diet containing 0.01% of DDC (that can be either commercially purchased or can be made in-house) for a period of 2-4 weeks [140]. By 2 weeks, significant ductular reaction is observed in the liver (Figure 2.3iv), which further increases by 4 weeks of DDC treatment. A prolonged DDC treatment such as 6 months results in the development of liver cancers. Similar to the 2-AAF+PH model, the major techniques to study HPC biology using DDC involve either immunohistochemistry or flow cytometry following collagenase perfusion of the livers [141].

A large number of proteins have been used as HPC markers in the DDC model without much consensus in the field. The HPC that arise in DDC model stain positive for CK-19 (Figure 2.3v) A6, OV6, Thy1, EpCAM, and Foxl1 among others [132,133,141]. Dorrell et al. have recently developed and validated an array of antibodies that stain for HPC in the DDC diet model including MIC1 and OC2-1D11 [141]. However, similar to 2-AAF+PH model, the issue of HPC markers remains unresolved and highly investigated in DDC model as well.

2.4.3 The CDE Diet Model

Another model to study HPC activation in mice, though less commonly used, is the CDE diet model. In this model, mice are fed specialized choline-deficient, ethionine-supplemented diet [143]. While more commonly used in mice, this model has been used for rats [144]. Other studies indicate that the CDE diet can also induce pancreatic fibrosis and thus this model may not be specific to liver [145]. In a typical experiment, animals are fed either a specialized choline-deficient diet supplemented with 0.15% (w/v) ethionine already mixed or a choline-deficient chow supplemented with 0.15% (w/v) ethionine drinking water up to 4 weeks [143,146].

2.5 ASSAYS USED TO ASSESS LIVER REGENERATION

There are several tools that investigators utilize to quantify liver regeneration. These assays include proliferating cell nuclear antigen (PCNA) immunostaining, 5-bromo-2′-deoxyuridine (BrdU) or tritiated thymidine (^{3}H-TdR) incorporation, Ki67 immunostaining, real-time polymerase chain reaction, and immunoblotting for cyclin gene expression as well as quantification of mitotic figures. When used in conjunction with liver-to-body weight ratio calculations, these techniques provide valuable information regarding the nature of the hepatic regenerative response in wild-type animals and reveal perturbations in the regenerative response in knockout or transgenic animals.

2.5.1 PCNA Immunolocalization

PCNA is a 36-kDa auxiliary protein of DNA polymerase delta, an enzyme required for DNA synthesis [147]. PCNA expression is induced in late G1, peaks in S-phase, and is reduced thereafter [148]. One of the most useful ways to analyze PCNA protein in cells during liver regeneration is though immunostaining; the clone most commonly used for this purpose is PC10, and it can detect PCNA in formalin-fixed tissues [149]. Using this method, each cell can be evaluated with respect to the phase of the cell cycle it was in at the time the liver was removed for study [150,151] (see Figure 2.4 for detailed description on how to determine cell-cycle phase using PCNA staining). Although time consuming, this strategy makes it possible to identify how different treatments, genetic backgrounds, and other manipulations affect cell-cycle progression and identify check point activation in many different liver regeneration studies [34,52,152–154]; others have used PCNA immunolocalization to quantify cells in S-phase alone which is also very useful [155].

2.5.2 ^{3}H-TdR and BrdU Incorporation

^{3}H-TdR and BrdU are utilized to specifically interrogate S-phase of the cell cycle. Both are thymidine analogs and are integrated into replicating DNA instead of thymidine during DNA synthesis. While the former technique is associated with some reports that ^{3}H-TdR induces cell-cycle arrest and apoptosis [156,157], others demonstrated reasonable parallels between that method and a flow cytometric approach to determining DNA content in regenerating hepatocytes [158]. About 50 μCi ^{3}H-TdR in saline is administered, i.v., at various time points after PH or chemical hepatectomy and 1 h later, animals are euthanized [159]. The liver is excised and homogenized in 5% trichloroacetic acid, and DNA content and incorporated radioactivity are determined [159].

Due to the potential toxicity and regulatory concerns surrounding use of radioactive isotopes, BrdU incorporation studies later gained appreciable momentum and are commonly used today. In addition to these concerns, it became clear that immunolocalization of BrdU was more sensitive and less time consuming than quantification of ^{3}H-TdR into proliferating cells [160]. The major advance in BrdU labeling technology is attributed to Howard G. Gratzner. Dr. Gratzner is best known for the development of the first monoclonal antibody to BrdU; this development improved the specificity of BrdU labeling which was plagued by the polyclonal antibodies used previously [161,162]. In addition to flow cytometric-based approaches to quantify BrdU-positive DNA, immunolocalization

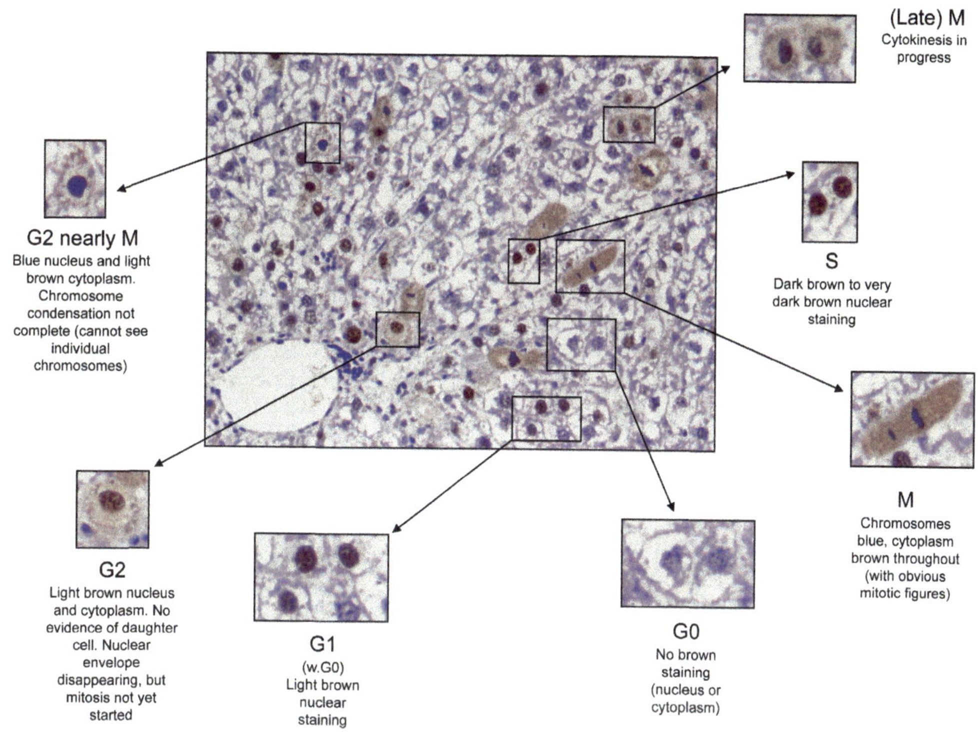

FIGURE 2.4 PCNA cell-cycle stage determination. Each phase of the cell cycle is listed in this image with a description of the visual characteristics of each phase. In brief, G0, or quiescent cells, lacks, any brown staining. G1 cells exhibit light brown nuclear staining, while S-phase cells exhibit dark brown to nearly black nuclear staining; G1 and S-phase cells have blue cytoplasm. Cells in G2 have brown cytoplasm and brown nuclei, while cells in mitosis have light brown cytoplasm but bright blue chromosomes. Please refer to the insets above for additional information.

of BrdU to hepatic nuclei is easily achieved using standard methodology.

Similar to ^{3}H-TdR, BrdU must be injected into an animal to be incorporated into DNA. This can be achieved by using the pulse-labeling technique. Specifically, BrdU in saline is administered, i.p., 1-2 h before euthanasia and liver harvest. Several different BrdU concentrations are utilized including 30, 50, and 150 μg/g body weight [34,163,164]. Some investigators utilize 5-fluoro-2′deoxyuridine, an inhibitor of thymidylate synthetase, which increases BrdU incorporation by lowering competition by endogenous thymidine during DNA synthesis. At the time of euthanasia, hepatectomy is performed and a portion of the liver is fixed in 10% neutral buffered formalin or stored for frozen tissue sectioning and later immunolocalized.

The main challenge with BrdU detection is getting the monoclonal antibody into the cell nucleus. Different methods exist to accomplish this including incubation with 2-4 N hydrochloric acid for 20-60 min at 37 °C (followed by acid neutralization in 0.1 M borate buffer), enzymatic digestion, and periodic acid (HIO_4, 60 °C, 30 min) exposure. Some investigators and companies prefer less harsh techniques such as using heat in combination with 10 mM citrate buffer, pH 6.0, antigen retrieval. Regardless, once the epitopes are revealed, the anti-BrdU monoclonal antibody can find its antigen after which, BrdU can be visualized using fluorescence or light microscopy depending on the detection system employed. Figure 2.2b shows an example of BrdU immunolocalization using the same procedure as found in Ref. [34].

2.5.3 Ki67 Immunolocalization

Another useful immunohistochemical technique to identify proliferating cells relies on the Ki67 monoclonal antibody. Its name originated from the city in which the antibody was developed (Kiel) and the number of the original clone in the cell culture plate from which it was derived [165]. This monoclonal antibody recognizes a nuclear antigen present in G1, S, G2, and M phases of the cell cycle; G0, quiescent cells do not express this antigen [165–167]. It was originally developed to determine the fraction of proliferating cells in tumors of the breast and prostate [165] but is easily applicable to any proliferating cell type, including the liver. Ki67 immunolocalization is most useful for determining the entire fraction of proliferating cells in a given sample. In contrast to PCNA immunolocalization, it is not easy to quantify in which stage of the cell cycle any given cell is found when using Ki67 staining, although subtle differences in nuclear staining pattern do exist [165].

2.6 ADDITIONAL METHODS

The methods described above rely heavily on immunohistochemical techniques. Additional assays, such as the measurement of cyclin mRNA and protein accumulation, liver-to-body weight ratio calculation, and mitotic figure enumeration, are also helpful assays to quantify cell proliferation. Flow cytometry is a powerful tool utilized to determine ploidy and therefore may reveal perturbations in cell-cycle progression [168]. In addition to thorough analysis of proliferation, it is important to assess cell death pathways when investigating liver regeneration. Assays to measure apoptosis by TUNEL staining, caspase 3 activation, and caspase 3-mediated cleavage of cytokeratin 18 are helpful in this regard. When used together, the assays described in this section are incredibly revealing and helpful to investigators who explore the complexities of liver regeneration.

References

[1] Brockes JP. Amphibian limb regeneration: rebuilding a complex structure. Science 1997;276(5309):81–7.

[2] Colwell AS, Longaker MT, Lorenz HP. Mammalian fetal organ regeneration. Adv Biochem Eng Biotechnol 2005;93:83–100.

[3] Glim JE, van Egmond M, Niessen FB, Everts V, Beelen RH. Detrimental dermal wound healing: what can we learn from the oral mucosa? Wound Repair Regen 2013;21(5):648–60.

[4] Fausto N, Campbell JS, Riehle KJ. Liver regeneration. Hepatology 2006;43(2 Suppl. 1):S45–53.

[5] Taub R. Liver regeneration: from myth to mechanism. Nat Rev 2004;5(10):836–47.

[6] Koniaris LG, McKillop IH, Schwartz SI, Zimmers TA. Liver regeneration. J Am Coll Surg 2003;197(4):634–59.

[7] Madrahimov N, Dirsch O, Broelsch C, Dahmen U. Marginal hepatectomy in the rat: from anatomy to surgery. Ann Surg 2006;244(1):89–98.

[8] Michalopoulos GK. Advances in liver regeneration. Expert Rev Gastroenterol Hepatol 2014;8(8):897–907.

[9] Apte U, Gkretsi V, Bowen WC, Mars WM, Luo JH, Donthamsetty S, et al. Enhanced liver regeneration following changes induced by hepatocyte-specific genetic ablation of integrin-linked kinase. Hepatology 2009;50(3):844–51.

[10] Higgins GM, Anderson RM. Experimental pathology of liver: Restoration of liver of the white rat following partial surgical removal. Arch Pathol 1931;12:186–202.

[11] Borude P, Edwards G, Walesky C, Li F, Ma X, Kong B, et al. Hepatocyte-specific deletion of farnesoid X receptor delays but does not inhibit liver regeneration after partial hepatectomy in mice. Hepatology 2012;56(6):2344–52.

[12] Lindroos PM, Zarnegar R, Michalopoulos GK. Hepatocyte growth factor (hepatopoietin A) rapidly increases in plasma before DNA synthesis and liver regeneration stimulated by partial hepatectomy and carbon tetrachloride administration. Hepatology 1991;13(4):743–50.

[13] Huang W, Ma K, Zhang J, Qatanani M, Cuvillier J, Liu J, et al. Nuclear receptor-dependent bile acid signaling is required for normal liver regeneration. Science 2006;312(5771):233–6.

[14] Apte U, Mehendale HM. Hepatic defenses against toxicology: liver regeneration and tissue repair. In: McQueen CA, editor. Comprehensive toxicology. 2nd ed. Philadelphia: Elsevier; 2009. p. 339–67.
[15] Lehmann V, Freudenberg MA, Galanos C. Lethal toxicity of lipopolysaccharide and tumor necrosis factor in normal and D-galactosamine-treated mice. J Exp Med 1987;165(3):657–63.
[16] Tiegs G, Wolter M, Wendel A. Tumor necrosis factor is a terminal mediator in galactosamine/endotoxin-induced hepatitis in mice. Biochem Pharmacol 1989;38 (4):627–31.
[17] Galanos C, Freudenberg MA, Reutter W. Galactosamine-induced sensitization to the lethal effects of endotoxin. Proc Natl Acad Sci USA 1979;76(11): 5939–43.
[18] Maezono K, Mawatari K, Kajiwara K, Shinkai A, Maki T. Effect of alanine on D-galactosamine-induced acute liver failure in rats. Hepatology 1996;24(5):1211–6.
[19] Chung H, Kim HJ, Jang KS, Kim M, Yang J, Kang KS, et al. Comprehensive analysis of differential gene expression profiles on D-galactosamine-induced acute mouse liver injury and regeneration. Toxicology 2006;227(1-2):136–44.
[20] Keppler D, Lesch R, Reutter W, Decker K. Experimental hepatitis induced by D-galactosamine. Exp Mol Pathol 1968;9(2):279–90.
[21] Decker K, Keppler D. Galactosamine hepatitis: key role of the nucleotide deficiency period in the pathogenesis of cell injury and cell death. Rev Physiol Biochem Pharmacol 1974;71:77–106.
[22] Keppler DO, Pausch J, Decker K. Selective uridine triphosphate deficiency induced by D-galactosamine in liver and reversed by pyrimidine nucleotide precursors. Effect on ribonucleic acid synthesis. J Biol Chem 1974;249(1):211–6.
[23] Liehr H, Grun M, Seelig HP, Seelig R, Reutter W, Heine WD. On the pathogenesis of galactosamine hepatitis. Indications of extrahepatocellular mechanisms responsible for liver cell death. Virchows Archiv B: Cell Pathol 1978;26(4):331–44.
[24] Kuhlmann WD, Peschke P. Hepatic progenitor cells, stem cells, and AFP expression in models of liver injury. Int J Exp Pathol 2006;87(5):343–59.
[25] Abdul-Hussain SK, Mehendale HM. Ongoing hepatocellular regeneration and resiliency toward galactosamine hepatotoxicity. Arch Toxicol 1992;66 (10):729–42.
[26] Apte UM, Limaye PB, Ramaiah SK, Vaidya VS, Bucci TJ, Warbritton A, et al. Upregulated promitogenic signaling via cytokines and growth factors: potential mechanism of robust liver tissue repair in calorie-restricted rats upon toxic challenge. Toxicol Sci 2002;69(2):448–59.
[27] Chilakapati J, Shankar K, Korrapati MC, Hill RA, Mehendale HM. Saturation toxicokinetics of thioacetamide: role in initiation of liver injury. Drug Metab Dispos 2005;33(12):1877–85.
[28] Mehendale HM. Tissue repair: an important determinant of final outcome of toxicant-induced injury. Toxicol Pathol 2005;33(1):41–51.
[29] Mehendale HM. Once initiated, how does toxic tissue injury expand? Trends Pharmacol Sci 2012;33 (4):200–6.
[30] Faroon O, Taylor J, Roney N, Fransen ME, Bogaczyk S, Diamond G. Toxicological profile for carbon tetrachloride. Atlanta, GA: Agency for Toxic Substances and Disease Registry (ATSDR), US Department of Health & Human Services, Public Health Service; 2005. p. 1–361.
[31] Tucker A. The history of the lava lamp. Smithsonian 2013.
[32] Weber LW, Boll M, Stampfl A. Hepatotoxicity and mechanism of action of haloalkanes: carbon tetrachloride as a toxicological model. Crit Rev Toxicol 2003;33 (2):105–36.
[33] Lindros KO. Zonation of cytochrome P450 expression, drug metabolism and toxicity in liver. Gen Pharmacol 1997;28(2):191–6.
[34] Pritchard MT, Malinak RN, Nagy LE. Early growth response (EGR)-1 is required for timely cell-cycle entry and progression in hepatocytes after acute carbon tetrachloride exposure in mice. Am J Physiol Gastrointest Liver Physiol 2011;300(6):G1124–31.
[35] Pritchard MT, Cohen JI, Roychowdhury S, Pratt BT, Nagy LE. Early growth response-1 attenuates liver injury and promotes hepatoprotection after carbon tetrachloride exposure in mice. J Hepatol 2010;53(4):655–62.
[36] Shi J, Aisaki K, Ikawa Y, Wake K. Evidence of hepatocyte apoptosis in rat liver after the administration of carbon tetrachloride. Am J Pathol 1998;153(2):515–25.
[37] Constandinou C, Henderson N, Iredale JP. Modeling liver fibrosis in rodents. Methods Mol Med 2005;117:237–50.
[38] Taub R, Greenbaum LE, Peng Y. Transcriptional regulatory signals define cytokine-dependent and -independent pathways in liver regeneration. Semin Liver Dis 1999;19(2):117–27.
[39] Peng Y, Du K, Ramirez S, Diamond RH, Taub R. Mitogenic up-regulation of the PRL-1 protein-tyrosine phosphatase gene by Egr-1. Egr-1 activation is an early event in liver regeneration. J Biol Chem 1999;274 (8):4513–20.
[40] Fitzhugh OG, Nelson AA. Liver tumors in rats fed thiourea or thioacetamide. Science 1948;108(2814):626–8.
[41] Mormone E, Lu Y, Ge X, Fiel MI, Nieto N. Fibromodulin, an oxidative stress-sensitive proteoglycan, regulates the fibrogenic response to liver injury in mice. Gastroenterology 2012;142(3):612–21, e615.

[42] Corton JC, Apte U, Anderson SP, Limaye P, Yoon L, Latendresse J, et al. Mimetics of caloric restriction include agonists of lipid-activated nuclear receptors. J Biol Chem 2004;279(44):46204–12.

[43] Abe H, Ogawa T, Wang L, Kimura M, Tanaka T, Morita R, et al. Promoter-region hypermethylation and expression downregulation of Yy1 (Yin yang 1) in preneoplastic liver lesions in a thioacetamide rat hepatocarcinogenesis model. Toxicol Appl Pharmacol 2014;280(3):467–74.

[44] Ramaiah SK, Apte U, Mehendale HM. Cytochrome P4502E1 induction increases thioacetamide liver injury in diet-restricted rats. Drug Metab Dispos 2001;29 (8):1088–95.

[45] Mangipudy RS, Mehendale HM. Temporal changes in tissue repair upon repeated exposure to thioacetamide. Toxicol Appl Pharmacol 1998;149(2):254–7.

[46] Shankar K, Vaidya VS, Wang T, Bucci TJ, Mehendale HM. Streptozotocin-induced diabetic mice are resistant to lethal effects of thioacetamide hepatotoxicity. Toxicol Appl Pharmacol 2003;188(2):122–34.

[47] Nourjah P, Ahmad SR, Karwoski C, Willy M. Estimates of acetaminophen (paracetamol)-associated overdoses in the United States. Pharmacoepidemiol Drug Saf 2006;15(6):398–405.

[48] Lee WM. Acute liver failure in the United States. Semin Liver Dis 2003;23(3):217–26.

[49] Larson AM, Polson J, Fontana RJ, Davern TJ, Lalani E, Hynan LS, et al. Acetaminophen-induced acute liver failure: results of a United States multicenter, prospective study. Hepatology 2005;42(6):1364–72.

[50] Gow PJ, Smallwood RA, Angus PW. Paracetamol overdose in a liver transplantation centre: an 8-year experience. J Gastroenterol Hepatol 1999;14(8):817–21.

[51] Jaeschke H, Williams CD, McGill MR, Xie Y, Ramachandran A. Models of drug-induced liver injury for evaluation of phytotherapeutics and other natural products. Food Chem Toxicol 2013;55:279–89.

[52] Bhushan B, Walesky C, Manley M, Gallagher T, Borude P, Edwards G, et al. Pro-regenerative signaling after acetaminophen-induced acute liver injury in mice identified using a novel incremental dose model. Am J Pathol 2014;184(11):3013–25.

[53] Jaeschke H, Williams CD, Ramachandran A, Bajt ML. Acetaminophen hepatotoxicity and repair: the role of sterile inflammation and innate immunity. Liver Int 2012;32(1):8–20.

[54] Jones DP, Lemasters JJ, Han D, Boelsterli UA, Kaplowitz N. Mechanisms of pathogenesis in drug hepatotoxicity putting the stress on mitochondria. Mol Interv 2010;10(2):98–111.

[55] Ni HM, Bockus A, Boggess N, Jaeschke H, Ding WX. Activation of autophagy protects against acetaminophen-induced hepatotoxicity. Hepatology 2012;55(1):222–32.

[56] Mehendale HM, Limaye PB. Calpain: a death protein that mediates progression of liver injury. Trends Pharmacol Sci 2005;26(5):232–6.

[57] Harrill AH, Ross PK, Gatti DM, Threadgill DW, Rusyn I. Population-based discovery of toxicogenomics biomarkers for hepatotoxicity using a laboratory strain diversity panel. Toxicol Sci 2009;110 (1):235–43.

[58] Park Y, Smith RD, Combs AB, Kehrer JP. Prevention of acetaminophen-induced hepatotoxicity by dimethyl sulfoxide. Toxicology 1988;52(1-2):165–75.

[59] Arndt K, Haschek WM, Jeffery EH. Mechanism of dimethylsulfoxide protection against acetaminophen hepatotoxicity. Drug Metab Rev 1989;20(2-4):261–9.

[60] Jaeschke H, Farhood A, Bautista AP, Spolarics Z, Spitzer JJ. Complement activates Kupffer cells and neutrophils during reperfusion after hepatic ischemia. Am J Physiol 1993;264(4 Part 1):G801–9.

[61] Jaeschke H, Farhood A. Neutrophil and Kupffer cell-induced oxidant stress and ischemia-reperfusion injury in rat liver. Am J Physiol 1991;260(3 Part 1): G355–62.

[62] Lentsch AB, Yoshidome H, Cheadle WG, Miller FN, Edwards MJ. Chemokine involvement in hepatic ischemia/reperfusion injury in mice: roles for macrophage inflammatory protein-2 and KC. Hepatology 1998;27(4):1172–7.

[63] Jaeschke H, Farhood A, Smith CW. Neutrophils contribute to ischemia/reperfusion injury in rat liver in vivo. FASEB J 1990;4(15):3355–9.

[64] Hughes H, Farhood A, Jaeschke H. Role of leukotriene B4 in the pathogenesis of hepatic ischemia-reperfusion injury in the rat. Prostaglandins Leukot Essent Fat Acids 1992;45(2):113–9.

[65] Clarke CN, Kuboki S, Tevar A, Lentsch AB, Edwards M. CXC chemokines play a critical role in liver injury, recovery, and regeneration. Am J Surg 2009;198 (3):415–9.

[66] Paonessa G, Graziani R, De Serio A, Savino R, Ciapponi L, Lahm A, et al. Two distinct and independent sites on IL-6 trigger gp 130 dimer formation and signalling. EMBO J 1995;14(9):1942–51.

[67] Hassan W, Ding L, Gao RY, Liu J, Shang J. Interleukin-6 signal transduction and its role in hepatic lipid metabolic disorders. Cytokine 2014;66(2):133–42.

[68] Trautwein C, Rakemann T, Niehof M, Rose-John S, Manns MP. Acute-phase response factor, increased binding, and target gene transcription during liver regeneration. Gastroenterology 1996;110(6):1854–62.

[69] Cressman DE, Greenbaum LE, DeAngelis RA, Ciliberto G, Furth EE, Poli V, et al. Liver failure and

defective hepatocyte regeneration in interleukin-6-deficient mice. Science 1996;274(5291):1379–83.
[70] Selzner M, Camargo CA, Clavien PA. Ischemia impairs liver regeneration after major tissue loss in rodents: protective effects of interleukin-6. Hepatology 1999;30(2):469–75.
[71] Zipfel PF, Skerka C. Complement regulators and inhibitory proteins. Nat Rev Immunol 2009;9(10):729–40.
[72] Bohana-Kashtan O, Ziporen L, Donin N, Kraus S, Fishelson Z. Cell signals transduced by complement. Mol Immunol 2004;41(6-7):583–97.
[73] Duehrkop C, Rieben R. Ischemia/reperfusion injury: effect of simultaneous inhibition of plasma cascade systems versus specific complement inhibition. Biochem Pharmacol 2014;88(1):12–22.
[74] Strey CW, Markiewski M, Mastellos D, Tudoran R, Spruce LA, Greenbaum LE, et al. The proinflammatory mediators C3a and C5a are essential for liver regeneration. J Exp Med 2003;198(6):913–23.
[75] Markiewski MM, DeAngelis RA, Strey CW, Foukas PG, Gerard C, Gerard N, et al. The regulation of liver cell survival by complement. J Immunol 2009;182(9):5412–8.
[76] Halperin JA, Taratuska A, Nicholson-Weller A. Terminal complement complex C5b-9 stimulates mitogenesis in 3 T3 cells. J Clin Invest 1993;91(5):1974–8.
[77] Qin X, Gao B. The complement system in liver diseases. Cell Mol Immunol 2006;3(5):333–40.
[78] He S, Atkinson C, Qiao F, Cianflone K, Chen X, Tomlinson S. A complement-dependent balance between hepatic ischemia/reperfusion injury and liver regeneration in mice. J Clin Invest 2009;119(8):2304–16.
[79] Arumugam TV, Woodruff TM, Stocks SZ, Proctor LM, Pollitt S, Shiels IA, et al. Protective effect of a human C5a receptor antagonist against hepatic ischaemia-reperfusion injury in rats. J Hepatol 2004;40(6):934–41.
[80] Zhang J, Hu W, Xing W, You T, Xu J, Qin X, et al. The protective role of CD59 and pathogenic role of complement in hepatic ischemia and reperfusion injury. Am J Pathol 2011;179(6):2876–84.
[81] Saidi RF, Rajeshkumar B, Shariftabrizi A, Dresser K, Walter O. Human C1 inhibitor attenuates liver ischemia-reperfusion injury and promotes liver regeneration. J Surg Res 2014;187(2):660–6.
[82] Ratnoff OD, Lepow IH. Some properties of an esterase derived from preparations of the first component of complement. J Exp Med 1957;106(2):327–43.
[83] Ratnoff OD, Pensky J, Ogston D, Naff GB. The inhibition of plasmin, plasma kallikrein, plasma permeability factor, and the C'1r subcomponent of the first component of complement by serum C'1 esterase inhibitor. J Exp Med 1969;129(2):315–31.
[84] Sim RB, Reboul A, Arlaud GJ, Villiers CL, Colomb MG. Interaction of ^{125}I-labelled complement subcomponents C-1r and C-1s with protease inhibitors in plasma. FEBS Lett 1979;97(1):111–5.
[85] Ziccardi RJ, Cooper NR. Active disassembly of the first complement component, C-1, by C-1 inactivator. J Immunol 1979;123(2):788–92.
[86] Kerr FK, Thomas AR, Wijeyewickrema LC, Whisstock JC, Boyd SE, Kaiserman D, et al. Elucidation of the substrate specificity of the MASP-2 protease of the lectin complement pathway and identification of the enzyme as a major physiological target of the serpin, C1-inhibitor. Mol Immunol 2008;45(3):670–7.
[87] Jiang H, Wagner E, Zhang H, Frank MM. Complement 1 inhibitor is a regulator of the alternative complement pathway. J Exp Med 2001;194(11):1609–16.
[88] Davis 3rd AE, Mejia P, Lu F. Biological activities of C1 inhibitor. Mol Immunol 2008;45(16):4057–63.
[89] Colletti LM, Kunkel SL, Walz A, Burdick MD, Kunkel RG, Wilke CA, et al. The role of cytokine networks in the local liver injury following hepatic ischemia/reperfusion in the rat. Hepatology 1996;23(3):506–14.
[90] Colletti LM, Cortis A, Lukacs N, Kunkel SL, Green M, Strieter RM. Tumor necrosis factor up-regulates intercellular adhesion molecule 1, which is important in the neutrophil-dependent lung and liver injury associated with hepatic ischemia and reperfusion in the rat. Shock 1998;10(3):182–91.
[91] Ren X, Carpenter A, Hogaboam C, Colletti L. Mitogenic properties of endogenous and pharmacological doses of macrophage inflammatory protein-2 after 70% hepatectomy in the mouse. Am J Pathol 2003;163(2):563–70.
[92] Colletti LM, Green M, Burdick MD, Kunkel SL, Strieter RM. Proliferative effects of CXC chemokines in rat hepatocytes in vitro and in vivo. Shock 1998;10(4):248–57.
[93] Kuboki S, Shin T, Huber N, Eismann T, Galloway E, Schuster R, et al. Hepatocyte signaling through CXC chemokine receptor-2 is detrimental to liver recovery after ischemia/reperfusion in mice. Hepatology 2008;48(4):1213–23.
[94] Van Sweringen HL, Sakai N, Quillin RC, Bailey J, Schuster R, Blanchard J, et al. Roles of hepatocyte and myeloid CXC chemokine receptor-2 in liver recovery and regeneration after ischemia/reperfusion in mice. Hepatology 2013;57(1):331–8.
[95] Wilson GC, Freeman CM, Kuethe JW, Quillin RC, 3rd, Nojima H, Schuster R, et al. CXC chemokine receptor 4 signaling limits hepatocyte proliferation after hepatic ischemia/reperfusion in mice. Am J Physiol

Gastrointest Liver Physiol 2015;308(8):G702–9. http://dx.doi.org/10.1152/ajpgi.00257.2014.
[96] Septer S, Edwards G, Gunewardena S, Wolfe A, Li H, Daniel J, et al. Yes-associated protein is involved in proliferation and differentiation during postnatal liver development. Am J Physiol Gastrointest Liver Physiol 2012;302(5):G493–503.
[97] Kennedy S, Rettinger S, Flye MW, Ponder KP. Experiments in transgenic mice show that hepatocytes are the source for postnatal liver growth and do not stream. Hepatology 1995;22(1):160–8.
[98] Apte U, Zeng G, Thompson MD, Muller P, Micsenyi A, Cieply B, et al. Beta-catenin is critical for early postnatal liver growth. Am J Physiol Gastrointest Liver Physiol 2007;292(6):G1578–85.
[99] Cadoret A, Ovejero C, Saadi-Kheddouci S, Souil E, Fabre M, Romagnolo B, et al. Hepatomegaly in transgenic mice expressing an oncogenic form of beta-catenin. Cancer Res 2001;61(8):3245–9.
[100] Tan X, Apte U, Micsenyi A, Kotsagrelos E, Luo JH, Ranganathan S, et al. Epidermal growth factor receptor: a novel target of the Wnt/beta-catenin pathway in liver. Gastroenterology 2005;129(1):285–302.
[101] Harvey KF, Zhang X, Thomas DM. The Hippo pathway and human cancer. Nat Rev Cancer 2013;13(4):246–57.
[102] Low BC, Pan CQ, Shivashankar GV, Bershadsky A, Sudol M, Sheetz M. YAP/TAZ as mechanosensors and mechanotransducers in regulating organ size and tumor growth. FEBS Lett 2014;588(16):2663–70.
[103] Zhou D, Conrad C, Xia F, Park JS, Payer B, Yin Y, et al. Mst1 and Mst2 maintain hepatocyte quiescence and suppress hepatocellular carcinoma development through inactivation of the Yap1 oncogene. Cancer Cell 2009;16(5):425–38.
[104] Johnson R, Halder G. The two faces of Hippo: targeting the Hippo pathway for regenerative medicine and cancer treatment. Nat Rev Drug Discov 2014;13(1):63–79.
[105] Dong J, Feldmann G, Huang J, Wu S, Zhang N, Comerford SA, et al. Elucidation of a universal size-control mechanism in Drosophila and mammals. Cell 2007;130(6):1120–33.
[106] Happe H, de Heer E, Peters DJ. Polycystic kidney disease: the complexity of planar cell polarity and signaling during tissue regeneration and cyst formation. Biochim Biophys Acta 2011;1812(10):1249–55.
[107] Liu AM, Xu MZ, Chen J, Poon RT, Luk JM. Targeting YAP and Hippo signaling pathway in liver cancer. Expert Opin Ther Targets 2010;14(8):855–68.
[108] Thompson MD, Monga SP. WNT/beta-catenin signaling in liver health and disease. Hepatology 2007;45(5):1298–305.
[109] Bustamante JJ, Copple BL, Soares MJ, Dai G. Gene profiling of maternal hepatic adaptations to pregnancy. Liver Int 2010;30(3):406–15.
[110] Milona A, Owen BM, van Mil S, Dormann D, Mataki C, Boudjelal M, et al. The normal mechanisms of pregnancy-induced liver growth are not maintained in mice lacking the bile acid sensor FxR. Am J Physiol Gastrointest Liver Physiol 2010;298(2):G151–8.
[111] Smith RW. The effects of pregnancy and lactation on the activities in rat liver of some enzymes associated with glucose metabolism. Biochim Biophys Acta 1975;411(1):22–9.
[112] Smith RW, Walsh A. Composition of liver lipids of the rat during pregnancy and lactation. Lipids 1975;10(10):643–5.
[113] Hollister A, Okubara P, Watson JG, Chaykin S. Reproduction in mice: liver enlargement in mice during pregnancy and lactation. Life Sci 1987;40(1):11–8.
[114] Dai G, Bustamante JJ, Zou Y, Myronovych A, Bao Q, Kumar S, et al. Maternal hepatic growth response to pregnancy in the mouse. Exp Biol Med 2011;236(11):1322–32.
[115] Sigal SH, Rajvanshi P, Gorla GR, Sokhi RP, Saxena R, Gebhard Jr. DR, et al. Partial hepatectomy-induced polyploidy attenuates hepatocyte replication and activates cell aging events. Am J Physiol 1999;276(5 Part 1):G1260–72.
[116] Nadal C. Some aspects of the regulation of hepatocyte proliferation. Toxicol Pathol 1986;14(3):349–52.
[117] Yager JD, Zurlo J, Ni N. Sex hormones and tumor promotion in liver. Proc Soc Exp Biol Med 1991;198(2):667–74.
[118] Ni N, Yager JD. Comitogenic effects of estrogens on DNA synthesis induced by various growth factors in cultured female rat hepatocytes. Hepatology 1994;19(1):183–92.
[119] Ni N, Yager JD. The co-mitogenic effects of various estrogens for TGF-alpha-induced DNA synthesis in cultured female rat hepatocytes. Cancer Lett 1994;84(2):133–40.
[120] Shi YE, Yager JD. Effects of the liver tumor promoter ethinyl estradiol on epidermal growth factor-induced DNA synthesis and epidermal growth factor receptor levels in cultured rat hepatocytes. Cancer Res 1989;49(13):3574–80.
[121] Huang C, Snider F, Cross JC. Prolactin receptor is required for normal glucose homeostasis and modulation of beta-cell mass during pregnancy. Endocrinology 2009;150(4):1618–26.
[122] Kim H, Toyofuku Y, Lynn FC, Chak E, Uchida T, Mizukami H, et al. Serotonin regulates pancreatic beta cell mass during pregnancy. Nat Med 2010;16(7):804–8.
[123] Copple IM. The Keap1-Nrf2 cell defense pathway—a promising therapeutic target? Adv Pharmacol 2012;63:43–79.
[124] Bataille AM, Manautou JE. Nrf2: a potential target for new therapeutics in liver disease. Clin Pharmacol Ther 2012;92(3):340–8.

[125] Beyer TA, Xu W, Teupser D, auf dem, Keller U, Bugnon P, Hildt E, et al. Impaired liver regeneration in Nrf2 knockout mice: role of ROS-mediated insulin/IGF-1 resistance. EMBO J 2008;27(1):212–23.

[126] Beyer TA, Werner S. The cytoprotective Nrf2 transcription factor controls insulin receptor signaling in the regenerating liver. Cell Cycle 2008;7(7):874–8.

[127] Wakabayashi N, Shin S, Slocum SL, Agoston ES, Wakabayashi J, Kwak MK, et al. Regulation of notch1 signaling by nrf2: implications for tissue regeneration. Sci Signal 2010;3(130):ra52.

[128] Zou Y, Hu M, Bao Q, Chan JY, Dai G. Nrf2 participates in regulating maternal hepatic adaptations to pregnancy. J Cell Sci 2013;126(Part 7):1618–25.

[129] Castano G, Lucangioli S, Sookoian S, Mesquida M, Lemberg A, Di Scala M, et al. Bile acid profiles by capillary electrophoresis in intrahepatic cholestasis of pregnancy. Clin Sci 2006;110(4):459–65.

[130] Kim I, Morimura K, Shah Y, Yang Q, Ward JM, Gonzalez FJ. Spontaneous hepatocarcinogenesis in farnesoid X receptor-null mice. Carcinogenesis 2007;28 (5):940–6.

[131] Wang X, Foster M, Al-Dhalimy M, Lagasse E, Finegold M, Grompe M. The origin and liver repopulating capacity of murine oval cells. Proc Natl Acad Sci USA 2003;100(Suppl. 1):11881–8.

[132] Shin S, Kaestner KH. The origin, biology, and therapeutic potential of facultative adult hepatic progenitor cells. Curr Top Dev Biol 2014;107:269–92.

[133] Duncan AW, Dorrell C, Grompe M. Stem cells and liver regeneration. Gastroenterology 2009;137 (2):466–81.

[134] Kinosita R. Studies on the cancerogenic chemical substances. Trans Soc Pathol Jpn 1937;27:329–34.

[135] Bird TG, Lorenzini S, Forbes SJ. Activation of stem cells in hepatic diseases. Cell Tissue Res 2008;331 (1):283–300.

[136] Petersen BE, Zajac VF, Michalopoulos GK. Bile ductular damage induced by methylene dianiline inhibits oval cell activation. Am J Pathol 1997;151 (4):905–9.

[137] Solt DB, Medline A, Farber E. Rapid emergence of carcinogen-induced hyperplastic lesions in a new model for the sequential analysis of liver carcinogenesis. Am J Pathol 1977;88(3):595–618.

[138] Tatematsu M, Ho RH, Kaku T, Ekem JK, Farber E. Studies on the proliferation and fate of oval cells in the liver of rats treated with 2-acetylaminofluorene and partial hepatectomy. Am J Pathol 1984;114 (3):418–30.

[139] Petersen BE, Zajac VF, Michalopoulos GK. Hepatic oval cell activation in response to injury following chemically induced periportal or pericentral damage in rats. Hepatology 1998;27(4):1030–8.

[140] Apte U, Thompson MD, Cui S, Liu B, Cieply B, Monga SP. Wnt/beta-catenin signaling mediates oval cell response in rodents. Hepatology 2008;47 (1):288–95.

[141] Dorrell C, Erker L, Lanxon-Cookson KM, Abraham SL, Victoroff T, Ro S, et al. Surface markers for the murine oval cell response. Hepatology 2008;48 (4):1282–91.

[142] Tephly TR, Gibbs AH, De Matteis F. Studies on the mechanism of experimental porphyria produced by 3,5-diethoxycarbonyl-1,4-dihydrocollidine. Role of a porphyrin-like inhibitor of protohaem ferro-lyase. Biochem J 1979;180(1):241–4.

[143] Akhurst B, Croager EJ, Farley-Roche CA, Ong JK, Dumble ML, Knight B, et al. A modified choline-deficient, ethionine-supplemented diet protocol effectively induces oval cells in mouse liver. Hepatology 2001;34(3):519–22.

[144] Croager EJ, Smith PG, Yeoh GC. Ethanol interactions with a choline-deficient, ethionine-supplemented feeding regime potentiate pre-neoplastic cellular alterations in rat liver. Carcinogenesis 2002;23(10): 1685–93.

[145] Gray KD, Simovic MO, Blackwell TS, Christman JW, May AK, Parman KS, et al. Activation of nuclear factor kappa B and severe hepatic necrosis may mediate systemic inflammation in choline-deficient/ethionine-supplemented diet-induced pancreatitis. Pancreas 2006;33(3):260–7.

[146] Shin S, Upadhyay N, Greenbaum LE, Kaestner KH. Ablation of Foxl1-Cre-labeled hepatic progenitor cells and their descendants impairs recovery of mice from liver injury. Gastroenterology 2015;148(1):192–202, e193.

[147] Bravo R, Frank R, Blundell PA, Macdonald-Bravo H. Cyclin/PCNA is the auxiliary protein of DNA polymerase-delta. Nature 1987;326(6112):515–7.

[148] Kurki P, Vanderlaan M, Dolbeare F, Gray J, Tan EM. Expression of proliferating cell nuclear antigen (PCNA)/cyclin during the cell cycle. Exp Cell Res 1986;166(1):209–19.

[149] Hall PA, Woods AL. Immunohistochemical markers of cellular proliferation: achievements, problems and prospects. Cell Tissue Kinet 1990;23(6):505–22.

[150] Greenwell A, Foley JF, Maronpot RR. An enhancement method for immunohistochemical staining of proliferating cell nuclear antigen in archival rodent tissues. Cancer Lett 1991;59(3):251–6.

[151] Eldrige SR, Butterworth BE, Goldsworthy TL. Proliferating cell nuclear antigen: a marker for hepatocellular proliferation in rodents. Environ Health Perspect 1993;101(Suppl. 5):211–8.

[152] Wang T, Fontenot RD, Soni MG, Bucci TJ, Mehendale HM. Enhanced hepatotoxicity and toxic

outcome of thioacetamide in streptozotocin-induced diabetic rats. Toxicol Appl Pharmacol 2000;166(2):92–100.
[153] Donthamsetty S, Bhave VS, Mitra MS, Latendresse JR, Mehendale HM. Nonalcoholic fatty liver sensitizes rats to carbon tetrachloride hepatotoxicity. Hepatology 2007;45(2):391–403.
[154] von Montfort C, Beier JI, Kaiser JP, Guo L, Joshi-Barve S, Pritchard MT, et al. PAI-1 plays a protective role in CCl_4-induced hepatic fibrosis in mice: role of hepatocyte division. Am J Physiol Gastrointest Liver Physiol 2010;298(5):G657–66.
[155] Bruccoleri A, Gallucci R, Germolec DR, Blackshear P, Simeonova P, Thurman RG, et al. Induction of early-immediate genes by tumor necrosis factor alpha contribute to liver repair following chemical-induced hepatotoxicity. Hepatology 1997;25(1):133–41.
[156] Maurer HR. Potential pitfalls of [^{3}H]thymidine techniques to measure cell proliferation. Cell Tissue Kinet 1981;14(2):111–20.
[157] Hu VW, Black GE, Torres-Duarte A, Abramson FP. ^{3}H-thymidine is a defective tool with which to measure rates of DNA synthesis. FASEB J 2002;16(11):1456–7.
[158] Digernes V, Bronstad G, Sand TE, Christoffersen T. The proliferation response of rat liver parenchymal cells after partial hepatectomy. A methodological study comparing flow cytometry of nuclear DNA content and in vivo and in vitro uptake of thymidine. Cell Tissue Kinet 1982;15(5):521–8.
[159] Russell WE, Bucher NL. Vasopressin modulates liver regeneration in the Brattleboro rat. Am J Physiol 1983;245(2):G321–4.
[160] Raza A, Ucar K, Bhayana R, Kempski M, Preisler HD. Utility and sensitivity of anti BrdU antibodies in assessing S-phase cells compared to autoradiography. Cell Biochem Funct 1985;3(2):149–53.
[161] Gratzner HG. Monoclonal antibody to 5-bromo- and 5-iododeoxyuridine: a new reagent for detection of DNA replication. Science 1982;218(4571):474–5.
[162] Leif RC, Stein JH, Zucker RM. A short history of the initial application of anti-5-BrdU to the detection and measurement of S phase. Cytometry A 2004;58 (1):45–52.
[163] Luedde T, Rodriguez ME, Tacke F, Xiong Y, Brenner DA, Trautwein C. p18(INK4c) collaborates with other CDK-inhibitory proteins in the regenerating liver. Hepatology 2003;37(4):833–41.
[164] Drixler TA, Vogten MJ, Ritchie ED, van Vroonhoven TJ, Gebbink MF, Voest EE, et al. Liver regeneration is an angiogenesis-associated phenomenon. Ann Surg 2002;236(6):703–11, discussion 711–702.
[165] Scholzen T, Gerdes J. The Ki-67 protein: from the known and the unknown. J Cell Physiol 2000;182 (3):311–22.
[166] Gerdes J, Schwab U, Lemke H, Stein H. Production of a mouse monoclonal antibody reactive with a human nuclear antigen associated with cell proliferation. Int J Cancer 1983;31(1):13–20.
[167] Gerdes J, Lemke H, Baisch H, Wacker HH, Schwab U, Stein H. Cell cycle analysis of a cell proliferation-associated human nuclear antigen defined by the monoclonal antibody Ki-67. J Immunol 1984;133 (4):1710–5.
[168] Darzynkiewicz Z, Halicka HD, Zhao H. Analysis of cellular DNA content by flow and laser scanning cytometry. Adv Exp Med Biol 2010;676:137–47.

CHAPTER

3

Liver Regeneration in Zebrafish

*Wolfram Goessling**,†

*Brigham and Women's Hospital/Dana-Farber Cancer Institute, Harvard Medical School, Boston, Massachusetts, USA

†Harvard Stem Cell Institute and Broad Institute of MIT and Harvard, Cambridge, Massachusetts, USA

3.1 INTRODUCTION

Regenerative medicine holds the keys to regaining organ function after injury. Regenerative therapies can promote, enhance, and reestablish organ-specific repair processes and recover and repair organ structure and function after injury. In clinical practice, regenerative medicine has been largely limited to a few specialized although highly successful interventions, such as partial liver transplantation [1], autologous bone marrow transplantation [2], or skin grafting [3]. Many studies have focused on the discovery and isolation of specialized cell types (e.g., embryonic or adult stem cells) or small molecules that can boost the innate ability of the human body to achieve endogenous regenerative repair in the liver.

Zebrafish, native to the river basins in East India, were originally discovered in the Ganges River by Hamilton in 1822 [4] and are a popular aquarium species. They were introduced as a research animal model system by Streisinger and colleagues in the 1970s [5], followed by ground-breaking genetic screening efforts in Boston and Tübingen [6]. Over the last three decades, zebrafish have been very successfully studied to enhance our understanding of vertebrate developmental biology and organogenesis. One important finding was the identification and iterative use of key signaling pathways involved in germ layer development, axis formation, and specification and differentiation of mature organs. Fish, however, were first used as a scientific model organism in the field of regenerative medicine: Broussonet described regeneration of the pectoral fins of goldfish in 1786, and Thomas Hunt Morgan studied regeneration of amputated fish fins in 1901, as described in his book *Regeneration* [7], even prior to his seminal studies in fruit flies that founded the field of modern genetics.

In recent years, zebrafish have been used to discover mechanisms of liver repair both after physical and toxic injury. These findings have enhanced our understanding of the cellular and molecular mechanisms involved in organ repair, showing striking conservation of genetic regulation across organ systems. More recently, chemical genetic screens applied in conjunction

Liver Regeneration
http://dx.doi.org/10.1016/B978-0-12-420128-6.00003-8

with clinically relevant injury models have led to translational efforts aimed at the introduction of an array of novel therapeutic options for the field of regenerative medicine [8–11]. This chapter examines the utility of the zebrafish model in the elucidation of cellular and molecular mechanisms of liver regeneration.

3.2 THE THERAPEUTIC IMPACT OF ZEBRAFISH RESEARCH

The impact of the zebrafish as a model for organ development and disease has been demonstrated in many organ systems, including the liver. Large chemical mutagenesis screens using the potent mutagen *N*-ethyl-*N*-nitrosourea (ENU) identified essential genes and gave insight into the physiology and pathophysiology of liver development and function. Other forward genetic screening methods, such as mutagenesis via viral insertions [12] or use of transposon-mediated gene disruption [13], have yielded equally important and clinically applicable findings, which have been reviewed in detail [14,15]. These studies are relevant to liver regeneration because they provided clues into the genes, signaling cascades, and cellular networks that are important for creating both the structure and function of the liver. Therefore, studies of liver development using zebrafish mutants or chemical genetics can provide insights to re-establish or augment cellular differentiation or tissue function to aid regenerative repair *in vivo*, or even for the step-wise production and expansion of stem and progenitor populations *in vitro* for therapeutic cellular replacement strategies. One example for this latter approach is the use of the soluble factor Activin A to induce definitive endoderm *in vitro* from pluripotent stem cells, which is required to induce hepatocyte-like cells. Activin A is used because of its established role in nodal signaling in endoderm specification during embryogenesis discovered in zebrafish [16].

Modern genome-editing methods allow focused studies to validate mechanism or disease relevance for particular genes of interest. Prior to the advent of direct genome editing made available in recent years, an approach known as targeting-induced local lesions in genomes [17], exploited the potent mutagenic activity of ENU, used in the forward genetic screens and the availability of the zebrafish genome [18]. These studies delivered zebrafish mutants with striking phenotypic correlations to the associated clinical manifestations of the human disease mutation. For example, zebrafish heterozygous for a mutation in *Adenomatous Polyposis Coli* (*APC*) gene, a central regulator of β-catenin stability and Wnt signaling, develop intestinal tumors comparable to the intestinal polyposis in humans with a corresponding *APC* mutation at the genotypic and phenotypic level [19]. Further studies revealed roles for APC in liver development and liver regeneration [20]. More recently, direct gene targeting using zinc fingers [21], TAL effector nucleases [55], and CRISPR/Cas9 [22] in combination with the zebrafish mutation project [23] has brought the opportunity to mutate any gene in the genome, thereby widely expanding the experimental opportunities in the field.

Chemical screens are now widely used to interrogate biological processes from the earliest stages of development to organogenesis and organ repair. They provide insight into disease pathophysiology and led to the discovery and clinical application of novel treatment strategies [9,24]. As such, they represent a *direct* opportunity to translate findings from the "fish tank to the bedside." Chemical screens in the zebrafish were pioneered initially to identify compounds that could improve a mutant phenotype, usually relevant to a human disease state [25–27]. Subsequent studies were designed to identify novel compound regulators of well-conserved aspects organ formation, typically using chemical libraries composed of bioactive compounds [28].

3.3 ADULT LIVER ANATOMY AND PHYSIOLOGY

The adult zebrafish liver has a trilobar structure and is positioned at the cranial end of the intestinal tract [29,30]. Hepatocytes are arranged in bilayered cords separated by endothelial sinusoids distributed radially around a central vein. As in the mammalian liver, zebrafish hepatocytes are polarized, with the basal membrane facing the sinusoidal endothelial cells and the apical membrane forming the bile canaliculi. In contrast to the mammalian liver, the zebrafish liver is not zonated into liver lobules with a central vein and portal fields.

3.4 LIVER REGENERATION FOLLOWING PARTIAL HEPATECTOMY

The liver regenerates robustly after surgical injury, and, as outlined elsewhere, rodent models for partial hepatectomy have existed for more than 80 years [31]. An equivalent surgical resection approach can be effectively applied to adult zebrafish: a single liver lobe can be directly removed using microdissection scissors and an anterolateral surgical approach in the anesthetized fish [20,32]. No sutures or other forms of wound closure are necessary. The procedure is quick, highly efficient, allowing for dozens of fish to be resected within a few hours with high survival rates approaching 100%. In contrast to the compensatory hyperplasia observed in mammals, local regrowth is observed after resection of a single lobe with recovery of the original lobular structure which can be quantified easily by lobe length and recovery of liver:body weight ratios. Typically, 1/3 of the zebrafish liver is removed in most published studies. About 2/3 partial hepatectomy in zebrafish causes overwhelming mortality, with no animals surviving for 2 days [33] after resection; it remains unclear if that discrepancy represents a functional difference between fish and mammalian livers or simply a technical problem. These surgical approaches have identified important novel regulators of liver regrowth: innovative studies by Sadler et al. discovered that the cell-cycle regulator *uhrf1*, mutated via viral insertion, regulates both embryonic liver development and liver regeneration after partial hepatectomy: adult *uhrf1*$^{+/-}$ fish have reduced regrowth in the resected lobe [32]. We describe a similar surgical technique and employ ultrasound volumetric analysis in addition to length measurements to quantify regrowth to reveal similar regenerative kinetics between the liver of zebrafish and mice, each needing approximately 5-7 days for full recovery after surgical resection [20]. Through the use of *APC* mutant zebrafish and mice, this study also illustrates the conserved functional importance of Wnt signaling during embryonic and adult liver growth across vertebrate species [34,35]. Subsequent investigations demonstrate the maintenance of regulatory interactions between prostaglandin E2 (PGE2) and the Wnt signaling pathways in vertebrate liver regeneration [36]. In particular, the COX2 inhibitor indomethacin limits liver regeneration in both *apc*$^{+/-}$ fish and *Apc*$^{+/min}$ mice. Subsequent studies from our laboratory reveal that PGE2 exposure or inhibition of the PGE2-degrading enzyme prostaglandin dehydrogenase increases liver regeneration after partial hepatectomy, which is mediated by the pger4 receptor [37]. Subsequent resection studies performed by several groups have shown a role for topoisomerase 2a [38] and for both fibroblast growth factor and bone morphogenetic protein signaling in optimal liver repair [33]. Recent studies have also demonstrated that a fibrotic scar forms at the resection site [39]. Immunohistochemical analysis demonstrated that *def*-deficient zebrafish produce a prolonged inflammatory response with impaired resolution of the fibrotic scar.

Similar to the published work on uhrf1, WNT signaling, or PGE2, other studies exploit the strong conservation between signaling pathways affecting liver organogenesis and regeneration: building on developmental studies implicating nitric oxide and *S*-nitrosothiol signaling in hepatic growth, we examined the impact of *S*-glutathione reductase (GSNOR) on liver regrowth after partial hepatectomy: chemical inhibition of GSNOR accelerated hepatic regrowth [40]. In contrast, impaired NO signaling after to the NO synthase inhibitor L-NAME reduced liver repair. These studies have provided new insights into the regulation of hepatic growth following partial hepatectomy, which can provide a foundation for screening for novel chemical and genetic modifiers of liver regeneration.

3.5 DRUG-INDUCED HEPATOTOXICITY

Other zebrafish models of liver regeneration have focused on the induction of hepatocyte injury and necrosis, rather than surgical removal of tissue. The first regenerative response in the zebrafish liver was observed after toxic injury: exposure to 4-chloroaniline affected hepatic cytoarchitecture and ultrastructure and survival in a dose-dependent manner [41]. We have pioneered a clinically relevant model of acetaminophen (APAP)-induced liver injury [8,40,42]. APAP is the most common cause of acute liver failure, and only one FDA-approved therapy, *N*-acetylcysteine (NAC), is available to limit oxidative cellular damage [43]. A chemical modifier screen in APAP-exposed zebrafish larvae identified PGE2 also as a compound to act synergistically with NAC to limit liver damage and increase fish survival [8]. PGE2 activates the Wnt pathway to stimulate a regenerative response. In separate work, we discovered that nitric oxide and *S*-nitrosothiol signaling enhances not only liver growth after partial hepatectomy but also survival after APAP-induced liver injury [40]. GSNOR inhibition provided a hepatoprotective effect through sustained activation of the Nrf2 oxidative stress response pathway. GSNOR inhibition worked synergistically with NAC and enhanced survival even after delayed treatment. GSNOR-deficient mice were resistant to APAP-induced liver injury, confirming the conservation of hepatoprotective properties provided by *S*-nitrosothiol signaling across vertebrates. These studies further demonstrate similarities between organ development, physical injury, and toxic injury and support the hypothesis that inflammatory mediators (PGE2 and NO), which are generated in response to liver injury in the endothelium, can act to initiate and orchestrate regeneration [44].

3.6 GENETIC HEPATOCYTE ABLATION

A novel methodology to induce liver injury in zebrafish employs nitroreductase (NTR)-mediated hepatocyte ablation. Originally, envisioned for cancer gene therapy studies that used NTR as a prodrug gene suicide system to eliminate cancer cells [45], this approach has been utilized in *in vivo* studies in transgenic mice expressing NTR to efficiently ablate specific cell compartments [46,47]; however, its use in mammals is limited due to substantial bystander effects, resulting in the death of neighboring cells [48]. In the zebrafish, pioneering studies by Stainier [49,50] and Parsons [51,52] led to the adaptation of the NTR system to target specific organs and cell types, employing an alternative prodrug metronidazole. This compound can be easily applied by aqueous exposure, is well tolerated and has substantially reduced off-target and bystander effects. Transgenic fish were developed for use in liver regeneration studies that expressed a cyan fluorescent protein-NTR fusion driven by the hepatocyte promoter fatty acid-binding protein (fabp10a) [49,50]. Recent studies have utilized this approach to reveal that biliary epithelial cells (BECs) transdifferentiate to mature hepatocytes during liver regeneration following hepatocyte ablation [53,54]. These studies used a Cre/lox approach to label BECs and follow their fate during liver regeneration after hepatocyte ablation, thereby exploiting several advantages of the zebrafish system, namely, its optical clarity during embryo and larval stages to enable high-resolution visualization of liver regeneration and cell-fate mapping combined with targeted cell ablation. BECs first dedifferentiated into hepatic progenitors and then became hepatocytes. These studies demonstrate that *sox9b* was required for BEC transdifferentiation [54], while *wnt2bb* stimulates hepatocyte proliferation [53]. This cell ablation approach has also been used by Huang et al. to investigate the impact of fibrosis on regeneration, by exposing zebrafish larvae to ethanol to induce a response by hepatic stellate cells, resulting in the deposition of laminin and collagen. These studies illustrate the impact of toxin-mediated genetically directed ablation system that can be utilized in zebrafish larvae and adults, enabling *in vivo* imaging, cell-fate mapping, genetic epistasis, and drug screening approaches.

3.7 SUMMARY

Zebrafish have substantially contributed to liver research in the last two decades, beginning with the identification of essential signaling molecules during endoderm development, liver specification and liver differentiation, to the development of important, clinically relevant approaches to induce physical, toxic, or metabolic liver injury. Many pathways involved in liver organogenesis are conserved and reutilized and reactivated during liver regrowth after injury. Further, virtually all signals examined in zebrafish regeneration to date are conserved in mammals, indicating their direct relevance and immediate applicability to human disease and clinical problems. The combination of advanced genome-editing methods, high-resolution *in vivo* imaging, chemical genetics, and cutting-edge transcriptomic, genomic, and epigenomic techniques will further solidify the impact and innovative power of the zebrafish for our understanding of liver regeneration and regrowth with the aim to improve the outcome for patients with acute or chronic liver failure.

Acknowledgments

The author thanks Dr. Trista North and all members of the Goessling laboratory for ongoing and stimulating discussions. This work was supported by NIH NIDDK 1R01DK090311 (W. G.). W. G. is a Pew Scholar in the Biomedical Sciences.

References

[1] Vagefi PA, Parekh J, Ascher NL, Roberts JP, Freise CE. Outcomes with split liver transplantation in 106 recipients: the University of California, San Francisco, experience from 1993 to 2010. Arch Surg 2011;146(9):1052–9.

[2] Clift RA, Thomas ED. Seattle Marrow Transplant T. Follow-up 26 years after treatment for acute myelogenous leukemia. N Engl J Med 2004;351(23):2456–7.

[3] Markeson D, Pleat JM, Sharpe JR, Harris AL, Seifalian AM, Watt SM. Scarring, stem cells, scaffolds and skin repair. J Tissue Eng Regen Med 2013 http://dx.doi.org/10.1002/term.1841[Epub ahead of print].

[4] Hamilton F. An account of the fishes found in the river Ganges and its branches. Edinburgh: Archibald Constable and Company; 1822.

[5] Streisinger G, Walker C, Dower N, Knauber D, Singer F. Production of clones of homozygous diploid zebra fish (*Brachydanio rerio*). Nature 1981;291(5813):293–6.

[6] Mullins MC, Hammerschmidt M, Haffter P, Nusslein-Volhard C. Large-scale mutagenesis in the zebrafish: in search of genes controlling development in a vertebrate. Curr Biol 1994;4(3):189–202.

[7] Sunderland ME. Regeneration: Thomas Hunt Morgan's window into development. J Hist Biol 2010;43(Summer (2)):325–61.

[8] North TE, Babu IR, Vedder LM, et al. PGE2-regulated wnt signaling and *N*-acetylcysteine are synergistically hepatoprotective in zebrafish acetaminophen injury. Proc Natl Acad Sci USA 2010;107(40):17315–20.

[9] North TE, Goessling W, Walkley CR, et al. Prostaglandin E2 regulates vertebrate haematopoietic stem cell homeostasis. Nature 2007;447(7147):1007–11.

[10] Esterberg R, Coffin AB, Ou H, Simon JA, Raible DW, Rubel EW. Fish in a dish: drug discovery for hearing

habilitation. Drug Discov Today Dis Models 2013;10(1) http://dx.doi.org/10.1016/j.ddmod.2012.02.001.
[11] Sanker S, Cirio MC, Vollmer LL, et al. Development of high-content assays for kidney progenitor cell expansion in transgenic zebrafish. J Biomol Screen 2013;18 (10):1193–202.
[12] Amsterdam A, Burgess S, Golling G, et al. A large-scale insertional mutagenesis screen in zebrafish. Genes Dev 1999;13(20):2713–24.
[13] Kotani T, Nagayoshi S, Urasaki A, Kawakami K. Transposon-mediated gene trapping in zebrafish. Methods 2006;39(3):199–206.
[14] Patton EE, Zon LI. The art and design of genetic screens: zebrafish. Nat Rev Genet 2001;2(12):956–66.
[15] Jing L, Zon LI. Zebrafish as a model for normal and malignant hematopoiesis. Dis Model Mech 2011;4(4):433–8.
[16] Schier AF. Nodal signaling in vertebrate development. Annu Rev Cell Dev Biol 2003;19:589–621.
[17] Wienholds E, van Eeden F, Kosters M, Mudde J, Plasterk RH, Cuppen E. Efficient target-selected mutagenesis in zebrafish. Genome Res 2003;13(12):2700–7.
[18] Howe K, Clark MD, Torroja CF, et al. The zebrafish reference genome sequence and its relationship to the human genome. Nature 2013;496(7446):498–503.
[19] Haramis AP, Hurlstone A, van der Velden Y, et al. Adenomatous polyposis coli-deficient zebrafish are susceptible to digestive tract neoplasia. EMBO Rep 2006;7 (4):444–9.
[20] Goessling W, North TE, Lord AM, et al. APC mutant zebrafish uncover a changing temporal requirement for wnt signaling in liver development. Dev Biol 2008;320(1):161–74.
[21] Meng X, Noyes MB, Zhu LJ, Lawson ND, Wolfe SA. Targeted gene inactivation in zebrafish using engineered zinc-finger nucleases. Nat Biotechnol 2008;26(6):695–701.
[22] Hwang WY, Fu Y, Reyon D, et al. Efficient genome editing in zebrafish using a CRISPR-Cas system. Nat Biotechnol 2013;31(3):227–9.
[23] Kettleborough RN, Busch-Nentwich EM, Harvey SA, et al. A systematic genome-wide analysis of zebrafish protein-coding gene function. Nature 2013;496 (7446):494–7.
[24] White RM, Cech J, Ratanasirintrawoot S, et al. DHODH modulates transcriptional elongation in the neural crest and melanoma. Nature 2011;471(7339):518–22.
[25] Zhong TP, Rosenberg M, Mohideen MA, Weinstein B, Fishman MC. gridlock, an HLH gene required for assembly of the aorta in zebrafish. Science 2000;287:1820–4.
[26] Peterson RT, Shaw SY, Peterson TA, et al. Chemical suppression of a genetic mutation in a zebrafish model of aortic coarctation. Nat Biotechnol 2004;22 (5):595–9.
[27] Stern HM, Murphey RD, Shepard JL, et al. Small molecules that delay S phase suppress a zebrafish bmyb mutant. Nat Chem Biol 2005;1(7):366–70.
[28] Garnaas MK, Cutting CC, Meyers A, et al. Rargb regulates organ laterality in a zebrafish model of right atrial isomerism. Dev Biol 2012;372(2):178–89.
[29] Yao Y, Lin J, Yang P, et al. Fine structure, enzyme histochemistry, and immunohistochemistry of liver in zebrafish. Anat Rec (Hoboken) 2012;295(4):567–76.
[30] Menke AL, Spitsbergen JM, Wolterbeek AP, Woutersen RA. Normal anatomy and histology of the adult zebrafish. Toxicol Pathol 2011;39(5):759–75.
[31] Higgins GM, Anderson RM. Experimental pathology of the liver. Arch Pathol 1931;12:186–202.
[32] Sadler KC, Krahn KN, Gaur NA, Ukomadu C. Liver growth in the embryo and during liver regeneration in zebrafish requires the cell cycle regulator, uhrf1. Proc Natl Acad Sci USA 2007;104(5):1570–5.
[33] Kan NG, Junghans D, Izpisua Belmonte JC. Compensatory growth mechanisms regulated by BMP and FGF signaling mediate liver regeneration in zebrafish after partial hepatectomy. FASEB J 2009;23 (10):3516–25.
[34] Decaens T, Godard C, de Reynies A, et al. Stabilization of beta-catenin affects mouse embryonic liver growth and hepatoblast fate. Hepatology 2008;47(1):247–58.
[35] Tan X, Behari J, Cieply B, Michalopoulos GK, Monga SP. Conditional deletion of beta-catenin reveals its role in liver growth and regeneration. Gastroenterology 2006;131(5):1561–72.
[36] Goessling W, North TE, Loewer S, et al. Genetic interaction of PGE2 and Wnt signaling regulates developmental specification of stem cells and regeneration. Cell 2009;136(6):1136–47.
[37] Nissim S, Sherwood RI, Wucherpfennig J, et al. Prostaglandin E2 regulates liver versus pancreas cell-fate decisions and endodermal outgrowth. Dev Cell 2014;28 (4):423–37.
[38] Dovey M, Patton EE, Bowman T, et al. Topoisomerase II alpha is required for embryonic development and liver regeneration in zebrafish. Mol Cell Biol 2009;29 (13):3746–53.
[39] Zhu Z, Chen J, Xiong JW, Peng J. Haploinsufficiency of Def activates p53-dependent TGFbeta signalling and causes scar formation after partial hepatectomy. PLoS One 2014;9(5):e96576.
[40] Cox AG, Saunders DC, Kelsey Jr PB, et al. *S*-nitrosothiol signaling regulates liver development and improves outcome following toxic liver injury. Cell Rep 2014;6 (1):56–69.
[41] Burkhardt-Holm P, Oulmi Y, Schroeder A, Storch V, Braunbeck T. Toxicity of 4-chloroaniline in early life stages of zebrafish (Danio rerio): II. Cytopathology

and regeneration of liver and gills after prolonged exposure to waterborne 4-chloroaniline. Arch Environ Contam Toxicol 1999;37(1):85–102.
[42] Liu LY, Fox CS, North TE, Goessling W. Functional validation of GWAS gene candidates for abnormal liver function during zebrafish liver development. Dis Model Mech 2013;6(5):1271–8.
[43] Bunchorntavakul C, Reddy KR. Acetaminophen-related hepatotoxicity. Clin Liver Dis 2013;17 (4):587–607, viii.
[44] Cox AG, Goessling W. Regenerative biology: take the brakes off for liver repair. Nature 2014;506(7488): 299–300.
[45] Bridgewater JA, Springer CJ, Knox RJ, Minton NP, Michael NP, Collins MK. Expression of the bacterial nitroreductase enzyme in mammalian cells renders them selectively sensitive to killing by the prodrug CB1954. Eur J Cancer 1995;31A(13-14):2362–70.
[46] Drabek D, Guy J, Craig R, Grosveld F. The expression of bacterial nitroreductase in transgenic mice results in specific cell killing by the prodrug CB1954. Gene Ther 1997;4(2):93–100.
[47] Clark AJ, Iwobi M, Cui W, et al. Selective cell ablation in transgenic mice expression *E. coli* nitroreductase. Gene Ther 1997;4(2):101–10.
[48] Bridgewater JA, Knox RJ, Pitts JD, Collins MK, Springer CJ. The bystander effect of the nitroreductase/CB1954 enzyme/prodrug system is due to a cell-permeable metabolite. Hum Gene Ther 1997;8(6):709–17.
[49] Curado S, Anderson RM, Jungblut B, Mumm J, Schroeter E, Stainier DY. Conditional targeted cell ablation in zebrafish: a new tool for regeneration studies. Dev Dyn 2007;236(4):1025–35.
[50] Curado S, Stainier DY, Anderson RM. Nitroreductase-mediated cell/tissue ablation in zebrafish: a spatially and temporally controlled ablation method with applications in developmental and regeneration studies. Nat Protoc 2008;3(6):948–54.
[51] Pisharath H, Rhee JM, Swanson MA, Leach SD, Parsons MJ. Targeted ablation of beta cells in the embryonic zebrafish pancreas using *E. coli* nitroreductase. Mech Dev 2007;124(3):218–29.
[52] Pisharath H, Parsons MJ. Nitroreductase-mediated cell ablation in transgenic zebrafish embryos. Methods Mol Biol 2009;546:133–43.
[53] Choi TY, Ninov N, Stainier DY, Shin D. Extensive conversion of hepatic biliary epithelial cells to hepatocytes after near total loss of hepatocytes in zebrafish. Gastroenterology 2014;146(3):776–88.
[54] He J, Lu H, Zou Q, Luo L. Regeneration of liver after extreme hepatocyte loss occurs mainly via biliary transdifferentiation in zebrafish. Gastroenterology 2014;146 (3):789–800, e788.
[55] Bedell VM, Wang Y, Campbell JM, Poshusta TL, Starker CG, Krug 2nd RG, et al. In vivo genome editing using a high-efficiency TALEN system. Nature 2012;491 (7422):114–8.

MOLECULAR MECHANISMS OF LIVER REGENERATION

WHY DON'T YOU GIVE ME A BREAK AND LEAVE ME ALONE TODAY?
I'M JUST DOING THE JOB ZEUS COMMANDED ME TO DO.
GEEZ! IF IT'S NOT YOU COMPLAINING IT'S THE WIFE AND KIDS!
© Mark Weinstein
LIVER AGAIN?

CHAPTER

4

The Priming and Progression Theory of Liver Regeneration

*Kimberly J. Riehle**,†,‡, Kevin Riggle†, Jean S. Campbell*,‡,#*

*Northwest Liver Research Program, University of Washington, Seattle, Washington, USA

†Department of Surgery, University of Washington, Seattle, Washington, USA

‡Department of Pathology, University of Washington, Seattle, Washington, USA

4.1 OVERVIEW OF STUDIES OF LIVER REGENERATION

The "modern" era of liver regeneration research began in the 1930s with the classic description of successful hepatectomy in rats by Higgins and Anderson [1]. The amazing physiological process wherein liver mass is restored after a portion of the liver is surgically removed or chemically ablated has captivated basic scientists and clinicians for decades. To gain a sense of the progress that has been made in the field of liver regeneration, the reader is directed to a number of excellent reviews spanning the last five decades [2–8]. In rodents and humans, liver regeneration is technically not "regeneration" in the true sense of the word; compensatory hyperplasia is a more precise description of the process, as the original liver mass is restored, but not the original lobar architecture. Whether by surgical resection or chemically induced cell death, loss of liver mass stimulates synchronous deoxyribonucleic acid (DNA) synthesis in hepatocytes and nonparenchymal cells (NPCs), and liver mass is restored in 10-14 days in rodents. Partial hepatectomy, surgical removal of two-thirds of the liver (2/3PH), stimulates quiescent (G_0) hepatocytes to enter the cell cycle, replicate their genome, undergo cytokinesis, and once the original liver mass has been restored, exit the cell cycle and return to the resting state where they are poised to proliferate again. *In vivo*, normal uninjured adult hepatocytes have the capacity to self-renew without dedifferentiating. Unfortunately, this robust proliferative capacity is lost if the liver is steatotic or fibrotic, or if hepatocytes are cultured *in vitro* for extended periods of time.

Rodents have been used to help identify key signaling pathways that control various aspects of liver regeneration, as the basic components of liver regeneration appear to be similar in all mammals [9–11]. The power of genetic

Current address: OncoSec Medical Inc., 454 N. 34th Street, Seattle, WA 98103, USA.

Liver Regeneration
http://dx.doi.org/10.1016/B978-0-12-420128-6.00004-X

manipulation in particular, whether systemic or cell type-specific, constitutive or inducible, overexpression of a transgene or targeted deletion, allows investigators to study the importance of specific proteins and pathways. Genetic approaches are best complimented by pharmacological approaches, such as neutralizing antibodies or small-molecule inhibitors, to acutely block pathways that regulate specific aspects of liver regeneration.

About 50 years ago, the regulatory mechanisms of liver regeneration were unknown. Although the rates of hepatocyte mitoses and liver mass restoration had been characterized, the specifics of how hepatocytes replicated and whether other liver cells were necessary for liver regeneration were just beginning to be investigated. Mechanistic questions such as how cell replication is initiated or terminated, or what governs cell replication and progression through the stages of cell cycle were unclear. It is this scientific context from which Dr. Fausto developed the "priming and progression" theory of regeneration. A variety of independent research approaches and nonhepatic experimental systems sculpted this hypothesis (Figure 4.1). We speculate that these influences included concurrent investigations in fetal liver development, hormonal influences on liver regeneration [12], chemically induced hepatocellular carcinogenesis [13], investigations of DNA replication in cells in culture [14], and the use of *Saccharomyces cerevisiae* as a genetically tractable model organism to identify cell division cycle (CDC) genes [15].

It was known that hepatocytes in an adult liver rarely turnover and thus were usually "quiescent," yet after surgical removal of two-thirds of the liver, liver mass was rapidly restored. The problem that intrigued Professor Fausto was how hepatocyte DNA synthesis was initiated, ultimately leading to restoration of liver mass. In this chapter, we focus on the evolution of the "priming and progression" model of liver regeneration that was hypothesized and tested by Professor Fausto and members of his laboratory from the late 1970s to 2010. With the benefit of hindsight, we review the scientific knowledge available when Dr. Fausto began his career and speculate on conceptual influences that lead to his hypothesis of "priming and progress."

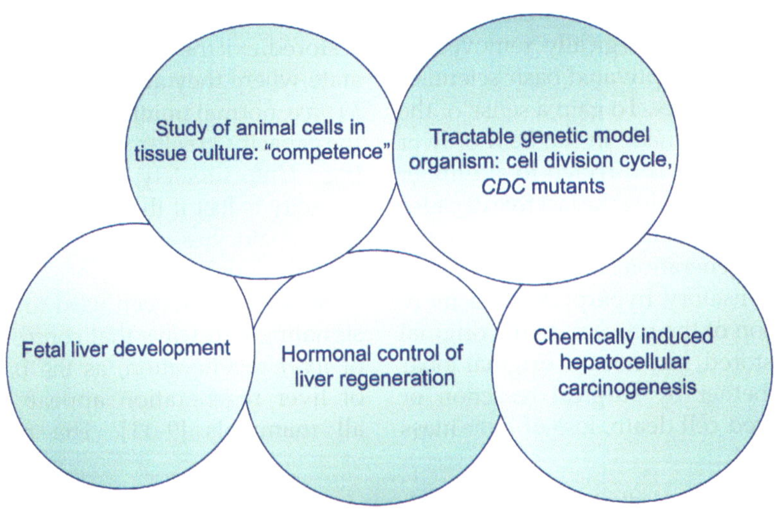

FIGURE 4.1 Developmental influences on "priming and progression" model of liver regeneration.

4.2 SALIENT FEATURES OF LIVER REGENERATION PRIOR TO 1970

Removal of two-thirds of the liver by ligation of the vascular pedicles (2/3PH) as described by Higgins and Anderson [1] is the best-studied experimental model of liver regeneration and is thus the primary focus of this chapter. The fact that liver mass is tightly regulated relative to lean body mass was illustrated by elegant liver transplant studies in dogs, in which donated livers decreased in mass when they were slightly too large for the recipient [16]. The precise mechanisms by which too much or too little functional liver mass are sensed are still not well understood.

To understand how nucleic acid synthesis was involved in restorative growth after 2/3PH, Grisham's laboratory employed a relatively new reagent to investigate the spatial relationship of proliferating hepatocytes within the liver lobule. Using DNA synthesized with trace amounts of ^{3}H-thymidine and emulsion-based autoradiography, Grisham showed that hepatocyte DNA synthesis began 12-18 h after 2/3PH in rats and that ultimately labeled hepatocytes were found in all residual lobes and lobules [17]. The presence of mitotic figures in hepatocytes temporally lagged behind ^{3}H-thymidine incorporation by approximately 4-8 h. A gradual increase in DNA synthesis in nonhepatocytes, including ductal cells and sinusoidal endothelial cells, was observed between 24 and 48 h. These studies illustrated a number of salient points regarding the hepatocyte cell cycle *in vivo*. One, a temporal lag between removal of liver tissue and subsequent DNA synthesis and mitoses was observed. Two, DNA synthesis was tightly synchronized, especially in hepatocytes. Three hepatocytes that were undergoing DNA synthesis were uniformly distributed throughout the lobules. Finally, these early investigations demonstrated that there is a sequential order of cellular events that lead to restoration of liver mass after 2/3PH.

The lag time between loss of liver mass and hepatocyte DNA synthesis was called the "pre-replication phase" and researchers focused on cellular changes that occurred during this time. Among these studies was seminal work by Nancy Bucher's group using a parabotic system in which two anesthetized rats had a shared blood supply; one animal underwent 2/3PH and the other animal did not. Her observation that hepatocytes in the animal that did not undergo 2/3PH still underwent DNA synthesis suggested that systemic circulating factors could stimulate hepatocyte replication [18,19]. These studies laid the experimental groundwork for the identification of soluble mitogens and growth factors that enhance primary hepatocyte growth in primary cultures and *in vivo*, known as hepatomitogens [3,20–22].

4.3 IDENTIFYING HEPATOMITOGENS IN THE 1970s-1980s

As the liver regeneration field matured, Nancy Bucher's group continued to examine hormonal factors that simulate hepatocyte DNA synthesis *in vivo* [3]. In parallel, experiments using established cell lines and primary cells in culture were used to define "factors" in serum that simulated cell proliferation [23–25]. In the early 1970s, commitment of cells to enter the cell cycle and replicate their DNA was considered to be "the result of complex interactions." For example, it was initially suggested that cell-cycle entry was random and characterized by a first-order rate constant, the transition probability [26]. However, a number of labs demonstrated that the commitment process could be separated into two stages. Pledger and coworkers described the stage between adding serum and committing to DNA replication (later termed "G_1") as becoming "competent," while the second phase of beginning to synthesize DNA was termed "progression" [23–25].

Importantly, these investigators noted a temporal sequence of events from quiescence induced by serum removal, to cell replication after addition of serum. While G_0 induced by serum-withdrawal is likely to differ somewhat from G_0 in resting hepatocytes *in vivo*, key phenotypic aspects were similar, supporting the validity of this terminology for hepatocytes during liver regeneration.

In this same decade, geneticists were investigating the stages of the cell cycle using a budding yeast, *S. cerevisiae*. The use of a genetically tractable model organism to identify CDC genes accelerated our fundamental understanding of stages of the cell cycle and regulatory pathways. Leland Harwell began these groundbreaking studies at the University of Washington in 1970 [15]; Drs. Nurse and Hunt used different model systems (sea urchin and *Schizosaccharomyces pombe*) to identify additional key regulators of the cell cycle. These studies, which resulted in these three investigators receiving the Nobel Prize in Physiology or Medicine in 2001, together with studies of mammalian cells in culture, helped define the phases of the cell cycle in the 1970s and 1980s.

The lag time between 2/3PH and the onset of hepatocyte replication was also reminiscent of ongoing studies using chemicals to initiate liver carcinogenesis. This work in the 1960s and 1970s lead to the concept of sequential phases or stages of cancer development, with each being necessary for the next step to proceed [13,27,28]. Drs. Farber and Solt at the University of Toronto, Drs. Van Potter, Pitot, and Drinkwater at the McArdle laboratory for Cancer Research at the University of Wisconsin in Madison [29,30], and many other investigators provided the experimental framework for the multi-staged carcinogenesis model, and developed the terms "initiation, promotion, and progression." In this experimental paradigm, diethyl nitrosamine was found to be an "initiating agent," and phenobarbital a promoting agent that allows progression to hepatocellular carcinoma. Dr. Fausto arrived in Wisconsin in the late 1960s and performed his liver regeneration studies in a research building adjacent to the McArdle labs, where Pitot's group was investigating mechanisms of hepatocarcinogenesis. Subsequent studies showed that 2/3PH is a surgical "promotion agent" [29,31]. The conceptual influence of these studies in close proximity to the initial phases of Dr. Fausto's training may have lead to his use of the terms "priming" or "competence" to describe the initial step in liver regeneration. Thus, an influential senior investigator and role model would likely have had a powerful influence on a young investigator such as Nelson.

4.4 1990s: DEVELOPMENT OF THE PRIMING AND PROGRESSION MODEL, WITH A FOCUS ON INFLAMMATORY STIMULI DURING REGENERATION

In the normal adult liver, the vast majority of hepatocytes are in a quiescent state, i.e., the G_0 phase of the cell cycle. In fact, at rest less than 1% of healthy murine hepatocytes are dividing at any given time, resulting in a 200-400 day lifespan [32,33]. Similarly, hepatocytes in culture are notorious for being difficult to maintain, in part due to their low proliferation rate. Despite this baseline low-proliferation rate, at 36 h after a 2/3PH, up to 40% of all hepatocytes are simultaneously undergoing DNA synthesis [5,34]. In the 1990s, a number of growth factors that stimulate hepatocyte proliferation in culture were identified, including hepatocyte growth factor (HGF) and epidermal growth factor (EGF) receptor ligands such as transforming growth factor-α (TGFα), EGF, and heparin-binding EGF-like growth factor (HB-EGF). These ligands also stimulate a modest degree of hepatocyte proliferation in rodents. A breakthrough discovery was the finding that pretreatment of cultured hepatocytes or rodents with a single dose of the inflammatory cytokine tumor necrosis factor (TNF) markedly amplifies the proliferative response of these cells to TGFα or HGF [35,36].

These experimental findings led to the idea that early inflammatory signals after hepatectomy allow hepatocytes to transition from G_0 into G_1, where they gain proliferative "competence," as defined by Drs. Mead and Fausto in 1990. The process of conferring an enhanced ability to proliferate without directly stimulating proliferation itself was termed priming. Numerous subsequent studies focused on the priming phase of regeneration and identified the proto-oncogenes *c-jun*, *c-fos*, and *c-myc* as critical immediate early genes activated early after PH [37,38]. Additionally, there were hints that one of the earliest responses after 2/3PH was the activation of nuclear factor kappa B (NFκB) [39,40]. Importantly, the priming process is reversible, as if hepatocytes are not exposed to growth factors or necessary metabolic components [41] within early G_1, they do not progress through the cell cycle but instead return to quiescence. Of note, priming after 2/3PH bears several similarities to the liver's acute-phase response, as described by Professors Decker and Heinrich in Germany, among others [42,43].

Integration of key aspects of the priming and progression model for hepatocytes *in vivo* and the understanding of the stages of mammalian cell cycle at the time the model was developed are shown in Figure 4.2. Quiescent (G_0) hepatocytes are poised to enter into the cell cycle after loss of liver mass. "Priming" is considered to be the molecular changes that allow quiescent hepatocytes to be *competent* to respond to factors that drive progression through the G_1 restriction point.

Further *in vitro* studies explored potential mechanisms by which cytokines and growth factors interact to link the priming and progression phases of proliferation and found that TNF activates TGFα-converting enzyme (TACE) in hepatocytes. TACE then cleaves TGFα to allow its release, subsequent activation of EGFR, and thus stimulates proliferation [44]. Contrary to these *in vitro* data, however, was the finding of normal regeneration in mice deficient for *Tace* in hepatocytes, though these negative findings may be a simple testament to the redundancy of pathways critical to regeneration [45]. George

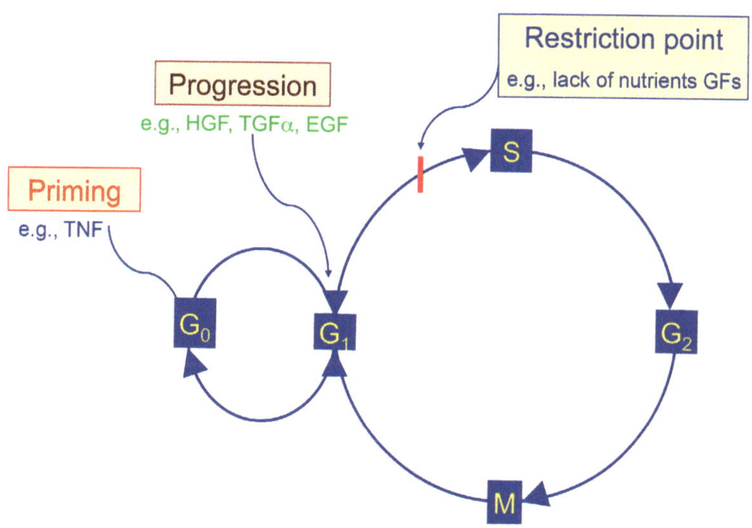

FIGURE 4.2 Schematic representation of the priming and progression theory.

Michalopoulos's group similarly documented that HGF and TGFα stimulate hepatocyte proliferation *in vivo* and focused on the ability of proteases to prime hepatocytes. Specifically, they discovered that intraportal collagenase infusion prior to stimulation with growth factors lead to a two- to fourfold increase in bromodeoxyuridine (BrdU) labeling compared to growth factor stimulation alone [46]. Serandour et al. then demonstrated that in cocultures of hepatocytes with biliary epithelial cells, stimulation with TNF and EGF leads to repeated waves of hepatocyte proliferation that cannot be obtained by EGF stimulation alone [47]. Further, they found that TNF activated matrix metalloproteinases, which degrade extracellular matrix components and allow hepatocytes to better respond to the growth factor. These studies lent further credence to the priming/progression theory and provided additional evidence of significant cross talk between cytokine and growth factor pathways within hepatocytes.

While resting or quiescent, hepatocytes may share some similarities to serum-starved cells in culture, but should not be considered identical [48,93]. Hepatocytes appear to have unlimited capacity to replicate *in vivo*, while hepatocytes in culture do not. It is not known whether hepatocytes have an "*in vivo* Hayflick limit" [49]. The G_1 phase contains a restriction (R) point, a point in mid to late G_1 where cells commit to completing the rest of the cell cycle. The time from priming to restriction point appears to vary between species [50], and a lack of nutrients or cofactors needed for metabolic pathways seems to lengthen the time between 2/3PH and DNA synthesis [5,41,51,52]. Amino acids in particular may also be involved in the transition from the priming to the progression phase of regeneration. As mentioned above, the Bucher and Fausto labs used nutritional manipulation to induce a state of replicative competence in hepatocytes *in vivo* [53]. Additionally, leucine has been shown to lead to the release of TGFα in culture [54], and infusion of branched chain amino acids hastens the onset of liver regeneration after PH in rats [55]. More recently, it has been shown that rapid release of ATP, even within a minute of hepatectomy, can itself trigger expression of immediate early genes and prime hepatocytes for proliferation after PH [56]. Together, these data suggest that a variety of pathways converge to drive the G_0 to G_1 transition, preparing hepatocytes to maximally proliferate.

4.5 PRIMING ALONE VERSUS PRIMING AND PROGRESSION: 1/3 VERSUS 2/3 HEPATECTOMY

The differential responses to 1/3 and 2/3PH highlight the distinction between the priming and progression phases of regeneration. Removal of 2/3 of the liver leads to a robust synchronous proliferative response involving the vast majority of the remaining hepatocytes. Conversely, removal of 1/3 of the liver does not elicit a coordinated wave of DNA replication, and resections involving between 1/3 and 2/3 of the liver cause a proportionally increasing peak of proliferation. Importantly, infusion of a growth factor (TGFα or HGF) after a 1/3 hepatectomy will elicit a more robust proliferative response than growth factor infusion alone, suggesting that 1/3PH leads to a priming response without progression through the cell cycle [36,57]. Mitchell et al. compared molecular events after 1/3 and 2/3 hepatectomies to identify factors that linked the priming and progression phases of liver regeneration and identified HB-EGF to be critical to this process [58]. Specifically, they found that expression of HB-EGF significantly differs after 1/3 compared to 2/3PH, while expression levels of TGFα and HGF do not. Further, infusion of HB-EGF after a 1/3PH induces a far more robust proliferative response than infusion of the other two growth factors, and HB-EGF-knockout (KO) mice have a significant impairment in cell-cycle progression during regeneration. Li et al. continued these investigations by performing gene expression profiling in livers at various time points after 1/3 and 2/3 hepatectomies [59]. These studies confirmed that at early time points after either operation, there are similar changes in gene expression, suggesting that both operations induce the priming phase. At 12 h after surgery, however, there is a shift in transcriptional programs, which pushes cells to massively replicate after a 2/3PH but not after 1/3PH, consistent with distinct priming and progression phases of regeneration.

4.6 CONTROVERSIES REGARDING THE IMPORTANCE OF PRIMING IN REGENERATION

Despite ample evidence that a plethora of factors can prime hepatocytes to become competent to proliferate, the precise role of specific mediators has been widely debated. A focus on components of the innate immune system originated with early studies by Cornell et al., demonstrating that rats with restricted lipopolysaccharide (LPS) production and LPS hyporesponsive mice have impaired proliferation after PH [60]. Later, when toll-like receptor 4 was identified as the primary receptor for LPS, a newer series of experiments called into question the importance of LPS in regeneration, as multiple strains of TLR KO mice regenerate normally. Interestingly, additional studies indicated that the TLR adaptor protein MyD88 might be critical to the priming response [61,62]. An absolute requirement for the inflammatory cytokines TNFα and IL-6 to initiate regeneration has also been brought into question. Initial work in this area demonstrated that anti-TNF antibodies inhibit DNA proliferation after PH [40] and that TNFR1-KO and IL6-KO mice have impaired regeneration [63,64]. Contrary to these findings, TNF KO mice regenerate normally [65,66]. Similarly, different groups have reported conflicting results on the outcome of 2/3PH in IL6-KO mice, varying from impaired proliferation to decreased survival to no phenotype at all [67–69]. It has been calculated that nearly 40% of the immediate early genes expressed in the regenerating liver are IL-6 dependent, suggesting a significant impact of this pathway on the priming phase [70]. Most conclusive, perhaps, were studies by Vaquero et al., wherein serum IL-6 levels at 4 h after 2/3PH were compared in various TLR KO strains on a mouse-by-mouse basis to the extent of hepatocyte proliferation, and no correlation was seen between the two [62].

4.7 RECENT INSIGHT INTO CELL-CYCLE "COMPETENCY"

As an extension of the initial sequential "priming" and "progression" phase model, recent experimental evidence indicates the existence of two distinct phases of G_1, which are defined in part by the activation of specific intracellular signaling pathways. Elegant work by Baffet's lab in 1999 demonstrated the growth-factor dependency of hepatocytes in mid-G_1 and linked MEK/ERK signaling to cyclin D1 accumulation. Additionally, Talarmin and coworkers showed that there were two waves of MEK/ERK signaling after 2/3PH [71], an observation also made by our lab [72]. The initial wave occurs immediately after surgery, peaking at 2 h and returning to baseline by 4 h; the second wave occurs between 10 and 11 h in the rat [71] and 20 h after 2/3PH in mice [72]. This initial burst of intracellular signaling coincides with induction of significant gene expression changes after both 1/3 and 2/3PH [59]. However, a majority of genes whose expression is induced by the removal of 1/3 of the liver return to baseline levels with an apparent absence of further signaling [59]. One interpretation of these data is that a second set of intracellular signals is needed to pass the G_1 restriction point.

In the early 1990s, experiments by Jones and Kazlauskas using growth factors in a discontinuous stimulation assay paradigm that evaluated various signaling pathways in nonhepatocyte cultures supported this hypothesis [73–75]. Two pulses of growth factors were necessary to traverse the G_1 restriction point. Specifically, when serum-starved cells in culture are exposed to growth factors, the first signaling pulse is quite short and involves MEK/ERK activation. The second "pulse" of signaling after growth-factor stimulation is much longer and appears to be dependent on PI3K signaling pathways.

More recent studies, including those by Yarden's group using human epithelial cells in culture, further explored the kinetics of growth factor exposure and commitment to cell-cycle progression. The first pulse of growth factors rapidly activated MEK/ERK signaling, induction of p53, and expression of metabolic enzymes, including steroid, cholesterol, and lipid metabolism [76]. The induction of p53 is believed to act as an "inhibitory brake on the cell cycle" and a molecular component of the G_1 restriction point. The second growth factor "pulse" stimulates PI3K and MEK/ERK activities, which among other actions, facilitates p53 degradation. The essential purpose of the proposed priming phase is to ensure that G_0 hepatocytes are competent to traverse G_1 and cross over the restriction point, where they commit to S phase. However, it is not known whether similar cell-cycle controls occur in hepatocytes *in vivo* after 2/3PH.

These recent cell-cycle studies in culture may provide some insight into the elegant work of Weglarz and Sandgren. These investigators demonstrated that mouse hepatocytes transplanted into rat livers maintain their species-specific regenerative timing after PH, despite being exposed the same endocrine and paracrine signals as the host rat hepatocytes, which divide on their own time line [50]. The species-specific circadian rhythms that control the timing of hepatocyte proliferation may be governed at least in part by WEE1 kinase [77]. Assuming that rat and mouse hepatocytes have similar intracellular signaling pathways and that cross-species differences in ligands are minimal, one interpretation of these findings is that mouse hepatocytes keep their own cell-cycle timing by maintaining their longer G_1 phase. It is not clear why hepatocytes from different species have variations in the length of G_1. Regulation of the G_1 restriction point may allow hepatocytes to prolong their G_1 phase until checkpoints have been validated or restoration of nutrients has been achieved [41,51,52].

4.8 THE ROLE OF NPCs IN PRIMING AND PROGRESSION

The priming phase of liver regeneration begins with the differential expression of a wide variety of genes, resulting in the production and release of cytokines. Kupffer cells are known to produce a variety of growth factors and cytokines, thus it has been postulated that they play an important role in liver regeneration [78]. Indeed, Meijer et al. showed that depletion of Kupffer cells delays liver regeneration after 2/3PH [79]. As mentioned above, NFκB is activated early after PH, and Abshagen et al. further evaluated NFκB signaling after 2/3PH in Kupffer cell-deficient mice. They found that there was a decrease in NFκB activation in these animals, an absence of elevation in TNFα and IL-6, and decreased regenerative capacity [80]. Conversely, Taub and coworkers observed that mice deficient for the p50 subunit of NFκB did not have impaired liver regeneration [81], or was it impaired in the hepatocyte-specific KO strain developed by Chaisson and colleagues [82].

More recently, Shahin Rafii's laboratory has highlighted the importance of liver sinusoidal endothelial cells (LSECs) in supporting normal hepatocyte proliferation, as well as in restoring functional vasculature to the regenerating liver. They show that activation of vascular endothelial growth factor receptor-2 promotes early hepatocyte regeneration in the first 3 days following 2/3PH through upregulation of Wnt2 and HGF. LSECs are further involved in regeneration by promoting angiogenesis to vascularize and sustain the expanding liver mass [83]. Importantly, hepatocytes do not grow well if removed from a normal microenvironment, as is evidenced clinically by the regenerative failure that is seen in humans with significant liver fibrosis or cirrhosis [4]. In an elegant series of experiments, Ira Fox's laboratory demonstrated that hepatocytes from livers of rodents with uncompensated cirrhosis do not replicate for 2 months after transplantation into a normal liver. Hepatocytes from normal livers or animals with compensated cirrhosis, however, proliferate normally immediately after transplantation into healthy mice. These data confirm that signals from other cell types are critical in supporting hepatocyte proliferation, though further studies are needed to elucidate the underlying mechanisms and how this process can be manipulated to improve outcomes of patients with liver disease undergoing hepatectomy.

4.9 FUTURE DIRECTIONS

The past 50 years of investigation into the mechanisms underlying liver regeneration has brought about an exponential increase in our understanding of this unique process. Multiple models have been developed to use liver regeneration as a means to study signal transduction and the cell cycle *in vivo* [10,84]. These studies have in turn uncovered further mechanistic and clinical questions, which need to be answered in order to fill critical knowledge gaps that persist despite years of research. For instance, the use of fate mapping by Willenbring and colleagues in the 2/3PH model of liver regeneration has confirmed that adult hepatocytes are the source of new liver cells following resection in normal liver [32]. In a healthy microenvironment, hepatocytes have a remarkable proliferative capacity, and can undergo up to 100 population doublings [85,86]. However, we have yet to translate our sophisticated cell-cycle knowledge into the clinic. Other experimental systems may provide further insight into the mechanisms utilized by hepatocytes to commit to cell-cycle progress, allowing regenerative medicine to be translated to the care of patients with liver failure, and to even potentially improve our management of liver cancer. We know that hepatocytes have a near-infinite capacity to regenerate, based on the early studies in which serial partial hepatectomies were performed, as well as more recent serial transplantation studies by Markus Grompe's group [87]. Do NPCs, which are prominent actors in liver regeneration, also have this ability? Further, why is this regenerative capacity so adversely affected in livers that are steatotic or cirrhotic [88,89]? Accordingly, in models of hepatic injury and necrosis, liver progenitor cells seem to be instrumental in regenerating the liver. It will be interesting to use new technologies to study the source of new liver cells after liver resections in inflamed, fibrotic, and fatty livers, as progenitor cells may play a more prominent role after resection in these disease states [90].

It is likely that epigenetic changes and microRNAs (miRNAs) further alter the regenerative capacity of the liver. We have hypothesized that chromatin remodeling leads to increased hepatocyte proliferation after partial hepatectomy [59]. Other groups have shown that histone deacetylase (HDAC) activity increases following PH, and that HDAC inhibition delays liver regeneration [91]. Microarray studies focusing on miRNA regulation after 2/3PH demonstrated that early on, up to 40% of all miRNAs are upregulated, followed by an overall 70% downregulation at 24 h. These data suggest that miRNA regulation is involved in the priming phase of liver regeneration, but more work must be done to determine the mechanisms that underlie this regulation [92].

As our understanding of regenerative mechanisms has expanded, so has our repertoire of experimental models. Initially, 2/3PH and models of hepatocellular injury brought about understanding of the basic molecular and cellular mechanisms of regeneration. Further dissection of the specific molecules and pathways involved in regeneration were elucidated following the advent of genetically modified mice, and knowledge regarding global genetic changes was uncovered using whole-genome sequencing. The future may bring animal models that are better suited to reflect clinical problems, such as regeneration in livers that are cirrhotic, steatotic, or have been previously treated with chemotherapeutic drugs. Liver-cell-specific-KO models will further our understanding of the specific role each particular cell type plays in regeneration, particularly different fibroblast populations, which have not yet been studied in detail. Finally, new technologies using animal models and 3D decellularized scaffolds as *in vivo* vectors for human hepatocyte expansion will provide a unique and powerful model for studying regeneration.

From a clinical perspective, a more thorough understanding of regenerative mechanisms will lead to therapies for acute and chronic liver failure. Given the shortage of viable organs for transplantation and increased demand [4], increasing the regenerative capacity of donor livers would immediately increase the supply of this precious resource. A more detailed understanding of regeneration in diseased livers may also facilitate resection in those who otherwise would not have been surgical candidates. Whether hepatocytes residing in diseased livers are "incompetent" in that they lack the ability to be primed, or have prolonged G_1 phases and never traverse the restriction point are questions that remain to be answered. The framework for these questions reflects the richness of the scientific contributions and spirited debates instigated by Dr. Fausto's "priming and progression theory."

Acknowledgments

The authors would like to thank Drs. Jordi Bruix and Robert H. Pierce for helpful discussions related to Nelson's early career. Many scientists who trained with Nelson wish to thank him for his scientific guidance and generosity that extended beyond the laboratory.

References

[1] Higgins G, Anderson R. Restoration of liver of white rat following partial surgical removal. Arch Pathol 1931;12:186–202.
[2] Bucher NL. Regeneration of mammalian liver. Int Rev Cytol 1963;15:245–300.
[3] Bucher NL, Patel U, Cohen S. Hormonal factors concerned with liver regeneration. Ciba Found Symp 1977;55:95–107.
[4] Clavien PA, Petrowsky H, DeOliveira ML, Graf R. Strategies for safer liver surgery and partial liver transplantation. N Engl J Med 2007;356:1545–59.
[5] Fausto N, Campbell JS, Riehle KJ. Liver regeneration. Hepatology 2006;43:S45–53.
[6] Michalopoulos GK. Liver regeneration. J Cell Physiol 2007;213:286–300.
[7] Taub R. Liver regeneration: from myth to mechanism. Nat Rev Mol Cell Biol 2004;5:836–47.
[8] Steer CJ. Liver regeneration. FASEB J 1995;9:1396–400.
[9] Bucher NL. Liver regeneration: an overview. J Gastroenterol Hepatol 1991;6:615–24.
[10] Fausto N, Riehle KJ. Mechanisms of liver regeneration and their clinical implications. J Hepatobiliary Pancreat Surg 2005;12:181–9.
[11] Fausto N, Webber EM. Control of liver growth. Crit Rev Eukaryot Gene Expr 1993;3:117–35.
[12] Weinbren K. Regeneration of the liver. Gastroenterology 1959;37:657–78.
[13] Solt DB, Medline A, Farber E. Rapid emergence of carcinogen-induced hyperplastic lesions in a new model for the sequential analysis of liver carcinogenesis. Am J Pathol 1977;88:595–618.
[14] Sisken JE, Morasca L. Intrapopulation kinetics of the mitotic cycle. J Cell Biol 1965;25:179–89.
[15] Hartwell LH, Culotti J, Reid B. Genetic control of the cell-division cycle in yeast. I. Detection of mutants. Proc Natl Acad Sci USA 1970;66:352–9.
[16] Kam I, Lynch S, Svanas G, Todo S, Polimeno L, Francavilla A, et al. Evidence that host size determines

liver size: studies in dogs receiving orthotopic liver transplants. Hepatology 1987;7:362–6.
[17] Grisham JW. A morphologic study of deoxyribonucleic acid synthesis and cell proliferation in regenerating rat liver; autoradiography with thymidine-H3. Cancer Res 1962;22:842–9.
[18] Moolten FL, Bucher NL. Regeneration of rat liver: transfer of humoral agent by cross circulation. Science 1967;158:272–4.
[19] Sakai A. Humoral factor triggering DNA synthesis after partial hepatectomy in the rat. Nature 1970;228: 1186–7.
[20] Bucher NL, Scott JF, Aub JC. Regeneration of the liver in parabiotic rats. Cancer Res 1951;11:457–65.
[21] Sigel B. The extracellular regulation of liver regeneration. J Surg Res 1969;9:387–94.
[22] Sigel B, Baldia LB, Dunn MR, Menduke H. Humoral control of liver regeneration. Surg Gynecol Obstet 1967;124:1023–31.
[23] Pardee AB. A restriction point for control of normal animal cell proliferation. Proc Natl Acad Sci USA 1974;71:1286–90.
[24] Pledger WJ, Stiles CD, Antoniades HN, Scher CD. Induction of DNA synthesis in BALB/c 3T3 cells by serum components: reevaluation of the commitment process. Proc Natl Acad Sci USA 1977;74:4481–5.
[25] Pledger WJ, Stiles CD, Antoniades HN, Scher CD. An ordered sequence of events is required before BALB/c-3T3 cells become committed to DNA synthesis. Proc Natl Acad Sci USA 1978;75:2839–43.
[26] Robinson JH, Smith JA, Totty NF, Riddle PN. Transition probability and the hormonal and density-dependent regulation of cell proliferation. Nature 1976;262: 298–300.
[27] Farber E, Sarma DS. Chemical carcinogenesis: the liver as a model. Pathol Immunopathol Res 1986;5:1–28.
[28] Pitot HC, Sirica AE. The stages of initiation and promotion in hepatocarcinogenesis. Biochim Biophys Acta 1980;605:191–215.
[29] Drinkwater NR. Experimental models and biological mechanisms for tumor promotion. Cancer Cells 1990;2:8–15.
[30] Drinkwater NR, Ginsler JJ. Genetic control of hepatocarcinogenesis in C57BL/6J and C3H/HeJ inbred mice. Carcinogenesis 1986;7:1701–7.
[31] Hanigan MH, Winkler ML, Drinkwater NR. Partial hepatectomy is a promoter of hepatocarcinogenesis in C57BL/6J male mice but not in C3H/HeJ male mice. Carcinogenesis 1990;11:589–94.
[32] Malato Y, Naqvi S, Schurmann N, Ng R, Wang B, Zape J, et al. Fate tracing of mature hepatocytes in mouse liver homeostasis and regeneration. J Clin Invest 2011;121:4850–60.
[33] Magami Y, Azuma T, Inokuchi H, Kokuno S, Moriyasu F, Kawai K, et al. Cell proliferation and renewal of normal hepatocytes and bile duct cells in adult mouse liver. Liver 2002;22:419–25.
[34] Fausto N, Campbell JS, Riehle KJ. Liver regeneration. J Hepatol 2012;57:692–4.
[35] Kirillova I, Chaisson M, Fausto N. Tumor necrosis factor induces DNA replication in hepatic cells through nuclear factor kappaB activation. Cell Growth Differ 1999;10:819–28.
[36] Webber EM, Bruix J, Pierce RH, Fausto N. Tumor necrosis factor primes hepatocytes for DNA replication in the rat. Hepatology 1998;28:1226–34.
[37] Thompson NL, Mead JE, Braun L, Goyette M, Shank PR, Fausto N. Sequential protooncogene expression during rat liver regeneration. Cancer Res 1986;46:3111–7.
[38] Mohn KL, Laz TM, Melby AE, Taub R. Immediate-early gene expression differs between regenerating liver, insulin-stimulated H-35 cells, and mitogen-stimulated Balb/c 3T3 cells. Liver-specific induction patterns of gene 33, phosphoenolpyruvate carboxykinase, and the jun, fos, and egr families. J Biol Chem 1990;265:21914–21.
[39] Tewari M, Dobrzanski P, Mohn KL, Cressman DE, Hsu JC, Bravo R, et al. Rapid induction in regenerating liver of RL/IF-1 (an I kappa B that inhibits NF-kappa B, RelB-p50, and c-Rel-p50) and PHF, a novel kappa B site-binding complex. Mol Cell Biol 1992;12:2898–908.
[40] Akerman P, Cote P, Yang SQ, McClain C, Nelson S, Bagby GJ, et al. Antibodies to tumor necrosis factor-alpha inhibit liver regeneration after partial hepatectomy. Am J Physiol 1992;263:G579–85.
[41] Huang J, Rudnick DA. Elucidating the metabolic regulation of liver regeneration. Am J Pathol 2014;184: 309–21.
[42] Andus T, Geiger T, Hirano T, Kishimoto T, Tran-Thi TA, Decker K, et al. Regulation of synthesis and secretion of major rat acute-phase proteins by recombinant human interleukin-6 (BSF-2/IL-6) in hepatocyte primary cultures. Eur J Biochem 1988;173:287–93.
[43] Heinrich PC, Castell JV, Andus T. Interleukin-6 and the acute phase response. Biochem J 1990;265:621–36.
[44] Argast GM, Campbell JS, Brooling JT, Fausto N. Epidermal growth factor receptor transactivation mediates tumor necrosis factor-induced hepatocyte replication. J Biol Chem 2004;279:34530–6.
[45] McMahan RS, Riehle KJ, Fausto N, Campbell JS. A disintegrin and metalloproteinase 17 regulates TNF and TNFR1 levels in inflammation and liver regeneration in mice. Am J Physiol Gastrointest Liver Physiol 2013;305:G25–34.
[46] Liu ML, Mars WM, Zarnegar R, Michalopoulos GK. Collagenase pretreatment and the mitogenic effects of hepatocyte growth factor and transforming growth

factor-alpha in adult rat liver. Hepatology 1994;19:1521–7.

[47] Serandour AL, Loyer P, Garnier D, Courselaud B, Theret N, Glaise D, et al. TNFalpha-mediated extracellular matrix remodeling is required for multiple division cycles in rat hepatocytes. Hepatology 2005;41:478–86.

[48] Corlu A, Loyer P. Regulation of the g1/s transition in hepatocytes: involvement of the cyclin-dependent kinase cdk1 in the DNA replication. Int J Hepatol 2012;2012:689324.

[49] Rubin H. The disparity between human cell senescence in vitro and lifelong replication in vivo. Nat Biotechnol 2002;20:675–81.

[50] Weglarz TC, Sandgren EP. Timing of hepatocyte entry into DNA synthesis after partial hepatectomy is cell autonomous. Proc Natl Acad Sci USA 2000;97: 12595–600.

[51] Riehle KJ, Haque J, McMahan RS, Kavanagh TJ, Fausto N, Campbell JS. Sustained glutathione deficiency interferes with the liver response to TNF-alpha and liver regeneration after partial hepatectomy in mice. J Liver Disease Transplant 2013;1(2):1–14.

[52] Vail ME, Chaisson ML, Thompson J, Fausto N. Bcl-2 expression delays hepatocyte cell cycle progression during liver regeneration. Oncogene 2002;21:1548–55.

[53] Mead JE, Braun L, Martin DA, Fausto N. Induction of replicative competence ("priming") in normal liver. Cancer Res 1990;50:7023–30.

[54] Kimura M, Ogihara M. Effects of branched-chain amino acids on DNA synthesis and proliferation in primary cultures of adult rat hepatocytes. Eur J Pharmacol 2005;510:167–80.

[55] Holecek M, Simek J, Palicka V, Zadak Z. Effect of glucose and branched chain amino acid (BCAA) infusion on onset of liver regeneration and plasma amino acid pattern in partially hepatectomized rats. J Hepatol 1991;13:14–20.

[56] Crumm S, Cofan M, Juskeviciute E, Hoek JB. Adenine nucleotide changes in the remnant liver: an early signal for regeneration after partial hepatectomy. Hepatology 2008;48:898–908.

[57] Webber EM, Godowski PJ, Fausto N. In vivo response of hepatocytes to growth factors requires an initial priming stimulus. Hepatology 1994;19:489–97.

[58] Mitchell C, Nivison M, Jackson LF, Fox R, Lee DC, Campbell JS, et al. Heparin-binding epidermal growth factor-like growth factor links hepatocyte priming with cell cycle progression during liver regeneration. J Biol Chem 2005;280:2562–8.

[59] Li J, Campbell JS, Mitchell C, McMahan RS, Yu X, Riehle KJ, et al. Relationships between deficits in tissue mass and transcriptional programs after partial hepatectomy in mice. Am J Pathol 2009;175:947–57.

[60] Cornell RP, Liljequist BL, Bartizal KF. Depressed liver regeneration after partial hepatectomy of germ-free, athymic and lipopolysaccharide-resistant mice. Hepatology 1990;11:916–22.

[61] Campbell JS, Riehle KJ, Brooling JT, Bauer RL, Mitchell C, Fausto N. Proinflammatory cytokine production in liver regeneration is Myd88-dependent, but independent of Cd14, Tlr2, and Tlr4. J Immunol 2006;176:2522–8.

[62] Vaquero J, Campbell JS, Haque J, McMahan RS, Riehle KJ, Bauer RL, et al. Toll-like receptor 4 and myeloid differentiation factor 88 provide mechanistic insights into the cause and effects of interleukin-6 activation in mouse liver regeneration. Hepatology 2011;54:597–608.

[63] Cressman DE, Greenbaum LE, DeAngelis RA, Ciliberto G, Furth EE, Poli V, et al. Liver failure and defective hepatocyte regeneration in interleukin-6-deficient mice. Science 1996;274:1379–83.

[64] Yamada Y, Kirillova I, Peschon JJ, Fausto N. Initiation of liver growth by tumor necrosis factor: deficient liver regeneration in mice lacking type I tumor necrosis factor receptor. Proc Natl Acad Sci USA 1997;94:1441–6.

[65] Fujita J, Marino MW, Wada H, Jungbluth AA, Mackrell PJ, Rivadeneira DE, et al. Effect of TNF gene depletion on liver regeneration after partial hepatectomy in mice. Surgery 2001;129:48–54.

[66] Hayashi H, Nagaki M, Imose M, Osawa Y, Kimura K, Takai S, et al. Normal liver regeneration and liver cell apoptosis after partial hepatectomy in tumor necrosis factor-alpha-deficient mice. Liver Int 2005;25:162–70.

[67] Blindenbacher A, Wang X, Langer I, Savino R, Terracciano L, Heim MH. Interleukin 6 is important for survival after partial hepatectomy in mice. Hepatology 2003;38:674–82.

[68] Wuestefeld T, Klein C, Streetz KL, Betz U, Lauber J, Buer J, et al. Interleukin-6/glycoprotein 130-dependent pathways are protective during liver regeneration. J Biol Chem 2003;278:11281–8.

[69] Sakamoto T, Liu Z, Murase N, Ezure T, Yokomuro S, Poli V, et al. Mitosis and apoptosis in the liver of interleukin-6-deficient mice after partial hepatectomy. Hepatology 1999;29:403–11.

[70] Li W, Liang X, Leu JI, Kovalovich K, Ciliberto G, Taub R. Global changes in interleukin-6-dependent gene expression patterns in mouse livers after partial hepatectomy. Hepatology 2001;33:1377–86.

[71] Talarmin H, Rescan C, Cariou S, Glaise D, Zanninelli G, Bilodeau M, et al. The mitogen-activated protein kinase kinase/extracellular signal-regulated kinase cascade activation is a key signalling pathway involved in the regulation of G(1) phase progression in proliferating hepatocytes. Mol Cell Biol 1999;19:6003–11.

[72] Campbell JS, Argast GM, Yuen SY, Hayes B, Fausto N. Inactivation of p38 MAPK during liver regeneration. Int J Biochem Cell Biol 2011;43:180–8.
[73] Jones SM, Kazlauskas A. Growth-factor-dependent mitogenesis requires two distinct phases of signalling. Nat Cell Biol 2001;3:165–72.
[74] Jones SM, Klinghoffer R, Prestwich GD, Toker A, Kazlauskas A. PDGF induces an early and a late wave of PI 3-kinase activity, and only the late wave is required for progression through G1. Curr Biol 1999;9:512–21.
[75] Kazlauskas A. The priming/completion paradigm to explain growth factor-dependent cell cycle progression. Growth Factors 2005;23:203–10.
[76] Zwang Y, Sas-Chen A, Drier Y, Shay T, Avraham R, Lauriola M, et al. Two phases of mitogenic signaling unveil roles for p53 and EGR1 in elimination of inconsistent growth signals. Mol Cell 2011;42:524–35.
[77] Matsuo T, Yamaguchi S, Mitsui S, Emi A, Shimoda F, Okamura H. Control mechanism of the circadian clock for timing of cell division in vivo. Science 2003; 302:255–9.
[78] Takeishi T, Hirano K, Kobayashi T, Hasegawa G, Hatakeyama K, Naito M. The role of Kupffer cells in liver regeneration. Arch Histol Cytol 1999;62:413–22.
[79] Meijer C, Wiezer MJ, Diehl AM, Schouten HJ, Schouten HJ, Meijer S, et al. Kupffer cell depletion by CI2MDP-liposomes alters hepatic cytokine expression and delays liver regeneration after partial hepatectomy. Liver 2000;20:66–77.
[80] Abshagen K, Eipel C, Kalff JC, Menger MD, Vollmar B. Loss of NF-kappaB activation in Kupffer cell-depleted mice impairs liver regeneration after partial hepatectomy. Am J Physiol Gastrointest Liver Physiol 2007;292:G1570–7.
[81] DeAngelis RA, Kovalovich K, Cressman DE, Taub R. Normal liver regeneration in p50/nuclear factor kappaB1 knockout mice. Hepatology 2001;33:915–24.
[82] Chaisson ML, Brooling JT, Ladiges W, Tsai S, Fausto N. Hepatocyte-specific inhibition of NF-kappaB leads to apoptosis after TNF treatment, but not after partial hepatectomy. J Clin Invest 2002;110:193–202.
[83] Ding BS, Nolan DJ, Butler JM, James D, Babazadeh AO, Rosenwaks Z, et al. Inductive angiocrine signals from sinusoidal endothelium are required for liver regeneration. Nature 2010;468:310–5.
[84] Michalopoulos GK. Liver regeneration after partial hepatectomy: critical analysis of mechanistic dilemmas. Am J Pathol 2010;176:2–13.
[85] Rhim JA, Sandgren EP, Degen JL, Palmiter RD, Brinster RL. Replacement of diseased mouse liver by hepatic cell transplantation. Science 1994;263: 1149–52.
[86] Rhim JA, Sandgren EP, Palmiter RD, Brinster RL. Complete reconstitution of mouse liver with xenogeneic hepatocytes. Proc Natl Acad Sci USA 1995;92:4942–6.
[87] Simpson GE, Finckh ES. The pattern of regeneration of rat liver after repeated partial hepatectomies. J Pathol Bacteriol 1963;86:361–70.
[88] Selzner M, Clavien PA. Fatty liver in liver transplantation and surgery. Semin Liver Dis 2001;21:105–13.
[89] Vetelainen R, van Vliet A, Gouma DJ, van Gulik TM. Steatosis as a risk factor in liver surgery. Ann Surg 2007;245:20–30.
[90] Greenbaum LE, Wells RG. The role of stem cells in liver repair and fibrosis. Int J Biochem Cell Biol 2011;43:222–9.
[91] Huang J, Barr E, Rudnick DA. Characterization of the regulation and function of zinc-dependent histone deacetylases during rodent liver regeneration. Hepatology 2013;57:1742–51.
[92] Shu J, Kren BT, Xia Z, Wong PY, Li L, Hanse EA, et al. Genomewide microRNA down-regulation as a negative feedback mechanism in the early phases of liver regeneration. Hepatology 2011;54:609–19.
[93] Coller HA. Cell biology. The essence of quiescence Science 2011;334(6059):1074–5[PMID: 22116874].

CHAPTER

5

Extracellular Signals Involved in Liver Regeneration: Direct and Auxiliary Mitogens

*Udayan Apte**, *Pallavi B. Limaye*[†], *George K. Michalopoulos*[‡]

[*]Department of Pharmacology, Toxicology and Therapeutics, University of Kansas Medical Center, Kansas City, Kansas, USA

[†]Xenometrics LLC, Stilwell, Kansas, USA

[‡]Department of Pathology, University of Pittsburgh School of Medicine, Pittsburgh, Pennsylvania, USA

The ability to regenerate upon surgical resection or chemical and other types of injury is a unique characteristic of liver [1,2]. Liver regeneration has been extensively studied and a number of signaling pathways comprising a large variety of extracellular and intracellular proteins have been implicated in stimulation of liver regeneration [3]. At its heart, liver regeneration involves proliferation of hepatocytes, the parenchymal cells of the liver that form 60-70% of the organ, followed by proliferation of the other cell types, aimed to form a complete tissue at the end of the regenerative process. Liver regeneration is a complex event tightly regulated by multiple redundant signaling mechanisms that are activated in a time-dependent fashion. Proregenerative signaling can be detected within minutes after partial hepatectomy (PH), the most widely used model to study liver regeneration, where approximately 70% of the liver is surgically resected and the remaining liver is allowed to regenerate. Whereas liver regeneration after PH takes almost 2 weeks to complete in rodents, the signaling begins very early and signals such as activation of urokinase, activation of pro-HGF, activation of transcriptional regulators such as β-catenin, Notch, and NF-κB, and many other signals can be detected with 30 min to 1 h after surgery [1]. Similar rapid activation of intracellular promitogenic signaling in hepatocytes is observed after drug-induced liver injury [4]. Extracellular signaling factors such as growth factors, cytokines, chemokine, hormones, and other ligands associated with liver regeneration orchestrate and coordinate the intracellular signaling triggered in hepatocytes in response to PH (Figure 5.1).

While a role for extracellular circulating signals in liver regeneration was hypothesized before, the first evidence for the role of extracellular factors stimulating liver regeneration came in mid-1960s by seminal experiments performed

Liver Regeneration
http://dx.doi.org/10.1016/B978-0-12-420128-6.00005-1

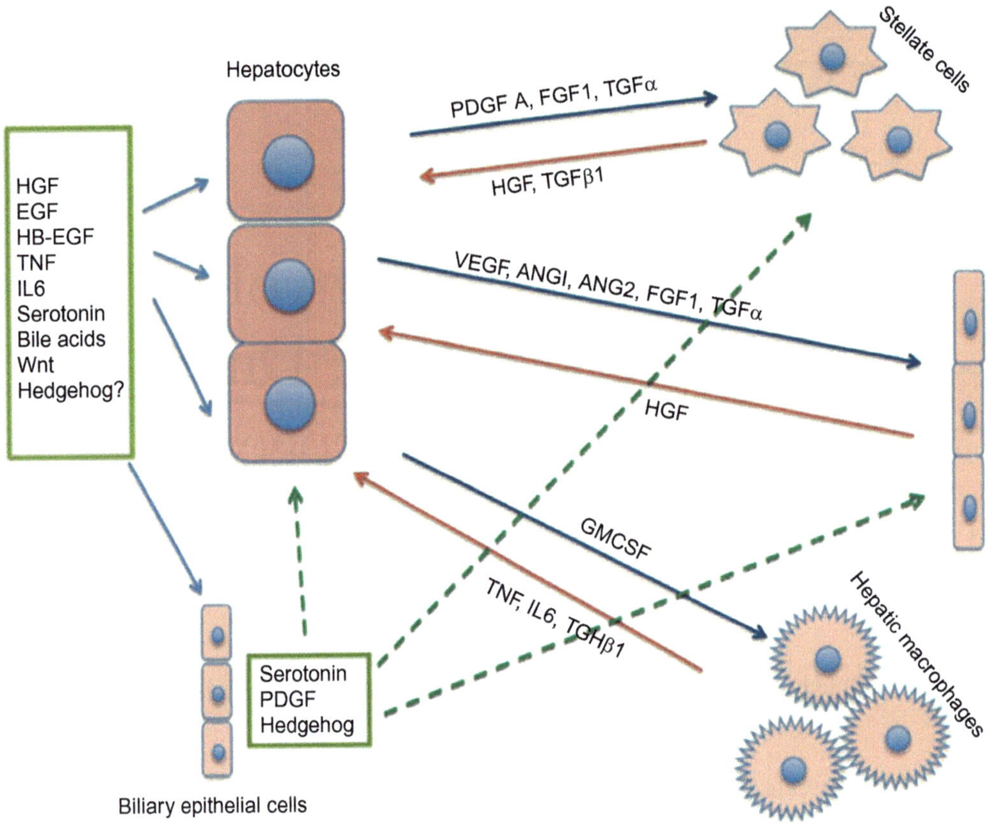

FIGURE 5.1 Scheme showing various direct and auxiliary mitogens involved in stimulation of liver regeneration after partial hepatectomy (PH).

by Bucher and colleagues [5]. In these famous parabiosis experiments in rats, where cross-circulation was established between a normal and a hepatectomized rat resulting in increased DNA synthesis in the normal rat liver. Similar results were obtained in other studies involving heterotrophic hepatic tissue fragment grafts and hepatocytes implanted in subcutaneous fat pad [6]. These studies demonstrated that liver regeneration is stimulated by blood-borne or humoral signals and led to extensive studies to identify these extracellular circulating signals. These studies culminated in discovery of hepatocyte growth factor (HGF), which was discovered originally by Michalopoulos and colleagues as hepatopoietin A and later fully sequenced in humans by Nakamura and colleagues in 1980s [7–9]. Subsequent studies resulted in identification of a new role in liver regeneration for many known and newly isolated proteins including growth factors (epidermal growth factor (EGF), transforming growth factor-alpha (TGFα), and other EGFR-ligand family members), cytokines (TNFα, IL6), hormones (insulin), and other circulating ligands such as leptin, bile acids, norepinephrine, etc. [1].

Whereas the number of signaling factors known to be involved in stimulation of liver regeneration has increased over the years, additional questions have emerged. One of the most important questions is which of the numerous circulating proteins known to stimulate regeneration can directly activate hepatocyte proliferation without requiring additional support? In other words, are all extracellular signals created equal? Studies in the last decade after advent of transgenic and knockout mouse technology demonstrated a significant redundancy in the proregenerative signaling system. Genetic deletion of growth factors, cytokines, or inhibition of their signaling via deletion of their cognate receptors in mice rarely resulted in partial inhibition of liver regeneration after PH. Mice that exhibited near total inhibition of liver regeneration invariably also exhibited extensive cell death in the regenerating lobes due to the loss of cytoprotective functions of the deleted factor (e.g., farnesoid X receptor deletion results in increased bile acid-induced injury and inhibited regeneration) [10]. Whereas the *in vivo* studies using transgenic and knockout mice have been very useful, they have presented additional questions about whether hepatocytes respond to all the factors similarly or prefer certain stimuli to others when they need to proliferate.

5.1 HEPATOCYTE GROWTH ASSAY

One way to address this question would be to do *in vitro* studies to test the ability of specific factors to induce hepatocyte growth. Interestingly, despite the innate ability of proliferation *in vivo*, the adult mature hepatocytes, when cultured *in vitro*, exhibit active but limited proliferative activity [11]. The assay termed hepatocyte growth assay (HGA) used primary rat hepatocytes isolated by employing standard collagenase perfusion and has been adapted for mouse [12] and even human hepatocytes [13]. The goal of the assay is to elucidate the conditions that allow hepatocytes to enter into sustained clonal growth and to identify the relative effects of specific mitogens, nutrients, and matrix. This was systematically studied by individually subtracting these factors from culture media (Table 5.1) [11]. Isolated rat hepatocytes were cultured in collagen-coated six-well cluster plates in serum-free chemically defined media (called hepatocyte growth medium or HGM). The DNA per culture after 14 days of growth was measured as an endpoint of hepatocyte proliferation.

TABLE 5.1 Composition of Modified Hepatocyte Growth Medium (HGM)

Reagent	Concentration
DMEM 4.5 g/L glucose	500 mL
DMEM no glucose	500 mL
Bovine serum albumin	2 g/L
D-(+)-Galactose	2 g/L
L-Ornithine monohydrochloride	0.1 g/L
L-Proline	0.3 g/L
HEPES	2.38 g/L
Nicotinamide	0.3 g/L
Zinc chloride	0.544 mg/L
Copper(II) sulfate pentahydrate	0.2 mg/L
Zinc sulfate heptahydrate	0.75 mg/L
Manganese(II) sulfate monohydrate	0.025 mg/L
L-Glutamine	0.15 mg/L
Penicillin-streptomycin	1%
Dexamethasone[a]	10^{-7} M
Insulin-transferrin-sodium selenite supplement[b]	10 mg/L

Dissolve the ingredients included in the table in 1 L of a 1:1 mixture of DMEM high glucose (4.5 g/L) and DMEM no glucose. Filter sterilize using a 0.22-μm filter before adding dexamethasone and ITS supplement.

[a] *Add 100 μL from 10^{-3} M stock dissolved in ethanol.*

[b] *Add to final concentration 5 μg/mL insulin, 5 μg/mL transferrin, and 5 ng/mL sodium selenite.*

A number of different growth factors were tested for their ability to stimulate hepatocyte growth. HGF, EGF, TGFα, keratinocyte growth factor (KGF), stem cell factor (SCF), and acidic fibroblast growth factor (aFGF) were added individually to culture medium. The series of experiments showed that in rat hepatocytes, only HGF, EGF, and TGFα caused significant cell proliferation, as measured by the total amount of DNA per culture whereas KGF, SCF, when added alone or in combination had no effect. It also showed that HGF and EGF added together had the strongest proliferative effect than any other single mitogen or combination, whereas TGFα when added alone had the strongest effect than any other mitogen added alone. Interestingly, addition of all the growth factors together had no more effect than the combined HGF and EGF.

In addition to growth factors, the initial studies were focused on investigating the role of nutrients on hepatocyte growth. Among the nutrients, glucose, galactose, ornithine, albumin, dexamethasone, transferrin and selenium, nicotinamide, and trace elements were tested. Removal of dexamethasone, insulin, and nicotinamide dramatically reduced hepatocyte growth. Addition of iron containing diferric transferrin was effective in promoting growth. Removal of glucose, galactose, and ornithine together led to complete termination of proliferation. Addition of glucose alone was successful in restoring most of the proliferative response, whereas addition of galactose alone was less effective and ornithine alone had very little effect. Additional studies tested the effect of various matrix proteins on hepatocyte growth. Among the matrix substances tested, dry coating of collagens type IV (mouse), type I (bovine), fibronectin, and laminin was comparable to each other in promoting cell growth.

Recent studies have further modified the HGA, the mouse hepatocyte growth medium, and adapted the assay for using mouse hepatocytes given that mouse is becoming more common animal model to study liver regeneration [12]. In this assay, ^{3}H-thymidine incorporation is used to evaluate hepatocyte DNA synthesis and growth. The results in mouse hepatocytes mostly corroborated observations in rat hepatocytes. Additionally, FGF1 and FGF2 exhibited a modest mitogenic effect in mouse hepatocytes [12]. Other methods to measure cell proliferation such as proliferating cell nuclear antigen staining can also be used. Recent studies using mouse hepatocytes and MHGM have revealed innate differences in mouse and rat hepatocytes and demonstrated that process of collagenase perfusion activates growth pathways in the mouse cells [12].

A modified version of the assay called hepatocyte organoid cultures has also been used not only to study hepatocyte growth but also to demonstrate plasticity of hepatocytes in culture enabling them to undergo hepatocyte to biliary transdifferentiation [14]. One of the criticisms to the HGA was that hepatocytes may not express cognate receptors in culture and thus may not respond to growth factors. The hepatocyte organoid culture assays confirmed that the hepatocytes in culture do express the membrane receptors for various tyrosine kinase receptor ligands and are activated on exposure to their respective ligands such as Flt1 (VEGF receptor), PDGFR-α, and PDGFR-β (PDGF receptors), c-kit (SCF receptor), Ron (MSP receptor), and FGFR 1, 2, 3, and 4 (FGF receptors). However, except HGF and EGF, all the other GFs failed to stimulate hepatocytes proliferation despite expression and activation of their cognate receptors [14].

5.2 COMPLETE VERSUS AUXILIARY MITOGENS

The HGA has been very effective in determining ability of various circulating ligands in stimulating hepatocyte growth. A new system was developed to classify the ligands based on their ability to stimulate hepatocyte proliferation in culture [1]. The ligands have been classified in to two categories—(1) direct or complete or primary mitogens and (2) auxiliary or secondary mitogens. The direct or primary mitogens can stimulate hepatocyte proliferation and growth in conditioned HGM medium by themselves. There are only few proteins that fall in to this category including HGF, EGF, and other EGF family members such as TGFα, heparin-binding EGF, and amphiregulin. Most other extracellular ligands known to play a role in liver regeneration fail to stimulate hepatocyte growth in the HGA independently and are termed as auxiliary or secondary mitogens. The auxiliary mitogens can increase hepatocyte proliferation in HGA when added in combination of one of the primary mitogens such as HGF and EGF. The list of auxiliary mitogens is long and includes cytokines such as TNFα, IL6, growth factors such as PDGF, FGF, chemokines such as CXC chemokines, neurotransmitters such as norepinephrine, and hormones such as estrogen. Table 5.2 summarizes a role of a number of factors classified as direct and auxiliary mitogens for hepatocytes. Here, we discuss some of the main direct and auxiliary mitogens.

TABLE 5.2 Extracellular Factors Acting as Direct and Auxiliary Mitogens for Hepatocytes

Factor	Effect	References
DIRECT MITOGENS		
HGF	Clonal expansion of hepatocytes in culture	[11]
	HGF injection in portal vein of normal rats and mice causes proliferation of hepatocytes and enlargement of the liver	[15,16]
	Targeted genetic elimination of cMet from the liver is associated with very diminished or absent regenerative response	[17]
	RNA interference after PH *in vivo* against cMet is associated with complete blockade of the cell cycle	[18]
EGF	EGF given in intact animals causes hepatocyte proliferation	[19]
	Clonal expansion of hepatocytes in culture	[11]
TGF-α	Mice with targeted transgenic expression of TGFα in hepatocytes have dramatic liver enlargement and develop tumors	[20]
Amphiregulin	Mice deficient in amphiregulin have deficient liver regeneration	[21]
HB-EGF	HB-EGF transgenic mice with liver-targeted production have enhanced regeneration	[22]
	HB-EGF knockout mice have deficient regenerative response	[23]
AUXILIARY MITOGENS		
TNFα	Antibodies against TNFα administered at the time of hepatectomy decrease the regenerative response	[24]
	Mice with genetic deletions of the TNF receptor 1 (TNFR1) have slow and deficient response following PH	[25]
	It does not induce DNA synthesis in primary cultures of hepatocytes in serum-free media nor does it induce hepatocytes DNA synthesis when injected in whole animals. It does, however, enhance the mitogenic effects of direct mitogens such as HGF, both *in vivo* and in cell culture	[26]
IL6	Mice overexpressing both IL6 and its soluble receptor gp130 have areas of periportal hepatocyte hyperplasia	[27]
	IL6 is not a direct mitogen for hepatocytes and does not enhance the mitogenic effect of other growth factors	[11]

Continued

TABLE 5.2 Extracellular Factors Acting as Direct and Auxiliary Mitogens for Hepatocytes—cont'd

Factor	Effect	References
Norepinephrine	Norepinephrine substantially enhances the mitogenic effects of EGF and HGF in hepatocyte cultures and it decreases the mito-inhibitory effects of TGFβ1	[28]
	Addition of norepinephrine triggers high-level hepatocyte DNA synthesis	[29]
Serotonin	In thrombocytopenic mice, a serotonin agonist reconstituted liver proliferation	[30]
	Serotonin is not a mitogen for hepatocytes in culture, thus its effects on this process are likely to be indirect	
iNOS	Mice with deficiency in iNOS have defective liver regeneration	[31]
Bile acids	Depletion of bile acids leads to decreased regeneration	[32]
Insulin	Diversion of the portal vein flow to vena cava (portacaval shunt) forces insulin to bypass the liver. Liver atrophies to about 1/3 of its size. Administration of insulin directly to the liver in animals with experimental portacaval shunts reverses hepatic atrophy and is associated with rapid hepatocyte proliferation	[33,34]
	Hepatocytes in culture have diminished response to mitogens in the absence of insulin	[28]
Notch and Jagged (involvement is not clear)	Treatment with RNAi against either Notch-1 or Jagged-1 partially suppressed regeneration and recombinant Jagged-1 induced DNA synthesis in hepatocyte cultures	
	Elimination of Notch using a Cre/Lox system soon after birth from the entire body resulted in an adaptive hepatic nodular hyperplasia	[35]
Components of complement	Mice deficient in components of complement C3 and C5 have defective regenerative responses to both PH and to recovery from centrilobular necrosis following injury with CCl4. Administration of the missing components restores the efficiency of the regenerative response	[36]

5.3 COMPLETE OR DIRECT MITOGENS

5.3.1 Hepatocyte Growth Factor

HGF was originally identified as Hepatopoietin A by Michalopoulos and colleagues using serum from hepatectomized rats and was partially sequenced [7,8]. Nakamura et al. published a full sequence of human HGF [9]. HGF is made up of two subunits, a 70-KDa heavy chain and a 30-KDa light chain. HGF shares homology with plasminogen and has several Kringle domains in its structure. In the liver, HGF is bound to the extracellular matrix in an inactive form and is activated rapidly upon PH by urokinase. It is also found in matrix of other organs including lungs, spleen, placenta, brain, etc. [37]. HGF in liver is produced predominantly not only by the stellate cells [38] but also by sinusoidal endothelial cells (SECs) [1]. In SECs, HGF is produced in response to of VEGF via the nonmitogenic VEGF receptor I. Interestingly, the liver sequesters systemically injected HGF more than any other organ [39,40]. Studies in early 1990s identified Met oncogene as the cognate cell-surface receptor for HGF [41,42]. Met is expressed in most epithelial cells, endothelial cells, as well as neurons and mediates all the effects of HGF. Genetic elimination of either HGF or Met is associated with embryonic lethality involving abnormalities in many organs, most notably in placenta [1].

Livers of the embryos are smaller than the wild-type controls.

Activation of the matrix-bound HGF by urokinase is observed very rapidly after PH within first hour resulting in an increase in serum HGF [15]. This is followed by increase in synthesis of new HGF mRNA at 3 h after PH and peaks at 24 h after PH. Increase in HGF mRNA after PH is observed not only in the liver but also in lung and spleen [43]. The signals triggering this extrahepatic participation are not clear. Norepinephrine and IL6, both rising in the plasma after PH, are known to stimulate production of HGF in responding cells and are possible candidates for mediating this effect [44]. Studies with hepatocytes in culture have suggested that part of the effects of HGF is caused by stimulation of production of TGFα [45]. However, liver regeneration proceeds normally in TGFα-deficient mice, whereas the effects of inactivation of the HGF receptor cannot be compensated by other mitogenic receptors. Studies in rats using ShRNA against Met showed a significant inhibition of liver regeneration after PH. Met inhibition also resulted in substantial activation of caspase-3 and increased apoptosis [18]. Recent studies with conditional HGF-knockout mice generated using Mx1-Cre further reinforced role of HGF in liver regeneration. Interestingly, the HGF-KO mice generated using the Mx1-cre exhibited persistence of HGF mRNA in small quantities in the liver. However, these mice had substantially decreased liver regeneration after CCl_4 treatment [46].

The view of HGF as an initiator of liver regeneration is bolstered by the fact that it is a direct mitogen for hepatocytes, it activates its receptor very early, and it can induce most of the changes occurring during liver regeneration (including massive hepatic enlargement) by administration in intact mice and rats. Several events occur before HGF becomes demonstrably available (e.g., increase in urokinase, migration of NICD). Currently, there is no evidence that one single initiating change after PH exists which alone leads to liver regeneration. Of all the signals participating in the very early events after PH, however, given its properties, actions, and impact of elimination of its receptor, the signaling by HGF appears as the most irreplaceable contributor.

5.3.2 Ligands of EGFR

EGFR is a member of ErbB family of cell-surface receptors [47]. There are a total of six ligands for EGFR including EGF, TGFα, HB-EGF, amphiregulin, betacellulin, and epiregulin out of which EGF, TGFα, amphiregulin, and HB-EGF have been tested and shown to stimulate hepatocyte growth when used in the HGA. EGFR undergoes tyrosine phosphorylation upon binding to its ligands and transmits the signal downstream via a cascade of mitogen-activated protein kinases. After PH, similar to Met, EGFR activation is observed within 1 h [48]. The liver-specific EGFR knockout mice exhibit higher mortality and delayed liver regeneration after PH. The delay in liver regeneration is due to inhibited cell-cycle progression from G1 to S phase accompanied by decreased Cyclin D1 expression. Similarly, shRNA-mediated knockdown of EGFR results in delayed liver regeneration in rats. However, treatment of EGFR shRNA results in compensatory upregulation of other members of ErbB family members as well as Met [49].

Four EGFR ligands including EGF, HB-EGF, amphiregulin, and TGFα are complete or direct mitogens for hepatocytes. *In vivo* infusion of EGF results in increased hepatocyte DNA synthesis [1]. EGF is produced in salivary glands as well as Brunner's glands in intestine, which provide a constant supply for EGF to liver via portal circulation. Studies have indicated that EGF production in Brunner's glands is stimulated by norepinephrine after PH [50]. HB-EGF is produced in the liver by SECs and Kupffer cells and its levels increase after PH. Mice deficient in HB-EGF have decreased liver regeneration [23]. Interestingly, minimal liver regeneration is observed with 30% PH, which is significantly increased with HB-EGF treatment after 30% PH. Hepatocytes produce TGFα and amphiregulin, the two other EGFR ligands that can independently induce hepatocyte DNA synthesis in HGA [51]. Amphiregulin is a target of Yap and amphiregulin knockout mice show decreased liver regeneration. TGFα has the strongest promitogenic effect of all the EGFR ligands in the HGA [1]. It is a transmembrane protein produced as a 28 KDa precursor and is cleaved by proteases to release an 8 KDa active protein, which may act in either autocrine or paracrine fashion. Interestingly, TGFα knockout mice have normal liver regeneration, presumably due to compensation by other EGFR ligands [52].

5.4 AUXILIARY MITOGENS

There are a number of extracellular circulating proteins that can stimulate hepatocyte proliferation when present along with the complete or direct mitogens and most often additively or synergistically augment the effect of direct mitogens. These include various cytokines such as TNFα and IL6, some chemokines, bile acids, hormones, and neurotransmitters [1]. The role of bile acids, Hedgehog- and Wnt-stimulated pathways, and TGFβ has been discussed in detail elsewhere in this book. Here, we will discuss some of the most important auxiliary mitogens.

5.5 TNFα

TNFα is produced by Kupffer cells in the liver and functions in a context-dependent manner [53]. TNFα activation is associated with hepatic cell-death models such as LPS (bacterial endotoxin) alone, LPS + galactosamine, and the anti-Fas agonistic antibody-induced apoptotic cell death. However, many studies have shown that TNFα plays a progrowth role in liver regeneration after PH [54,55]. Similar role for TNFα has also been identified in drug-induced liver injury [4]. TNFα levels increase in plasma after PH and antibody-mediated neutralization of TNFα results in delayed liver regeneration [24]. Furthermore, TNF receptor-knockout mice show delayed liver regeneration [25]. The downstream proregenerative signaling of TNFα is mediated by NF-κB. TNFα cannot independently stimulate hepatocyte growth in HGA, and infusion of TNFα *in vivo* does not stimulate hepatocyte proliferation [1].

5.6 IL6

IL6 is a proinflammatory cytokine secreted by Kupffer cells under the influence of TNFα. It is best known for its role in innate immunity and in stimulation of acute-phase response [56]. IL6 signals by binding to a soluble circulating receptor, which then binds to a cell-surface receptor called gp130 to initiate downstream signaling culminating in activation of STAT3 [27,57]. IL6 levels increase in serum after PH. IL6 knockout mice have a significantly delayed liver regeneration due to delayed STAT3 activation [57]. However, like TNFα, IL6 cannot independently stimulate hepatocyte growth in HGA and infusion of IL6 cannot stimulate hepatocyte proliferation *in vivo*.

5.7 NOREPINEPHRINE

Norepinephrine is a catecholamine produced mainly by adrenal medulla, noradrenergic neurons in the locus coeruleus in the central nervous system, and by hepatic stellate cells in the liver [58]. It is activated by monoamine oxidase and functions as a sympathetic agonist. Whereas norepinephrine cannot stimulate hepatocyte growth independently in HGA, it augments promitogenic effects of HGF and EGF [28]. Circulating norepinephrine (and epinephrine) levels increase following PH. Blocking norepinephrine signaling by Prazosine-mediated blockage of α1 adrenergic receptors results in delays of liver regeneration in rats [59]. Interestingly, norepinephrine can activate HGF and EGF production and is also involved in STAT3 as well as NF-κB signaling. In addition, it stimulates production of HGF by mesenchymal cells [44,59] and production of EGF from Brunner's glands of the duodenum [50].

5.8 INSULIN

Insulin, the main glucose-regulating hormone in the body, is produced by pancreatic beta cells and enters liver directly via portal circulation. Presence of insulin in hepatocyte cultures is essential for survival of hepatocytes [11]. However, insulin does not stimulate hepatocyte growth independent of HGF and EGF in HGA, and insulin infusion does not stimulate liver regeneration *in vivo*. Nevertheless, the insulin is required for proper liver regeneration as shown by studies in rats with portacaval shunt, which diverts portal blood supply away from the liver, results in decreased liver regeneration after PH [1].

5.9 SUMMARY

The *in vitro* HGA studies along with *in vivo* evidence have revealed role of plethora of extracellular and extrahepatic factors in stimulation of liver regeneration. Classifying them into complete or direct mitogens and auxiliary mitogens provides a system to identify the most important factor and develops a hierarchical system that could be used in future for application in regenerative medicine. Increasing evidence suggests that lack of proper liver regeneration is a contributing factor in development of chronic liver diseases and the final outcome of acute liver failure. Finding regenerative therapies based on stimulating regeneration of existing hepatocytes holds particular promise. Defining the role of humoral signaling and identifying their sequential or simultaneous roles in liver regeneration will provide a strong basis for further investigations in their therapeutic use as stimulators of liver regeneration to treat liver diseases.

References

[1] Michalopoulos GK. Principles of liver regeneration and growth homeostasis. Compr Physiol 2013;3:485–513.
[2] Fausto N, Campbell JS, Riehle KJ. Liver regeneration. Hepatology 2006;43:S45–53.
[3] Michalopoulos GK, DeFrances MC. Liver regeneration. Science 1997;276:60–6.
[4] Bhushan B, Walesky C, Manley M, Gallagher T, Borude P, Edwards G, et al. Pro-regenerative signaling after acetaminophen-induced acute liver injury in mice identified using a novel incremental dose model. Am J Pathol 2014;184:3013–25.
[5] Moolten FL, Bucher NL. Regeneration of rat liver: transfer of humoral agent by cross circulation. Science 1967;158:272–4.
[6] Leong GF, Grisham JW, Hole BV, Albright ML. Effect of partial hepatectomy on DNA synthesis and mitosis in heterotopic partial autografts of rat liver. Cancer Res 1964;24:1496–501.
[7] Zarnegar R, Michalopoulos G. Purification and biological characterization of human hepatopoietin A, a polypeptide growth factor for hepatocytes. Cancer Res 1989;49:3314–20.
[8] Zarnegar R, DeFrances MC, Oliver L, Michalopoulos G. Identification and partial characterization of receptor binding sites for HGF on rat hepatocytes. Biochem Biophys Res Commun 1990;173:1179–85.
[9] Nakamura T, Nishizawa T, Hagiya M, Seki T, Shimonishi M, Sugimura A, et al. Molecular cloning and expression of human hepatocyte growth factor. Nature 1989;342:440–3.
[10] Borude P, Edwards G, Walesky C, Li F, Ma X, Kong B, et al. Hepatocyte-specific deletion of farnesoid X receptor delays but does not inhibit liver regeneration after partial hepatectomy in mice. Hepatology 2012;56:2344–52.
[11] Block GD, Locker J, Bowen WC, Petersen BE, Katyal S, Strom SC, et al. Population expansion, clonal growth, and specific differentiation patterns in primary cultures of hepatocytes induced by HGF/SF, EGF and TGF alpha in a chemically defined (HGM) medium. J Cell Biol 1996;132:1133–49.

[12] Bowen WC, Michalopoulos AW, Orr A, Ding MQ, Stolz DB, Michalopoulos GK. Development of a chemically defined medium and discovery of new mitogenic growth factors for mouse hepatocytes: mitogenic effects of FGF1/2 and PDGF. PLoS One 2014;9:e95487.
[13] Runge DM, Runge D, Dorko K, Pisarov LA, Leckel K, Kostrubsky VE, et al. Epidermal growth factor- and hepatocyte growth factor-receptor activity in serum-free cultures of human hepatocytes. J Hepatol 1999;30:265–74.
[14] Limaye PB, Bowen WC, Orr AV, Luo J, Tseng GC, Michalopoulos GK. Mechanisms of hepatocyte growth factor-mediated and epidermal growth factor-mediated signaling in transdifferentiation of rat hepatocytes to biliary epithelium. Hepatology 2008;47:1702–13.
[15] Liu ML, Mars WM, Zarnegar R, Michalopoulos GK. Uptake and distribution of hepatocyte growth factor in normal and regenerating adult rat liver. Am J Pathol 1994;144:129–40.
[16] Patijn GA, Lieber A, Schowalter DB, Schwall R, Kay MA. Hepatocyte growth factor induces hepatocyte proliferation in vivo and allows for efficient retroviral-mediated gene transfer in mice. Hepatology 1998;28:707–16.
[17] Borowiak M, Garratt AN, Wustefeld T, Strehle M, Trautwein C, Birchmeier C. Met provides essential signals for liver regeneration. Proc Natl Acad Sci USA 2004;101:10608–13.
[18] Paranjpe S, Bowen WC, Bell AW, Nejak-Bowen K, Luo JH, Michalopoulos GK. Cell cycle effects resulting from inhibition of hepatocyte growth factor and its receptor c-Met in regenerating rat livers by RNA interference. Hepatology 2007;45:1471–7.
[19] Bucher NL, Patel U, Cohen S. Hormonal factors concerned with liver regeneration. Ciba Found Symp 1977;55:95–107.
[20] Webber EM, Wu JC, Wang L, Merlino G, Fausto N. Overexpression of transforming growth factor-alpha causes liver enlargement and increased hepatocyte proliferation in transgenic mice. Am J Pathol 1994;145:398–408.
[21] Berasain C, Garcia-Trevijano ER, Castillo J, Erroba E, Lee DC, Prieto J, et al. Amphiregulin: an early trigger of liver regeneration in mice. Gastroenterology 2005;128:424–32.
[22] Kiso S, Kawata S, Tamura S, Inui Y, Yoshida Y, Sawai Y, et al. Liver regeneration in heparin-binding EGF-like growth factor transgenic mice after partial hepatectomy. Gastroenterology 2003;124:701–7.
[23] Mitchell C, Nivison M, Jackson LF, Fox R, Lee DC, Campbell JS, et al. Heparin-binding epidermal growth factor-like growth factor links hepatocyte priming with cell cycle progression during liver regeneration. J Biol Chem 2005;280:2562–8.
[24] Akerman P, Cote P, Yang SQ, McClain C, Nelson S, Bagby GJ, et al. Antibodies to tumor necrosis factor-alpha inhibit liver regeneration after partial hepatectomy. Am J Physiol 1992;263:G579–85.
[25] Yamada Y, Kirillova I, Peschon JJ, Fausto N. Initiation of liver growth by tumor necrosis factor: deficient liver regeneration in mice lacking type I tumor necrosis factor receptor. Proc Natl Acad Sci USA 1997;94:1441–6.
[26] Webber EM, Bruix J, Pierce RH, Fausto N. Tumor necrosis factor primes hepatocytes for DNA replication in the rat. Hepatology 1998;28:1226–34.
[27] Wuestefeld T, Klein C, Streetz KL, Betz U, Lauber J, Buer J, et al. Interleukin-6/glycoprotein 130-dependent pathways are protective during liver regeneration. J Biol Chem 2003;278:11281–8.
[28] Cruise JL, Houck KA, Michalopoulos GK. Induction of DNA synthesis in cultured rat hepatocytes through stimulation of alpha 1 adrenoreceptor by norepinephrine. Science 1985;227:749–51.
[29] Houck KA, Michalopoulos GK. Altered responses of regenerating hepatocytes to norepinephrine and transforming growth factor type beta. J Cell Physiol 1989;141:503–9.
[30] Lesurtel M, Graf R, Aleil B, Walther DJ, Tian Y, Jochum W, et al. Platelet-derived serotonin mediates liver regeneration. Science 2006;312:104–7.
[31] Rai RM, Lee FY, Rosen A, Yang SQ, Lin HZ, Koteish A, et al. Impaired liver regeneration in inducible nitric oxide synthasedeficient mice. Proc Natl Acad Sci USA 1998;95:13829–34.
[32] Huang W, Ma K, Zhang J, Qatanani M, Cuvillier J, Liu J, et al. Nuclear receptor-dependent bile acid signaling is required for normal liver regeneration. Science 2006;312:233–6.
[33] Bucher NL, Weir GC. Insulin, glucagon, liver regeneration, and DNA synthesis. Metabolism 1976;25:1423–5.
[34] Evarts RP, Raab M, Marsden E, Thorgeirsson SS. Histochemical changes in livers from portacaval-shunted rats. J Natl Cancer Inst 1986;76:731–8.
[35] Croquelois A, Blindenbacher A, Terracciano L, Wang X, Langer I, Radtke F, et al. Inducible inactivation of Notch1 causes nodular regenerative hyperplasia in mice. Hepatology 2005;41:487–96.
[36] Strey CW, Markiewski M, Mastellos D, Tudoran R, Spruce LA, Greenbaum LE, et al. The proinflammatory mediators C3a and C5a are essential for liver regeneration. J Exp Med 2003;198:913–23.
[37] Matsumoto K, Nakamura T. Hepatocyte growth factor: molecular structure, roles in liver regeneration, and other biological functions. Crit Rev Oncog 1992;3:27–54.
[38] Schirmacher P, Geerts A, Jung W, Pietrangelo A, Rogler CE, Dienes HP. The role of Ito cells in the biosynthesis of HGF-SF in the liver. EXS 1993;65:285–99.

[39] Appasamy R, Tanabe M, Murase N, Zarnegar R, Venkataramanan R, Van Thiel DH, et al. Hepatocyte growth factor, blood clearance, organ uptake, and biliary excretion in normal and partially hepatectomized rats. Lab Invest 1993;68:270–6.
[40] Michalopoulos GK, Appasamy R. Metabolism of HGF-SF and its role in liver regeneration. EXS 1993;65:275–83.
[41] Bottaro DP, Rubin JS, Faletto DL, Chan AM, Kmiecik TE, Vande Woude GF, et al. Identification of the hepatocyte growth factor receptor as the c-met proto-oncogene product. Science 1991;251:802–4.
[42] Naldini L, Vigna E, Narsimhan RP, Gaudino G, Zarnegar R, Michalopoulos GK, et al. Hepatocyte growth factor (HGF) stimulates the tyrosine kinase activity of the receptor encoded by the proto-oncogene c-MET. Oncogene 1991;6:501–4.
[43] Stolz DB, Mars WM, Petersen BE, Kim TH, Michalopoulos GK. Growth factor signal transduction immediately after two-thirds partial hepatectomy in the rat. Cancer Res 1999;59:3954–60.
[44] Broten J, Michalopoulos G, Petersen B, Cruise J. Adrenergic stimulation of hepatocyte growth factor expression. Biochem Biophys Res Commun 1999;262:76–9.
[45] Tomiya T, Ogata I, Yamaoka M, Yanase M, Inoue Y, Fujiwara K. The mitogenic activity of hepatocyte growth factor on rat hepatocytes is dependent upon endogenous transforming growth factor-alpha. Am J Pathol 2000;157:1693–701.
[46] Nejak-Bowen K, Orr A, Bowen Jr WC, Michalopoulos GK. Conditional genetic elimination of hepatocyte growth factor in mice compromises liver regeneration after partial hepatectomy. PLoS One 2013;8:e59836.
[47] Michalopoulos GK, Khan Z. Liver regeneration, growth factors, and amphiregulin. Gastroenterology 2005; 128:503–6.
[48] Natarajan A, Wagner B, Sibilia M. The EGF receptor is required for efficient liver regeneration. Proc Natl Acad Sci USA 2007;104:17081–6.
[49] Paranjpe S, Bowen WC, Tseng GC, Luo JH, Orr A, Michalopoulos GK. RNA interference against hepatic epidermal growth factor receptor has suppressive effects on liver regeneration in rats. Am J Pathol 2010;176:2669–81.
[50] Olsen PS, Poulsen SS, Kirkegaard P. Adrenergic effects on secretion of epidermal growth factor from Brunner's glands. Gut 1985;26:920–7.
[51] Luetteke NC, Lee DC. Transforming growth factor alpha: expression, regulation and biological action of its integral membrane precursor. Semin Cancer Biol 1990;1:265–75.
[52] Russell WE, Kaufmann WK, Sitaric S, Luetteke NC, Lee DC. Liver regeneration and hepatocarcinogenesis in transforming growth factor-alpha-targeted mice. Mol Carcinog 1996;15:183–9.
[53] Tacke F, Luedde T, Trautwein C. Inflammatory pathways in liver homeostasis and liver injury. Clin Rev Allergy Immunol 2009;36:4–12.
[54] Diehl AM, Yang SQ, Yin M, Lin HZ, Nelson S, Bagby G. Tumor necrosis factor-alpha modulates CCAAT/enhancer binding proteins-DNA binding activities and promotes hepatocyte-specific gene expression during liver regeneration. Hepatology 1995;22:252–61.
[55] Diehl AM, Yin M, Fleckenstein J, Yang SQ, Lin HZ, Brenner DA, et al. Tumor necrosis factor-alpha induces c-jun during the regenerative response to liver injury. Am J Physiol 1994;267:G552–61.
[56] Ramadori G, Christ B. Cytokines and the hepatic acute-phase response. Semin Liver Dis 1999;19: 141–55.
[57] Cressman DE, Greenbaum LE, DeAngelis RA, Ciliberto G, Furth EE, Poli V, et al. Liver failure and defective hepatocyte regeneration in interleukin-6-deficient mice. Science 1996;274:1379–83.
[58] Oben JA, Roskams T, Yang S, Lin H, Sinelli N, Li Z, et al. Sympathetic nervous system inhibition increases hepatic progenitors and reduces liver injury. Hepatology 2003;38:664–73.
[59] Cruise JL, Knechtle SJ, Bollinger RR, Kuhn C, Michalopoulos G. Alpha 1-adrenergic effects and liver regeneration. Hepatology 1987;7:1189–94.

CHAPTER

6

Developmental Pathways in Liver Regeneration-I

Kari Nichole Nejak-Bowen, Satdarshan Pal Singh Monga

Department of Pathology, University of Pittsburgh, School of Medicine, Pittsburgh, Pennsylvania, USA

6.1 INTRODUCTION

Many of the pathways involved in the initiation and progression of liver regeneration are also essential to embryogenesis and liver development. The four signaling pathways discussed in this section—Wnt/β-catenin, Notch, Hippo, and NF-κB—are highly evolutionarily conserved, attesting to their importance both in embryogenesis and in developmental processes, such as regeneration, that occur in adult tissues. The precise regulation of these pathways requires coordination and exchange of signals with neighboring cells. An additional characteristic of these pathways is that signal transduction involves regulated proteolysis. In the Wnt/β-catenin, Hippo, and NF-κB pathways, activation occurs when an inhibitor is degraded, freeing the downstream effector from repression and proteasomal targeting. In the case of the Notch-signaling pathway, cleavage of the Notch receptor itself activates signaling. The roles of these pathways in liver regeneration have been elucidated by a variety of mechanisms, including genetic knockout mouse studies using global, conditional, or tissue-specific deletion of genes, transient inhibition through siRNA, morpholinos, or chemical inhibitors, or through overexpression of pathway components and targets both *in vivo* and *in vitro*.

6.2 WNT/β-CATENIN SIGNALING

Wnt signaling has been shown to play an important role in both development and regulating tissue homeostasis in adults [1]. Wnt proteins are secreted proteins that are required to undergo certain posttranslational modifications in order to be biologically active [2] (Figure 6.1). Porcupine, which localizes to the endoplasmic reticulum (ER), is responsible for both acylation in the form of palmitoylation and glycosylation of the Wnt proteins in the ER [3,4]. Another highly conserved protein called the G-protein receptor-177 (Gpr177) or Evenness Interrupted (Evi) or Wntless (Wls) is essential and specific for Wnt secretion from a cell [5]. It has been shown to cargo Wnt from Golgi to cell membrane for secretion. Biologically active Wnt proteins bind to

Liver Regeneration
http://dx.doi.org/10.1016/B978-0-12-420128-6.00006-3

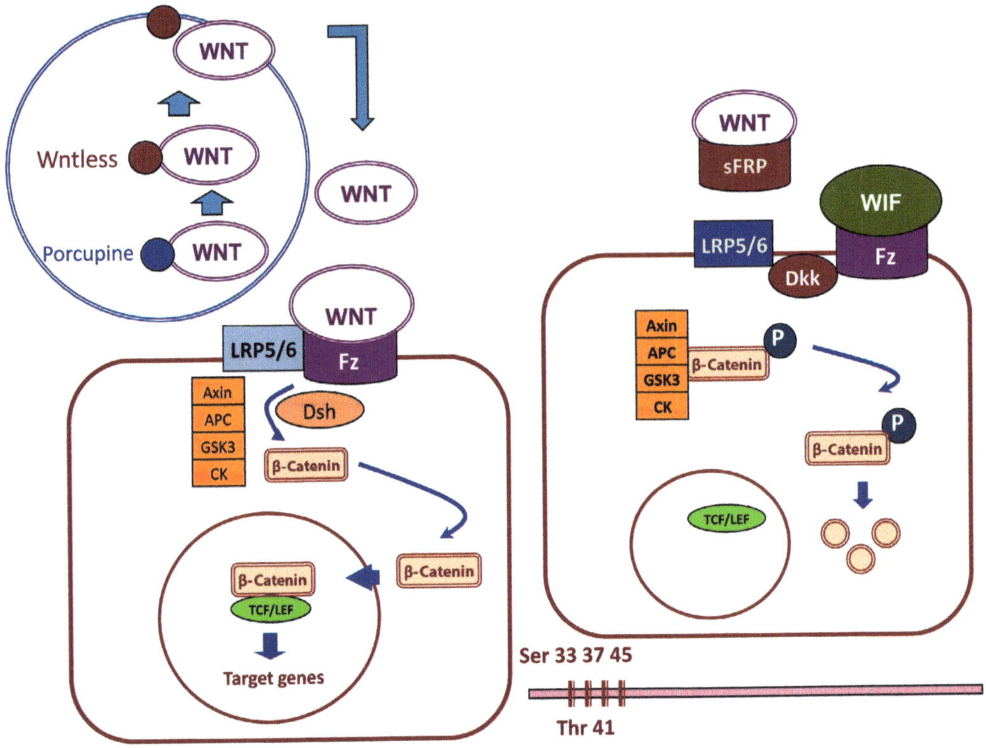

FIGURE 6.1 Wnt/β-catenin signaling. Left panel shows active β-catenin signaling. Wnt proteins undergo posttranslational modifications in endoplasmic reticulum, which is brought about by Porcupine protein. Wntless protein is a cargo receptor for Wnt proteins and allows its transport from Golgi to cell membrane and eventually facilitates its secretion from a cell. Upon binding of Wnt to frizzled receptor and coreceptors LRP5 or LRP6, the signal is transduced through Disheveled protein to β-catenin degradation complex comprising glycogen synthase kinase 3β (GSK3β), casein kinase 1 (CK1), Axin, and adenomatous polyposis coli gene product (APC) to inactivate it. β-Catenin cannot be phosphorylated by GSK3β and CK1 and in its hypophosphorylated form dissociates from the multimeric complex to eventually accumulate in cytoplasm and eventually translocates to the nucleus where it acts as a cofactor for TCF/LEF family of transcription factors. The eventual result is an increase in expression of several target genes of this pathway. As shown in the right panel, Wnt signaling can be disrupted by the presence of inhibitors that would sequester Wnt in extracellular space such as soluble frizzled-related protein (sFRP) or bind to frizzled to make it unavailable for Wnt proteins to bind such as Wnt inhibitory factor (WIF) or by protein such as Dickoppf-1 (Dkk1) that prevents association of frizzled and LRP5/6. Disruption of Wnt signaling allows for β-catenin degradation complex to be active. CK1 phosphorylates β-catenin at serine 45, followed by GSK3β phosphorylating it at serine 33, 27, and threonine 41 (lower panel), and this allows recognition of β-catenin by β-transducin repeat containing protein for its degradation by ubiquitin proteosome. In certain tumors of the liver, these phosphorylation sites (right lower panel) or adjacent residues in exon-3 are mutated and hence β-catenin cannot be phosphorylated or degraded and hence becomes stable and constitutively active.

frizzled receptors and LDL-related protein 5 or 6 (LRP5/6) to induce a signaling cascade that ultimately converges on to β-catenin degradation complex composed of glycogen synthase kinase 3β (GSK3β), casein kinase 1 (CK1), adenomatous polyposis coli gene product or APC protein, and Axin (Figure 6.1). In the presence of Wnt protein, this complex is inactive, and β-catenin in its hypophosphorylated form is released from the complex to eventually translocate to the nucleus where it acts as a cofactor for transcription factors, the most well studied being the T-cell factor/lymphoid enhancement factor (TCF/LEF) family of transcription factors. β-Catenin is also known to function in collaboration with other transcription factors such as FoxO and HIF. Several target genes of β-catenin/TCF are known including c-Myc, cyclin-D1, and others although tissue specificity is clearly evident. In the absence of Wnt, β-catenin is phosphorylated by GSK3β and CK1 at specific serine and threonine sites in exon-3 of the protein, which allow its identification by an E3 ubiquitin ligase component β-transducin repeat containing protein (β-TrCP) for its degradation (Figure 6.1). Mutations in exon-3 of β-catenin gene or components of its degradation complex such as Axin can disrupt its degradation leading to activation and increased expression of an array of target genes with diverse functions in tumorigenesis [6].

6.2.1 Wnt/β-Catenin Signaling in Liver Growth

There are many physiological and pathological examples of hepatic growth. Liver growth is evident in various hepatic pathologies such as benign and malignant tumors. Liver growth also occurs during specific development stages especially in hepatic bud once endoderm is specified to form hepatoblasts, the bipotential progenitors that will yield hepatocytes and cholangiocytes. A postnatal growth spurt is also well known and is especially evident during the first month after birth after which liver is mostly quiescent and turns over very slowly renewing almost once a year. β-Catenin activation has been reported in almost all of these kinds of hepatic growth. There are several excellent reviews on the role of β-catenin in liver cancer and postnatal liver growth. This section is more focused on the role of Wnt/β-catenin signaling in liver regeneration.

6.2.2 β-Catenin Signaling During Liver Regeneration After Partial Hepatectomy

Liver is a unique adult organ that despite being mostly quiescent and extremely slow in turning over possesses a remarkable ability to regenerate when its substantive mass is either removed surgically or occurs due to loss of hepatocytes to a toxicant [7,8]. Another novel aspect of liver regeneration is that the cells that participate in the process of tissue regeneration are in fact mature functioning resident cells of an adult liver. Only under specific conditions where mature hepatocytes are unable to proliferate and restore the organ homeostasis in a timely fashion will hepatic progenitors be called into play [9]. Such situations usually evident in clinic as chronic active hepatitis of many etiologies have been replicated in animal models by specific diets or combinations of diet and surgery or radiation and surgery that lead to blockade of hepatocyte proliferation while creating a need for hepatocyte proliferation. This leads to activation of the so-called oval cells that initially expand in number and eventually differentiate to hepatocytes or cholangiocytes and hence considered as bipotential progenitors [10]. Understanding the cellular and molecular basis of this process will not only enable improved strategies to treat acute or chronic liver failure in the clinic, but also may provide useful information that may be applicable to the broad discipline of tissue regeneration and regenerative medicine. In both adult-cell-driven liver regeneration in situations of PH or toxicant-induced liver injury as well as in oval cell-driven regeneration, Wnt/β-catenin signaling has been shown to play a critical role [11]. Here, we will focus on the role of β-catenin signaling during the process of hepatic regeneration after PH.

6.2.2.1 β-Catenin Activation and Downstream Signaling After Partial Hepatectomy

The first study that examined the role of Wnt signaling in liver regeneration did so in a rat model of partial hepatectomy (PH) [12]. After removal of three of the four lobes, the remaining lobe displayed a transient stabilization of β-catenin along with its nuclear translocation within minutes after the surgery. Nuclear β-catenin was evident for up to 24 h followed by its normalization and localization to hepatocyte membrane. Since RNA expression of β-catenin was induced only 6 h after PH, transient immediate early increase in total β-catenin protein within minutes was due to posttranslational regulation leading to its decreased degradation. The upstream mechanism leading to β-catenin activation in rats' livers after PH remains poorly understood as levels of total Wnt1, adenomatous polyposis coli gene product, or GSK3β did not change during first few minutes to have explained β-catenin activation. A follow-up study addressed the role of β-catenin in rat liver regeneration process more mechanistically [13]. Upon phospho-morpholino antisense-mediated knockdown of β-catenin at the time of PH, a dramatic decrease in hepatocyte proliferation was clearly evident and lead to decreased liver weight to body weight ratio. The animals with β-catenin knockdown also showed decreased expression of key cell-cycle genes and exhibited morbidity, prolonged period of inactivity, and hence, further substantiated the role of Wnt signaling in liver regeneration.

It should also be stated that immediately after its transient stabilization in rats after PH, the total levels of β-catenin significantly decreased [12]. However, nuclear translocation was evident at this time even though its total levels were lower than those at baseline. Decrease in total β-catenin protein during LR beyond 15 min after PH coincided with decrease in total Wnt1 levels and increase in adenomatous polyposis coli gene product. These data suggest that a very tight regulation of Wnt/β-catenin signaling during early rat liver regeneration may in fact be contributing to this regulated growth of the liver in a way that positive signals are immediately followed by negative signals to check or monitor any excessive induction of progrowth pathways and thus prevent any prolonged or undue hepatic growth.

The kinetics of β-catenin activation is a little better understood in mice. After PH, nuclear translocation of β-catenin in hepatocytes occurs within 1-3 h after surgery [14–16]. A β-catenin-TCF4 complex is observed as early as 4-6 h [16]. This transcriptional complex is recruited to the promoter of *Ccnd1* at 12 h and an increase in cyclin-D1 protein is appreciable at 24 h, which is maintained for additionally up to 48 h. This leads to G1- to S-phase transition of hepatocytes that is evident quite clearly at 40-72 h as seen by positivity to markers such as Ki-67 or proliferating cell nuclear antigen and followed immediately by increased mitosis [15,17]. This cascade of events was disrupted when β-catenin was conditionally deleted from hepatocytes and cholangiocytes in the KO obtained by interbreeding floxed β-catenin mice and *Albumin-cre* transgenic mice. While these animals showed a mild but significant basal liver weight to body weight ratio and lacked any overt phenotype, these mice showed a notable lag in normal liver regeneration after being subjected to PH [15,17,18]. In fact there were 50% fewer hepatocytes in S-phase at 40 h after PH in KO than their control littermates. There was a concurrent decrease in total levels of cyclin-D1, which also led to decreases in other cyclins such as cyclin-A and cyclin-E, which all culminated into a blunted regenerative response [15]. However, by 72 h the KO displayed comparable cell proliferation to controls due to presently unknown redundant mechanisms.

Similar studies have also been performed in zebrafish to reveal the role of β-catenin signaling [19]. When subjected to one-third hepatectomy, β-catenin activation was evident in remnant livers. In addition, there was a considerable decrease in hepatocyte proliferative response in zebrafish expressing dominant-negative T-cell factor-4 in hepatocytes [19]. These data further affirm a critical role of β-catenin signaling in regulating liver growth *in vivo* in surgical resection model.

6.2.2.2 β-Catenin Signaling in Toxicant-Induced Liver Injury and Regeneration

Additional nonsurgical models that lead to hepatic regeneration have also demonstrated crucial roles of β-catenin signaling. Acetaminophen (APAP) overdose is an important cause of acute liver failure. Suprapharmacological doses of APAP lead to hepatotoxicity characterized by glutathione depletion and centrilobular necrosis resulting in acute liver failure. APAP is metabolized to its reactive metabolite called *N*-acetyl-*p*-benzoquinone imine (NAPQI) by cytochrome P450 2E1 (Cyp2e1) [20–22]. If sublethal doses of acetaminophen are administered to a mouse, it causes early hepatocyte necrosis in zone 3, which sets the midzonal hepatocytes into proliferation to compensate for loss of hepatic mass and eventually liver recovers. In fact, spontaneous regeneration in patients after acetaminophen dose has eliminated the need for liver transplant [23].

Studies show an important role of β-catenin signaling in this model of liver regeneration as well. In fact, β-catenin stabilization due to inactivation of GSK3β is evident as early as an hour after administration of sublethal doses of acetaminophen [24]. It is important to note that lethal doses of this agent did not induce β-catenin activation in the liver. β-Catenin activation led to increased cyclin-D1 expression that stimulated G1- to S-phase transition of midzonal hepatocytes and eventually hepatocyte proliferation and recovery of liver. Liver-specific β-catenin KO could not be directly used to address role of β-catenin in acetaminophen-induced liver regeneration since β-catenin is the chief regulator of pericentral gene expression and hence a contributor to metabolic zonation in the liver [24]. Interestingly, β-catenin KO mice lack Cyp2e1 and Cyp1a2 and hence are unable to metabolize acetaminophen to NAPQI and remain protected [18]. However, after induction of these xenobiotic metabolism enzymes by exogenous stimuli, β-catenin KO mice become susceptible to acetaminophen. At equitoxic doses (at doses where KO and WT mice have similar injury), the KO showed a significant decrease in hepatocyte proliferation when compared to the controls thus verifying the role of β-catenin signaling in toxicant-induced liver regeneration as well [24].

6.2.3 Enhanced β-Catenin Activation Stimulates Regeneration of Liver

Conversely, enhanced β-catenin activation has been shown to impart a regenerative advantage to liver across different species as well. Transgenic mice expressing serine 45-mutated and hence stable β-catenin in hepatocytes showed accelerated liver regeneration after PH [25]. Recent studies have also shown enhanced liver regeneration in β-catenin transgenic mice in response to acetaminophen overdose. Mice expressing relatively stable β-catenin due to mutation affecting serine 45 residue in hepatocytes showed a pronounced hepatocyte proliferation in response to sublethal doses of acetaminophen. This was also due to increased downstream signaling and hence increased cyclin-D1 expression [26]. Furthermore, zebrafish heterozygous for loss of adenomatous polyposis coli gene in hepatocytes led to notable β-catenin activation and enhanced liver regeneration after one-third hepatectomy [19]. A similar phenotype was also evident in transgenic zebrafish that overexpressed Wnt8 in the liver, which was also due to β-catenin activation [19]. Similarly, mice heterozygous for loss of adenomatous polyposis coli gene in hepatocytes showed a significantly increased hepatocyte proliferation after PH [19].

β-Catenin activation has also been observed in patients during toxicant-induced liver injury. Two cohorts of patient livers from acetaminophen overdose were assessed for nuclear translocation of β-catenin and proliferating cell nuclear antigen [24]. The livers of patients that showed high nuclear β-catenin also showed greater number of hepatocytes in S-phase of cell cycle. Most patients in this group did not require a liver transplant as they displayed spontaneous liver regeneration. Conversely, explanted livers from patients that did not spontaneously regenerate and thus required transplant, showed absent nuclear β-catenin and only occasional hepatocytes in S-phase. These observations suggest that β-catenin activation may be an important component driving cell-cycle machinery in hepatocytes that may eventually determine tissue regeneration response after hepatic insult.

6.2.4 Exogenous β-Catenin Activation and Liver Regeneration

Exogenous activation of β-catenin has now also been shown to positively regulate tissue regeneration. Hydrodynamic delivery of naked Wnt-1 plasmid through tail vein led to its enhanced expression in hepatocytes that led to β-catenin activation [25]. These animals, when subjected to PH, displayed a significantly earlier hepatocyte proliferation due to enhanced cyclin-D1 expression. More recently, an alternate mechanism of stimulating β-catenin activity has been identified, which may be of high translational significance in cases of impaired liver regeneration. Triiodothyronine or T3 hormone has been shown to induce hepatocyte mitogenesis even in the absence of hepatic resection through upregulation of cyclin-D1 expression [27–29]. Because β-catenin is known to regulate cyclin-D1 during liver regeneration, the requirement for β-catenin in T3-driven cyclin-D1 gene expression and stimulation of hepatocyte proliferation was investigated. When T3 was fed to hepatocyte-specific β-catenin KO mice, there was no increase of cyclin-D1 and hepatocytes lacked entry to S-phase and eventually showed impaired proliferative response [30]. Further studies identified T3's ability to activate β-catenin to be protein kinase A dependent. Thus, T3 may be a more physiological and clinically relevant modality to activate β-catenin for inducing tissue regeneration that may be beneficial in a subset of end-stage liver disease patients [28].

6.2.5 Cell-Molecule Circuitry of Wnt Signaling in Liver Regeneration After Hepatectomy

With ample evidence of a positive role of β-catenin activation and signaling in several models of liver regeneration in various species, it is highly relevant to mechanistically address the regulation of this pathway. This may lead to better understanding of the biology, which in turn may result in exploitation of this signaling cascade for regenerative therapies and hepatic tissue engineering. This is specifically relevant since β-catenin can be regulated not only by the upstream Wnt signaling but also by other factors such as receptor tyrosine kinases, protein kinase A, transforming growth factor-β, Flt3, and others [31–40]. Furthermore, liver regeneration after PH involves multiple cell types in an adult liver that can both send and receive signals to eventually participate in the hepatic restoration process.

To further elucidate the mechanisms that regulate β-catenin activity during liver regeneration after PH, novel approaches to specifically disrupt Wnt "communication" to β-catenin are being used. This enables preserving β-catenin to be available for participating downstream of other signaling mechanisms that are known to activate it. LRP5/6 is a Wnt coreceptor that transduces Wnt/frizzled signaling to Disheveled protein [41–43]. We used LRP5/LRP6 double-floxed mice to conditionally KO these Wnt coreceptors in hepatocytes by interbreeding to *Albumin-cre* transgenic mice. These mice have been recently reported and disrupt canonical Wnt signaling only and hence are invaluable resource to distinguish Wnt-dependent from Wnt-independent activation of β-catenin [44]. Liver-specific LRP5/6 double KO, which lack these two proteins from both hepatocytes and cholangiocytes, when subjected to PH, phenocopied liver-specific β-catenin KO mice in that there was a significant diminution of β-catenin activation that was observed as lack of β-catenin-TCF4 complex formation at 4 h after PH, which led to reduced expression of cyclin-D1 [16]. A delay in liver regeneration ensued and was due to notably fewer hepatocytes in S-phase at both 40 h as well as 72 h. Since β-catenin levels are intact in LRP5/6 double

KO mice in contrast to β-catenin KO, and still these two strains of mice show dampened tissue regeneration, it is safe to conclude that Wnt signaling is the sole upstream effector of β-catenin during liver regeneration after PH.

Based on observation in LRP5/6 double KO mice, the next relevant question is the source and identity of Wnt proteins that play a role in β-catenin activation during the liver regeneration process. This is a cumbersome question to answer due to the multiplicity of Wnt proteins. In fact, there are 19 Wnt proteins and a significant subset of these are expressed in multiple cell types of the liver [45]. Thus, isolation of various cell types at different times after PH to detect changes in the expression of multiple Wnt is an arduous task. A novel approach to circumvent this problem is to inhibit all Wnt secretion from a cell. Wntless protein is a highly conserved molecule that is specific to Wnt secretion only [46]. It is also called G protein-coupled receptor 177 or Evi and is known to interact in Wnt proteins within a cell especially in ER, Golgi apparatus, and plasma membrane [47,48]. It has been shown that Wntless is essential in secretion of biologically active, hydrophobic Wnt proteins from a cell. In the absence of Wntless, Wnt signaling is disrupted as has been shown in embryonic lethal phenotypes that also demonstrate complete lack of β-catenin activation [48,49]. This demonstrates a critical role of this highly Wnt-specific protein in Wnt secretion from a cell. Recently, Wntless-floxed mice have been generated, and these mice are also an invaluable resource to generate loss of function of all Wnt genes in a tissue- or cell-specific manner [50,51].

In order to directly address the role of cells in the liver that may be a source of Wnt proteins to induce β-catenin activation in hepatocytes, conditional Wntless knockout mice in four major cells of the liver were generated [16]. *Albumin-cre* was used to generate hepatocyte- and cholangiocyte-specific Wntless KO. These mice are viable and do not show any overt hepatic phenotype or otherwise. When subjected to PH, the livers regenerate normally and hepatocytes show expected increase in cyclin-D1 expression and hepatocyte proliferation [16]. These studies suggest that Wnt proteins are neither originating from hepatocytes for autocrine activation of β-catenin nor emanating from cholangiocytes to induce hepatocyte β-catenin activation in a paracrine manner during liver regeneration process. Tie2-cre enabled successful deletion of Wntless from endothelial cells, however, this led to embryonic lethality, further precluding studies on liver regeneration in these animals [16,52]. Mice expressing only one allele of Wntless in endothelial cells did not show any discernible phenotype after PH. Thus, the role of endothelial cells in liver regeneration process will have to be determined using inducible conditional KO mice. It should be noted that a recent study has shown hepatic sinusoidal endothelial cells (SECs) to be a major source of Wnt2 and hepatocyte growth factor after PH and hence an inducible endothelial cell-specific cre-mediated deletion of Wntless would be valuable in directly addressing the role of these cells as a source of proregenerative Wnt proteins [53]. Macrophage-specific *Wntless* KO mice were generated by interbreeding floxed-Wntless mice with *LysM-Cre* transgenic mice [54]. These animals also exhibited normal liver morphology and lacked any overt phenotype. However, after PH these mice showed a decrease in β-catenin-TCF complex formation within first few hours of liver regeneration that eventually led to lower cyclin-D1 expression [16]. Eventually, a consistent decrease in numbers of hepatocytes in S-phase by 20% was evident at 40 h after PH in these animals. This suggests that Kupffer cells (KCs) are one of the cell sources for Wnt secretion that would act in a paracrine manner to induce hepatocyte-specific β-catenin activation. At this time, the identity of Wnt protein/s that originates from macrophages after PH remains elusive and future studies are necessary.

In summary, the PH studies in liver-specific LRP5/6 and liver-specific β-catenin KO mice

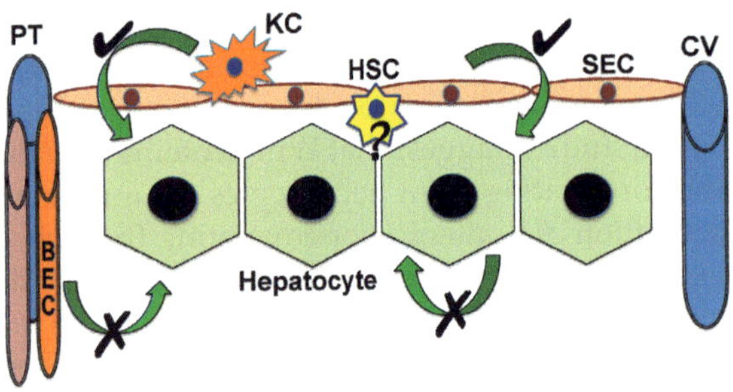

FIGURE 6.2 Cell circuitry of Wnt signaling during liver regeneration. Based on the partial hepatectomy studies in liver-specific LRP5/6 and liver-specific β-catenin KO mice, it is clear that β-catenin activation in hepatocytes during liver regeneration is indeed Wnt dependent. Using Wntless conditional KO, we have shown that the Wnt proteins that are directing β-catenin activation in hepatocytes during liver regeneration are not originating from biliary epithelial cells (BECs) or hepatocytes (based on the results in *Albumin-cre*-mediated *Wntless* deletion). However, based on the previously published studies and our studies in macrophage-Wntless KO mice, we show that β-catenin activation in hepatocytes after partial hepatectomy is at least partially dependent on Wnt proteins released from hepatic sinusoidal endothelial cells (SECs) and Kupffer cells (KCs). At this time, the role of hepatic stellate cells (HSCs) in Wnt secretion during liver regeneration remains unknown.

indicate that β-catenin activation in hepatocytes during liver regeneration is Wnt dependent (Figure 6.2). The studies in cell-specific Wntless KO mice suggest that the Wnt proteins directing β-catenin activation in hepatocytes during liver regeneration are not originating from biliary epithelial cells (BECs) or hepatocytes but are at least partially dependent on Wnt proteins released from hepatic SECs [53]and KCs [16] (Figure 6.2). At this time, the role of hepatic stellate cells (HSCs) in Wnt secretion during liver regeneration remains unknown.

6.3 NOTCH SIGNALING

The Notch receptor is part of a highly conserved signaling pathway that plays a role in many key cellular functions, including cell fate decisions, proliferation, differentiation, and survival [55,56]. In mammals, there are four different Notch receptors (Notch 1-4), which interact with ligands from either the Jagged (JAG 1 and 2) or Delta-like (DLL 1, 3, and 4) gene families [57,58]. The Notch signaling cascade is initiated by binding to its ligand, which is also a membrane-bound cell-surface receptor. Cleavage by ADAM/TACE at S2 releases a form of the Notch intracellular domain (NICD) tethered to the membrane, which is then cleaved by γ-secretase at S3, allowing the NICD to translocate to the nucleus [55,59,60]. In the nucleus, NICD recruits a coactivator, mastermind-like 1 (Maml1) and forms a complex with the DNA-binding protein CSL. This complex allows for transcription of Hes-1 and other Notch pathway targets, including Myc and cyclin-D1 [59,61] (Figure 6.3).

6.3.1 Notch Signaling in Hepatic Pathophysiology

The first major observation linking the role of liver Jagged/Notch signaling pathway to liver was in hepatic pathology. Mutations in Jagged are known to cause Alagille syndrome, a disorder that is characterized by paucity of interlobular bile ducts [62,63]. Further studies have identified that Jagged-1 in the portal mesenchyme regulates bile duct development, providing a mechanistic basis for the defect evident in Alagille syndrome [64]. The authors demonstrate that loss of Jag1 expression in specific cells within the portal vein mesenchyme, not endothelial cells, results in organizational defects of cytokeratin (CK)-19-positive cells. While the ductal plate is formed, the CK-19-positive cells do not form bile ducts, which demonstrates the importance of the mesenchymal compartment in shaping and regulating biliary morphogenesis.

6.3.2 Notch Signaling in Liver Regeneration After PH

Both Notch-1 and Jagged-1 are expressed on normal hepatocytes and bile duct cells in liver.

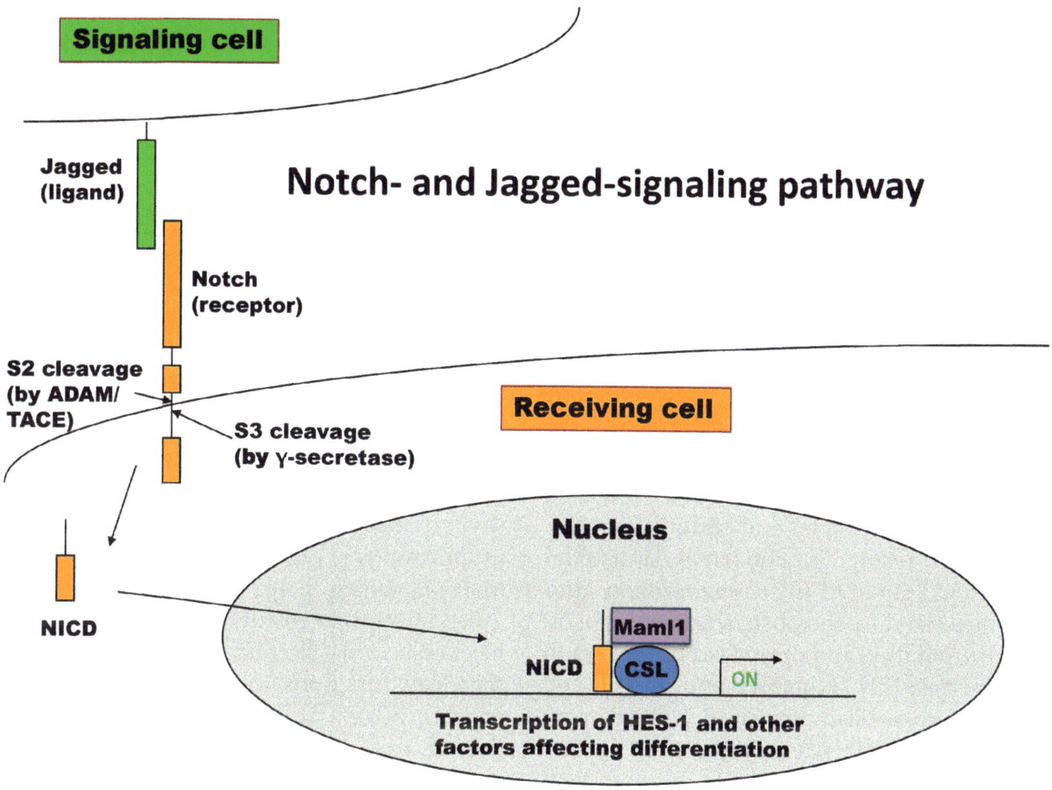

FIGURE 6.3 Notch/Jagged signaling. Notch signaling is activated through interaction of membrane-bound Jagged ligands and Notch receptors on adjacent cells. This interaction triggers sequential cleavage of Notch by a disintegrin and metalloproteinase (ADAM)/TNF-α-converting enzyme (TACE) and γ-secretase, resulting in release of the Notch intracellular domain (NICD), which translocates to the nucleus. In the nucleus, the NICD interacts with the transcription factor CBF1/Su (H)/Lag-1 (CSL), and along with its coactivator, mastermind-like 1 (Maml1), activates transcription of hairy and enhancer of split-1 (Hes-1) and other Notch pathway targets.

After PH, both Notch and Jagged proteins are upregulated for up to 4 days. Additionally, rapid translocation of NICD to the nucleus was observed within 15 min [65], which preceded Hes-1 expression. Jagged-1 was also shown to be a mitogen for hepatocytes in culture. Further, treatment with either Notch or Jagged-1 siRNA suppressed proliferation during liver regeneration [65]. Another study employing the rat 2-acetylaminofluorine-PH model, which induces oval cells, demonstrated that interrupting Notch signaling with a γ-secretase inhibitor delayed oval cell differentiation 11 days post-PH [66]. Interestingly, conditional deletion of Notch-1 in the liver resulted in an adaptive nodular regenerative hyperplasia, characterized by an increase in proliferating hepatocytes [58]. However, restoration of liver mass after PH was markedly delayed due to a reduction in the number of proliferating hepatocytes. Although Notch signaling has been implicated in bile duct formation and BEC survival [67,68], regeneration of the biliary compartment did not appear to be impaired in any of these studies.

In addition to its involvement in hepatocyte proliferation, activation of the Jagged/Notch pathway has several other functions in regenerating liver. Constitutive expression of the NICD in hepatocytes leads to hepatomegaly and intrahepatic bile duct density. These mice were also protected against acetaminophen (APAP)-induced hepatotoxicity through activation of p45 nuclear factor erythroid-derived 2-related factor 2 (Nrf2), a downstream target of Notch that may act in part to increase the rate of regeneration after APAP injury [69]. Further, disruption of Nrf2 in mice impaired liver regeneration after PH, which can be rescued by expression of the NICD [70]. Notch signaling is also essential for revascularlizing the regenerating liver, as shown by the deletion of recombination signal-binding protein-J (RBP-J), a CSL component, in hepatocytes and endothelial cells. This deletion results in abnormal sinusoidal structure and impaired liver regeneration after PH, characterized by the obstruction of sinusoidal microcirculation and decreased proliferation of hepatocytes [71]. Notably, endothelial cell proliferation was also impaired, which may have resulted in a decrease or absence of mitogens that normally emanate from this compartment [57]. Thus, Notch signaling appears to be important for proliferation and differentiation of both hepatocytes and endothelial cells during regeneration.

6.4 HIPPO SIGNALING

Yes-associated protein, or Yap, is a nuclear transcription factor and the effector of the Hippo signaling pathway. Yap activation is associated with many cellular functions, most notably cell differentiation and proliferation, and as such is a key regulator of growth regulation and organ size [72,73]. In the cell, upstream signals induce mammalian sterile 20 like (Mst) 1/2 kinase to activate large tumor suppressor (Lats) 1/2 kinase, which in turn phosphorylates and inactivates Yap, targeting it for ubiquitination and degradation. When dephosphorylated, Yap interacts with other coactivators and transcription factors, including transcriptional coactivator with PDZ-binding motif (TAZ) and TEA domain proteins (TEADs), to activate downstream targets [74,75].

6.4.1 Hippo Signaling in Hepatic Pathophysiology

Several studies have shown that activation of the Yap-signaling pathway preferentially affects liver growth, suggesting a predominant hepatic function. A liver-specific conditional mouse model transgenic for human Yap showed that induction of Yap causes massive hepatomegaly, which was due to increased cell number (hyperplasia) as opposed to increased cell size (hypertrophy) [72,76]. Similarly, deletion of Mst1/2, which helps to maintain hepatocyte quiescence, also results in enlarged livers due to excessive proliferation [73,77,78], indicating that Yap functions to maintain liver size and restrain postnatal liver growth [74,79]. Activation of Yap also protects against apoptosis, as demonstrated by the increased cell death seen in Yap-deficient livers and decreased apoptosis upon activation of Yap [80,81], suggesting that Yap is required for hepatocyte survival as well.

6.4.2 Hippo Signaling in Liver Regeneration After PH

Although studies that directly address the role of the Yap pathway in liver regeneration are lacking, one group found that Yap protein was increased in the later stages of regenerating rat liver almost threefold, due to posttranscriptional regulation [82]. Yap also promotes both hepatocyte and cholangiocyte proliferation after bile duct ligation [83]. However, more work is needed to elucidate the downstream targets of Yap that regulate liver regeneration and proliferation [84].

During liver regeneration, hepatocyte-specific deletion of integrin-linked kinase

(ILK), a component of extracellular matrix (ECM)/integrin signaling whose loss results in hepatomegaly, causes an increase in nuclear YAP and a decrease in the inactivated degraded form of Yap, indicating that Yap signaling and activation is at least partly under the control of ILK during regeneration [85]. Several other studies have also shown that nuclear localization of Yap increases in hepatocyte-specific ILK knockouts (KO) under conditions that stimulate enhanced liver growth, such as treatment with 1,4-bis [2-(3,5-dichaloropyridyloxy)] benzene or phenobarbitol [86,87]. Conversely, overexpression of glypican-3, a heparin-bound proteoglycan on the cell surface, results in suppressed proliferation after PH, a finding that also correlates with decrease in nuclear Yap [88,89]. Thus, it appears that regulation of the Yap pathway is at least partly under the control of ECM and integrin components, and its inhibition may be part of the signaling cascade that results in termination of regeneration [90].

6.5 NF-κB SIGNALING

Nuclear factor-κB (NF-κB) is a dimeric transcription factor that consists of five members: p50, p52, p65 (also known as RelA), c-Rel, and RelB, with the prototypical NF-κB complex consisting of a heterodimer between p50 and p65 [91,92]. NF-κB, which is robustly activated by tumor necrosis factor (TNF)-α [93], has pleiotropic effects in a wide variety of cellular programs, including inflammation and injury, cell survival, proliferation, and apoptosis [94–97]. In the absence of signal, NF-κB is retained in the cytoplasm by members of the inhibitory IκB complex [92]. Activation of the NF-κB pathway occurs when the IκBs undergo phosphorylation, ubiquitination, and degradation [98]. The kinases responsible for this phosphorylation are IκB kinases (IKKs). Loss of the IκB complex after IKK phosphorylation liberates NF-κB, which then translocates to the nucleus [99]. Full activation of NF-κB requires further posttranslational modification by phosphorylation of its various subunits through the action of kinases such as protein kinase A, casein kinase II, PKCζ, and GSK3β [91,100–104]. NF-κB is then able to provide signals to the cell through transcription of inflammatory genes, antiapoptotic genes, or hepatomitogens such as TNF-α and IL-6 [96,97,105,106] (Figure 6.4).

6.5.1 Signaling Upstream of NF-κB

TNF-α is a strong inducer of NF-κB activation. Although the role of TNF-α signaling in inflammation and apoptosis has been well described, TNF-α is also capable of stimulating hepatocyte proliferation [107]. Because of this, and also due to its very early activation after PH, TNF-α is considered to be a critical component in initiating the early phase of liver regeneration [108]. Treatment with antibodies to TNF-α or inhibition of TNF-α production results in decreased hepatocyte proliferation after PH [109–111]. Additionally, TNF-R1-deficient livers have impaired NF-κB activation and delayed regeneration after carbon tetrachloride and PH, demonstrating the importance of TNF-α in inducing NF-κB and a subsequent proliferative response [106,112]. However, deletion of TNF-α itself does not seem to impair regeneration [113,114]. This indicates that there may be an alternate mechanism for NF-κB activation and subsequent IL-6 production in NPCs that is not dependent on TNF-α, or that TNF-α is not required because other ligands can signal through TNF-R1 [115].

Lipopolysaccharide (LPS), which is released by gut-derived enteric bacteria into the portal circulation, is a ligand for Toll-like receptor (TLR)-4. LPS was thought to be an important stimulator of liver regeneration through activation of NF-κB activity in NPCs [116–118]. In support of this hypothesis, studies have demonstrated that deletion of myeloid differentiation factor (MyD) 88, an adaptor molecule required for TLR-mediated signaling, resulted

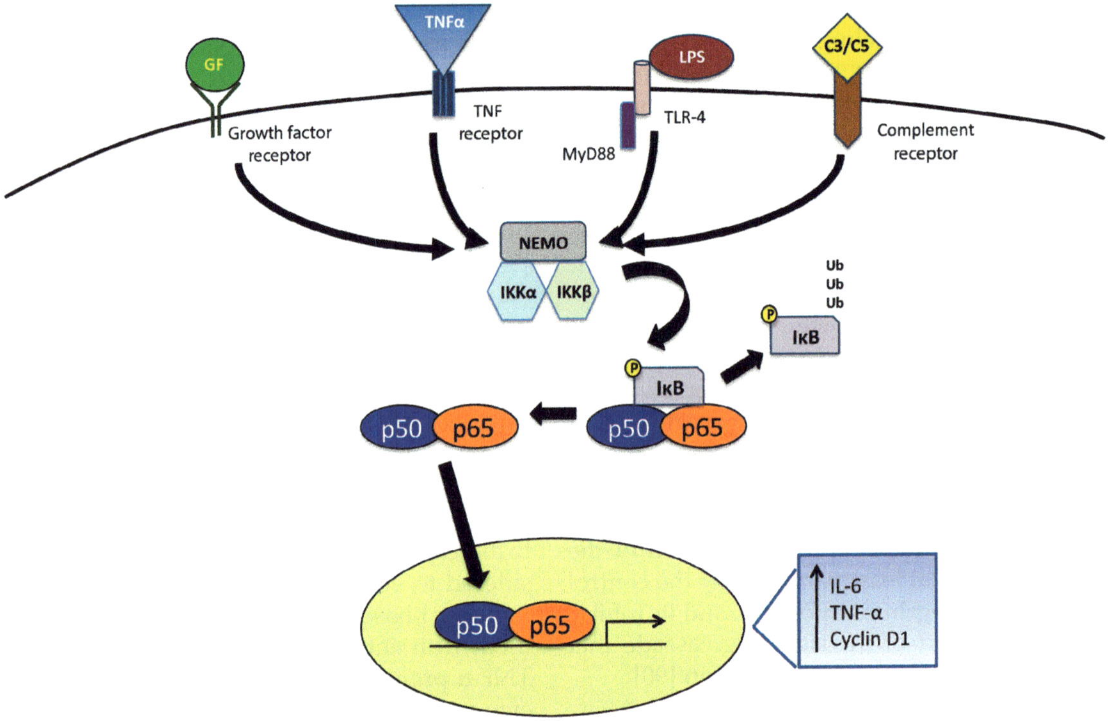

FIGURE 6.4 NF-κB signaling. Nuclear factor-kappaB (NF-κB) signaling is most commonly induced by binding of tumor necrosis factor (TNF) to its cell-surface receptor. This results in activation of the IκB kinase (IKK), consisting of NEMO (NF-κB essential modulator), a master regulatory subunit, as well as the catalytic IKKα and IKKβ subunits. IKK phosphorylates inhibitor of κB (IκB), which under unstimulated conditions inhibits NF-κB by sequestering it in the cytoplasm. Once IκB is phosphorylated, it is rapidly ubiquitinated (ub) and degraded, which allows NF-κB to translocate to the nucleus and activate expression of genes such as interleukin (IL)-6, TNF-α, and cyclin-D1. Other extracellular signaling pathways are also known to activate NF-κB during liver regeneration, including complement components C3 and C5, and several different growth factors (GF), such as hepatocyte growth factor (HGF), or epidermal growth factor (EGF). Signaling through the lipopolysaccharide (LPS)/Toll-like receptor (TLR)/myeloid differentiation factor 88 (MyD88) pathway has also been postulated to play a role in activation of NF-κB during regeneration, although it is still unknown exactly which pathway members are required for this process.

in abnormal NF-κB activation and IL-6 production [119,120]. However, in TLR, TLR4, and TLR9 KO mice, regeneration proceeded normally, suggesting that another ligand and/or receptor that can signal through MyD88 is responsible for NF-κB activation and subsequent regeneration [119–122].

Complement proteins, which are mediators of innate immunity, also appear to play an important role in inducing NF-κB during liver regeneration. Mice deficient in C3 and/or C5, critical components of the complement cascade, display suboptimal regeneration, which is in part due to loss of NF-κB activation and subsequent decreased levels of TNF-α and IL-6 [123,124].

Finally, several studies have also shown that the HGF/Met signaling pathway is capable of stimulating NF-κB activity in multiple cell types, including hepatic-derived cells [125–127]. Addition of HGF induced NF-κB activation in HepG2 cells [128]. Similarly, HGF-induced proliferation in cultured hepatic stem cells was blocked using an NF-κB inhibitor [129]. Further, addition of HGF to a liver-derived cell line stimulates NF-κB binding and activation [130]. Transmission of epidermal growth factor (EGF) signaling, which is also an important driver of cell-cycle progression during liver regeneration, can also be mediated through activation of NF-κB [131]. These data implicate a role for NF-κB as an effector in growth factor signal transduction and proliferation during regeneration.

6.5.2 Signaling Downstream of NF-κB

NF-κB promotes production of IL-6, which activates the Jak/STAT pathway, an important stimulator of early liver regeneration [132,133]. In fact, it is estimated that nearly 40% of the immediate early genes activated during liver regeneration are regulated by IL-6 signaling [134]. In the absence of IL-6, hepatocyte proliferation is impaired, causing a delay but not cessation of the regenerative response [135,136]. Additionally, overexpression of IL-6 and its soluble receptor causes an acceleration of liver regeneration after PH or during acute liver injury [137]. Other reports, however, indicate that IL-6 signaling plays more of a cytoprotective, antiapoptotic role during regeneration [138,139], although even in these studies the proliferative response was delayed. Deletion of glycoprotein 130 (gp130), the common signal transducer of all IL-6 family members, phenocopied IL-6 KOs in terms of decreased survival and proliferation after PH [140,141]. Still other studies, however, have found that the IL-6-signaling pathway can delay hepatocyte proliferation via Socs3 [122,140,142]. Therefore, it appears that IL-6 may have a pleiotropic effect in liver regeneration that is dependent on context, timing, and activation of additional signaling pathways [133,143].

In addition to its indirect role in stimulating proliferation through IL-6, NF-κB activation is believed to be a key contributor to the proliferative response through activation of cyclin-D1. NF-κB activates transcription of cyclin-D1 through NF-κB binding sites on the cyclin-D1 promoter [144–146]. Accordingly, inhibition of NF-κB caused a dramatic reduction of cyclin D1 in several different cell types [131, 144,146], thus suggesting that NF-κB-mediated activation of cyclin-D1 might participate in the regenerative process through regulation of cell-cycle progression.

6.5.3 Role of NF-κB Signaling in Liver Pathophysiology

The essential role of NF-κB in liver was demonstrated by the finding that homozygous deletion of RelA (p65) in mice leads to embryonic lethality due to massive hepatocyte apoptosis [147]. This is due to the requirement for NF-κB in preventing TNF-α-induced cell death [148,149]. Indeed, mice deficient in both RelA and TNF-R1 survive until shortly after birth, suggesting that RelA knockout mice have an enhanced susceptibility to TNF-α, which is expressed in high amounts by hematopoietic progenitors [150]. Mice harboring deletions of IKK complex members, which control NF-κB activation, also die from hepatocyte apoptosis during embryonal development [151–153]. Thus, proper function of the NF-κB pathway is critical to hepatocyte maturation, and activation of NF-κB is indispensable for protection from TNF-α-mediated apoptosis [108,154].

6.5.4 Role of NF-κB Signaling in Liver Regeneration

NF-κB is activated very early during liver regeneration, translocating to the nucleus and binding to DNA within 30 min to 1 h after PH [155–158]. However, the location, function, and relative importance of NF-κB activation in different liver cell populations have long been a source of debate. Early reports demonstrated that NF-κB binding occurs primarily in hepatocytes during regeneration [155,156]. NF-κB binding activities were stronger in nonparenchymal cells (NPCs) than hepatocytes at baseline but were not further increased after PH [156]. In contrast, Sakuda et al. found that although NF-κB translocates to the nucleus in both hepatocytes and NPCs after PH, activation in the NPC population occurs prior to activation in hepatocytes [159]. This earlier activation may be reflective of the ability of NPCs to produce and respond to growth factors and cytokines such as TNF-α and IL-6 during the immediate early phase of liver regeneration. Furthermore, since NF-κB requires further modification and phosphorylation in order to become fully active, evidence of NF-κB nuclear translocation may not be sufficient to indicate activation status [160].

Subsequent studies attempting to elucidate the role of NF-κB during liver regeneration have employed several different techniques to inhibit NF-κB, including adenoviral infection with NF-κB pathway repressors, cell-specific gene deletion, chemical inhibition, and conditional genetic elimination, often with contradictory results. Many of these methodologies are context- and species-specific, and thus the specific cell types targeted are of particular relevance as this can directly affect the interpretation of the results. Therefore, depending on the model used, NF-κB activation during liver regeneration can be interpreted as increasing, decreasing, or having no effect on proliferation and/or apoptosis. The results of the NF-κB inhibition studies are summarized in Table 6.1.

One of the earliest studies to employ inhibitors of the NF-κB pathway during liver regeneration concluded that NF-κB activation primarily occurs in hepatocytes and is essential for initiating proliferation (Figure 6.5a). Rats expressing a phosphorylation- and degradation-resistant form of IκB by adenoviral vector exhibited suppressed NF-κB binding and activity, which resulted in massive apoptosis after PH and failure to progress through the cell cycle [161]. Further experiments in mice using an adenoviral vector expressing a mutated form of IκBα also resulted in decreased hepatocyte proliferation; however, unlike in the previous report, no increase in apoptosis was observed, which may reflect species-specific differences with respect to NF-κB activation and cell-cycle length [162]. Subsequent reports seemed to support the hypothesis that NF-κB activates proliferation in hepatocytes. Mice fed a high-fat diet had impaired liver regeneration after PH that was attributed to increased IκBα expression, which reduced NF-κB activity [163]. Although the cell type overexpressing IκBα was not specifically identified, the metabolic abnormalities induced by high-fat diet occur primarily in hepatocytes, and might contribute to reduced degradation of IκBα in these cells [163].

Additional work, however, suggested that the role of NF-κB in liver regeneration is more complex, and that it may not have a direct effect on hepatocyte proliferation. Hepatocyte-specific deletion of NEMO (IKKγ), the regulatory subunit of IKK, also significantly attenuated the proliferative response after PH in mice [164]. Lack of hepatocyte proliferation corresponded with massive apoptosis, which were both attributed to the absence of NF-κB activation. However, the mechanism of impaired regeneration in this model was found to be increased oxidative stress due to lack of antioxidants, which are regulated by NF-κB. Thus, although NF-κB activation in hepatocytes was shown to be essential for regeneration, its primary role was in preventing oxidative stress after liver injury, thus allowing hepatocytes to proliferate [164].

Further attempts to elucidate the relationship between NF-κB activation and hepatocyte proliferation through inhibition of NF-κB have been complicated by questions of cell specificity. Treatment with gliotoxin, a fungal metabolite that causes potent immunosuppression through NF-κB inhibition [165], prior to PH resulted in decreased NF-κB activity and a dramatic decrease in DNA synthesis in hepatocytes,

TABLE 6.1 Summary of Reports Describing Role of NF-κB Signaling in Liver Regeneration

First Author	Type of NF-κB Inhibition	Cell Type(s) Involved	Species	Effect on Proliferation	Effect on Apoptosis	Notes
Limuro	IκBα superrepressor (adenovirus)	Hepatocytes, maybe NPCs	Rat	Decrease	Increase	
Yang	Mutated form of IκBα (adenovirus)	Hepatocytes, maybe NPCs	Mouse	Decrease	No change	NF-κB primarily found in Kupffer cells
DeAngelis	High-fat diet (increases IκBα)	Hepatocytes	Mouse	Decrease	Increase	
Plumpe	Gliotoxin	Hepatocytes, maybe NPCs	Mouse	Decrease	Increase	
Abshagen	Dichloromethylene diphosphonate	Kupffer cells	Mouse	Decrease	Unknown	Decrease in NF-κB secondary to Kupffer cell depletion
Malato	Hep-IKKγ KO	Hepatocytes only	Mouse	Decrease	Increase	
Maeda	Mx-Cre IKKβ KO	All liver cell types	Mouse	Decrease	Unknown	
Laurent	Pyrrolidine dithiocarbamate (PDTC)	All liver cell types	Rat	No change	No change	
Malato	IKKβ inhibitor AS602868	All liver cell types	Mouse	No change	Unknown	
DeAngelis	p50 KO	All cell types of the body	Mouse	No change	Unknown	
Chaisson	Hep-IκBα KO	Hepatocytes only	Mouse	No change	No change	
Ringelhan	Mx-cre p65 KO	All liver cell types	Mouse	No change	No change	
Maeda/ Koch	Hep-IKKβ KO	Hepatocytes only	Mouse	Earlier/ enhanced	No change	
Malato	Hep-IKKβ KO	Hepatocytes only	Mouse	Earlier/ enhanced	Unknown	
Ringelhan	Hep-p65 KO	Hepatocytes only	Mouse	Earlier	No change	

concomitant with increased apoptosis [166]. The authors demonstrated that gliotoxin has a direct effect on hepatocytes; however, subsequent studies have demonstrated that gliotoxin nonselectively targets the liver NPC population [167–170]. Therefore, whether impaired hepatocyte proliferation after PH is due to direct suppression of hepatocyte NF-κB activity or due to the absence of growth and proliferation signals such as HGF and IL-6 that would normally be provided by NPCs during liver regeneration is unknown.

In contrast to the studies mentioned above, hepatocyte-specific inhibition of NF-κB by

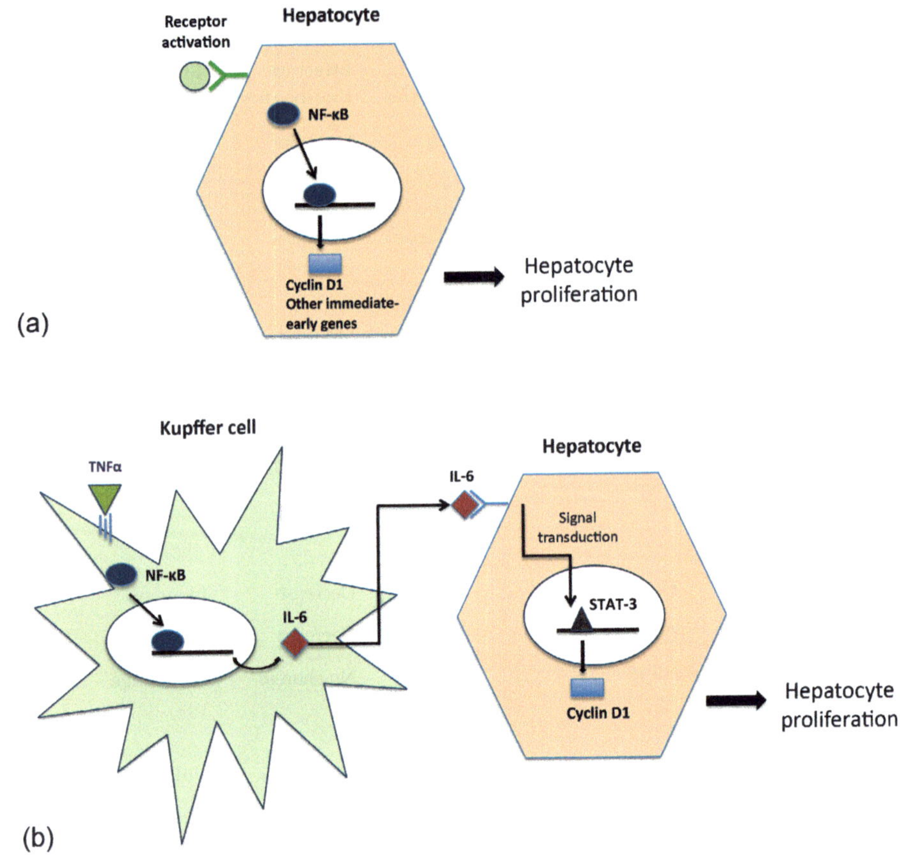

FIGURE 6.5 Two theories for the role of NF-κB in liver regeneration. (a) Activation of NF-κB occurs primarily in hepatocytes, which drives regeneration. Activation of cell-surface receptors such as TNF-R1 occurs immediately after PH; this results in derepression of NF-κB, which translocates to the nucleus and turns on expression of cyclin-D1 and/or immediate early genes essential for hepatocyte proliferation. (b) NF-κB is activated in Kupffer cells and other nonparenchymal cells, which results in production of cytokines and growth factors that stimulate hepatocyte proliferation and liver regeneration. Binding of TNF-α, LPS, or other cytokines to the surface of Kupffer cells induces NF-κB activation, resulting in production of IL-6 and TNF-α. These cytokines then stimulate proliferation of adjacent hepatocytes through activation of signal transducer and activator of transcription (STAT)-3.

expression of a mutated form of IκBα does not lead to either apoptosis or inhibited proliferation after PH [171]. The authors speculated that NPCs, which have intact NF-κB signaling, are responsible for providing antiapoptotic and proliferative signals to hepatocytes during LR. The use of adenovirus, which can infect other liver cell types including KCs, may have confounded the interpretation of results in the previous studies [172]. Indeed, Yang et al. demonstrated through use of a transgenic mouse expressing enhanced green fluorescent protein under the transcriptional control of NF-κB that activation primarily occurs in KCs rather than hepatocytes [162].

The ambiguity of previous findings has lead researchers to directly test the hypothesis that

NF-κB activation in NPCs is essential for liver regeneration. Hepatocyte-specific removal of IKKβ, which is upstream of IκB and required for NF-κB activation, causes no changes in proliferation after PH; in fact, regeneration was slightly accelerated. However, when IKKβ is removed from all cells of the liver by interferon-induced Mx1-cre recombination, a significant reduction in proliferation is observed [173]. Depletion of KCs in liver by liposome-encapsulated dichloromethylene diphosphonate prior to PH also resulted in loss of NF-κB activation and impaired liver regeneration [174]. These studies highlight the role of the NPC compartment in NF-κB activation and regeneration.

Over time, the concept that activation of NF-κB occurs primarily in KCs, not hepatocytes, has gradually evolved and has emerged as the most likely function for NF-κB signaling during regeneration. In this model, binding of TNF-α, LPS, or other cytokines to the surface of KCs induces NF-κB activation, resulting in production of IL-6 and TNF-α, which stimulates hepatocyte proliferation in a paracrine manner [115,175] (Figure 6.5b).

However, other studies attempting to elucidate the role of NF-κB signaling during liver regeneration seem to contradict this hypothesis. Hepatocytes lacking IKKβ due to expression of cre recombinase under an albumin promoter and α-fetoprotein enhancer (AFP-cre) had delayed NF-κB activation during regeneration. However, there was paradoxical earlier NF-κB activation in the NPC compartment, which preceded earlier entry of hepatocytes into the cell cycle and subsequent proliferation [176]. This was attributed to enhanced expression of TNF-α and other cytokines in these livers, which induced a stronger inflammatory response in NPCs of Alfp-cre IKKβ KOs. Interestingly, in contrast to the studies utilizing Mx-cre to delete IKKβ from all liver cells, pharmacological inhibition of IKKβ in all cell compartments had no effect on proliferation kinetics during regeneration, reversing the proliferative advantage seen in Alfp-cre IKKβ KO mice, which have enhanced NF-κB activation in NPCs only [176]. Koch et al. further analyzed the regenerative properties of albumin (Alb)-cre IKKβ KO mice after PH, which displayed similar kinetics to the Alfp-cre IKKβ KO mice above. They found that the accelerated liver regeneration in Alb-cre IKKβ KO was due to earlier activation of cyclin-D1 and resulting mitosis [177]. Further, *ex vivo* deletion of IKKβ in cultured hepatocytes with cre-expressing adenovirus stimulated proliferation as well. These effects were attributed to increased sensitivity of IKKβ KO hepatocytes to TNF-α. Thus, in these models, inactivation of NF-κB in hepatocytes has a paradoxical promotional effect on proliferation, while inhibition of NF-κB in all liver cells has no effect on regeneration.

Seeking to resolve the contradictions and discrepancies of the previous studies, Ringelhan et al. directly inhibited p65 in either hepatocytes or all liver cells [178]. Deletion of p65 from hepatocytes only results in accelerated cell-cycle progression, but ultimately did not change the mass of regenerating liver. There was also no change in liver injury after PH in Hep-p65 KO. Interestingly, deletion of p65 from both parenchymal and NPCs using Mx-cre neutralized the proliferative advantage seen after hepatocyte-specific deletion of p65. Therefore, the authors concluded that p65 in all cell compartments is dispensable for proper liver regeneration.

Other studies have also demonstrated that NF-κB activation in either hepatocytes or NPCs is not required for liver regeneration. Treatment with either the antioxidant pyrrolidine dithiocarbamate (PDTC), an inhibitor of NF-κB, or TNF-α inhibitor prior to PH in rats resulted in inhibition of NF-κB activation 30 min after PH; however, DNA synthesis and cell proliferation were equivalent to controls at 24 h after PH. Thus, the authors concluded that activation of NF-κB is not necessary to initiate liver regeneration after PH [179]. Similarly, deletion of the NF-κB subunit p50 does not impair overall liver regeneration. However, this may be due to compensatory increase in the p65 subunit, rendering p50 dispensable during regeneration [180].

In summary, NF-κB, a pleiotropic transcription factor that plays a role in many diverse biological processes, such as inflammation, proliferation, and especially survival, has been shown to be activated during liver regeneration; however, its overall contribution to regeneration is yet to be definitively determined [181].

6.6 CONCLUSIONS

In the past several decades, many of the genes and signaling events that are involved in initiation and perpetuation of liver regeneration have been elucidated. Interestingly, several of these signaling pathways are also critical to embryonic development and hepatogenesis. Some pathways, such as Wnt/β-catenin and NF-κB, have been extensively studied in the process of liver regeneration and have been shown to play key roles in hepatocyte proliferation and survival. Additionally, there is emerging evidence to suggest that other key developmental pathways, such as the Hippo/YAP and Jagged/Notch pathways, participate in these processes as well.

Extensive studies over many years have demonstrated the inherent redundancy of liver regeneration, such that there is no known pathway whose absence completely abrogates this process. Indeed it appears that liver regeneration is the result of a complex convergence of many signaling pathways, and no one pathway alone can drive this response. Therefore, the pathways discussed here likely function as part of a coordinated and regulated process that involves intercellular and intermolecular crosstalks. Moreover, communication between these developmental pathways may be essential to the efficient execution of liver regeneration, and thus studies that characterize how activation of one pathway directly or indirectly affects other pathways will be highly informative.

Acknowledgments

This study was funded by NIH Grants 1R01DK62277, 1R01DK100287, and 1R01DK095498 to SPSM.

References

[1] Cadigan KM, Peifer M. Wnt signaling from development to disease: insights from model systems. Cold Spring Harb Perspect Biol 2009;1:a002881.

[2] Bartscherer K, Boutros M. Regulation of Wnt protein secretion and its role in gradient formation. EMBO Rep 2008;9:977–82.

[3] Willert K, Brown JD, Danenberg E, Duncan AW, Weissman IL, Reya T, et al. Wnt proteins are lipid-modified and can act as stem cell growth factors. Nature 2003;423:448–52.

[4] Zhai L, Chaturvedi D, Cumberledge S. Drosophila wnt-1 undergoes a hydrophobic modification and is targeted to lipid rafts, a process that requires porcupine. J Biol Chem 2004;279:33220–7.

[5] Bartscherer K, Pelte N, Ingelfinger D, Boutros M. Secretion of Wnt ligands requires Evi, a conserved transmembrane protein. Cell 2006;125:523–33.

[6] Moon RT, Kohn AD, De Ferrari GV, Kaykas A. WNT and beta-catenin signalling: diseases and therapies. Nat Rev Genet 2004;5:691–701.

[7] Michalopoulos GK. Liver regeneration. J Cell Physiol 2007;213:286–300.

[8] Michalopoulos GK. Principles of liver regeneration and growth homeostasis. Compr Physiol 2013;3:485–513.

[9] Lade AG, Monga SP. Beta-catenin signaling in hepatic development and progenitors: which way does the WNT blow? Dev Dyn 2011;240:486–500.

[10] Oh SH, Hatch HM, Petersen BE. Hepatic oval 'stem' cell in liver regeneration. Semin Cell Dev Biol 2002;13:405–9.

[11] Nejak-Bowen KN, Monga SP. Beta-catenin signaling, liver regeneration and hepatocellular cancer: sorting the good from the bad. Semin Cancer Biol 2011;21:44–58.

[12] Monga SP, Pediaditakis P, Mule K, Stolz DB, Michalopoulos GK. Changes in WNT/beta-catenin pathway during regulated growth in rat liver regeneration. Hepatology 2001;33:1098–109.

[13] Sodhi D, Micsenyi A, Bowen WC, Monga DK, Talavera JC, Monga SP. Morpholino oligonucleotide-triggered beta-catenin knockdown compromises normal liver regeneration. J Hepatol 2005;43:132–41.

[14] Nakamura I, Fernandez-Barrena MG, Ortiz-Ruiz MC, Almada LL, Hu C, Elsawa SF, et al. Activation of the transcription factor GLI1 by WNT signaling underlies the role of SULFATASE 2 as a regulator of tissue regeneration. J Biol Chem 2013;288:21389–98.
[15] Tan X, Behari J, Cieply B, Michalopoulos GK, Monga SP. Conditional deletion of beta-catenin reveals its role in liver growth and regeneration. Gastroenterology 2006;131:1561–72.
[16] Yang J, Mowry LE, Nejak-Bowen KN, Okabe H, Diegel CR, Lang RA, et al. Beta-catenin signaling in murine liver zonation and regeneration: a Wnt-Wnt situation! Hepatology 2014;60:964–76.
[17] Sekine S, Gutierrez PJ, Lan BY, Feng S, Hebrok M. Liver-specific loss of beta-catenin results in delayed hepatocyte proliferation after partial hepatectomy. Hepatology 2007;45:361–8.
[18] Sekine S, Lan BY, Bedolli M, Feng S, Hebrok M. Liver-specific loss of beta-catenin blocks glutamine synthesis pathway activity and cytochrome p450 expression in mice. Hepatology 2006;43:817–25.
[19] Goessling W, North TE, Lord AM, Ceol C, Lee S, Weidinger G, et al. APC mutant zebrafish uncover a changing temporal requirement for wnt signaling in liver development. Dev Biol 2008;320:161–74.
[20] Corbett MD, Corbett BR, Hannothiaux MH, Quintana SJ. Metabolic activation and nucleic acid binding of acetaminophen and related arylamine substrates by the respiratory burst of human granulocytes. Chem Res Toxicol 1989;2:260–6.
[21] Patten CJ, Thomas PE, Guy RL, Lee M, Gonzalez FJ, Guengerich FP, et al. Cytochrome P450 enzymes involved in acetaminophen activation by rat and human liver microsomes and their kinetics. Chem Res Toxicol 1993;6:511–8.
[22] Raucy JL, Lasker JM, Lieber CS, Black M. Acetaminophen activation by human liver cytochromes P450IIE1 and P450IA2. Arch Biochem Biophys 1989;271:270–83.
[23] Mehendale HM. Tissue repair: an important determinant of final outcome of toxicant-induced injury. Toxicol Pathol 2005;33:41–51.
[24] Apte U, Singh S, Zeng G, Cieply B, Virji MA, Wu T, et al. Beta-catenin activation promotes liver regeneration after acetaminophen-induced injury. Am J Pathol 2009;175:1056–65.
[25] Nejak-Bowen KN, Thompson MD, Singh S, Bowen Jr WC, Dar MJ, Khillan J, et al. Accelerated liver regeneration and hepatocarcinogenesis in mice overexpressing serine-45 mutant beta-catenin. Hepatology 2010;51:1603–13.
[26] Bhushan B, Walesky C, Manley M, Gallagher T, Borude P, Edwards G, et al. Pro-regenerative signaling after acetaminophen-induced acute liver injury in mice identified using a novel incremental dose model. Am J Pathol 2014;184:3013–25.
[27] Kowalik MA, Perra A, Pibiri M, Cocco MT, Samarut J, Plateroti M, et al. TRbeta is the critical thyroid hormone receptor isoform in T3-induced proliferation of hepatocytes and pancreatic acinar cells. J Hepatol 2010;53:686–92.
[28] Columbano A, Simbula M, Pibiri M, Perra A, Deidda M, Locker J, et al. Triiodothyronine stimulates hepatocyte proliferation in two models of impaired liver regeneration. Cell Prolif 2008;41:521–31.
[29] Ledda-Columbano GM, Pibiri M, Loi R, Perra A, Shinozuka H, Columbano A. Early increase in cyclin-D1 expression and accelerated entry of mouse hepatocytes into S phase after administration of the mitogen 1,4-bis[2-(3,5-Dichloropyridyloxy)] benzene. Am J Pathol 2000;156:91–7.
[30] Fanti M, Singh S, Ledda-Columbano GM, Columbano A, Monga SP. Triiodothyronine induces hepatocyte proliferation by protein kinase A-dependent β-catenin activation in rodents. Hepatology 2014; 59: 2309–20.
[31] Apte U, Zeng G, Muller P, Tan X, Micsenyi A, Cieply B, et al. Activation of Wnt/beta-catenin pathway during hepatocyte growth factor-induced hepatomegaly in mice. Hepatology 2006;44:992–1002.
[32] Bonvini P, An WG, Rosolen A, Nguyen P, Trepel J, Garcia de Herreros A, et al. Geldanamycin abrogates ErbB2 association with proteasome-resistant beta-catenin in melanoma cells, increases beta-catenin-E-cadherin association, and decreases beta-catenin-sensitive transcription. Cancer Res 2001;61:1671–7.
[33] Guturi KK, Mandal T, Chatterjee A, Sarkar M, Bhattacharya S, Chatterjee U, et al. Mechanism of beta-catenin-mediated transcriptional regulation of epidermal growth factor receptor expression in glycogen synthase kinase 3 beta-inactivated prostate cancer cells. J Biol Chem 2012;287:18287–96.
[34] Hino S, Tanji C, Nakayama KI, Kikuchi A. Phosphorylation of beta-catenin by cyclic AMP-dependent protein kinase stabilizes beta-catenin through inhibition of its ubiquitination. Mol Cell Biol 2005;25:9063–72.
[35] Hoshida Y, Nijman SM, Kobayashi M, Chan JA, Brunet JP, Chiang DY, et al. Integrative transcriptome analysis reveals common molecular subclasses of human hepatocellular carcinoma. Cancer Res 2009;69:7385–92.
[36] Kajiguchi T, Katsumi A, Tanizaki R, Kiyoi H, Naoe T. Y654 of beta-catenin is essential for FLT3/ITD-related tyrosine phosphorylation and nuclear localization of beta-catenin. Eur J Haematol 2012;88:314–20.
[37] Monga SP, Mars WM, Pediaditakis P, Bell A, Mule K, Bowen WC, et al. Hepatocyte growth factor induces

Wnt-independent nuclear translocation of beta-catenin after Met-beta-catenin dissociation in hepatocytes. Cancer Res 2002;62:2064–71.
[38] Taurin S, Sandbo N, Qin Y, Browning D, Dulin NO. Phosphorylation of beta-catenin by cyclic AMP-dependent protein kinase. J Biol Chem 2006;281:9971–6.
[39] van Veelen W, Le NH, Helvensteijn W, Blonden L, Theeuwes M, Bakker ER, et al. Beta-catenin tyrosine 654 phosphorylation increases Wnt signalling and intestinal tumorigenesis. Gut 2011;60:1204–12.
[40] Zeng G, Apte U, Micsenyi A, Bell A, Monga SP. Tyrosine residues 654 and 670 in beta-catenin are crucial in regulation of Met-beta-catenin interactions. Exp Cell Res 2006;312:3620–30.
[41] Cadigan KM, Liu YI. Wnt signaling: complexity at the surface. J Cell Sci 2006;119:395–402.
[42] Zeng X, Huang H, Tamai K, Zhang X, Harada Y, Yokota C, et al. Initiation of Wnt signaling: control of Wnt coreceptor Lrp6 phosphorylation/activation via frizzled, dishevelled and axin functions. Development 2008;135:367–75.
[43] Zeng X, Tamai K, Doble B, Li S, Huang H, Habas R, et al. A dual-kinase mechanism for Wnt co-receptor phosphorylation and activation. Nature 2005;438:873–7.
[44] Zhong Z, Baker JJ, Zylstra-Diegel CR, Williams BO. Lrp5 and Lrp6 play compensatory roles in mouse intestinal development. J Cell Biochem 2012;113:31–8.
[45] Zeng G, Awan F, Otruba W, Muller P, Apte U, Tan X, et al. Wnt'er in liver: expression of Wnt and frizzled genes in mouse. Hepatology 2007;45:195–204.
[46] Banziger C, Soldini D, Schutt C, Zipperlen P, Hausmann G, Basler K. Wntless, a conserved membrane protein dedicated to the secretion of Wnt proteins from signaling cells. Cell 2006;125:509–22.
[47] Belenkaya TY, Wu Y, Tang X, Zhou B, Cheng L, Sharma YV, et al. The retromer complex influences Wnt secretion by recycling wntless from endosomes to the trans-Golgi network. Dev Cell 2008;14:120–31.
[48] Franch-Marro X, Wendler F, Guidato S, Griffith J, Baena-Lopez A, Itasaki N, et al. Wingless secretion requires endosome-to-Golgi retrieval of Wntless/Evi/Sprinter by the retromer complex. Nat Cell Biol 2008;10:170–7.
[49] Fu J, Jiang M, Mirando AJ, Yu HM, Hsu W. Reciprocal regulation of Wnt and Gpr177/mouse Wntless is required for embryonic axis formation. Proc Natl Acad Sci USA 2009;106:18598–603.
[50] Carpenter AC, Rao S, Wells JM, Campbell K, Lang RA. Generation of mice with a conditional null allele for Wntless. Genesis 2010;48:554–8.
[51] Zhong Z, Zylstra-Diegel CR, Schumacher CA, Baker JJ, Carpenter AC, Rao S, et al. Wntless functions in mature osteoblasts to regulate bone mass. Proc Natl Acad Sci USA 2012;109:E2197–204.
[52] Kisanuki YY, Hammer RE, Miyazaki J, Williams SC, Richardson JA, Yanagisawa M. Tie2-Cre transgenic mice: a new model for endothelial cell-lineage analysis in vivo. Dev Biol 2001;230:230–42.
[53] Ding BS, Nolan DJ, Butler JM, James D, Babazadeh AO, Rosenwaks Z, et al. Inductive angiocrine signals from sinusoidal endothelium are required for liver regeneration. Nature 2010;468:310–5.
[54] Clausen BE, Burkhardt C, Reith W, Renkawitz R, Forster I. Conditional gene targeting in macrophages and granulocytes using LysMcre mice. Transgenic Res 1999;8:265–77.
[55] Baron M. An overview of the Notch signalling pathway. Semin Cell Dev Biol 2003;14:113–9.
[56] Rampal R, Luther KB, Haltiwanger RS. Notch signaling in normal and disease States: possible therapies related to glycosylation. Curr Mol Med 2007;7:427–45.
[57] Bohm F, Kohler UA, Speicher T, Werner S. Regulation of liver regeneration by growth factors and cytokines. EMBO Mol Med 2010;2:294–305.
[58] Croquelois A, Blindenbacher A, Terracciano L, Wang X, Langer I, Radtke F, et al. Inducible inactivation of Notch1 causes nodular regenerative hyperplasia in mice. Hepatology 2005;41:487–96.
[59] Chillakuri CR, Sheppard D, Lea SM, Handford PA. Notch receptor-ligand binding and activation: insights from molecular studies. Semin Cell Dev Biol 2012; 23:421–8.
[60] Bray SJ. Notch signalling: a simple pathway becomes complex. Nat Rev Mol Cell Biol 2006;7:678–89.
[61] Ranganathan P, Weaver KL, Capobianco AJ. Notch signalling in solid tumours: a little bit of everything but not all the time. Nat Rev Cancer 2011;11:338–51.
[62] Oda T, Elkahloun AG, Pike BL, Okajima K, Krantz ID, Genin A, et al. Mutations in the human Jagged1 gene are responsible for Alagille syndrome. Nat Genet 1997;16:235–42.
[63] Li L, Krantz ID, Deng Y, Genin A, Banta AB, Collins CC, et al. Alagille syndrome is caused by mutations in human Jagged1, which encodes a ligand for Notch1. Nat Genet 1997;16:243–51.
[64] Hofmann JJ, Zovein AC, Koh H, Radtke F, Weinmaster G, Iruela-Arispe ML. Jagged1 in the portal vein mesenchyme regulates intrahepatic bile duct development: insights into Alagille syndrome. Development 2010;137:4061–72.
[65] Kohler C, Bell AW, Bowen WC, Monga SP, Fleig W, Michalopoulos GK. Expression of Notch-1 and its ligand Jagged-1 in rat liver during liver regeneration. Hepatology 2004;39:1056–65.
[66] Darwiche H, Oh SH, Steiger-Luther NC, Williams JM, Pintilie DG, Shupe TD, et al. Inhibition of Notch signaling affects hepatic oval cell response in rat model of 2AAF-PH. Hepat Med 2011;3:89–98.

[67] McCright B, Lozier J, Gridley T. A mouse model of Alagille syndrome: Notch2 as a genetic modifier of Jag1 haploinsufficiency. Development 2002;129:1075–82.
[68] Nijjar SS, Crosby HA, Wallace L, Hubscher SG, Strain AJ. Notch receptor expression in adult human liver: a possible role in bile duct formation and hepatic neovascularization. Hepatology 2001;34:1184–92.
[69] Wakabayashi N, Skoko JJ, Chartoumpekis DV, Kimura S, Slocum SL, Noda K, et al. Notch-Nrf2 axis: regulation of Nrf2 gene expression and cytoprotection by notch signaling. Mol Cell Biol 2014;34:653–63.
[70] Wakabayashi N, Shin S, Slocum SL, Agoston ES, Wakabayashi J, Kwak MK, et al. Regulation of notch1 signaling by nrf2: implications for tissue regeneration. Sci Signal 2010;3:ra52.
[71] Wang L, Wang CM, Hou LH, Dou GR, Wang YC, Hu XB, et al. Disruption of the transcription factor recombination signal-binding protein-Jkappa (RBP-J) leads to veno-occlusive disease and interfered liver regeneration in mice. Hepatology 2009;49:268–77.
[72] Dong J, Feldmann G, Huang J, Wu S, Zhang N, Comerford SA, et al. Elucidation of a universal size-control mechanism in Drosophila and mammals. Cell 2007;130:1120–33.
[73] Lu L, Li Y, Kim SM, Bossuyt W, Liu P, Qiu Q, et al. Hippo signaling is a potent in vivo growth and tumor suppressor pathway in the mammalian liver. Proc Natl Acad Sci USA 2010;107:1437–42.
[74] Halder G, Johnson RL. Hippo signaling: growth control and beyond. Development 2011;138:9–22.
[75] Zhao B, Li L, Lei Q, Guan KL. The Hippo-YAP pathway in organ size control and tumorigenesis: an updated version. Genes Dev 2010;24:862–74.
[76] Tordjmann T. Hippo signalling: liver size regulation and beyond. Clin Res Hepatol Gastroenterol 2011; 35:344–6.
[77] Song H, Mak KK, Topol L, Yun K, Hu J, Garrett L, et al. Mammalian Mst1 and Mst2 kinases play essential roles in organ size control and tumor suppression. Proc Natl Acad Sci USA 2010;107:1431–6.
[78] Zhou D, Conrad C, Xia F, Park JS, Payer B, Yin Y, et al. Mst1 and Mst2 maintain hepatocyte quiescence and suppress hepatocellular carcinoma development through inactivation of the Yap1 oncogene. Cancer Cell 2009;16:425–38.
[79] Avruch J, Zhou D, Fitamant J, Bardeesy N. Mst1/2 signalling to Yap: gatekeeper for liver size and tumour development. Br J Cancer 2011;104:24–32.
[80] Zhang N, Bai H, David KK, Dong J, Zheng Y, Cai J, et al. The Merlin/NF2 tumor suppressor functions through the YAP oncoprotein to regulate tissue homeostasis in mammals. Dev Cell 2010;19:27–38.
[81] Camargo FD, Gokhale S, Johnnidis JB, Fu D, Bell GW, Jaenisch R, et al. YAP1 increases organ size and expands undifferentiated progenitor cells. Curr Biol 2007;17:2054–60.
[82] Wang C, Zhang L, He Q, Feng X, Zhu J, Xu Z, et al. Differences in Yes-associated protein and mRNA levels in regenerating liver and hepatocellular carcinoma. Mol Med Rep 2012;5:410–4.
[83] Bai H, Zhang N, Xu Y, Chen Q, Khan M, Potter JJ, et al. Yes-associated protein regulates the hepatic response after bile duct ligation. Hepatology 2012;56: 1097–107.
[84] Shafritz D, Oertel M, Dabeva M, Grompe M. Liver repopulation by cell transplantation and the role of stem cells. In: Arias I, Wolkoff A, Boyer J, Shafritz D, Fausto N, Alter H, et al., The liver: biology and pathobiology. 5th ed. Hoboken, NJ: John Wiley and Sons; 2011. p. 577–95.
[85] Apte U, Gkretsi V, Bowen WC, Mars WM, Luo JH, Donthamsetty S, et al. Enhanced liver regeneration following changes induced by hepatocyte-specific genetic ablation of integrin-linked kinase. Hepatology 2009;50:844–51.
[86] Donthamsetty S, Bhave VS, Kliment CS, Bowen WC, Mars WM, Bell AW, et al. Excessive hepatomegaly of mice with hepatocyte-targeted elimination of integrin linked kinase following treatment with 1,4-bis [2-(3,5-dichaloropyridyloxy)] benzene. Hepatology 2011; 53:587–95.
[87] Donthamsetty S, Bowen W, Mars W, Bhave V, Luo JH, Wu C, et al. Liver-specific ablation of integrin-linked kinase in mice results in enhanced and prolonged cell proliferation and hepatomegaly after phenobarbital administration. Toxicol Sci 2010;113:358–66.
[88] Lin CW, Mars WM, Paranjpe S, Donthamsetty S, Bhave VS, Kang LI, et al. Hepatocyte proliferation and hepatomegaly induced by phenobarbital and 1,4-bis [2-(3,5-dichloropyridyloxy)] benzene is suppressed in hepatocyte-targeted glypican 3 transgenic mice. Hepatology 2011;54:620–30.
[89] Liu B, Bell AW, Paranjpe S, Bowen WC, Khillan JS, Luo JH, et al. Suppression of liver regeneration and hepatocyte proliferation in hepatocyte-targeted glypican 3 transgenic mice. Hepatology 2010;52:1060–7.
[90] Michalopoulos G, et al. Liver regeneration. In: Monga SP, editor. Molecular pathology of liver diseases. New York, NY: Springer; 2010. p. 261–78.
[91] Chen LF, Greene WC. Shaping the nuclear action of NF-kappaB. Nat Rev Mol Cell Biol 2004;5:392–401.
[92] Wajant H, Pfizenmaier K, Scheurich P. Tumor necrosis factor signaling. Cell Death Differ 2003;10:45–65.
[93] Guicciardi ME, Gores GJ. Life and death by death receptors. FASEB J 2009;23:1625–37.
[94] Bellas RE, FitzGerald MJ, Fausto N, Sonenshein GE. Inhibition of NF-kappa B activity induces apoptosis in murine hepatocytes. Am J Pathol 1997;151:891–6.

[95] Chen S, Guttridge DC, Tang E, Shi S, Guan K, Wang CY. Suppression of tumor necrosis factor-mediated apoptosis by nuclear factor kappaB-independent bone morphogenetic protein/Smad signaling. J Biol Chem 2001;276:39259–63.
[96] Tak PP, Firestein GS. NF-kappaB: a key role in inflammatory diseases. J Clin Invest 2001;107:7–11.
[97] Karin M, Lin A. NF-kappaB at the crossroads of life and death. Nat Immunol 2002;3:221–7.
[98] Tacke F, Luedde T, Trautwein C. Inflammatory pathways in liver homeostasis and liver injury. Clin Rev Allergy Immunol 2009;36:4–12.
[99] Perkins ND. The Rel/NF-kappa B family: friend and foe. Trends Biochem Sci 2000;25:434–40.
[100] Zhong H, SuYang H, Erdjument-Bromage H, Tempst P, Ghosh S. The transcriptional activity of NF-kappaB is regulated by the IkappaB-associated PKAc subunit through a cyclic AMP-independent mechanism. Cell 1997;89:413–24.
[101] Wang D, Westerheide SD, Hanson JL, Baldwin Jr AS. Tumor necrosis factor alpha-induced phosphorylation of RelA/p65 on Ser529 is controlled by casein kinase II. J Biol Chem 2000;275:32592–7.
[102] Leitges M, Sanz L, Martin P, Duran A, Braun U, Garcia JF, et al. Targeted disruption of the zetaPKC gene results in the impairment of the NF-kappaB pathway. Mol Cell 2001;8:771–80.
[103] Hoeflich KP, Luo J, Rubie EA, Tsao MS, Jin O, Woodgett JR. Requirement for glycogen synthase kinase-3beta in cell survival and NF-kappaB activation. Nature 2000;406:86–90.
[104] O'Mahony AM, Montano M, Van Beneden K, Chen LF, Greene WC. Human T-cell lymphotropic virus type 1 tax induction of biologically active NF-kappaB requires IkappaB kinase-1-mediated phosphorylation of RelA/p65. J Biol Chem 2004;279:18137–45.
[105] Lawrence T. The nuclear factor NF-kappaB pathway in inflammation. Cold Spring Harb Perspect Biol 2009;1: a001651.
[106] Yamada Y, Kirillova I, Peschon JJ, Fausto N. Initiation of liver growth by tumor necrosis factor: deficient liver regeneration in mice lacking type I tumor necrosis factor receptor. Proc Natl Acad Sci USA 1997;94:1441–6.
[107] Kirillova I, Chaisson M, Fausto N. Tumor necrosis factor induces DNA replication in hepatic cells through nuclear factor kappaB activation. Cell Growth Differ 1999;10:819–28.
[108] Wullaert A, van Loo G, Heyninck K, Beyaert R. Hepatic tumor necrosis factor signaling and nuclear factor-kappaB: effects on liver homeostasis and beyond. Endocr Rev 2007;28:365–86.
[109] Akerman P, Cote P, Yang SQ, McClain C, Nelson S, Bagby GJ, et al. Antibodies to tumor necrosis factor-alpha inhibit liver regeneration after partial hepatectomy. Am J Physiol 1992;263:G579–85.
[110] Diehl AM, Yin M, Fleckenstein J, Yang SQ, Lin HZ, Brenner DA, et al. Tumor necrosis factor-alpha induces c-jun during the regenerative response to liver injury. Am J Physiol 1994;267:G552–61.
[111] Kimura T, Sakaida I, Terai S, Matsumura Y, Uchida K, Okita K. Inhibition of tumor necrosis factor-alpha production retards liver regeneration after partial hepatectomy in rats. Biochem Biophys Res Commun 1997;231:557–60.
[112] Yamada Y, Fausto N. Deficient liver regeneration after carbon tetrachloride injury in mice lacking type 1 but not type 2 tumor necrosis factor receptor. Am J Pathol 1998;152:1577–89.
[113] Fujita J, Marino MW, Wada H, Jungbluth AA, Mackrell PJ, Rivadeneira DE, et al. Effect of TNF gene depletion on liver regeneration after partial hepatectomy in mice. Surgery 2001;129:48–54.
[114] Hayashi H, Nagaki M, Imose M, Osawa Y, Kimura K, Takai S, et al. Normal liver regeneration and liver cell apoptosis after partial hepatectomy in tumor necrosis factor-alpha-deficient mice. Liver Int 2005;25:162–70.
[115] Fausto N, Campbell JS, Riehle KJ. Liver regeneration. Hepatology 2006;43:S45–53.
[116] Cornell RP, Liljequist BL, Bartizal KF. Depressed liver regeneration after partial hepatectomy of germ-free, athymic and lipopolysaccharide-resistant mice. Hepatology 1990;11:916–22.
[117] Cornell RP. Gut-derived endotoxin elicits hepatotrophic factor secretion for liver regeneration. Am J Physiol 1985;249:R551–62.
[118] Poltorak A, He X, Smirnova I, Liu MY, Van Huffel C, Du X, et al. Defective LPS signaling in C3H/HeJ and C57BL/10ScCr mice: mutations in Tlr4 gene. Science 1998;282:2085–8.
[119] Seki E, Tsutsui H, Iimuro Y, Naka T, Son G, Akira S, et al. Contribution of Toll-like receptor/myeloid differentiation factor 88 signaling to murine liver regeneration. Hepatology 2005;41:443–50.
[120] Campbell JS, Riehle KJ, Brooling JT, Bauer RL, Mitchell C, Fausto N. Proinflammatory cytokine production in liver regeneration is Myd88-dependent, but independent of Cd14, Tlr2, and Tlr4. J Immunol 2006;176:2522–8.
[121] Iimuro Y, Fujimoto J. TLRs, NF-kappaB, JNK, and liver regeneration. Gastroenterol Res Pract 2010; 2010:Article ID 598109, 7 p. http://dx.doi.org/10.1155/2010/598109.
[122] Vaquero J, Campbell JS, Haque J, McMahan RS, Riehle KJ, Bauer RL, et al. Toll-like receptor 4 and myeloid differentiation factor 88 provide mechanistic insights into the cause and effects of interleukin-6

activation in mouse liver regeneration. Hepatology 2011; 54:597–608.
[123] Mastellos D, Papadimitriou JC, Franchini S, Tsonis PA, Lambris JD. A novel role of complement: mice deficient in the fifth component of complement (C5) exhibit impaired liver regeneration. J Immunol 2001;166:2479–86.
[124] Strey CW, Markiewski M, Mastellos D, Tudoran R, Spruce LA, Greenbaum LE, et al. The proinflammatory mediators C3a and C5a are essential for liver regeneration. J Exp Med 2003;198:913–23.
[125] Block GD, Locker J, Bowen WC, Petersen BE, Katyal S, Strom SC, et al. Population expansion, clonal growth, and specific differentiation patterns in primary cultures of hepatocytes induced by HGF/SF, EGF and TGF alpha in a chemically defined (HGM) medium. J Cell Biol 1996;132:1133–49.
[126] Fan S, Gao M, Meng Q, Laterra JJ, Symons MH, Coniglio S, et al. Role of NF-kappaB signaling in hepatocyte growth factor/scatter factor-mediated cell protection. Oncogene 2005;24:1749–66.
[127] Maroni P, Bendinelli P, Matteucci E, Desiderio MA. HGF induces CXCR4 and CXCL12-mediated tumor invasion through Ets1 and NF-kappaB. Carcinogenesis 2007;28:267–79.
[128] Tacchini L, Dansi P, Matteucci E, Desiderio MA. Hepatocyte growth factor signal coupling to various transcription factors depends on triggering of Met receptor and protein kinase transducers in human hepatoma cells HepG2. Exp Cell Res 2000;256:272–81.
[129] Yao P, Zhan Y, Xu W, Li C, Yue P, Xu C, et al. Hepatocyte growth factor-induced proliferation of hepatic stem-like cells depends on activation of NF-kappaB. J Hepatol 2004;40:391–8.
[130] Muller M, Morotti A, Ponzetto C. Activation of NF-kappaB is essential for hepatocyte growth factor-mediated proliferation and tubulogenesis. Mol Cell Biol 2002;22:1060–72.
[131] Biswas DK, Cruz AP, Gansberger E, Pardee AB. Epidermal growth factor-induced nuclear factor kappa B activation: a major pathway of cell-cycle progression in estrogen-receptor negative breast cancer cells. Proc Natl Acad Sci USA 2000;97:8542–7.
[132] Libermann TA, Baltimore D. Activation of interleukin-6 gene expression through the NF-kappa B transcription factor. Mol Cell Biol 1990;10:2327–34.
[133] Taub R. Liver regeneration: from myth to mechanism. Nat Rev Mol Cell Biol 2004;5:836–47.
[134] Li W, Liang X, Leu JI, Kovalovich K, Ciliberto G, Taub R. Global changes in interleukin-6-dependent gene expression patterns in mouse livers after partial hepatectomy. Hepatology 2001;33:1377–86.
[135] Cressman DE, Greenbaum LE, DeAngelis RA, Ciliberto G, Furth EE, Poli V, et al. Liver failure and defective hepatocyte regeneration in interleukin-6-deficient mice. Science 1996;274:1379–83.
[136] Sakamoto T, Liu Z, Murase N, Ezure T, Yokomuro S, Poli V, et al. Mitosis and apoptosis in the liver of interleukin-6-deficient mice after partial hepatectomy. Hepatology 1999;29:403–11.
[137] Galun E, Zeira E, Pappo O, Peters M, Rose-John S. Liver regeneration induced by a designer human IL-6/sIL-6R fusion protein reverses severe hepatocellular injury. FASEB J 2000;14:1979–87.
[138] Blindenbacher A, Wang X, Langer I, Savino R, Terracciano L, Heim MH. Interleukin 6 is important for survival after partial hepatectomy in mice. Hepatology 2003;38:674–82.
[139] Kovalovich K, DeAngelis RA, Li W, Furth EE, Ciliberto G, Taub R. Increased toxin-induced liver injury and fibrosis in interleukin-6-deficient mice. Hepatology 2000;31:149–59.
[140] Dierssen U, Beraza N, Lutz HH, Liedtke C, Ernst M, Wasmuth HE, et al. Molecular dissection of gp130-dependent pathways in hepatocytes during liver regeneration. J Biol Chem 2008;283:9886–95.
[141] Wuestefeld T, Klein C, Streetz KL, Betz U, Lauber J, Buer J, et al. Interleukin-6/glycoprotein 130-dependent pathways are protective during liver regeneration. J Biol Chem 2003;278:11281–8.
[142] Riehle KJ, Campbell JS, McMahan RS, Johnson MM, Beyer RP, Bammler TK, et al. Regulation of liver regeneration and hepatocarcinogenesis by suppressor of cytokine signaling 3. J Exp Med 2008;205:91–103.
[143] Zimmers TA, Pierce RH, McKillop IH, Koniaris LG. Resolving the role of IL-6 in liver regeneration. Hepatology 2003;38:1590–1 [author reply 1591].
[144] Hinz M, Krappmann D, Eichten A, Heder A, Scheidereit C, Strauss M. NF-kappaB function in growth control: regulation of cyclin D1 expression and G0/G1-to-S-phase transition. Mol Cell Biol 1999;19:2690–8.
[145] Guttridge DC, Albanese C, Reuther JY, Pestell RG, Baldwin Jr AS. NF-kappaB controls cell growth and differentiation through transcriptional regulation of cyclin D1. Mol Cell Biol 1999;19:5785–99.
[146] Joyce D, Bouzahzah B, Fu M, Albanese C, D'Amico M, Steer J, et al. Integration of Rac-dependent regulation of cyclin D1 transcription through a nuclear factor-kappaB-dependent pathway. J Biol Chem 1999;274: 25245–9.
[147] Beg AA, Sha WC, Bronson RT, Ghosh S, Baltimore D. Embryonic lethality and liver degeneration in mice lacking the RelA component of NF-kappa B. Nature 1995;376:167–70.
[148] Beg AA, Baltimore D. An essential role for NF-kappaB in preventing TNF-alpha-induced cell death. Science 1996;274:782–4.

[149] Van Antwerp DJ, Martin SJ, Verma IM, Green DR. Inhibition of TNF-induced apoptosis by NF-kappa B. Trends Cell Biol 1998;8:107–11.

[150] Rosenfeld ME, Prichard L, Shiojiri N, Fausto N. Prevention of hepatic apoptosis and embryonic lethality in RelA/TNFR-1 double knockout mice. Am J Pathol 2000;156:997–1007.

[151] Li Q, Van Antwerp D, Mercurio F, Lee KF, Verma IM. Severe liver degeneration in mice lacking the IkappaB kinase 2 gene. Science 1999;284:321–5.

[152] Rudolph D, Yeh WC, Wakeham A, Rudolph B, Nallainathan D, Potter J, et al. Severe liver degeneration and lack of NF-kappaB activation in NEMO/IKK-gamma-deficient mice. Genes Dev 2000;14:854–62.

[153] Tanaka M, Fuentes ME, Yamaguchi K, Durnin MH, Dalrymple SA, Hardy KL, et al. Embryonic lethality, liver degeneration, and impaired NF-kappa B activation in IKK-beta-deficient mice. Immunity 1999;10:421–9.

[154] Papa S, Bubici C, Zazzeroni F, Franzoso G. Mechanisms of liver disease: cross-talk between the NF-kappaB and JNK pathways. Biol Chem 2009;390:965–76.

[155] Cressman DE, Greenbaum LE, Haber BA, Taub R. Rapid activation of post-hepatectomy factor/nuclear factor kappa B in hepatocytes, a primary response in the regenerating liver. J Biol Chem 1994;269:30429–35.

[156] FitzGerald MJ, Webber EM, Donovan JR, Fausto N. Rapid DNA binding by nuclear factor kappa B in hepatocytes at the start of liver regeneration. Cell Growth Differ 1995;6:417–27.

[157] Mohn KL, Laz TM, Melby AE, Taub R. Immediate-early gene expression differs between regenerating liver, insulin-stimulated H-35 cells, and mitogen-stimulated Balb/c 3T3 cells. Liver-specific induction patterns of gene 33, phosphoenolpyruvate carboxykinase, and the jun, fos, and egr families. J Biol Chem 1990;265:21914–21.

[158] Tewari M, Dobrzanski P, Mohn KL, Cressman DE, Hsu JC, Bravo R, et al. Rapid induction in regenerating liver of RL/IF-1 (an I kappa B that inhibits NF-kappa B, RelB-p50, and c-Rel-p50) and PHF, a novel kappa B site-binding complex. Mol Cell Biol 1992;12:2898–908.

[159] Sakuda S, Tamura S, Yamada A, Miyagawa J, Yamamoto K, Kiso S, et al. NF-kappaB activation in non-parenchymal liver cells after partial hepatectomy in rats: possible involvement in expression of heparin-binding epidermal growth factor-like growth factor. J Hepatol 2002;36:527–33.

[160] Yang WJ, Zhang QY, Yu ZP, Song QT, Liang HP, Xu X, et al. Effects of nuclear factor-kappaB on rat hepatocyte regeneration and apoptosis after 70% portal branch ligation. World J Gastroenterol 2005;11:6775–9.

[161] Iimuro Y, Nishiura T, Hellerbrand C, Behrns KE, Schoonhoven R, Grisham JW, et al. NFkappaB prevents apoptosis and liver dysfunction during liver regeneration. J Clin Invest 1998;101:802–11.

[162] Yang L, Magness ST, Bataller R, Rippe RA, Brenner DA. NF-kappaB activation in Kupffer cells after partial hepatectomy. Am J Physiol Gastrointest Liver Physiol 2005;289:G530–8.

[163] DeAngelis RA, Markiewski MM, Taub R, Lambris JD. A high-fat diet impairs liver regeneration in C57BL/6 mice through overexpression of the NF-kappaB inhibitor, IkappaBalpha. Hepatology 2005;42:1148–57.

[164] Malato Y, Ehedego H, Al-Masaoudi M, Cubero FJ, Bornemann J, Gassler N, et al. NF-kappaB essential modifier is required for hepatocyte proliferation and the oval cell reaction after partial hepatectomy in mice. Gastroenterology 2012;143:1597–608.

[165] Pahl HL, Krauss B, Schulze-Osthoff K, Decker T, Traenckner EB, Vogt M, et al. The immunosuppressive fungal metabolite gliotoxin specifically inhibits transcription factor NF-kappaB. J Exp Med 1996;183:1829–40.

[166] Plumpe J, Malek NP, Bock CT, Rakemann T, Manns MP, Trautwein C. NF-kappaB determines between apoptosis and proliferation in hepatocytes during liver regeneration. Am J Physiol Gastrointest Liver Physiol 2000;278:G173–83.

[167] Nejak-Bowen KN, Orr AV, Bowen Jr WC, Michalopoulos GK. Gliotoxin-induced changes in rat liver regeneration after partial hepatectomy. Liver Int 2013;33:1044–55.

[168] Anselmi K, Stolz DB, Nalesnik M, Watkins SC, Kamath R, Gandhi CR. Gliotoxin causes apoptosis and necrosis of rat Kupffer cells in vitro and in vivo in the absence of oxidative stress: exacerbation by caspase and serine protease inhibition. J Hepatol 2007;47:103–13.

[169] Hagens WI, Olinga P, Meijer DK, Groothuis GM, Beljaars L, Poelstra K. Gliotoxin non-selectively induces apoptosis in fibrotic and normal livers. Liver Int 2006;26:232–9.

[170] Wright MC, Issa R, Smart DE, Trim N, Murray GI, Primrose JN, et al. Gliotoxin stimulates the apoptosis of human and rat hepatic stellate cells and enhances the resolution of liver fibrosis in rats. Gastroenterology 2001;121:685–98.

[171] Chaisson ML, Brooling JT, Ladiges W, Tsai S, Fausto N. Hepatocyte-specific inhibition of NF-kappaB leads to apoptosis after TNF treatment, but not after partial hepatectomy. J Clin Invest 2002;110:193–202.

[172] Wheeler MD, Yamashina S, Froh M, Rusyn I, Thurman RG. Adenoviral gene delivery can inactivate Kupffer cells: role of oxidants in NF-kappaB activation and cytokine production. J Leukoc Biol 2001;69:622–30.

[173] Maeda S, Kamata H, Luo JL, Leffert H, Karin M. IKK-beta couples hepatocyte death to cytokine-driven compensatory proliferation that promotes chemical hepatocarcinogenesis. Cell 2005;121:977–90.
[174] Abshagen K, Eipel C, Kalff JC, Menger MD, Vollmar B. Loss of NF-kappaB activation in Kupffer cell-depleted mice impairs liver regeneration after partial hepatectomy. Am J Physiol Gastrointest Liver Physiol 2007;292:G1570–7.
[175] Taub R. Blocking NF-kappaB in the liver: the good and bad news. Hepatology 1998;27:1445–6.
[176] Malato Y, Sander LE, Liedtke C, Al-Masaoudi M, Tacke F, Trautwein C, et al. Hepatocyte-specific inhibitor-of-kappaB-kinase deletion triggers the innate immune response and promotes earlier cell proliferation during liver regeneration. Hepatology 2008;47:2036–50.
[177] Koch KS, Maeda S, He G, Karin M, Leffert HL. Targeted deletion of hepatocyte Ikkbeta confers growth advantages. Biochem Biophys Res Commun 2009;380:349–54.
[178] Ringelhan M, Schmid RM, Geisler F. The NF-kappaB subunit RelA/p65 is dispensable for successful liver regeneration after partial hepatectomy in mice. PLoS One 2012;7:e46469.
[179] Laurent S, Horsmans Y, Starkel P, Leclercq I, Sempoux C, Lambotte L. Disrupted NF- kappa B activation after partial hepatectomy does not impair hepatocyte proliferation in rats. World J Gastroenterol 2005;11:7345–50.
[180] DeAngelis RA, Kovalovich K, Cressman DE, Taub R. Normal liver regeneration in p50/nuclear factor kappaB1 knockout mice. Hepatology 2001;33:915–24.
[181] Robinson SM, Mann DA. Role of nuclear factor kappaB in liver health and disease. Clin Sci (Lond) 2010;118:691–705.

CHAPTER

7

Mechanisms of Termination of Liver Regeneration

Chad Walesky, Udayan Apte†*

***Department of Medicine—Genetics Division, Brigham and Women's Hospital, Harvard Medical School, Boston, Massachusetts, USA**

†Department of Pharmacology, Toxicology and Therapeutics, University of Kansas Medical Center, Kansas City, Kansas, USA

7.1 INTRODUCTION

Mechanisms surrounding the control of liver regeneration are of great interest to the scientific community. Understanding these complex mechanisms not only adds to the knowledge of basic liver biology but also provides direction for developing new strategies to combat disease and/or toxic exposure that lead to liver injury and ultimately death. Whereas there are several models to study liver regeneration, partial hepatectomy in rodents remains the model of choice for many investigators. In this model, approximately 60-70% of the rodent (rat or mouse) liver is surgically removed and the remaining lobes are allowed to regenerate. It is known that liver regenerates back to it presurgical mass within approximately 7-10 days. One of the most interesting aspects of this process is that cell proliferation is terminated and liver regeneration stops precisely when the liver has attained it presurgical mass. Much of the work on liver regeneration has focused on mechanisms of initiation of liver regeneration. However, the mechanisms that control the termination of liver regeneration are relatively understudied and remain an unexplored frontier in liver biology. Equally intriguing is the signaling connection between mechanisms of termination of liver regeneration and the mechanisms of pathogenesis of liver cancers. Recent studies indicate that these factors may be common and developing a better understanding of the signaling crosstalk between termination of regeneration and cancer pathogenesis is likely to lead to identification of new therapeutic targets for liver cancers. Studies in the last two decades have identified some pathways contributing to the termination phase of liver regeneration, but there are likely many more that are yet to be discovered (Figure 7.1).

7.2 TRANSFORMING GROWTH FACTOR β

Transforming growth factor β (TGF-β) is a multifunctional cytokine with a broad range of

Liver Regeneration
http://dx.doi.org/10.1016/B978-0-12-420128-6.00007-5

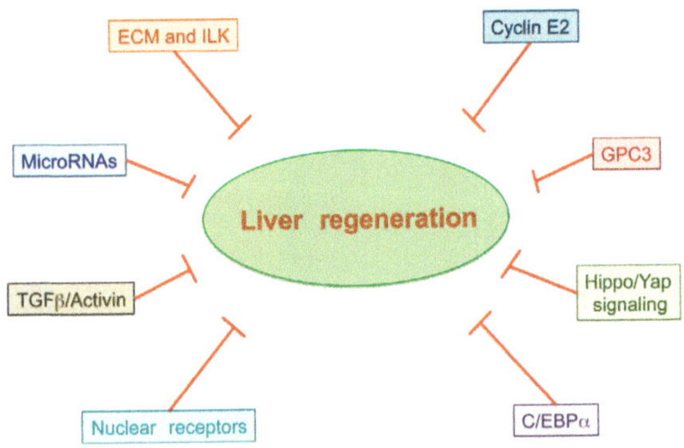

FIGURE 7.1 Various pathways and protein involved in termination of liver regeneration.

effects on development, growth, and homeostasis. TGF-β protein exists in three isoforms (TGF-β 1, 2, and 3) that are secreted as precursor proteins that are later cleaved into active forms and dimerize with a second TGF-β molecule. Three forms of the TGF-β receptor exist on the cell surface, where binding with the active form of TGF-β leads to autophosphorylation of the receptor. Activation of the receptor then leads to phosphorylation of Smad-2 and Smad-3 transcription factors, followed by their dimerization with Smad-4 and translocation to the nucleus [1].

In most cases, TGF-β is a potent inhibitor of cell proliferation. Signaling results in cell-cycle arrest in the G1 phase by stimulating production of the cyclin-dependent kinase (CDK) inhibitor p15 and by inhibiting other cell-cycle regulators, such as CDKs 2 and 4 and cyclins D and E [2]. This leads to cell-cycle arrest at the G1/S transition due to direct effects on Rb phosphorylation leading to E2F sequestration. Because of this, TGF-β was a logical candidate to contribute to the termination of liver regeneration, and early studies suggested that this might be the case. *In vitro* studies utilizing adult rat hepatocytes revealed that treatment with TGF-β in culture inhibited DNA synthesis in a dose-dependent manner [3]. Further, expression of TGF-β mRNA in the liver is increased early after partial hepatectomy in the rat [4,5]. Taken together, these studies suggested that TGF-β might contribute to the termination of liver regeneration. More recent studies, however, suggest otherwise. Hepatocyte-specific disruption of TGF-β II receptor (TβIIR) in conditional knockout mice revealed that disruption of TGF-β signaling enhances the proliferative phase of regeneration but has no effect on the termination of liver regeneration [6]. Liver regeneration after partial hepatectomy (PHX) was faster in TβIIR-KO mice characterized by faster entry of cells in to S-phase. However, by 120 h after PHX both the WT and TβIIR-KO mice showed decrease in cell proliferation. Interestingly, inhibition of Activin A [7] with follistatin resulted in prolonged cell proliferation in TβIIR-KO mice, indicating that TGFβ-Activin A complex may play a role in termination of liver regeneration.

7.3 EXTRACELLULAR MATRIX AND INTEGRIN-LINKED KINASE

The breakdown and remodeling of the extracellular matrix (ECM) is well characterized in regards with the beginning and mid-phase of liver regeneration, including the enzymes that digest components of the ECM leading to the release of various matrix-bound growth factors and initiation of signaling cascades via integrins. At the end of regeneration, there is synthesis of new ECM components to return the liver to normal architecture and reset the process.

There is strong evidence to support that ECM is involved in the termination of liver regeneration though the mechanisms are not well understood. Hepatocytes in culture in the absence of ECM have decreased expression of hepatocyte-specific genes and enhanced proliferation in response to hepatocyte growth factor (HGF) and epidermal growth factor (EGF) [8,9]. Further, addition of ECM, in the form of matrigel and collagen gels, restores hepatocyte-specific gene expression and inhibits the response to growth factors [10]. Therefore, the presence of ECM is important to maintain hepatic identity and growth control.

Communication between hepatocytes and the ECM is thought to be achieved through integrins and their associated integrin-proximal adhesion molecules [11]. The integrin-linked kinase (ILK) signaling complex consists of ILK and its partners Parvin, PINCH, and the PINCH-associated protein RSU1. The ILK-signaling complex is activated by interacting with integrins present in the ECM and is thought to play a key role in suppressing hepatocyte growth [12,13]. Hepatocyte-specific ILK-KO mice (hepILK-KO) lack the ability to transduce extracellular signals from integrins resulting in a constitutively inactive pathway. The hepILK-KO exhibit increased proliferation of hepatocytes and biliary epithelial cells resulting in hepatomegaly [9]. In WT mice, majority of hepatocytes are in quiescence after the postnatal growth period from postnatal days 10 to 30. However, in the hepILK-KO mice, hepatocytes and other cells in the liver remain in cell cycle resulting in a substantially higher liver to body weight ratio. There is also a significant change in the integrin pattern of the liver. Interestingly, after partial hepatectomy, the hepILK-KO mice regenerate their livers much faster and attain the prehepatectomy size by 5 days after PHX. Furthermore, cell proliferation continues in the hepILK-KO mice resulting in almost 60% higher liver size at 14 days after PHX. Thus, deletion of ILK in the liver results in a termination defect in liver regeneration. The increase is due to sustained proliferation at least partially explained by maintained HGF/Met, β-catenin, and Hippo kinase signaling [13].

The above findings highlight the importance of ECM and its components, such as integrins, in the regulation of liver regeneration. Beyond signal transduction mechanisms, ECM is also thought to play an important role in the regulation of liver regeneration by acting as a growth factor and cytokine reservoir. ECM achieves this function by binding inactive forms of growth factors (HGF, HB-EGF, FGFs, etc.) [14].

7.4 GLYPICAN-3

Glypican-3 (GPC3) belongs to a family of cell-surface heparan sulfate proteoglycans and is highly expressed during embryogenesis and organ development [15,16]. GPC3 has limited expression in adult tissues, including lung, ovaries, mammary epithelium, and mesothelium [15,17]. Normally, the liver does not express detectable levels of GPC3, but it has been observed to be highly upregulated in hepatocellular carcinomas and hepatoblastomas [18,19]. Loss-of-function mutations in GPC3 result in overgrowth of many organs (Simpson-Golabi-Behmel syndrome), including the liver, which has been confirmed using GPC3-deficient mice [20–22]. This observation suggests that GPC3 plays a role in growth suppression and organ size control.

A recent study shows that GPC3 protein and mRNA levels are increased in the termination phase of liver regeneration following partial hepatectomy in rats [23]. Further, GPC3 was also upregulated during the end of the growth phase of hepatocytes in culture. Finally, overexpression of GPC3 resulted in decreased proliferation after PHX and delayed liver regeneration. Because of its lack of a signaling domain, GPC3 is thought to function through interactions with various cofactors, such as Hedgehog family members and/or CD81 [23–26].

7.5 ACTIVIN

Activins are members of the TGF-β superfamily having similar structure to TGF-β receptors and sharing signaling pathways. The Activin A type II receptor signals through Smad proteins similar to TβIIR and is known to be upregulated during liver regeneration [7]. Blockage of Activin A pharmacologically (follistatin) results in increased liver regeneration rates and hepatomegaly following partial hepatectomy in rats. These results suggest that Activin signaling plays an important role in the termination of liver regeneration [27].

7.6 C/EBPα

The CCAAT/enhancer-binding protein (C/EBP) family of transcription factors has a well-accepted function in regulating cell-cycle arrest [28]. C/EBPα-KO mice have impaired liver function in terms of glucose homeostasis and die soon after birth [29]. Along with impaired liver function, C/EBPα-KO livers have increased expression of *Jun* and *Myc*, along with the marker for enhanced proliferation, proliferating cell nuclear antigen, though livers do not exhibit hepatomegaly [30]. It is possible that decreased availability of energy because of impaired glucose homeostasis results a decreased ability for hepatocytes to proliferate although they are showing signs of cell-cycle activation. Further studies would need to be conducted to investigate this possibility.

C/EBPα is thought to induce cell-cycle arrest in a variety of ways: stimulation of p21, inhibition of CDK activity, regulation of RB-E2F complexes, inhibition of E2F-mediated transcription [28]. A recent study performed using mice with mutated C/EBPα (S193A; blocks phosphorylation at this site) found that phosphorylation of C/EBPα is necessary for complex formation with histone deacetylase 1 and p300. These mice exhibit dysregulation of liver differentiation/proliferation after birth, liver proliferation after partial hepatectomy, and following liver injury (CCl_4-induced). A key observation was that mice with C/EBPα-S193A do not have proper termination of liver regeneration leading to hepatomegaly following partial hepatectomy or injury induced with CCl_4 [31].

7.7 CYCLIN E1 AND E2

E-type cyclins play an important role in the transition of quiescent cells into the cell cycle [32]. Two E-type cyclins have been previously described, Cyclin E1 and Cyclin E2 [33–35]. Cyclin E1 is known to activate CDK2 inducing S-phase transition and DNA synthesis associated with mitosis. Because of the critical role that E-type cyclins play in cell-cycle progression, it makes sense that they would play an important role in mechanisms initiating liver regeneration. Recent studies using KO mice for Cyclins E1 and E2, as well as CDK2, have helped to reveal a potential role for E-type cyclins in the termination of liver regeneration [36,37].

Mice deficient for Cyclin E1 displayed only a mild delay in liver regeneration following PH. This is due to compensation of Cyclin E1 by Cyclin A, which can also activate CDK2. In contrast, depletion of Cyclin E2 results in accelerated and amplified DNA synthesis along with hepatomegaly following PHX. This is due to an upregulation of Cyclin E1 and increased and prolonged activation of CDK2. The observed increased liver weight is not due to increased mitosis, but rather increased DNA synthesis and hypertrophy. Further, there is an observed increase in polyploidization in regenerating livers deficient in Cyclin E2 suggesting that S-phase is the primarily affected phase instead of the complete cell cycle [36]. Deletion of CDK2 along with Cyclin E2 blocks the observed increase in DNA synthesis suggesting that increased S-phase activity is due to a Cyclin E1/CDK2-dependent effect [37]. It is also observed that Cyclin E2 expression is dependent on the presence of Cyclin E1 and CDK2. Taken together, these data suggest a possible negative feedback loop where Cyclin E1/CDK2 causes cell-cycle progression at the S-phase transition along with expression of Cyclin E2, which may then function to inhibit cell-cycle progression by inhibition of Cyclin E1/CDK2 activity. Therefore, a complex balance of E-type cyclin expression and CDK2 activation may play an important role in liver regeneration where Cyclin E1 promotes regeneration and Cyclin E2 is important in termination.

7.8 NUCLEAR RECEPTORS

As mentioned previously, the ECM is thought to serve as a reservoir for growth factors and other signaling molecules. Many receptor ligands, both endogenous and exogenous, can stimulate increased liver growth and hepatomegaly. It is likely that signaling through transcription factors, such as nuclear hormone receptors, plays a key role in balancing the mechanisms that both initiate and terminate the liver regenerative response [14]. Examples of nuclear receptors thought to potentially play a role in the termination phase of liver regeneration can be found in Table 7.1. Further studies regarding the complex relationship that nuclear receptors play in hepatic proliferation and liver regeneration, especially in regards with the termination phase, will reveal further detail in the interworking of these mechanisms.

TABLE 7.1 Nuclear Receptors Thought to Play a Role in Hepatocyte Proliferation and Mechanisms Involving Different Phases of Liver Regeneration

Name	Symbol	References
Constitutive androstane receptor	CAR	[38,39]
Estrogen receptor	ER	[40]
Peroxisome proliferation receptor alpha	PPARα	[38,41,42]
Peroxisome proliferation receptor gamma	PPARγ	[38,41,42]
Glucocorticoid receptor	GR	[38,43–45]
Liver X rector	LXR	[38,41,42]
Farnesoid X receptor	FXR	[46]
Pregnane X receptor	PXR	[47]
Retinoid X receptor	RXR	[48,49]
Hepatocyte nuclear factor 4 alpha	HNF4α	[50–52]
Thyroid hormone receptor	TR	[38,53,54]
Nuclear receptor REV-ERB	REV-ERB	[38,55]

7.9 HIPPO/YAP SIGNALING PATHWAY

The Hippo kinase pathway is known to be an important growth-suppressive pathway in both *Drosophila* and mammals [56]. Activation of the Hippo kinase pathway results in phosphorylation of Yes-associated protein (Yap). Yap functions as a transcription factor where it is involved in the expression of multiple genes related to cell differentiation and proliferation. Recent studies have highlighted a role of Hippo/Yap signaling in liver growth control, liver regeneration, and cancer pathogenesis. Mice deficient in Mst1/2 show a loss in phosphorylation of Yap resulting in massive overgrowth and ultimately hepatocellular carcinoma in the liver [57,58]. Further, overactivation of GPC3, either by overexpression or elimination of ILK, results in an increase in Yap protein within the nuclei of hepatocytes [13,59–61]. Taken together, these results suggest that Hippo/Yap signaling may play a central role in the determination of liver size control and by extension may be involved in termination of regeneration.

7.10 MicroRNAs 34a AND 23b

Recent evidence suggests that miRNAs play a role in the initiation phase, as well as the termination phase of liver regeneration [62–65]. One study found that miR-23b was downregulated in regenerating rat liver following 70% partial hepatectomy suggesting that it is important in quiescent liver and may play a role in the termination phase of regeneration [64]. Investigators found that Smad3, important in TGF-β and Activin pathway, is a target for miR-23b. They conclude that downregulation of miR-23b during liver regeneration results in the activation of Smad3, which is known to induce apoptosis in response to TGF-β [66–68]. Therefore, miR-23b/Smad3 signaling cascade may play an important role in apoptosis that is present during the termination phase of liver regeneration.

In a separate study, miR-34a was found to be upregulated following partial hepatectomy of rat livers [65]. Therefore, investigators hypothesized that miR-34a is a suppressor of hepatocyte proliferation and might regulate signals important for the termination phase of liver regeneration. It was confirmed that *Inhibin βB (INHBB)*, a gene that encodes for a monomer of the Activin B complex, is a target gene for miR-34a. In this study, researchers show that miR-34a can inhibit *INHBB* directly in the context of liver regeneration. They speculate that this results in an imbalance in the expression of Activin A versus Activin B, which they argue have shown opposite effects on hepatocyte proliferation. Further, they also provide evidence that Met, the receptor for HGF, is another target of miR-34a. Met is inhibited in the presence of miR-34a, which may contribute to the suppression of hepatocyte proliferation during liver regeneration. Regardless, these studies highlight the importance of further investigation into the role of miRs in the mechanisms involved in the termination of liver regeneration. It is not surprising that a delicate balance between signaling pathways is important in this process leaving a great possibility for miRs to contribute to these mechanisms.

7.11 CONCLUSIONS

The mechanisms surrounding liver regeneration are extremely complex and involve orchestrated cascades of events controlled by mitogenic growth factors, cytokines, transcription factors, metabolic changes, and microRNAs [14]. The initiation and proliferation phases of liver regeneration are relatively well studied, while the termination phase of liver regeneration has only recently seen much attention. Several other yet to be tested mechanisms including nutritional signaling, metabolic signaling, and hemostatic signaling may be also involved in termination of regeneration. The exact role that each of the factors outlined in this chapter plays on the regulation of liver regeneration, especially in terms of termination, is not fully clear. Understanding these mechanisms will help us to further understand human diseases such as cancer, as well as give us new ideas for treating acute liver injury and liver failure.

References

[1] Bierie B, Moses HL. Transforming growth factor beta (TGF-beta) and inflammation in cancer. Cytokine Growth Factor Rev 2010;21:49–59.
[2] Alexandrow MG, Moses HL. Transforming growth factor beta and cell cycle regulation. Cancer Res 1995;55:1452–7.
[3] Nakamura T, Tomita Y, Hirai R, Yamaoka K, Kaji K, Ichihara A. Inhibitory effect of transforming growth factor-beta on DNA synthesis of adult rat hepatocytes in primary culture. Biochem Biophys Res Commun 1985;133:1042–50.
[4] Braun L, Mead JE, Panzica M, Mikumo R, Bell GI, Fausto N. Transforming growth factor beta mRNA increases during liver regeneration: a possible paracrine mechanism of growth regulation. Proc Natl Acad Sci USA 1988;85:1539–43.
[5] Nakatsukasa H, Evarts RP, Hsia CC, Thorgeirsson SS. Transforming growth factor-beta 1 and type I procollagen transcripts during regeneration and early fibrosis of rat liver. Lab Invest 1990;63:171–80.
[6] Oe S, Lemmer ER, Conner EA, Factor VM, Leveen P, Larsson J, et al. Intact signaling by transforming growth factor beta is not required for termination of liver regeneration in mice. Hepatology 2004;40:1098–105.
[7] Yasuda H, Mine T, Shibata H, Eto Y, Hasegawa Y, Takeuchi T, et al. Activin A: an autocrine inhibitor of initiation of DNA synthesis in rat hepatocytes. J Clin Invest 1993;92:1491–6.
[8] Block GD, Locker J, Bowen WC, Petersen BE, Katyal S, Strom SC, et al. Population expansion, clonal growth, and specific differentiation patterns in primary cultures of hepatocytes induced by HGF/SF, EGF and TGF alpha in a chemically defined (HGM) medium. J Cell Biol 1996;132:1133–49.
[9] Gkretsi V, Bowen WC, Yang Y, Wu C, Michalopoulos GK. Integrin-linked kinase is involved in matrix-induced hepatocyte differentiation. Biochem Biophys Res Commun 2007;353:638–43.
[10] Rana B, Mischoulon D, Xie Y, Bucher NL, Farmer SR. Cell-extracellular matrix interactions can regulate the

switch between growth and differentiation in rat hepatocytes: reciprocal expression of C/EBP alpha and immediate-early growth response transcription factors. Mol Cell Biol 1994;14:5858–69.

[11] Hynes RO. Integrins: bidirectional, allosteric signaling machines. Cell 2002;110:673–87.

[12] Gkretsi V, Apte U, Mars WM, Bowen WC, Luo JH, Yang Y, et al. Liver-specific ablation of integrin-linked kinase in mice results in abnormal histology, enhanced cell proliferation, and hepatomegaly. Hepatology 2008;48:1932–41.

[13] Apte U, Gkretsi V, Bowen WC, Mars WM, Luo JH, Donthamsetty S, et al. Enhanced liver regeneration following changes induced by hepatocyte-specific genetic ablation of integrin-linked kinase. Hepatology 2009;50: 844–51.

[14] Michalopoulos GK. Principles of liver regeneration and growth homeostasis. Compr Physiol 2013;3:485–513.

[15] Song HH, Filmus J. The role of glypicans in mammalian development. Biochim Biophys Acta 2002;1573:241–6.

[16] Filmus J. Glypicans in growth control and cancer. Glycobiology 2001;11:19R–23R.

[17] Iglesias BV, Centeno G, Pascuccelli H, Ward F, Peters MG, Filmus J, et al. Expression pattern of glypican-3 (GPC3) during human embryonic and fetal development. Histol Histopathol 2008;23:1333–40.

[18] Nakatsura T, Kageshita T, Ito S, Wakamatsu K, Monji M, Ikuta Y, et al. Identification of glypican-3 as a novel tumor marker for melanoma. Clin Cancer Res 2004;10:6612–21.

[19] Luo JH, Ren B, Keryanov S, Tseng GC, Rao UN, Monga SP, et al. Transcriptomic and genomic analysis of human hepatocellular carcinomas and hepatoblastomas. Hepatology 2006;44:1012–24.

[20] Hughes-Benzie RM, Pilia G, Xuan JY, Hunter AG, Chen E, Golabi M, et al. Simpson-Golabi-Behmel syndrome: genotype/phenotype analysis of 18 affected males from 7 unrelated families. Am J Med Genet 1996;66:227–34.

[21] Pellegrini M, Pilia G, Pantano S, Lucchini F, Uda M, Fumi M, et al. Gpc3 expression correlates with the phenotype of the Simpson-Golabi-Behmel syndrome. Dev Dyn 1998;213:431–9.

[22] Cano-Gauci DF, Song HH, Yang H, McKerlie C, Choo B, Shi W, et al. Glypican-3-deficient mice exhibit developmental overgrowth and some of the abnormalities typical of Simpson-Golabi-Behmel syndrome. J Cell Biol 1999;146:255–64.

[23] Liu B, Paranjpe S, Bowen WC, Bell AW, Luo JH, Yu YP, et al. Investigation of the role of glypican 3 in liver regeneration and hepatocyte proliferation. Am J Pathol 2009;175:717–24.

[24] Beckett K, Franch-Marro X, Vincent JP. Glypican-mediated endocytosis of Hedgehog has opposite effects in flies and mice. Trends Cell Biol 2008;18:360–3.

[25] Capurro MI, Xu P, Shi W, Li F, Jia A, Filmus J. Glypican-3 inhibits Hedgehog signaling during development by competing with patched for Hedgehog binding. Dev Cell 2008;14:700–11.

[26] Kim MS, Saunders AM, Hamaoka BY, Beachy PA, Leahy DJ. Structure of the protein core of the glypican Dally-like and localization of a region important for hedgehog signaling. Proc Natl Acad Sci USA 2011;108: 13112–7.

[27] Kogure K, Omata W, Kanzaki M, Zhang YQ, Yasuda H, Mine T, et al. A single intraportal administration of follistatin accelerates liver regeneration in partially hepatectomized rats. Gastroenterology 1995;108:1136–42.

[28] Johnson PF. Molecular stop signs: regulation of cell-cycle arrest by C/EBP transcription factors. J Cell Sci 2005;118:2545–55.

[29] Wang ND, Finegold MJ, Bradley A, Ou CN, Abdelsayed SV, Wilde MD, et al. Impaired energy homeostasis in C/EBP alpha knockout mice. Science 1995;269:1108–12.

[30] Flodby P, Barlow C, Kylefjord H, Ahrlund-Richter L, Xanthopoulos KG. Increased hepatic cell proliferation and lung abnormalities in mice deficient in CCAAT/enhancer binding protein alpha. J Biol Chem 1996;271: 24753–60.

[31] Jin J, Hong IH, Lewis K, Iakova P, Breaux M, Jiang Y, et al. Cooperation of C/EBP family proteins and chromatin remodeling proteins is essential for termination of liver regeneration. Hepatology 2015;61:315–25.

[32] Geng Y, Yu Q, Sicinska E, Das M, Schneider JE, Bhattacharya S, et al. Cyclin E ablation in the mouse. Cell 2003;114:431–43.

[33] Gudas JM, Payton M, Thukral S, Chen E, Bass M, Robinson MO, et al. Cyclin E2, a novel G1 cyclin that binds Cdk2 and is aberrantly expressed in human cancers. Mol Cell Biol 1999;19:612–22.

[34] Koff A, Cross F, Fisher A, Schumacher J, Leguellec K, Philippe M, et al. Human cyclin E, a new cyclin that interacts with two members of the CDC2 gene family. Cell 1991;66:1217–28.

[35] Lauper N, Beck AR, Cariou S, Richman L, Hofmann K, Reith W, et al. Cyclin E2: a novel CDK2 partner in the late G1 and S phases of the mammalian cell cycle. Oncogene 1998;17:2637–43.

[36] Nevzorova YA, Tschaharganeh D, Gassler N, Geng Y, Weiskirchen R, Sicinski P, et al. Aberrant cell cycle progression and endoreplication in regenerating livers of mice that lack a single E-type cyclin. Gastroenterology 2009;137:691–703, e691-6.

[37] Hu W, Nevzorova YA, Haas U, Moro N, Sicinski P, Geng Y, et al. Concurrent deletion of cyclin E1 and cyclin-dependent kinase 2 in hepatocytes inhibits DNA replication and liver regeneration in mice. Hepatology 2014;59:651–60.

[38] Vacca M, D'Amore S, Graziano G, D'Orazio A, Cariello M, Massafra V, et al. Clustering nuclear receptors in liver regeneration identifies candidate modulators of hepatocyte proliferation and hepatocarcinoma. PLoS One 2014;9:e104449.
[39] Columbano A, Ledda-Columbano GM, Pibiri M, Concas D, Reddy JK, Rao MS. Peroxisome proliferator-activated receptor-alpha mice show enhanced hepatocyte proliferation in response to the hepatomitogen 1,4-bis [2-(3,5-dichloropyridyloxy)] benzene, a ligand of constitutive androstane receptor. Hepatology 2001;34:262–6.
[40] Liddle C, Farrell GC. Role of the oestrogen receptor in liver regeneration in the male rat. J Gastroenterol Hepatol 1993;8:524–9.
[41] Vacca M, Degirolamo C, Massafra V, Polimeno L, Mariani-Costantini R, Palasciano G, et al. Nuclear receptors in regenerating liver and hepatocellular carcinoma. Mol Cell Endocrinol 2013;368:108–19.
[42] Gazit V, Huang J, Weymann A, Rudnick DA. Analysis of the role of hepatic PPARgamma expression during mouse liver regeneration. Hepatology 2012;56: 1489–98.
[43] Lien HC, Lu YS, Shun CT, Yao YT, Chang WC, Cheng AL. Differential expression of glucocorticoid receptor in carcinomas of the human digestive system. Histopathology 2008;52:314–24.
[44] Karabelyos C, Dobozy O, Szalai C, Klenjanszki K, Varju K, Hadhazi A, et al. Elevated hepatic glucocorticoid receptor expression during liver regeneration in rats. Pathol Oncol Res 1999;5:107–9.
[45] Mueller KM, Kornfeld JW, Friedbichler K, Blaas L, Egger G, Esterbauer H, et al. Impairment of hepatic growth hormone and glucocorticoid receptor signaling causes steatosis and hepatocellular carcinoma in mice. Hepatology 2011;54:1398–409.
[46] Huang W, Ma K, Zhang J, Qatanani M, Cuvillier J, Liu J, et al. Nuclear receptor-dependent bile acid signaling is required for normal liver regeneration. Science 2006;312:233–6.
[47] Dai G, He L, Bu P, Wan YJ. Pregnane X receptor is essential for normal progression of liver regeneration. Hepatology 2008;47:1277–87.
[48] Imai T, Jiang M, Kastner P, Chambon P, Metzger D. Selective ablation of retinoid X receptor alpha in hepatocytes impairs their lifespan and regenerative capacity. Proc Natl Acad Sci USA 2001;98:4581–6.
[49] Yang X, Guo M, Wan YJ. Deregulation of growth factor, circadian clock, and cell cycle signaling in regenerating hepatocyte RXRalpha-deficient mouse livers. Am J Pathol 2010;176:733–43.
[50] Bonzo JA, Ferry CH, Matsubara T, Kim JH, Gonzalez FJ. Suppression of hepatocyte proliferation by hepatocyte nuclear factor 4alpha in adult mice. J Biol Chem 2012;287:7345–56.
[51] Walesky C, Edwards G, Borude P, Gunewardena S, O'Neil M, Yoo B, et al. Hepatocyte nuclear factor 4 alpha deletion promotes diethylnitrosamine-induced hepatocellular carcinoma in rodents. Hepatology 2013;57:2480–90.
[52] Walesky C, Gunewardena S, Terwilliger EF, Edwards G, Borude P, Apte U. Hepatocyte-specific deletion of hepatocyte nuclear factor-4alpha in adult mice results in increased hepatocyte proliferation. Am J Physiol Gastrointest Liver Physiol 2013;304: G26–37.
[53] Lopez-Fontal R, Zeini M, Traves PG, Gomez-Ferreria M, Aranda A, Saez GT, et al. Mice lacking thyroid hormone receptor Beta show enhanced apoptosis and delayed liver commitment for proliferation after partial hepatectomy. PLoS One 2010;5:e8710.
[54] Pibiri M, Ledda-Columbano GM, Cossu C, Simbula G, Menegazzi M, Shinozuka H, et al. Cyclin D1 is an early target in hepatocyte proliferation induced by thyroid hormone (T3). FASEB J 2001;15:1006–13.
[55] Meier V, Tron K, Batusic D, Elmaouhoub A, Ramadori G. Expression of AFP and Rev-Erb A/Rev-Erb B and N-CoR in fetal rat liver, liver injury and liver regeneration. Comp Hepatol 2006;5:2.
[56] Dong J, Feldmann G, Huang J, Wu S, Zhang N, Comerford SA, et al. Elucidation of a universal size-control mechanism in Drosophila and mammals. Cell 2007;130:1120–33.
[57] Zhou D, Conrad C, Xia F, Park JS, Payer B, Yin Y, et al. Mst1 and Mst2 maintain hepatocyte quiescence and suppress hepatocellular carcinoma development through inactivation of the Yap1 oncogene. Cancer Cell 2009;16:425–38.
[58] Avruch J, Zhou D, Fitamant J, Bardeesy N. Mst1/2 signalling to Yap: gatekeeper for liver size and tumour development. Br J Cancer 2011;104:24–32.
[59] Donthamsetty S, Bhave VS, Kliment CS, Bowen WC, Mars WM, Bell AW, et al. Excessive hepatomegaly of mice with hepatocyte-targeted elimination of integrin linked kinase following treatment with 1,4-bis [2-(3,5-dichaloropyridyloxy)] benzene. Hepatology 2011;53: 587–95.
[60] Donthamsetty S, Bowen W, Mars W, Bhave V, Luo JH, Wu C, et al. Liver-specific ablation of integrin-linked kinase in mice results in enhanced and prolonged cell proliferation and hepatomegaly after phenobarbital administration. Toxicol Sci 2010;113:358–66.
[61] Lin CW, Mars WM, Paranjpe S, Donthamsetty S, Bhave VS, Kang LI, et al. Hepatocyte proliferation and hepatomegaly induced by phenobarbital and 1,4-bis [2-(3,5-dichloropyridyloxy)] benzene is suppressed in

hepatocyte-targeted glypican 3 transgenic mice. Hepatology 2011;54:620–30.

[62] Song G, Sharma AD, Roll GR, Ng R, Lee AY, Blelloch RH, et al. MicroRNAs control hepatocyte proliferation during liver regeneration. Hepatology 2010;51:1735–43.

[63] Marquez RT, Wendlandt E, Galle CS, Keck K, McCaffrey AP. MicroRNA-21 is upregulated during the proliferative phase of liver regeneration, targets Pellino-1, and inhibits NF-kappaB signaling. Am J Physiol Gastrointest Liver Physiol 2010;298: G535–41.

[64] Yuan B, Dong R, Shi D, Zhou Y, Zhao Y, Miao M, et al. Down-regulation of miR-23b may contribute to activation of the TGF-beta1/Smad3 signalling pathway during the termination stage of liver regeneration. FEBS Lett 2011;585:927–34.

[65] Chen H, Sun Y, Dong R, Yang S, Pan C, Xiang D, et al. Mir-34a is upregulated during liver regeneration in rats and is associated with the suppression of hepatocyte proliferation. PLoS One 2011;6:e20238.

[66] He WX, Niu ZY, Zhao SL, Smith AJ. Smad protein mediated transforming growth factor beta1 induction of apoptosis in the MDPC-23 odontoblast-like cell line. Arch Oral Biol 2005;50:929–36.

[67] Kim BC, Mamura M, Choi KS, Calabretta B, Kim SJ. Transforming growth factor beta 1 induces apoptosis through cleavage of BAD in a Smad3-dependent mechanism in FaO hepatoma cells. Mol Cell Biol 2002;22:1369–78.

[68] Yamamura Y, Hua X, Bergelson S, Lodish HF. Critical role of Smads and AP-1 complex in transforming growth factor-beta-dependent apoptosis. J Biol Chem 2000;275:36295–302.

CHAPTER

8

Role of CXC Chemokines in Liver Repair and Regeneration

Christopher M. Freeman, Gregory C. Wilson, Hiroyuki Nojima, Alex B. Lentsch

Department of Surgery, University of Cincinnati Medical Center, Cincinnati, Ohio, USA

8.1 INTRODUCTION

The regenerative properties of the liver have been recognized at least since the eighth century BC. Greek mythology, as recorded and retold by Homer and Hesiod, possesses two stories—Prometheus and Tityus—which indicates that they had knowledge of this unique property of the liver [1–3]. Furthermore, even the Greek word for liver, "hēpar" (after hēpaomai) which means "to repair oneself," shows their knowledge of this regenerative process [4]. However, broader understanding of the full extent of hepatic regeneration did not occur until millennia later. In the 1830s, Cruveilhier and Andral provided the first experimental proof through liver resections in rabbits that was published by Tillmans in 1879 [3,5]. The understanding of liver regeneration was further advanced in 1931 with the introduction of the partial hepatectomy model first described by Higgins and Anderson [6]. The field progressed again in 1967 when Moolten and Bucher demonstrated that circulating humoral factors obtained from mice which underwent partial hepatectomy could induce liver regeneration in mice with normal livers [7]. At the same time our understanding of the cellular kinetics of regeneration improved with the discovery that DNA synthesis starts in hepatocytes at 12-16 h following partial hepatectomy and demonstrates a second peak at 22-24 h following partial hepatectomy [8,9]. These findings prompted a search to identify and isolate hepatocyte growth factors (HGFs). In the 1980s, with the development of techniques to isolate and culture hepatocytes, HGF was simultaneously identified by Nakamura and Michalopoulos in 1984 [10,11]. Shortly thereafter, HGF was purified from rat platelets, human plasma, and rabbit plasma [12–14]. In the 1990s, molecular studies of gene expression cascades in regenerating liver provided evidence that growth factors (HGF, TGFα, and the antiproliferative factor TGFβ) stimulate DNA replication which causes a mitogenic effect resulting in cultured hepatocyte proliferation [15–17]. Interestingly, it was shown that infusion of these growth factors (HGF and TGFα) *in vivo* had only a minimal increase in hepatocyte DNA synthesis which suggested the need for a priming

Liver Regeneration
http://dx.doi.org/10.1016/B978-0-12-420128-6.00008-7

agent [18]. In 1996, Cressman et al. demonstrated that mice deficient in a cytokine not derived from hepatic parenchyma, IL-6, demonstrated defective hepatocyte regeneration [19]. Supporting this finding, Yamada et al. showed that the sequential exposure of hepatocytes first to cytokines then to growth factors was necessary for liver regeneration [20].

While much is now known about the cytokine- and growth factor-mediated events contributing to liver regeneration, much less is known about the factors that fine-tune this response. This chapter will focus on the involvement of a class of small-protein mediators called chemokines and how they function as a rheostat to regulate liver repair and regeneration.

8.2 CLINICAL SCENARIOS AND THEIR ANALOGOUS INJURY MODELS

The liver, being the largest organ in the body, is subject to a myriad of insults and injuries. Furthermore, because of the critical role the liver plays in a multitude of homeostatic processes—detoxification, metabolism, synthesis, and storage—injury or dysfunction results in a supply and demands equation which, if unbalanced, can result in the death of the organism. Injury to the liver can be broadly categorized as physical loss of hepatic tissue or function loss of hepatic tissue.

In the clinical setting, the liver can be injured through trauma from both penetrating injury or from blunt injury, the latter due to the size, mass, and fixed position of the organ undergoing abrupt changes in speed relative to the surrounding tissue and organs. This often results in the need for hepatic resection to stop life-threatening hemorrhage. Additionally, the liver can be host to malignancy or other clinical entities—hemangioma, focal nodular hyperplasia, and adenomas—which can require surgical resection. These patterns of injury may be closely modeled by partial hepatectomy. Rodent models of partial hepatectomy are the most widely used and, in these models, two-thirds of the liver is ligated and surgically excised. Within 1 week the liver mass is restored to its original size and is capable of carrying on the complete list of functional capabilities of the original liver [6]. It is important to note, and will be discussed later, that the remnant liver has minimal inflammation and all cellular effort is directed to regeneration of appropriate functional liver mass. For this reason, partial hepatectomy model is the gold standard for studies of the regenerative process as they are "clean," having few confounding variables such as inflammation or damaged tissue that needs to be cleared.

The liver can further be injured in instances where blood flow is temporarily halted resulting in hepatic ischemia. This can occur in many clinical entities: organ transplantation in which the allograft liver often experiences significant ischemic time to allow for transport between donor and recipient; following periods of profound and prolonged systemic hypotension, the liver can develop "shock liver" from an organismal low flow state. This injury has been successfully reproduced using rodent models of hepatic ischemia/reperfusion (I/R). In these models, a temporary cessation of hepatic blood flow through both the portal vein and hepatic artery, typically to two-thirds of the liver, is induced for a period of time followed by restoration of blood flow and reperfusion of the ischemic liver. In this model, the insult results in significant inflammation, injury, and parenchymal death leading to a loss of functional mass while retaining total physical mass. Thus, the liver undergoes a complex series of events to clear dead tissue, repair any salvageable parenchyma, and regenerate functional mass.

Chemical and pharmacologic injury of the liver is another major clinical problem and is related to the liver being the primary site of detoxification and drug metabolism. Acetaminophen toxicity, as a result of intended or accidental overdose, results in more cases of fulminant hepatic failure (65%) than any other etiology in developed countries [21–24]. In one multicenter US study (22 centers over a 6-year period) with 275 patients fulfilling criteria for acute liver failure (encephalopathy and coagulopathy) from acetaminophen toxicity, 27% patients died, 8% required transplantation, but 65% survived [25]. Similar to I/R injury, acetaminophen toxicity results in loss of functional liver mass from hepatocyte death, and restoration of appropriate function must occur along with reparative processes.

The liver is also subject to injury from both inherited and acquired diseases of biliary stasis. Biliary atresia is a neonatal disease that results in the obliteration of the biliary tree and has an incidence of 1 in 5000 to 1 in 12,000 live births [26]. This injury can be modeled in mice in which rhesus rotavirus (RRV) infection of new born mice leads to an obstructive cholangiopathy [27]. Interestingly, maternal vaccination against RRV has been shown to prevent rotavirus-induced biliary atresia in the murine model [28]. At this time, there is a paucity of research using a biliary atresia model in the study of regeneration. Biliary stasis resulting in hepatic parenchymal injury can also occur outside of the perinatal period as the result of primary biliary cirrhosis which affects the intrahepatic biliary tree, primary sclerosing cholangitis which affects both the intra- and extrahepatic biliary tree, and further mechanical obstructions such as malignancy, operative injury, or benign biliary strictures. Interestingly, elevated levels of HGF have been reported in obstructive pathologies including biliary atresia and primary biliary cirrhosis [29,30]. The well-studied rodent model of bile duct ligation simulates this injury pattern [31,32]. So the processes by which the liver coordinates regeneration in this setting represent an unexplored area.

8.3 GENERAL PRINCIPLES OF LIVER REGENERATION

The process of liver regeneration is dependent on the unique ability of mature hepatocytes to return to the cell replication cycle and proliferate. The transition from mature quiescent hepatocyte to proliferating hepatocyte can be divided into three phases: priming, proliferation, and termination [33,34]. Each phase is discernible from the others by the distinct mediators that predominate during the specific phases: priming phase (cytokines), proliferation (growth factors), and termination (metabolic pathways). Each of these has been reviewed in detail elsewhere [17,33,35,36]. It is important to note that all three phases occur in a seamless and synchronized fashion.

Cytokines mediate the priming phase and provide the impetus for the quiescent hepatocyte to transition from G_0 to G_1 phase. The predominant cytokines for this phase are TNFα and IL-6, which activate the transcription factors, STAT3 and NF-κB [33,37]. The transition of hepatocytes out of the quiescent phase by cytokines then requires the presence of the mitogenic growth factors, HGF and epidermal growth factor (EGF) for the primed hepatocyte to undergo proliferation, that is, to transition from G_1 to the S phase of mitosis [33]. These mitogenic growth factors, most notably HGF, are further upregulated during liver regeneration and are the most important mediators of hepatocyte proliferation [10,11].

After the progression from G_1 to S phase mediated by the mitogens HGF/EGF, phospholipase Cb1 (PLCβ1), phospholipase Cg1 (PLCγ1), phospholipase D1, and phosphoinositide-3-kinase (PI3K) become the predominant mediators of hepatocyte proliferation [38–41]. PLCβ1 and PLCγ1 induce DNA replication, PLCγ1 contributes to cell-cycle progression from G_2 to M phase, and PI3K is an important downstream product of HGF-receptor activation that functions to mediate cyclin D expression and contributes to DNA replication [39,40,42]. Furthermore, HGF-receptor activation induces the upregulation of these prostaglandins through a mitogen-activated protein kinase-dependent production of arachidonic acid through phospholipase A_2 [38]. It is this upregulation of prostaglandins, particularly E2 and F2, that results in increased hepatocyte proliferation [43].

8.4 CHEMOKINES AND THEIR RECEPTORS

Chemokines are a group of chemotactic cytokines originally identified by their ability to attract immune cells to sites of inflammation and cellular injury [44,45]. They are small (8-10 kD), basic, heparin-binding proteins which are secreted by numerous cell types [45,46]. Subsequent to the discoveries of their chemotactic properties, it became apparent that they play important roles in a diverse range of cellular processes including apoptosis, fibrosis, angiogenesis, proliferation, and cytotoxicity [47–50]. There are four subgroups of chemokines—CXC, CC, CX_3C, and C—with the differentiating factor being the configuration of cysteine residues in the amino terminus, where X is any amino acid [51]. The CX_3C subgroup has only one member, fractalkin, and the C subgroup is represented by two members, lymphotactin-alpha and lymphotactin-beta [52], whereas there are numerous CC and CXC chemokines. To date, only members of the CXC subgroup have been found to have functional significance to the process of liver regeneration. Members of the CXC subgroup can be further classified by the presence or absence of a Glu-Leu-Arg (ELR) amino acid motif at the amino terminus of the peptide which determines receptor-binding specificity [53,54].

CXC chemokines exert their effects via G protein-coupled receptors. These receptors consist of an extracellular component for specific ligand binding, a transmembrane segment, and intracellular loops with a C-terminal domain responsible for receptor signaling and internalization [49,55]. There are seven known CXC chemokine receptors in mammals, CXCR1 through CXCR7. Importantly, the receptors CXCR1 and CXCR2 bind only to CXC chemokines containing the amino terminal ELR motif (ELR+). Both CXCR1 and CXCR2 have been linked with regulation of hepatocyte proliferation and regeneration after hepatectomy, acetaminophen-induced liver injury, and hepatic I/R injury [56–61]. Of the receptors that bind ELR− CXC chemokines, only CXCR4 has been implicated as a potential regulator of liver regeneration [61–65].

8.5 ROLES FOR CXC CHEMOKINES IN LIVER REGENERATION

8.5.1 ELR+ Chemokines

The ELR+ CXC chemokines are important mediators in both the immediate inflammatory cascade following hepatic injury and the subsequent reparative and regenerative stages [57]. Interestingly, these chemokines have demonstrated a dose-response relationship with respect to repair and regeneration following acute injury: low concentrations are associated with repair/regeneration, whereas high concentrations are hepatotoxic.

The apparent divergent roles for ELR+ CXC chemokines was initially identified when comparing studies of the hepatic response in two different model systems—partial hepatectomy and I/R injury. In the partial hepatectomy model, the hepatic remnant has minimal inflammation and ELR+ chemokines were found to be upregulated and blockade of the chemokines or the receptor, CXCR2, resulted in decreased regeneration [58]. Demonstrating a similar outcome, *in vitro* hepatocytes treated with ELR+ chemokines exhibited proliferation similar to those cells treated with HGF. In contrast to this, ELR+ chemokines were found to have detrimental effects on hepatocyte proliferation and liver regeneration after I/R injury [56,60]. In contrast to partial hepatectomy, I/R results in significant inflammation and parenchymal damage that require the liver to repair damaged tissue prior to regeneration. This injury response is manifested within 24 h of the insult and is biphasic in nature. The early phase consists of injury caused by oxidative stress and leads to the induction of proinflammatory mediators that lead to and participate in a later phase characterized by an inflammatory cascade and the recruitment of activated neutrophils and their release of reactive oxygen species and proteases which result in hepatocyte injury [66–74]. Liver repairs and regeneration begins at 48 h after the initial insult, lasts up to 5-7 days in the murine model, and is accompanied by increased expression of stathmin and concomitant initiation of hepatocyte proliferation [75]. During this period, the ELR+ CXC chemokines no longer play a proinflammatory role but instead appear to directly affect hepatocyte fate, either proliferation or cell death. *In vivo*, their level of expression increases hundreds- to thousands-fold. *In vitro* treatment of hepatocytes with varying concentrations of ELR+ CXC chemokines indicated a dose relationship—low concentrations reduced baseline cell death *in vitro*, whereas high concentrations increased cytotoxicity. *In vivo*, both genetic knockout and pharmacologic blockade of CXCR2, the primary ELR+ CXC chemokine receptor in rodents, resulted in accelerated liver recovery and increased hepatocyte proliferation associated with increased activation of NF-κB and STAT3 [60]. Thus, in the setting of I/R injury, in which these chemokines are expressed at very high levels, they appear to limit hepatocyte proliferation and the regenerative process.

More recent studies have directly addressed the phenomenon of concentration-dependent effects [76]. Following partial hepatectomy, there is a fivefold increase in ELR+ CXC chemokine expression [58]. In contrast, following I/R injury, ELR+ CXC chemokine expression increases hundreds- to thousands-fold [60]. Thus low concentrations of ELR+ CXC chemokines are associated with hepatic regeneration whereas high concentrations are not. Direct examination of this relationship found that in mice undergoing partial hepatectomy which were given exogenous CXC chemokines that replicated serum levels observed after I/R injury displayed a significant reduction in hepatocyte proliferation and regeneration [76]. Other studies are consistent with these findings. Adenoviral-mediated overexpression of a CXCL1 analog, keratinocyte chemoattractant (KC), in the liver (> 100-fold) resulted in hepatocellular necrosis [72]. In another, using a model of acetaminophen toxicity, low concentrations of ELR+ CXC chemokines were shown to be associated with improved liver regeneration [59]. So clearly there is a concentration or bioavailability effect that dictates the effects of ELR+ CXC chemokines on hepatocyte proliferation and cell death.

A potential explanation for the divergent function-related expression levels or concentrations may be the manner in which these ligands signal. Two receptors for ELR+ CXC chemokines exist, CXCR1 and CXCR2. While most studies have focused on CXCR2, recent studies have also examined the role of CXCR1 in these model systems. The results of these studies indicate that CXCR2 is indeed the primary receptor [56,61]. CXCR2 is constitutively expressed on hepatocytes and can be upregulated in the presence of specific cytokines [59,77]. In contrast, CXCR1 is not constitutively expressed in hepatocytes but can be induced following I/R injury [56,78]. Interestingly, CXCR1 has no impact on hepatocyte proliferation or liver regeneration after partial hepatectomy [79]. In contrast, in the model of I/R injury, genetic knockout or pharmacologic blockade of CXCR1 resulted in modest, but significant decreases in proliferation and liver repair, suggesting an alternative signaling function from CXCR2 [56]. However, when both receptors were blocked, the resulting phenotype was similar to CXCR2 knockout/blockade, indicating that CXCR2 is the dominant signaling pathway operant *in vivo*.

8.5.2 ELR− Chemokines

A developing area of study has centered on the function of ELR− CXC chemokines, which includes SDF-1 and its receptor CXCR4, in the hepatic regenerative response. Studies suggest that this chemokine/receptor system may play an important role in the hepatic response to injury. In a normal physiologic state, the SDF-1/CXCR4 axis functions to isolate progenitor cells to bone marrow endothelium [80]. Following hepatic injury, SDF-1 expression is upregulated resulting in recruitment of hematopoietic stem cells to the liver. This includes CXCR4-expressing hepatocyte progenitor cells, which migrate and subsequently differentiate when exposed to elevated levels of SDF-1 [61,64, 81,82]. Supporting this finding, conditional CXCR4-knockout mice have increased susceptibility to chronic liver injury [61]. In the context of liver repair and regeneration, initial studies suggest that SDF-1 signaling via CXCR4 is detrimental to liver recovery after I/R injury [83]. Pharmacologic blockade of CXCR4 after I/R injury resulted in increased hepatocyte proliferation. In contrast, treatment with SDF-1 decreased hepatocyte proliferation. These findings were not a result of direct signaling of SDF-1/CXCR4 in hepatocytes but found to be related to mobilization of endothelial progenitor cells from the bone marrow [83]. While more research is needed in this area, it appears that ELR− CXC chemokines, such as SDF-1/CXCR4 signaling, may indirectly regulate liver repair and regeneration.

8.6 CXC CHEMOKINES AND HEPATOCYTE EXOSOMES

While the role of cytokines, growth factors, and chemokines in the hepatic regenerative response has been studied for some time, recent work has identified another component of this response that is mediated by direct communication between hepatocytes through the release of exosomes. Exosomes are small (30-100 nm) membrane nanovesicles that form as part of the endosomal pathway, fuse with the plasma membrane, and released into the extracellular environment [84–89]. In this context, cells may sense their environment by taking in macromolecules from the extracellular environment via endocytosis and begin to process this information in early and late endosomes. There is also bidirectional vesicular traffic between endosomes and the Golgi that can result in the addition of proteins and other molecules to the endosome compartment. Subsequently, proteins, mRNA, microRNA (miRNA), and other molecules are sorted into smaller vesicles that pinch off into the lumen of late endosomes, forming multivesicular bodies (MVBs). Exosomes originate from the internal vesicles of MVBs [85,90,91] and are released into the extracellular compartment when the MVBs fuse with the plasma membrane. While the precise molecular mechanisms leading to the formation and release of exosomes is not well understood, there is strong evidence that the overall process is regulated by the lipid mediator ceramide and the neutral sphingomyelinase. Ceramide is produced from the hydrolysis of sphingomyelin by neutral and acid sphingomyelinases [92]. Neutral sphingomyelinase is localized to the cytoplasmic leaflet of plasma and organelle membranes and has been shown to be critical for exosome release, while the acid sphingomyelinase localizes within lysosomes and the extracellular leaflet of the plasma membrane [92]. Exosomes are enriched in ceramide and the release of exosomes is significantly reduced after inhibition of neutral sphingomyelinase [92,93]. Conversely, augmenting ceramide levels induces exosome production and release [94]. Ceramide has also been shown to regulate the sorting and export of miRNAs to exosomes [95–99].

Hepatocytes are known to produce exosomes and circulating hepatocyte exosomes have been implicated as a biomarker for liver injury [100–102]. More recent work has shown that the chemokine receptors, CXCR1 and CXCR2, regulate exosome production and release after hepatic I/R injury [103,104]. CXCR1-knockout mice had significantly fewer circulating exosomes after liver I/R compared to wild-type mice. Conversely, CXCR2-knockout mice had increased numbers of circulating exosomes after liver I/R compared to wild-type controls. Subsequent experiments showed that the observed changes in exosome production in CXCR1- and CXCR2-knockout mice after I/R injury were due to changes in the production and release of exosomes from hepatocytes and not other liver-resident cell types including Kupffer cells, sinusoidal endothelial cells, or liver-recruited neutrophils [104]. The mechanism for decreased exosome production by hepatocytes lacking expression of CXCR1 remains unknown. However, the increase in exosome production and release from hepatocytes lacking CXCR2 was found to be a consequence of increased neutral sphingomyelinase activity and higher intracellular ceramide conentrations [104]. Therefore, it seems that CXCR2 may have direct interactions with neutral sphingomyelinase in the plasma and/or Golgi membranes that regulates the production of ceramide and subsequent exosome formation.

Interestingly, it was discovered that hepatocyte exosomes have proliferative effects on hepatocytes both *in vitro* and *in vivo* [103,104]. The findings that CXCR1-knockout mice have lower production of hepatocyte exosomes and CXCR2-knockout mice have higher production of hepatocyte exosomes after I/R injury are consistent with previous studies demonstrating that knockout of CXCR1 is detrimental and knockout of CXCR2 is beneficial for hepatocyte proliferation and liver regeneration after I/R injury [56,69]. The mechanism of the proliferative effects of hepatocyte exosomes was found to be mediated by the delivery of the machinery necessary to synthesize sphingosine-1-phosphate (S1P) [104]. Several lines of evidence supported this conclusion. First, hepatocyte exosomes were found to be enriched in ceramide, neutral ceramidase, and sphingosine kinase, but not S1P or neutral sphingomyelinase. Second, inhibition of sphingosine kinase in exosomes completely inhibited the proliferative effects, and lastly, exosomes from CXCR1-nullizygous hepatocytes were found to lack any proliferative effects, and these exosomes lacked sphingosine kinase [103].

8.7 CONCLUSION

The liver possesses the unique ability to undergo regeneration. The process is regulated by a complex cascade of events that involves numerous soluble mediators and cellular participants. CXC chemokines, particularly ELR+ CXC chemokines have been found to be important modulators of the reparative and regenerative responses of the liver. Based on the current body of evidence, CXC chemokines appear to have direct effects on hepatocytes to induce proliferation as well as to regulate the production of exosomes which promote hepatocyte proliferation and liver regeneration in a paracrine fashion. Each of these mechanisms may have a great impact in the development of alternative therapeutic modalities for a variety of liver injuries and diseases.

References

[1] Tiniakos DG, Kandilis A, Geller SA. Tityus: a forgotten myth of liver regeneration. J Hepatol 2010;53(2):357–61.
[2] Gilgenkrantz H, Collin de l'Hortet A. New insights into liver regeneration. Clin Res Hepatol Gastroenterol 2011;35(10):623–9.
[3] Power C, Rasko JE. Whither prometheus' liver? Greek myth and the science of regeneration. Ann Intern Med 2008;149(6):421–6.
[4] Steinhoff G. Regenerative medicine: from protocol to patient. 2nd ed. New York City: Springer; 2013.

[5] Tillmanns H. Experimental und anatomische Untersuchgen uber Wundun der Leber und Niere. Virchows Arch 1879;78:437–65.
[6] Higgins GaA RM. Experimental pathology of the liver. I. Restoration of the liver of the white rat following partial surgical removal. Arch Pathol 1931;12:186–202.
[7] Moolten FL, Bucher NL. Regeneration of rat liver: transfer of humoral agent by cross circulation. Science 1967;158(3798):272–4.
[8] Grisham JW. A morphologic study of deoxyribonucleic acid synthesis and cell proliferation in regenerating rat liver; autoradiography with thymidine-H3. Cancer Res 1962;22:842–9.
[9] Rabes HM, Wirsching R, Tuczek HV, Iseler G. Analysis of cell cycle compartments of hepatocytes after partial hepatecomy. Cell Tissue Kinet 1976;9(6):517–32.
[10] Michalopoulos G, Houck KA, Dolan ML, Leutteke NC. Control of hepatocyte replication by two serum factors. Cancer Res 1984;44(10):4414–9.
[11] Nakamura T, Nawa K, Ichihara A. Partial purification and characterization of hepatocyte growth factor from serum of hepatectomized rats. Biochem Biophys Res Commun 1984;122(3):1450–9.
[12] Gohda E, Tsubouchi H, Nakayama H, et al. Purification and partial characterization of hepatocyte growth factor from plasma of a patient with fulminant hepatic failure. J Clin Invest 1988;81(2):414–9.
[13] Nakamura T, Nawa K, Ichihara A, et al. Purification and subunit structure of hepatocyte growth factor from rat platelets. FEBS Lett 1987;224(2):311–6.
[14] Zarnegar R, Michalopoulos G. Purification and biological characterization of human hepatopoietin A, a polypeptide growth factor for hepatocytes. Cancer Res 1989;49(12):3314–20.
[15] Fausto N, Mead JE. Regulation of liver growth: protooncogenes and transforming growth factors. Lab Invest 1989;60(1):4–13.
[16] Haber BA, Mohn KL, Diamond RH, Taub R. Induction patterns of 70 genes during nine days after hepatectomy define the temporal course of liver regeneration. J Clin Invest 1993;91(4):1319–26.
[17] Michalopoulos GK. Liver regeneration: molecular mechanisms of growth control. FASEB J 1990;4(2): 176–87.
[18] Webber EM, Godowski PJ, Fausto N. In vivo response of hepatocytes to growth factors requires an initial priming stimulus. Hepatology 1994;19(2):489–97.
[19] Cressman DE, Greenbaum LE, DeAngelis RA, et al. Liver failure and defective hepatocyte regeneration in interleukin-6-deficient mice. Science 1996;274(5291): 1379–83.
[20] Yamada Y, Kirillova I, Peschon JJ, Fausto N. Initiation of liver growth by tumor necrosis factor: deficient liver regeneration in mice lacking type I tumor necrosis factor receptor. Proc Natl Acad Sci USA 1997;94(4):1441–6.
[21] Bernal W, Auzinger G, Dhawan A, Wendon J. Acute liver failure. Lancet 2010;376(9736):190–201.
[22] Khashab M, Tector AJ, Kwo PY. Epidemiology of acute liver failure. Curr Gastroenterol Rep 2007;9(1):66–73.
[23] Ostapowicz G, Fontana RJ, Schiodt FV, et al. Results of a prospective study of acute liver failure at 17 tertiary care centers in the United States. Ann Intern Med 2002;137 (12):947–54.
[24] Polson J, Lee WM. Etiologies of acute liver failure: location, location, location!. Liver Transpl 2007;13(10):1362–3.
[25] Larson AM, Polson J, Fontana RJ, et al. Acetaminophen-induced acute liver failure: results of a United States multicenter, prospective study. Hepatology 2005;42(6): 1364–72.
[26] Hackam DJGT, Wang KS, Newman KD, Ford HR. Pediatric surgery. In: Brunicardi FC, editor. Schwartz's principles of surgery. New York City: McGraw-Hill; 2010. p. 1438–9.
[27] Tucker RM, Hendrickson RJ, Mukaida N, et al. Progressive biliary destruction is independent of a functional tumor necrosis factor-alpha pathway in a rhesus rotavirus-induced murine model of biliary atresia. Viral Immunol 2007;20(1):34–43.
[28] Bondoc AJ, Jafri MA, Donnelly B, et al. Prevention of the murine model of biliary atresia after live rotavirus vaccination of dams. J Pediatr Surg 2009;44(8):1479–90.
[29] Hsieh CS, Huang LT, Lee SY, et al. Evaluation of hepatocyte growth factor in patients with biliary atresia. J Pediatr Surg 2008;43(7):1333–7.
[30] Shiota G, Okano J, Kawasaki H, et al. Serum hepatocyte growth factor levels in liver diseases: clinical implications. Hepatology 1995;21(1):106–12.
[31] Kountouras J, Billing BH, Scheuer PJ. Prolonged bile duct obstruction: a new experimental model for cirrhosis in the rat. Br J Exp Pathol 1984;65(3):305–11.
[32] Rodriguez-Garay EA. Cholestasis: human disease and experimental animal models. Ann Hepatol 2003;2(4): 150–8.
[33] Fausto N, Campbell JS, Riehle KJ. Liver regeneration. Hepatology 2006;43(2 Suppl. 1):S45–53.
[34] Zimmermann A. Regulation of liver regeneration. Nephrol Dial Transplant 2004;19(Suppl. 4):iv6–iv10.
[35] Michalopoulos GK. Liver regeneration. J Cell Physiol 2007;213(2):286–300.
[36] Taub R. Liver regeneration: from myth to mechanism. Nat Rev Mol Cell Biol 2004;5(10):836–47.
[37] Webber EM, Bruix J, Pierce RH, Fausto N. Tumor necrosis factor primes hepatocytes for DNA replication in the rat. Hepatology 1998;28(5):1226–34.
[38] Adachi T, Nakashima S, Saji S, et al. Mitogen-activated protein kinase activation in hepatocyte growth

factor-stimulated rat hepatocytes: involvement of protein tyrosine kinase and protein kinase C. Hepatology 1996;23(5):1244–53.

[39] Albi E, Rossi G, Maraldi NM, et al. Involvement of nuclear phosphatidylinositol-dependent phospholipases C in cell cycle progression during rat liver regeneration. J Cell Physiol 2003;197(2):181–8.

[40] Coutant A, Rescan C, Gilot D, et al. PI3K-FRAP/mTOR pathway is critical for hepatocyte proliferation whereas MEK/ERK supports both proliferation and survival. Hepatology 2002;36(5):1079–88.

[41] Nebigil CG. Suppression of phospholipase C beta, gamma, and delta families alters cell growth and phosphatidylinositol 4,5-bisphosphate levels. Biochemistry 1997;36(50):15949–58.

[42] Watanabe A, Nakashima S, Adachi T, et al. Changes in the expression of lipid-mediated signal-transducing enzymes in the rat liver after partial hepatectomy. Surg Today 2000;30(7):622–30.

[43] Refsnes M, Dajani OF, Sandnes D, et al. On the mechanisms of the growth-promoting effect of prostaglandins in hepatocytes: the relationship between stimulation of DNA synthesis and signaling mediated by adenylyl cyclase and phosphoinositide-specific phospholipase C. J Cell Physiol 1995;164(3):465–73.

[44] Miller MD, Krangel MS. Biology and biochemistry of the chemokines: a family of chemotactic and inflammatory cytokines. Crit Rev Immunol 1992;12(1–2):17–46.

[45] Oppenheim JJ, Zachariae CO, Mukaida N, Matsushima K. Properties of the novel proinflammatory supergene "intercrine" cytokine family. Annu Rev Immunol 1991;9:617–48.

[46] Baggiolini M. Chemokines in pathology and medicine. J Intern Med 2001;250(2):91–104.

[47] Belperio JA, Keane MP, Arenberg DA, et al. CXC chemokines in angiogenesis. J Leukoc Biol 2000;68(1):1–8.

[48] Charo IF, Ransohoff RM. The many roles of chemokines and chemokine receptors in inflammation. N Engl J Med 2006;354(6):610–21.

[49] Mantovani A, Bonecchi R, Locati M. Tuning inflammation and immunity by chemokine sequestration: decoys and more. Nat Rev Immunol 2006;6(12):907–18.

[50] Strieter RM, Polverini PJ, Kunkel SL, et al. The functional role of the ELR motif in CXC chemokine-mediated angiogenesis. J Biol Chem 1995;270(45):27348–57.

[51] Luster AD. Chemokines–chemotactic cytokines that mediate inflammation. N Engl J Med 1998;338(7):436–45.

[52] Stievano L, Piovan E, Amadori A. C and CX3C chemokines: cell sources and physiopathological implications. Crit Rev Immunol 2004;24(3):205–28.

[53] Clark-Lewis I, Dewald B, Geiser T, et al. Platelet factor 4 binds to interleukin 8 receptors and activates neutrophils when its N terminus is modified with Glu-Leu-Arg. Proc Natl Acad Sci USA 1993;90(8):3574–7.

[54] Hebert CA, Vitangcol RV, Baker JB. Scanning mutagenesis of interleukin-8 identifies a cluster of residues required for receptor binding. J Biol Chem 1991;266(28):18989–94.

[55] Thorpe R, Bacon K, Baggiolini M, et al. Chemokine/chemokine receptor nomenclature. Cytokine 2003;21:48–9.

[56] Clarke C, Kuboki S, Sakai N, et al. CXC chemokine receptor-1 is expressed by hepatocytes and regulates liver recovery after hepatic ischemia/reperfusion injury. Hepatology 2011;53(1):261–71.

[57] Clarke CN, Kuboki S, Tevar A, et al. CXC chemokines play a critical role in liver injury, recovery, and regeneration. Am J Surg 2009;198(3):415–9.

[58] Colletti LM, Green M, Burdick MD, et al. Proliferative effects of CXC chemokines in rat hepatocytes in vitro and in vivo. Shock 1998;10(4):248–57.

[59] Hogaboam CM, Bone-Larson CL, Steinhauser ML, et al. Novel CXCR2-dependent liver regenerative qualities of ELR-containing CXC chemokines. FASEB J 1999;13(12):1565–74.

[60] Kuboki S, Shin T, Huber N, et al. Hepatocyte signaling through CXC chemokine receptor-2 is detrimental to liver recovery after ischemia/reperfusion in mice. Hepatology 2008;48(4):1213–23.

[61] Tsuchiya A, Imai M, Kamimura H, et al. Increased susceptibility to severe chronic liver damage in CXCR4 conditional knock-out mice. Dig Dis Sci 2012;57(11):2892–900.

[62] Dalakas E, Newsome PN, Harrison DJ, Plevris JN. Hematopoietic stem cell trafficking in liver injury. FASEB J 2005;19(10):1225–31.

[63] Ding BS, Cao Z, Lis R, et al. Divergent angiocrine signals from vascular niche balance liver regeneration and fibrosis. Nature 2014;505(7481):97–102.

[64] Hatch HM, Zheng D, Jorgensen ML, Petersen BE. SDF-1alpha/CXCR4: a mechanism for hepatic oval cell activation and bone marrow stem cell recruitment to the injured liver of rats. Cloning Stem Cells 2002;4(4):339–51.

[65] Khurana S, Mukhopadhyay A. Characterization of the potential subpopulation of bone marrow cells involved in the repair of injured liver tissue. Stem Cells 2007;25(6):1439–47.

[66] Colletti LM, Remick DG, Burtch GD, et al. Role of tumor necrosis factor-alpha in the pathophysiologic alterations after hepatic ischemia/reperfusion injury in the rat. J Clin Invest 1990;85(6):1936–43.

[67] Husted TL, Blanchard J, Schuster R, et al. Potential role for IL-23 in hepatic ischemia/reperfusion injury. Inflamm Res 2006;55(5):177–8.

[68] Jaeschke H. Reactive oxygen and ischemia/reperfusion injury of the liver. Chem Biol Interact 1991;79(2):115–36.

[69] Jaeschke H, Farhood A, Smith CW. Neutrophils contribute to ischemia/reperfusion injury in rat liver in vivo. FASEB J 1990;4(15):3355–9.
[70] Kato A, Gabay C, Okaya T, Lentsch AB. Specific role of interleukin-1 in hepatic neutrophil recruitment after ischemia/reperfusion. Am J Pathol 2002;161(5):1797–803.
[71] Mavier P, Guigui B, Preaux AM, et al. In vitro toxicity of hydrogen peroxide against normal vs. tumor rat hepatocytes: role of catalase and of the glutathione redox cycle. Hepatology 1988;8(6):1673–8.
[72] Stefanovic L, Stefanovic B. Mechanism of direct hepatotoxic effect of KC chemokine: sequential activation of gene expression and progression from inflammation to necrosis. J Interferon Cytokine Res 2006;26(10):760–70.
[73] Wanner GA, Ertel W, Muller P, et al. Liver ischemia and reperfusion induces a systemic inflammatory response through Kupffer cell activation. Shock 1996;5(1):34–40.
[74] Wu J, Weiss B. Two divergently transcribed genes, soxR and soxS, control a superoxide response regulon of *Escherichia coli*. J Bacteriol 1991;173(9):2864–71.
[75] Barone S, Okaya T, Rudich S, et al. Distinct and sequential upregulation of genes regulating cell growth and cell cycle progression during hepatic ischemia-reperfusion injury. Am J Physiol Cell Physiol 2005;289 (4):C826–35.
[76] Wilson GC, Kuboki S, Freeman CM, Nojima H, Schuster RM, Edwards MJ, et al. CXC chemokines function as a rheostat for hepatocyte proliferation and liver regeneration. PLoS ONE 2015;10(3):e0120092.
[77] Bone-Larson CL, Hogaboam CM, Evanhoff H, et al. IFN-gamma-inducible protein-10 (CXCL10) is hepatoprotective during acute liver injury through the induction of CXCR2 on hepatocytes. J Immunol 2001;167 (12):7077–83.
[78] Fu W, Zhang Y, Zhang J, Chen WF. Cloning and characterization of mouse homolog of the CXC chemokine receptor CXCR1. Cytokine 2005;31(1):9–17.
[79] Sakai N, Kuboki S, Van Sweringen HL, et al. CXCR1 deficiency does not alter liver regeneration after partial hepatectomy in mice. Transplant Proc 2011;43(5):1967–70.
[80] Peled A, Kollet O, Ponomaryov T, et al. The chemokine SDF-1 activates the integrins LFA-1, VLA-4, and VLA-5 on immature human CD34(+) cells: role in transendothelial/stromal migration and engraftment of NOD/SCID mice. Blood 2000;95(11):3289–96.
[81] Ceradini DJ, Kulkarni AR, Callaghan MJ, et al. Progenitor cell trafficking is regulated by hypoxic gradients through HIF-1 induction of SDF-1. Nat Med 2004;10 (8):858–64.
[82] Kollet O, Shivtiel S, Chen YQ, et al. HGF, SDF-1, and MMP-9 are involved in stress-induced human CD34+ stem cell recruitment to the liver. J Clin Invest 2003;112(2):160–9.
[83] Wilson GC, Freeman CM, Kuethe JW, Quillin RC, 3rd, Nojima H, Schuster R, et al. CXC chemokine receptor-4 signaling limits hepatocyte proliferation after hepatic ischemia-reperfusion in mice. Am J Physiol Gastrointest Liver Physiol 2015;308(8):702–9.
[84] Conde-Vancells J, Rodriguez-Suarez E, Embade N, et al. Characterization and comprehensive proteome profiling of exosomes secreted by hepatocytes. J Proteome Res 2008;7(12):5157–66.
[85] Denzer K, Kleijmeer MJ, Heijnen HF, et al. Exosome: from internal vesicle of the multivesicular body to intercellular signaling device. J Cell Sci 2000;113(Part 19):3365–74.
[86] Harding C, Heuser J, Stahl P. Endocytosis and intracellular processing of transferrin and colloidal gold-transferrin in rat reticulocytes: demonstration of a pathway for receptor shedding. Eur J Cell Biol 1984;35(2):256–63.
[87] Mathivanan S, Ji H, Simpson RJ. Exosomes: extracellular organelles important in intercellular communication. J Proteomics 2010;73(10):1907–20.
[88] Schorey JS, Bhatnagar S. Exosome function: from tumor immunology to pathogen biology. Traffic 2008;9(6):871–81.
[89] Thery C, Zitvogel L, Amigorena S. Exosomes: composition, biogenesis and function. Nat Rev Immunol 2002;2 (8):569–79.
[90] Corrado C, Raimondo S, Chiesi A, et al. Exosomes as intercellular signaling organelles involved in health and disease: basic science and clinical applications. Int J Mol Sci 2013;14(3):5338–66.
[91] Keller S, Sanderson MP, Stoeck A, Altevogt P. Exosomes: from biogenesis and secretion to biological function. Immunol Lett 2006;107(2):102–8.
[92] Trajkovic K, Hsu C, Chiantia S, et al. Ceramide triggers budding of exosome vesicles into multivesicular endosomes. Science 2008;319(5867):1244–7.
[93] Li J, Liu K, Liu Y, et al. Exosomes mediate the cell-to-cell transmission of IFN-alpha-induced antiviral activity. Nat Immunol 2013;14(8):793–803.
[94] Li R, Blanchette-Mackie EJ, Ladisch S. Induction of endocytic vesicles by exogenous C(6)-ceramide. J Biol Chem 1999;274(30):21121–7.
[95] Boon RA, Vickers KC. Intercellular transport of microRNAs. Arterioscler Thromb Vasc Biol 2013;33(2):186–92.
[96] Kogure T, Lin WL, Yan IK, et al. Intercellular nanovesicle-mediated microRNA transfer: a mechanism of environmental modulation of hepatocellular cancer cell growth. Hepatology 2011;54(4):1237–48.
[97] Kosaka N, Iguchi H, Yoshioka Y, et al. Secretory mechanisms and intercellular transfer of microRNAs in living cells. J Biol Chem 2010;285(23):17442–52.
[98] Mittelbrunn M, Gutierrez-Vazquez C, Villarroya-Beltri C, et al. Unidirectional transfer of microRNA-loaded exosomes from T cells to antigen-presenting cells. Nat Commun 2011;2:282.

[99] Wang K, Zhang S, Weber J, et al. Export of microRNAs and microRNA-protective protein by mammalian cells. Nucleic Acids Res 2010;38(20):7248–59.
[100] Freeman CM, Quillin 3rd RC, Wilson GC, et al. Characterization of microparticles after hepatic ischemia-reperfusion injury. PLoS One 2014;9(5):e97945.
[101] Ren XP, Wu J, Wang X, et al. MicroRNA-320 is involved in the regulation of cardiac ischemia/reperfusion injury by targeting heat-shock protein 20. Circulation 2009;119(17):2357–66.
[102] Rittner HL, Labuz D, Richter JF, et al. CXCR1/2 ligands induce p38 MAPK-dependent translocation and release of opioid peptides from primary granules in vitro and in vivo. Brain Behav Immun 2007;21 (8):1021–32.
[103] Nojima H, Wilson GC, Quillin RC, Freeman CM, Schuster RM, Blanchard JA, et al. Hepatocyte exosomes mediate liver repair and regeneration via sphingosine-1-phosphate. Submitted for publication.
[104] Nojima H, Freeman CM, Schuster RM, Lukasz J, Burkhard K, Edwards MJ, et al. CXC chemokine receptor 1 is required for the proliferative effects of the hepatocyte exosomes. Submitted for publication.

CHAPTER

9

Bile Acid Receptors and Liver Regeneration

Mingjie Fan, Lili Ding†,‡, Wendong Huang†*

*Institute of Life Science, Taishan Medical University, Taian, China

†Department of Diabetes and Metabolic Diseases Research, Beckman Research Institute, City of Hope National Medical Center, Duarte, California, USA

‡Shanghai Key Laboratory of Complex Prescriptions and MOE Key Laboratory for Standardization of Chinese Medicines, Institute of Chinese Materia Medica, Shanghai University of Traditional Chinese Medicine, Shanghai, China

9.1 INTRODUCTION

Liver is one of the few organs that can regenerate itself in response to partial ablation or liver injury. Liver regeneration has been widely studied as a paradigm for regenerative organ regrowth since the introduction of a rodent partial hepatectomy (PHx) model in 1931 [1]. Unlike a typically anatomic regeneration, regeneration of the liver is a compensatory hyperplasia of the remaining tissues and is driven by the functional deficit of the organism. Liver regeneration consists of several well-orchestrated phases, with rapid induction of proliferating factors activating the quiescent hepatocytes and priming their subsequent progression through the cell cycles, followed by re-establishment of original liver size and renewed quiescence [2–4]. Growth factors and cytokines are the important early signals to induce the expression of downstream target genes via activation of several key transcription factors [5]. In addition to growth factors and cytokines, metabolic signals are considered as the third major signals during liver regeneration, which is relatively less studied [6]. Recently, bile acids (BAs) and their receptors were identified as key metabolic signaling pathway during liver regeneration and their roles in promoting liver regeneration have received more and more attention [7,4]. In this review, the roles of BA signaling and BA receptors in liver regeneration will be summarized and discussed.

9.2 METABOLIC SIGNALS AND LIVER REGENERATION

Liver regeneration is an adaptive regrowth response induced by specific stimuli and the subsequently sequential changes in gene expression and morphologic reconstruction.

Liver Regeneration
http://dx.doi.org/10.1016/B978-0-12-420128-6.00009-9

It is generally accepted that the remaining hepatocytes are the major cell types that replicate to regenerate liver in the models of 70% PHx or liver injury. Only in specific injury models, activation and replication of liver progenitors are observed when the hepatocytes fail to replicate normally. In addition to hepatocytes, other cell types are actively involved in liver regeneration or repair. Recently, several excellent reviews also highlight the roles of liver stellate cells, liver sinusoidal endothelial cells, and liver stem/progenitor cells in liver regeneration and repair [8–11].

Liver regeneration includes a highly complex network of signal transductions. The essential circuitry required for this process is defined mainly by three major networks: cytokine, growth factor, and metabolic signaling [12]. These three networks subsequently activate specific genes and signaling pathways that are essential for liver regeneration. Compared to the cytokine and growth factor networks, little is known about the roles of metabolic signals in liver regeneration. The identification of several nuclear receptors as receptors for liver metabolites provides novel insight into the roles of metabolic signals in liver regeneration. Among them, the farnesoid X receptor (FXR, NR1H4) is identified as a primary BA receptor [13,14]. In addition to FXR, some other nuclear receptors have been implicated in regulating the liver metabolism, including constitutive androstane receptor (CAR, NR1I3), pregnane X receptor (PXR, NR1I2), vitamin D receptor (VDR, NR1I1), and liver X receptor alpha and beta (LXRα, NR1H3; LXRβ, NR1H2). All these receptors bind to DNA either as a monomer or as a heterodimer with a common partner for nuclear receptors, retinoid X receptor (RXR, NR2B1) to regulate the expression of various genes involved in BA, lipid, glucose, and drug metabolism [15]. Their roles in liver regeneration are also under active investigation. For example, upon PHx, liver regeneration is impaired in mice lacking RXRα in hepatocytes [16]. LXR may suppress liver regeneration after PHx through regulating the cholesterol levels in the liver [17]. CAR activation strongly induces hepatomegaly and may contribute to normal liver regeneration after 70% PHx [7,18]. Dai et al. indicated that PXR is required for normal progression of liver regeneration by modulating lipid homeostasis and regulating hepatocyte proliferation [19]. In contrast, PPARγ acts as a negative regulator of hepatocyte proliferation and may be responsible for the inhibition of liver growth in the late phase of liver regeneration [20]. Actually, direct activation of some of these receptors such as PPARα and CAR results in hepatomegaly quickly, which is distinct from the normal liver regeneration [21]. These results indicate that the increase of different endogenous metabolites and hormones during liver regeneration may activate their individual receptor, which help liver regeneration to different extent. In summary, metabolic signaling is an integrated component of normal liver regeneration. Multiple pathways are working in parallel to contribute to the overall process of liver regeneration (Figure 9.1) [22].

Liver is a major organ for metabolism. Therefore, there is an immense metabolic demand during liver regeneration. The requirement of metabolic signals for liver regeneration has been known for a long time. The question is whether all metabolic signals are required for liver regeneration. The answer may be dependent on the levels of endogenous ligands after 70% PH or liver injury. Although all these nuclear receptors have the potential to be activated and promote liver growth, only those whose endogenous ligands are strongly increased, such as BAs and FXR, will play major roles in contribution to normal liver regeneration. Activation of FXR by its ligand has been shown to promote liver regeneration in aged animals [23], suggesting that further understanding of the mechanism by which nuclear receptors stimulate liver regrowth may provide novel approaches to develop drugs for promotion of liver regeneration. Among

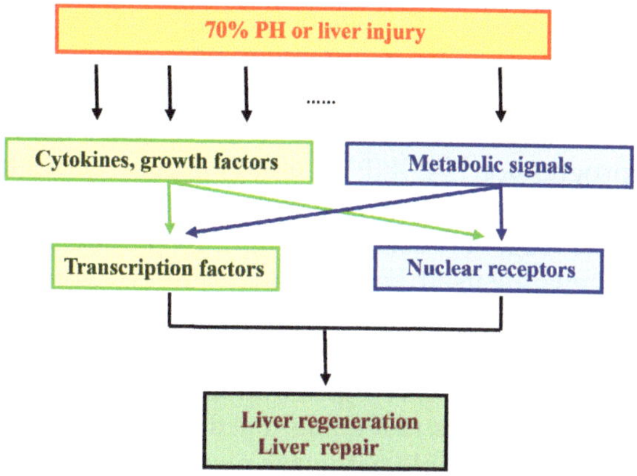

FIGURE 9.1 Multiple signaling pathways are engaged in promoting liver regeneration. In addition to cytokines and growth factors, metabolic signals are generated during liver regeneration, which activate the nuclear receptors to modulate the expression of a specific group of genes involved in liver regeneration and repair.

different metabolic signals, BAs are attractive signals for liver regeneration because the levels of BAs are tightly regulated. BAs are intrinsically toxic and cause liver injury if the levels are not controlled properly. As such, liver resection or injury will generate a BA overload in the liver, which is a potential driving force for liver regeneration [7,4].

9.3 BA SIGNALING AND LIVER REGENERATION

BAs are liver-specific metabolites. They are end products from cholesterol catabolism and are important for nutrition absorption in the intestine, including cholesterol, lipids, and fat-soluble vitamins. BAs are synthesized in the liver and stored in the gall bladder. They are secreted into the intestine when a meal is ingested, but 95% BAs are reabsorbed and transported back to the liver through the portal vein via enterohepatic circulation. Hepatic BAs comprise less than 5% of the total BA pool, and PHx increases bile influx, which rapidly generates a BA overload in the liver. Consistently with this, there is a sharp repression of Cyp7a1 gene expression after 70% PHx or liver injury [24]. Cyp7a1 is the rate-limiting enzyme required for the BA production from cholesterol catabolism. The importance for a stringent control of BA levels is illustrated by a delicate regulation of Cyp7a1 expression. There are many factors and pathways that can regulate the expression of Cyp7a1 gene. A negative feedback loop is identified to regulate BA levels, in which high levels of BA activate FXR to increase the mRNA levels of SHP, which is a negative regulator of *Cyp7a1* gene expression. Moreover, additional regulators of *Cyp7a1* expression are identified, including cytokines, growth factors, and nuclear receptors [25–27]. During liver regeneration, in addition to FXR-SHP axis, hepatocyte growth factor and JNK pathways are involved in suppressing *Cyp7a1* expression during the acute phases of liver regeneration [24]. Furthermore, during the early phase of liver regeneration, MAPK and other pathways may also participate to suppress the expression of *CYP7a1*.

The strong suppression of BA synthesis during liver regeneration indicates a BA overload stress in the liver. This also suggests that BAs may participate in the liver regeneration. Indeed, interruption of normal enterohepatic biliary circulation has been previously known to inhibit liver regeneration [28,29]. Moreover, there is also some direct evidence that BAs are able to stimulate hepatocyte proliferation [30–32]. In a study of 70% PHx mouse model, supplementation with a low dose of BAs promotes liver regeneration, while reduction of BA levels by a BA-binding resin delays liver regeneration [7]. Defective BA signaling causing delayed liver regeneration is also demonstrated in other animal models. In the absence of MRP3, a BA transporter, liver regeneration is delayed in mice due to lower BA concentration in the portal blood [33]. Similarly, deletion of *Cyp27*, an enzyme required for normal BA production and metabolism, results in lower BA pool and defective liver regeneration in mice [34]. In $FXR^{-/-}$ mice, the effect of BAs on promoting liver regeneration is lost [7]. Similarly, in $MRP3^{-/-}$ and $Cyp27^{-/-}$ mice, the delayed liver regeneration is due to impaired FXR activation, suggesting that FXR is the key player to mediate BA effect on liver regeneration. In conclusion, the identification of a novel role of FXR in liver regeneration is key to understand the molecular mechanism by which BAs promote liver regeneration.

9.4 FXR AND LIVER REGENERATION

FXR is highly expressed in the liver, intestine, kidney, and adrenals where the levels of BAs are relatively high [35], but with lower expression in the adipose tissues and heart. FXR is the primary sensor of BAs and both conjugated and unconjugated bile salts are able to activate FXR at physiological concentrations [36,37]. FXR regulates BA homeostasis by regulating genes involved in BA synthesis, secretion, transportation, absorption, conjugation, and detoxification [38–42]. As expected, FXR is also the BA receptor to mediate BA's effect on liver regeneration [7]. FXR is shown to promote liver regeneration after 70% PHx and stimulate liver repair after CCl_4-induced liver injury [43]. Interestingly, different from 70% PHx, there is massive cell death in liver injury model. It was shown that FXR activation specifically upregulated ERK pathways and protected liver cells from apoptosis induced by serum deprivation *in vitro* and fasting *in vivo* [44]. During liver repair after injury, this role of FXR in cell survival may be linked to the activation of STAT3, which is a key factor in cell survival [43]. Therefore, FXR is shown to have a dual role in promoting liver regeneration by both stimulating hepatocyte proliferation and protecting the hepatocyte from death [39,44]. However, the exact downstream events after FXR activation to prevent cell apoptosis are still unclear and needs further investigation.

In addition to metabolic genes, Foxm1b is identified as a FXR direct target gene during liver regeneration [7,23]. Foxm1b is a key cell cycle regulator essential for G1/S and G2/M progression [45,46]. Animal studies indicate that Foxm1b is a key transcription factor in liver regeneration. Although liver can fully regenerate itself, aging dramatically reduces this capacity of liver. Previous studies suggest that aging-induced suppression of liver regeneration is mediated by an epigenetic mechanism and this suppression is reversible. Further studies indicate that multiple pathways may work independently to promote liver regeneration [47,48]. The delayed and reduced proliferative response has been attributed to the decreased expression of some key transcription factors, such as c-Myc and Foxm1b and to the failure of aging-liver to diminish the age-specific C/EBPα-Brm-HDAC1 complex after PHx [49–51]. The complex suppressed Foxm1b induction after PHx through binding to Foxm1b promoter, which results in age-related proliferation defects upon PHx or liver injury [52]. These studies highlight Foxm1b as one of the key regulators in aging-liver regeneration. Loss of Foxm1b function in livers of young mice results in a significant reduction in hepatocyte DNA replication and inhibition of mitosis after PH [49]. More importantly, forced expression of Foxm1b in regenerating livers of old mice is sufficient to restore hepatocyte DNA replication and expression of necessary cell cycle regulatory genes to levels as seen in young animals [51,52]. Defective activation of FXR occurs in aged regenerating livers [23], which may account for the insufficient Foxm1b induction. Compared with young mice, aging mice did not have altered protein levels of FXR and RXR. Therefore, aging may affect the levels of endogenous FXR ligands such as BAs, which could result in defective activation of FXR during liver regeneration. Interestingly, in pregnant mice, loss of FXR results in reduced liver growth, indicating a similar function of FXR in mediating the liver growth during pregnancy [53].

Sirtuin1 (SIRT1) can modulate FXR activity and has an effect on liver regeneration through modulating FXR during liver regeneration [54]. The transgenic mice that overexpress *SIRT1* showed increased mortality, enhanced liver injury, and impaired hepatocyte proliferation after PHx. SIRT1 reduces FXR activities through persistent deacetylation and lower FXR expression. In summary, FXR is a key receptor and transcription factor that specifically mediates the effect of BA signaling to promote liver regeneration.

9.5 INTESTINE-FXR AND LIVER REGENERATION

FXR is highly expressed in both the liver and the intestine. Both hepatic- and intestine-FXR are involved in the regulation of BA homeostasis [55]. One critical FXR target gene in the intestine is *FGF15*. Indeed, several reports suggest that FGF15 secreted from ileum has profound effects on the suppression of *Cyp7a1* gene expression and liver metabolism through FGF receptor-mediated signaling pathways in the liver [56–58]. Suppressed *Cyp7a1* expression and decreased BA synthesis are known to be beneficial for liver regeneration. Therefore, FGF15 induction after liver damage may also contribute to the normal liver regeneration. Indeed, there is significantly delayed liver regeneration and increased liver injury in intestine-specific *FXR* knockout (*ΔIN-FXR*) mice compared to *FXR Fl/Fl* control mice after either 70% PHx or CCl_4 injection [59]. During liver regeneration, FXR also activates the expression of FGF15 in the intestine to suppress *Cyp7a1* transcription [59]. There results identify an unexpected role of intestine-FXR in regulating liver regeneration/repair. First, higher levels of BAs in *ΔIN-FXR* mice after liver injury may hamper the normal liver regeneration/repair. Second, the metabolic and mitogenic activities of FGF15 may contribute to liver regrowth. Third, the hydrophobic BA, deoxycholic acid (DCA), is significantly increased in fecal extracts from *ΔIN-FXR* mice but not from *FXR* KO or liver-*FXR* null mice, and DCA may cause hepatocyte apoptosis [60]. This may also be a protective function of intestine-FXR during liver regeneration/repair. Finally, exogenous delivery of FGF15 rescued the defect of liver repair in *ΔIN-FXR* and *FXR* KO mice [59], suggesting a direct role of FGF15 in promoting liver regeneration.

The direct effect of FGF15 on liver regeneration is also examined by comparing liver regeneration between WT and $FGF15^{-/-}$ mice [61]. As expected, in $FGF15^{-/-}$ mice, liver regeneration is delayed, and there is stronger liver injury in $FGF15^{-/-}$ mice. Furthermore, a recent report indicates that selective activation of intestinal FXR or treating mice with FGF19, a human homolog of murine FGF15, could reduce liver necrosis and inflammatory cell infiltration in cholestasis mouse models [62]. Taken together, intestinal FXR and its induction of FGF15 may have important roles in liver protection.

These data show that both liver- and intestine-FXR contribute to liver regeneration. Hepatic FXR directly induces Foxm1b expression and promotes liver regeneration. In contrast, intestine-FXR activates *FGF15* expression to promote liver regeneration. Therefore, both the cell-autonomous effect of hepatic FXR and the endocrine FGF15 pathway induced by intestine-FXR are required for normal liver regeneration (Figure 9.2).

9.6 TGR5 AND LIVER REGENERATION

TGR5 is a plasma membrane-bound G-protein-coupled BA receptor, which displays varied levels of expression in different tissues [27,63,64]. Hydrophobic BAs, such as lithocholic acid and DCA, are potent endogenous ligands of TGR5. TGR5 regulates BA homeostasis, glucose homeostasis, energy metabolism as well as inflammation [65–70].

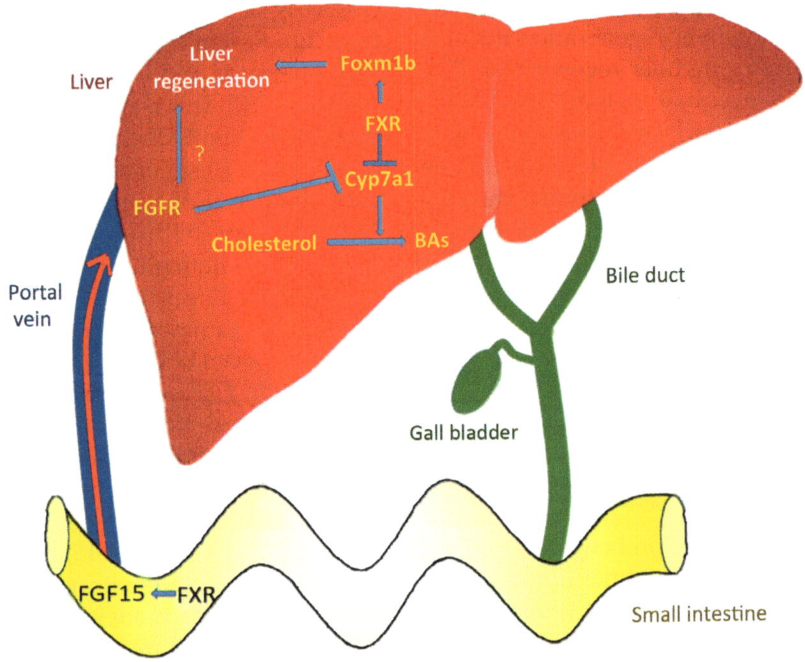

FIGURE 9.2 BA signaling activates FXR to promote liver regeneration. Liver resection or injury increases BA influx, which activates FXR in the liver and intestine. After activation, liver-FXR suppresses *Cyp7a1* to reduce BA production and activates Foxm1b to stimulate liver regeneration. Intestine-FXR is also activated to induce *FGF15* expression, which circulates back to liver through enterohepatic circulation. FGF15 binds to its receptor (FGFR) and initiates a signaling pathway to suppress *Cyp7a1* expression. At the same time, FGF15 also promotes liver regeneration through an unknown mechanism.

A role of TGR5 in liver regeneration was recently identified [71]. After 70% PHx, *TGR5*$^{-/-}$ mice displayed severe hepatocyte necrosis, prolonged cholestasis, exacerbated inflammatory response, and delayed liver regeneration [71]. TGR5 may primarily protect the BA-overloaded remnant liver by controlling BA hydrophobicity and suppressing inflammatory response. Moreover, TGR5 increases BA efflux in urine through kidney. The regulation of potential BA transporters in kidney needs further investigation. The defective BA clearance in the absence of TGR5 thus leads to exacerbated liver toxicity by BAs. These results highlight a distinct role of *TGR5* during liver regeneration.

Interestingly, a recent report indicates that serotonin also helps increase BA secretion through urine and reduces liver toxicity, suggesting that BA excretion through kidney may be an important mechanism for liver protection during liver regeneration [72]. Serotonin is not only a neurotransmitter but also a hormone involved in the initiation of liver regeneration [73,74].

9.7 FXR AND HCC DEVELOPMENT

Hepatocellular carcinoma (HCC) is the most common primary liver cancer and the incidence is rising worldwide largely due to hepatitis B virus and hepatitis C virus infection, alcohol abuse, and the epidemiological obesity-associated NASH [75,76]. *FXR* deficiency mice not only exhibit delayed LR after 70% PH but also demonstrate defective repair ability in the damaged liver [7,77]. Without the roles of FXR in promoting liver repair after injury, liver is prone to enter endless cycles of injury/repair that will keep producing inflammatory cytokines and other growth factors that are potential tumor promotors. Moreover, BAs are known to cause DNA damage and induce cell transformation if their levels are not controlled by FXR. Excessive accumulation of BAs has a cytotoxic effect and is considered an important etiology of tumorigenesis [78]. Therefore, FXR's role in promotion of liver repair could be an intrinsic mechanism to protect liver from carcinogenesis. In addition, FXR was found to participate in regulating hepatic fibrosis [79,80], cholestasis, hepatic inflammation [81,82], and immune response [83,84,43,85]. Studies found that disruption of BA metabolism is the major defect discovered in $FXR^{-/-}$ mice with spontaneous hepatocarcinogenesis [86,87]. Overload of BAs due to the depletion of the *FXR* gene is the causative factor for injury of liver cells, induction of chronic inflammation, enhancement of the cell proliferation, and development of liver tumor [86–88]. The persistently high levels of BA enhanced the inflammation and bile duct proliferation and led to the downregulation of *FXR* expression. Those data indicate that during hepatocarcinogenesis, BAs may function as tumor promoters as well as DNA-damaging initiators [86,89]. Therefore, further delineation of the link between FXR's roles in liver regeneration and hepatocarcinogenesis may provide novel insight into the mechanism of HCC development.

In summary, FXR is also a HCC suppressor. FXR exerts its anti-tumorigenic function via several mechanisms: (1) FXR maintains the normal liver metabolism of BAs, glucose, and lipid; (2) FXR promotes liver regeneration and repair after injury; (3) FXR protects the liver cells from death and enhances cell survival; (4) FXR suppresses hepatic inflammation, thereby preventing inflammatory damage; and (5) FXR can directly increase the expression of some tumor-suppressor genes and represse the transcription of several oncogenes (Figure 9.3) [90].

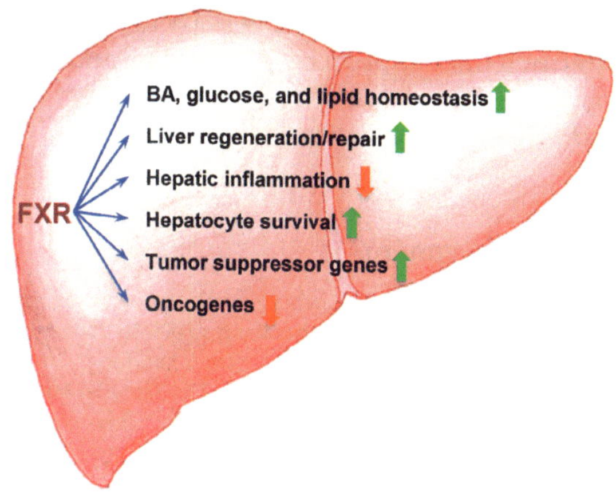

FIGURE 9.3 FXR and antihepatocarcinogenesis. FXR suppresses hepatocarcinogenesis via different mechanisms: (1) it maintains the normal liver homeostasis and metabolism of BAs, glucose, and lipid; (2) it promotes liver regeneration and repair after injury; (3) it suppresses hepatic inflammation; (4) it enhances hepatocyte cell survival; and (5) it increases the expression of some tumor-suppressor genes and represses the transcription of several oncogenes.

9.8 CONCLUSIONS AND PERSPECTIVE

BA signaling is now known as an important metabolic signal during liver regeneration. The novel roles of FXR and TGR5 in promoting liver regeneration are consistent with their defending roles against BA toxicity during liver regeneration. Moreover, there is a close relationship between aberrant liver regeneration and HCC in $FXR^{-/-}$ mice [91–93]. Similarly, abnormal BA homeostasis and cell proliferation in $SHP^{-/-}$ mice also result in HCC development [94,95]. Therefore, further studies on FXR and TGR5 in this new area will provide novel insights into the complex mechanism of liver regeneration, HCC, and other liver diseases. Encouragingly, there is already active research in searching for potent FXR and TGR5 ligands.

Thus, both FXR and TGR5 agonist ligands may offer potential approaches to prevent and treat insufficient liver regeneration as well as HCC and other liver diseases.

Acknowledgments

We apologize to colleagues whose work could not be cited due to space limitations. We thank our previous colleagues, especially Dr. Weidong Chen, Dr. Yandong Wang, Dr. Zhipeng Meng, Dr. Lisheng Zhang, Dr. Xiongfei Huang, Mr. Weiyu Zhao for their contributions. This work is supported by NCI 1R01-CA139158.

References

[1] Higgins GM, Anderson RM. Experimental pathology of the liver. I. Restoration of the liver of the white rat following partial surgical removal. Arch Pathol 1931;12:186–202.

[2] Fausto N. Liver regeneration. J Hepatol 2000;32:19–31.

[3] Taub R. Liver regeneration: from myth to mechanism. Nat Rev Mol Cell Biol 2004;5:836–47.

[4] Michalopoulos GK. Liver regeneration after partial hepatectomy: critical analysis of mechanistic dilemmas. Am J Pathol 2010;176:2–13.

[5] Costa RH, Kalinichenko VV, Holterman AX, Wang X. Transcription factors in liver development, differentiation, and regeneration. Hepatology 2003;38:1331–47.

[6] Mao SA, Glorioso JM, Nyberg SL. Liver regeneration. Transl Res 2014;163(4):352–62.

[7] Huang WD, Ma K, Zhang J, Qatanani M, Cuvillier J, Liu J, et al. Nuclear receptor-dependent bile acid signaling is required for normal liver regeneration. Science 2006;312:233–6.

[8] Diehl AM, Chute J. Underlying potential: cellular and molecular determinants of adult liver repair. J Clin Invest 2013;123:1858–60.

[9] Yin C, Evason KJ, Asahina K, Stainier DY. Hepatic stellate cells in liver development, regeneration, and cancer. J Clin Invest 2013;123:1902–10.

[10] DeLeve LD. Liver sinusoidal endothelial cells and liver regeneration. J Clin Invest 2013;123:1861–6.

[11] Kordes C, Häussinger D. Hepatic stem cell niches. J Clin Invest 2013;123:1874–80.

[12] Chen WD, Wang YD, Meng Z, Zhang L, Huang W. Nuclear bile acid receptor FXR in the hepatic regeneration. Biochim Biophys Acta 2011;1812:888–92.

[13] Makishima M, Okamoto AY, Repa JJ, Tu H, Learned RM, Luk A, et al. Identification of a nuclear receptor for bile acids. Science 1999;284:1362–5.

[14] Parks DJ, Blanchard SG, Bledsoe RK, Chandra G, Consler TG, Kliewer SA, et al. Bile acids: natural ligands for an orphan nuclear receptor. Science 1999;284:1365–8.

[15] Lefebvre P, Benomar Y, Staels B. Retinoid X receptors: common heterodimerization partners with distinct functions. Trends Endocrinol Metab 2010;21:676–83.

[16] Imai T, Jiang M, Kastner P, Chambon P, Metzger D. Selective ablation of retinoid X receptor alpha in hepatocytes impairs their lifespan and regenerative capacity. Proc Natl Acad Sci USA 2001;98:4581–6.

[17] Lo Sasso G, Celli N, Caboni M, Murzilli S, Salvatore L, Morgano A, et al. Down-regulation of the LXR transcriptome provides the requisite cholesterol levels to proliferating hepatocytes. Hepatology 2010;51: 1334–44.

[18] Huang W, Zhang J, Washington M, Liu J, Parant JM, Lozano G, et al. Xenobiotic stress induces hepatomegaly and liver tumors via the nuclear receptor constitutive androstane receptor. Mol Endocrinol 2005;19:1646–53.

[19] Dai G, He L, Bu P, Wan YJ. Pregnane X receptor is essential for normal progression of liver regeneration. Hepatology 2008;47:1277–87.

[20] Yamamoto Y, Ono T, Dhar DK, Yamanoi A, Tachibana M, Tanaka T, et al. Role of peroxisome proliferator-activated receptor-gamma (PPARgamma) during liver regeneration in rats. J Gastroenterol Hepatol 2008;23:930–7.

[21] Columbano A, Ledda-Columbano GM. Mitogenesis by ligands of nuclear receptors: an attractive model for the study of the molecular mechanisms implicated in liver growth. Cell Death Differ 2003;10:S19–21.

[22] Vacca M, Degirolamo C, Massafra V, Polimeno L, Mariani-Costantini R, Palasciano G, et al. Nuclear receptors in regenerating liver and hepatocellular carcinoma. Mol Cell Endocrinol 2013;368:108–19.

[23] Chen WD, Wang YD, Zhang L, Shiah S, Wang M, Yang F, et al. Farnesoid X receptor alleviates age-related proliferation defects in regenerating mouse livers by activating forkhead box m1b transcription. Hepatology 2010;51:953–62.

[24] Zhang L, Huang X, Meng Z, Dong B, Shiah S, Moore DD, et al. Significance and mechanism of CYP7a1 gene regulation during the acute phase of liver regeneration. Mol Endocrinol 2009;23:137–45.

[25] Qiao L, Han SI, Fang Y, Park JS, Gupta S, Gilfor D, et al. Bile acid regulation of C/EBPβ, CREB, and c-Jun function, via the extracellular signal-regulated kinase and c-Jun NH2-terminal kinase pathways, modulates the apoptotic response of hepatocytes. Mol Cell Biol 2003;23:3052–66.

[26] Maruyama T, Miyamoto Y, Nakamura T, Tamai Y, Okada H, Sugiyama E, et al. Identification of membrane-type receptor for bile acids (M-BAR). Biochem Biophys Res Commun 2002;298:714–9.

[27] Kawamata Y, Fujii R, Hosoya M, Harada M, Yoshida H, Miwa M, et al. A G protein-coupled receptor responsive to bile acids. J Biol Chem 2003;278:9435–40.
[28] Ueda J, Chijiiwa K, Nakano K, Zhao G, Tanaka M. Lack of intestinal bile results in delayed liver regeneration of normal rat liver after hepatectomy accompanied by impaired cyclin E-associated kinase activity. Surgery 2002;131:564–73.
[29] Otao R, Beppu T, Isiko T, Mima K, Okabe H, Hayashi H, et al. External biliary drainage and liver regeneration after major hepatectomy. Br J Surg 2012;99:1569–74.
[30] Barone M, Francavilla A, Polimeno L, Ierardi E, Romanelli D, Berloco P, et al. Modulation of rat hepatocyte proliferation by bile salts: in vitro and in vivo studies. Hepatology 1996;23:1159–66.
[31] Zhang J, Huang W, Qatanani M, Evans RM, Moore DD. The constitutive androstane receptor and pregnane X receptor function coordinately to prevent bile acid-induced hepatotoxicity. J Biol Chem 2004;279:49517–22.
[32] Columbano A, Ledda-Columbano GM. Mitogenesis by ligands of nuclear receptors: an attractive model for the study of the molecular mechanisms implicated in liver growth. Cell Death Differ 2003;1:S19–21.
[33] Fernández-Barrena MG, Monte MJ, Latasa MU, Uriarte I, Vicente E, Chang HC, et al. Lack of Abcc3 expression impairs bile-acid induced liver growth and delays hepatic regeneration after partial hepatectomy in mice. J Hepatol 2012;56:367–73.
[34] Meng Z, Liu N, Fu X, Wang X, Wang YD, Chen WD, et al. Insufficient bile acid signaling impairs liver repair in CYP27(−/−) mice. J Hepatol 2011;55:885–95.
[35] Forman BM, Goode E, Chen J, Oro AE, Bradley DJ, Perlmann T, et al. Identification of a nuclear receptor that is activated by farnesol metabolites. Cell 1995;81:687–93.
[36] Wang YD, Chen WD, Moore DD, Huang W. FXR: a metabolic regulator and cell protector. Cell Res 2008;18:1087–95.
[37] Claudel T, Staels B, Kuipers F. The farnesoid X receptor: a molecular link between bile acid and lipid and glucose metabolism. Arterioscler Thromb Vasc Biol 2005;25:2020–30.
[38] Cariou B, Staels B. FXR: a promising target for the metabolic syndrome? Trends Pharmacol Sci 2007;28:236–43.
[39] Sinal CJ, Tohkin M, Miyata M, Ward JM, Lambert G, Gonzalez FJ. Targeted disruption of the nuclear receptor FXR/BAR impairs bile acid and lipid homeostasis. Cell 2000;102:731–44.
[40] Edwards PA, Kast HR, Anisfeld AM. BAREing it all: the adoption of LXR and FXR and their roles in lipid homeostasis. J Lipid Res 2002;43:2–12.
[41] Zhang Y, Castellani LW, Sinal CJ, Gonzalez FJ, Edwards PA. Peroxisome proliferator-activated receptor-gamma coactivator 1alpha (PGC-1alpha) regulates triglyceride metabolism by activation of the nuclear receptor FXR. Genes Dev 2004;18:157–69.
[42] Li J, Pircher PC, Schulman IG, Westin SK. Regulation of complement C3 expression by the bile acid receptor FXR. J Biol Chem 2005;280:7427–34.
[43] Meng Z, Wang Y, Wang L, Jin W, Liu N, Pan H, et al. FXR regulates liver repair after CCl_4-induced toxic injury. Mol Endocrinol 2010;24:886–97.
[44] Wang YD, Yang F, Chen WD, Huang X, Lai L, Forman BM, et al. Farnesoid X receptor protects liver cells from apoptosis induced by serum deprivation in vitro and fasting in vivo. Mol Endocrinol 2008;22:1622–32.
[45] Costa RH. FoxM1 dances with mitosis. Nat Cell Biol 2005;7:108–10.
[46] Laoukili J, Stahl M, Medema RH. FoxM1: at the crossroads of ageing and cancer. Biochim Biophys Acta 2007;1775:92–102.
[47] Iakova P, Awad SS, Timchenko NA. Aging reduces proliferative capacities of liver by switching pathways of C/EBPalpha growth arrest. Cell 2003;113:495–506.
[48] Wang GL, Shi X, Salisbury E, Sun Y, Albrecht JH, Smith RG, et al. Growth hormone corrects proliferation and transcription of phosphoenolpyruvate carboxykinase in livers of old mice via elimination of CCAAT/enhancerbinding protein alpha-Brm complex. J Biol Chem 2007;282:1468–78.
[49] Wang X, Kiyokawa H, Dennewitz MB, Costa RH. The forkhead box m1b transcription factor is essential for hepatocyte DNA replication and mitosis during mouse liver regeneration. Proc Natl Acad Sci USA 2002;99:16881–6.
[50] Wang X, Krupczak-Hollis K, Tan Y, Dennewitz MB, Adami GR, Costa RH. Increased hepatic forkhead box M1B (FoxM1B) levels in old-aged mice stimulated liver regeneration through diminished p27Kip1 protein levels and increased Cdc25B expression. J Biol Chem 2002;277:44310–6.
[51] Wang X, Quail E, Hung NJ, Tan Y, Ye H, Costa RH. Increased levels of forkhead box M1B transcription factor in transgenic mouse hepatocytes prevent age-related proliferation defects in regenerating liver. Proc Natl Acad Sci USA 2001;98:11468–73.
[52] Krupczak-Hollis K, Wang X, Dennewitz MB, Costa RH. Growth hormone stimulates proliferation of old-aged regenerating liver through forkhead box m1b. Hepatology 2003;38:1552–62.
[53] Milona A, Owen BM, Van Mil S, Dormann D, Mataki C, Boudjelal M, et al. The normal mechanisms of pregnancy-induced liver growth are not maintained in mice lacking the bile acid sensor FXR. Am J Physiol Gastrointest Liver Physiol 2010;298:G151–8.
[54] García-Rodríguez JL, Barbier-Torres L, Fernández-Álvarez S, Juan VG, Monte MJ, Halilbasic E, et al. SIRT1 controls liver regeneration by regulating bile acid

metabolism through famesold X receptor and mammalian target of rapamycin signaling. Hepatology 2014;59 (5):1972–83.
[55] Inagaki T, Choi M, Moschetta A, Peng L, Cummins CL, McDonald JG, et al. Fibroblast growth factor 15 functions as an enterohepatic signal to regulate bile acid homeostasis. Cell Metab 2005;2:217–25.
[56] Holt JA, Luo G, Billin AN, Bisi J, McNeill YY, Kozarsky KF, et al. Definition of a novel growth factor-dependent signal cascade for the suppression of bile acid biosynthesis. Genes Dev 2003;17:1581–91.
[57] Gutierrez A, Ratliff EP, Andres AM, Huang X, McKeehan WL, Davis RA. Bile acids decrease hepatic paraoxonase 1 expression and plasma high-density lipoprotein levels via FXR-mediated signaling of FGFR4. Arterioscler Thromb Vasc Biol 2006;26:301–6.
[58] Miao J, Xiao Z, Kanamaluru D, Min G, Yau PM, Veenstra TD, et al. Bile acid signaling pathways increase stability of Small Heterodimer Partner (SHP) by inhibiting ubiquitin-proteasomal degradation. Genes Dev 2009;23:986–96.
[59] Zhang L, Wang YD, Chen WD, Wang X, Lou G, Liu N, et al. Promotion of liver regeneration/repair by farnesoid X receptor in both liver and intestine in mice. Hepatology 2012;56:2336–43.
[60] Kim I, Ahn SH, Inagaki T, Choi M, Ito S, Guo GL, et al. Differential regulation of bile acid homeostasis by the farnesoid X receptor in liver and intestine. J Lipid Res 2007;48:2664–72.
[61] Uriarte I, Fernandez-Barrena MG, Monte MJ, Latasa MU, Chang HC, Carotti S, et al. Identification of fibroblast growth factor 15 as a novel mediator of liver regeneration and its application in the prevention of post-resection liver failure in mice. Gut 2013;62:899–910.
[62] Modica S, Petruzzelli M, Bellafante E, Murzilli S, Salvatore L, Celli N, et al. Selective activation of nuclear bile acid receptor FXR in the intestine protects mice against cholestasis. Gastroenterology 2012;142:355–65.
[63] Keitel V, Reinehr R, Gatsios P, Rupprecht C, Gorg B. The G-protein coupled bile salt receptor TGR5 is expressed in liver sinusoidal endothelial cells. Hepatology 2003;45:695–704.
[64] Vassileva G, Golovko A, Markowitz L, Abbondanzo SJ, Zeng M. Targeted deletion of Gpbar1 protects mice from cholesterol gallstone formation. Biochem J 2006;398:423–30.
[65] Watanabe M, Houten SM, Mataki C, Christoffolete MA, Kim BW. Bile acids induce energy expenditure by promoting intracellular thyroid hormone activation. Nature 2006;439:484–9.
[66] Thomas C, Gioiello A, Noriega L, Strehle A, Oury J. TGR5-mediated bile acid sensing controls glucose homeostasis. Cell Metab 2009;10:167–77.
[67] Maruyama T, Tanaka K, Suzuki J, Miyoshi H, Harada N. Targeted disruption of G protein-coupled bile acid receptor 1 (Gpbar1/M-Bar) in mice. J Endocrinol 2006;191: 197–205.
[68] Keitel V, Donner M, Winandy S, Kubitz R, Haussinger D. Expression and function of the bile acid receptor TGR5 in Kupffer cells. Biochem Biophys Res Commun 2008;372:78–84.
[69] Wang YD, Chen WD, Yu D, Forman BM, Huang W. The G-protein-coupled bile acid receptor, Gpbar1 (TGR5), negatively regulates hepatic inflammatory response through antagonizing nuclear factor kappa light-chain enhancer of activated B cells (NF-kappaB) in mice. Hepatology 2011;54:1421–32.
[70] Pols TW, Nomura M, Harach T, Lo Sasso G, Oosterveer MH. TGR5 activation inhibits atherosclerosis by reducing macrophage inflammation and lipid loading. Cell Metab 2011;14:747–57.
[71] Péan N, Doignon I, Garcin I, Besnard A, Julien B, Liu B, et al. The receptor TGR5 protects the liver from bile acid overload during liver regeneration in mice. Hepatology 2013;58:1451–60.
[72] Jang JH, Rickenbacher A, Humar B, Weber A, Raptis DA, Lehmann K, et al. Serotonin protects mouse liver from cholestatic injury by decreasing bile salt pool after bile duct ligation. Hepatology 2012;56:209–18.
[73] Lesurtel M, Graf R, Aleil B, Walther DJ, Tian Y, Jochum W, et al. Platelet-derived serotonin mediates liver regeneration. Science 2006;312:104–7.
[74] Matondo RB, Punt C, Homberg J, Toussaint MJ, Kisjes R, Korporaal SJ, et al. Deletion of the serotonin transporter in rats disturbs serotonin homeostasis without impairing liver regeneration. Am J Physiol Gastrointest Liver Physiol 2009;296:G963–8.
[75] Fuchs M. Non-alcoholic fatty liver disease: the bile acid-activated farnesoid X receptor as an emerging treatment target. J lipids 2012;2012:934396.
[76] Elsharkawy AM, Mann DA. Nuclear factor-kappaB and the hepatic inflammation-fibrosis-cancer axis. Hepatology 2007;46(2):590–7.
[77] Borude P, Edwards G, Walesky C, Li F, Ma X, Kong B, et al. Hepatocyte-specific deletion of farnesoid X receptor delays but does not inhibit liver regeneration after partial hepatectomy in mice. Hepatology 2012;56(6):2344–52.
[78] Bernstein C, Holubec H, Bhattacharyya AK, Nguyen H, Payne CM, Zaitlin B, et al. Carcinogenicity of deoxycholate, a secondary bile acid. Arch Toxicol 2011;85 (8):863–71.
[79] Li J, Kuruba R, Wilson A, Gao X, Zhang Y, Li S. Inhibition of endothelin-1-mediated contraction of hepatic stellate cells by FXR ligand. PLoS One 2010;5(11):e13955.
[80] Fiorucci S, Antonelli E, Rizzo G, Renga B, Mencarelli A, Riccardi L, et al. The nuclear receptor SHP mediates

inhibition of hepatic stellate cells by FXR and protects against liver fibrosis. Gastroenterology 2004;127(5): 1497–512.
[81] Wang YD, Chen WD, Wang M, Yu D, Forman BM, Huang W. Farnesoid X receptor antagonizes nuclear factor kappaB in hepatic inflammatory response. Hepatology 2008;48(5):1632–43.
[82] Nijmeijer RM, Gadaleta RM, van Mil SW, van Bodegraven AA, Crusius JB, Dijkstra G, et al. Farnesoid X receptor (FXR) activation and FXR genetic variation in inflammatory bowel disease. PLoS One 2011;6(8):e23745.
[83] Vavassori P, Mencarelli A, Renga B, Distrutti E, Fiorucci S. The bile acid receptor FXR is a modulator of intestinal innate immunity. J Immunol 2009;183(10):6251–61.
[84] Mencarelli A, Renga B, Migliorati M, Cipriani S, Distrutti E, Santucci L, et al. The bile acid sensor farnesoid X receptor is a modulator of liver immunity in a rodent model of acute hepatitis. J Immunol 2009;183 (10):6657–66.
[85] Inagaki T, Moschetta A, Lee YK, Peng L, Zhao G, Downes M, et al. Regulation of antibacterial defense in the small intestine by the nuclear bile acid receptor. Proc Natl Acad Sci USA 2006;103(10):3920–5.
[86] Yang F, Huang X, Yi T, Yen Y, Moore DD, Huang W. Spontaneous development of liver tumors in the absence of the bile acid receptor farnesoid X receptor. Cancer Res 2007;67(3):863–7.
[87] Kim I, Morimura K, Shah Y, Yang Q, Ward JM, Gonzalez FJ. Spontaneous hepatocarcinogenesis in farnesoid X receptor-null mice. Carcinogenesis 2007;28(5):940–6.
[88] Wolfe A, Thomas A, Edwards G, Jaseja R, Guo GL, Apte U. Increased activation of the Wnt/beta-catenin pathway in spontaneous hepatocellular carcinoma observed in farnesoid X receptor knockout mice. J Pharmacol Exp Ther 2011;338(1):12–21.
[89] Lozano E, Sanchez-Vicente L, Monte MJ, Herraez E, Briz O, Banales JM, et al. Cocarcinogenic effects of intrahepatic bile acid accumulation in cholangiocarcinoma development. Mol Cancer Res 2014;12:91–100.
[90] Huang X, Zhao W, Huang W. FXR and liver carcinogenesis. Acta Pharmacol Sin 2014;36(2015):37–43.
[91] Liu N, Meng Z, Lou G, Zhou W, Wang X, Zhang Y, et al. Hepatocarcinogenesis in FXR −/− mice mimics human HCC progression that operates through HNF1α regulation of FXR expression. Mol Endocrinol 2012;26: 775–85.
[92] Pan H, Fu X, Huang W. Molecular mechanism of liver cancer. Anticancer Agents Med Chem 2011;11: 493–9.
[93] Wang X, Fu X, Van Ness C, Meng Z, Ma X, Huang W. Bile acid receptors and liver cancer. Curr Pathobiol Rep 2013;1:29–35.
[94] Song KH, Ellis E, Strom S, Chiang JY. Hepatocyte growth factor signaling pathway inhibits cholesterol 7alpha-hydroxylase and bile acid synthesis in human hepatocytes. Hepatology 2007;46:1993–2002.
[95] Zhang Y, Xu P, Park K, Choi Y, Moore DD, Wang L. Orphan receptor small heterodimer partner suppresses tumorigenesis by modulating cyclin D1 expression and cellular proliferation. Hepatology 2008;48:289–98.

WHAT SONGS ARE ON YOUR PLAYLIST?
"TORN BETWEEN TWO LIVERS," "CRY ME A LIVER," "LIVER OF DREAMS," "50 WAYS TO LEAVE YOUR LIVER," "MOON LIVER," "LIVER LET DIE..."
© Mark Weinstein

CHAPTER

10

Role of Developmental Morphogens in Liver Regeneration

Mariana Verdelho Machado, Anna Mae Diehl

Division of Gastroenterology, Duke University, Durham, North Carolina, USA

ABBREVIATIONS

α-SMA	α-smooth muscle actin
BBS	Bardet-Biedl syndrome proteins
Cdo	CAM-related downregulated by oncogenes
Ci	Cubitus interruptus
Dhh	Desert Hedgehog
EMT	epithelial to mesenchymal transition
Gli	glioma-associated oncogene homolog
GSK	glycogen synthase kinase
Hip	Hedgehog-interacting protein
HSC	hepatic stellate cell
IFP	intraflagellar transport proteins
Ihh	Indian Hedgehog
MSCs	mesenchymal stem cells
PC	primary cilium
Ptc	Patched
Shh	Sonic Hedgehog
Smo	Smoothened
TGF	transforming growth factor

10.1 INTRODUCTION

The liver is unique with respect to other tissues/organs because of its tremendous ability to regenerate. In rodents, the organ can recover its initial mass in less than a week after partial hepatectomy removes two thirds of the liver [1]. In humans, however, the liver is not always able to reacquire its preoperative size [2]. After removal of liver mass, the otherwise healthy liver reacts with compensatory hyperplasia, i.e., the remaining liver cells proliferate to replace the lost mass. The dogma is that liver regenerates by the replication of mature epithelial cells that were in replicative arrest before the insult [3]. However, more recently, it has been shown that progenitors also play a role in this "injury-free regeneration," [4] particularly when there is massive loss of liver parenchyma [5]. Other cell types have proved to be crucial in the regeneration process. For example, hepatic stellate cells (HSCs) not only send growth signals to progenitor cells, but may even act as progenitors themselves by transitioning from a more mesenchymal to a more epithelial phenotype [6,7] via a process dubbed mesenchymal to epithelial transition [8].

In the chronically injured liver, hepatocytes are not able to replicate, because they suffer "damage-induced senescence." [9] In this context, injury evokes a wound-healing response,

Liver Regeneration
http://dx.doi.org/10.1016/B978-0-12-420128-6.00010-5

and replacement of dying cells is highly dependent on progenitors. Progenitor cells are more resistant to senescence-inducing stimuli, such as transforming growth factor (TGF)-β [10,11] and more responsive to proliferative stimuli, such as Hedgehog ligands [12]. Dying hepatocytes release damage-associated molecules, including Hedgehog ligands, which orchestrate the healing process. The Hedgehog pathway has several functions in Hedgehog-responsive cells that are crucial for regenerative wound-healing. It exerts antiapoptotic and proliferative effects on progenitor cells. Indeed, progenitor populations expand or shrink according to the availability of Hedgehog ligands [12]. Hedgehog also acts on HSCs, promoting HSC activation to a more mesenchymal/myofibroblastic (MF) phenotype through epithelial to mesenchymal transition (EMT) [8]. MF-HSCs promote regeneration and confine the injurious stimuli by producing a collagen-enriched extracellular matrix. The Hedgehog pathway also promotes chemotaxis and activation of inflammatory cells, such as NKT cells and macrophages [13,14], directly or through induction of proinflammatory molecules such as osteopontin [15]. Inflammatory cells not only fight injurious agents but also remove damaged cells. The importance of these Hedgehog-driven responses for liver regeneration is demonstrated by evidence that liver regeneration is highly compromised when the Hedgehog pathway is inhibited [16]. However, activation of the Hedgehog pathway must be highly regulated to assure effective regeneration. When injurious stimuli become chronic or the pathway is deregulated, Hedgehog signaling does not shut down and defective repair ensues. Excessive HSC activation promotes liver fibrosis, over-activation of the immune cell response exacerbates liver injury, and overly exuberant progenitor growth may promote carcinogenesis. This chapter will summarize the Hedgehog pathway and its role in liver regeneration.

10.2 OVERVIEW OF THE HEDGEHOG PATHWAY

10.2.1 Understanding Hedgehog Ligands Physiology

Nüsslein-Volhard and Wieschaus were the first to identify Hedgehog using a genetic screen in *Drosophila melanogaster* [17]. The name Hedgehog was coined because Hedgehog-null fly embryos displayed a layer of disorganized hair-like bristles, resembling the homonymous mammal. Hedgehog is a morphogen, i.e., it regulates development and cell fate with differential responses according to its concentration. Hedgehog is released from producing cells and diffuses in the extracellular matrix. Cells closer to the ones that release Hedgehog are exposed to higher concentrations [18].

In mammals, there are three Hedgehog ligand counterparts: Sonic Hedgehog (Shh), Indian Hedgehog (Ihh), and Desert Hedgehog (Dhh). Dhh expression is restricted to the nervous tissue and testis, whereas Shh and Ihh are widely expressed. Shh is the predominant ligand expressed in the proximal gut and Ihh in the hindgut [19].

After translation, Hedgehog proteins experience several modifications, including cleavage and lipidations. The 45 kDa precursor undergoes auto-catalytic cleavage via auto-proteolytic activity of its C-terminal domain. The C-terminal domain also has cholesterol transferase activity which promotes covalent attachment of cholesterol to the newly generated N-terminal fragment. Afterward, the C-terminus is degraded in the proteasome, leaving the N-terminal peptide. The latter undergoes a second lipidation with attachment of palmitic acid via a reaction catalyzed by skinny Hedgehog (SKI), a member of the membrane-bound *O*-acyltransferase protein family [18]. These lipidations are very important for ligand activity. Cholesterol modification promotes retention of ligand in plasma membrane lipid rafts, and palmitoylation is required for full ligand full activity and long-distance movement [20]. Both lipid modifications are required for multimerization. Hedgehog can be released from producing cells through three mechanisms: Dispatched protein-facilitated process, assembly in very low-density lipoprotein particles, or released from the cell surface in exovesicles [20]. Diffusion of Hedgehog requires strategies to adapt to the hydrophilic extracellular matrix: assembly in soluble multimeric structures and formation of complexes with lipoproteins. Glypicans not only facilitate the latter, they also promote internalization by the receptor [21].

10.2.2 Hedgehog Pathway in Target Cells

In Hedgehog-responsive target cells, Hedgehog ligand binds to its receptor, Patched (Ptc). Ligand-receptor interaction is enhanced by several coreceptors: CAM-related downregulated by oncogenes (Cdo), brother of Cdo, and growth arrest-specific [18]. There are two Patch receptors, Patch 1 and 2; however, downstream signaling seems to be regulated mainly by Patch 1 [22]. Ptc is a transmembrane protein that constitutively represses a protein called Smoothened (Smo). When Hedgehog binds to Ptc, it eliminates Ptc repressor activity on Smo [18]. Activation of Smo regulates the activity of zinc-finger transcriptional factors, Cubitus interruptus (Ci) in *Drosophila* and the vertebrate equivalent, glioma-associated oncogene homologs (Gli) (Figure 10.1).

In the absence of Hedgehog ligands, Ci undergoes sequential serine phosphorylations by protein kinase A, glycogen synthase kinase (GSK)-3β, and different members of caseine kinase 1 family. Phosphorylation promotes binding of Ci to β-transducin repeat-containing protein that primes it for ubiquitination and targeting to the proteasome for partial degradation. This generates a Ci N-terminal, which translocate to the nucleus, repressing transcription (Ci-R). The mechanisms by which Ci/Gli-R suppress transcription of target genes is not fully understood. Recent studies in *Drosophila* and zebrafish suggest that this process may involve histone deacetylation. Ci/Gli-R binds to Atrophin, a transcriptional corepressor, allowing recruitment of histone deacetylase Rpd3 [23]. When Smo is activated by Hedgehog ligand-Ptc interaction, it blocks Ci phosphorylation and subsequent proteosomal degradation. This allows the full-length Ci protein to enter the nucleus and act as a transcriptional activator through the binding of Gli-binding motif [24] (Figure 10.1). Important target genes are vascular endothelial growth factor, angiopoietin 1 and 2 (endothelial cells); snail, twist-2, FoxF1, α-smooth muscle actin (α-SMA), vimentin, interleukin-6 (fibroblasts/myofibroblasts); and Sox-2 and -9 and Nanog (stem/progenitor cells) [19].

Mammals and flies exhibit some differences in the Hedgehog pathway. In mammals, there are three Gli protein counterparts to *Drosophila* Ci, Gli-1, Gli-2, and Gli-3. Also, for Gli-inhibitory phosphorylation to occur, Gli must couple to a suppressor protein complex composed by fused kinase, suppressor of fused, and Costal-2 [19,25]. The three Gli proteins also behave differently: Gli-1 does not have a repressor domain, and it is not proteolytically processed; Gli-3 acts mainly as a transcriptional repressor because the majority of full-length Gli-3 protein is proteolytically processed; and Gli-2 acts mainly as a transcriptional activator although it can have a weak repressing activity. Gli-2 proteolytical processing to form transcriptional repressor is extremely inefficient, and Gli-2 full-length protein is readily degraded. Hedgehog ligand signaling suppresses both processing and degradation of Gli-2. Thus, Hedgehog signaling typically results in accumulation of full-length Gli-2 [26]. The resistant full-length proteins undergo additional processing steps to convert them in transcription activators (Gli-A, CiA). The mechanisms involved are not fully understood but may require phosphorylation or

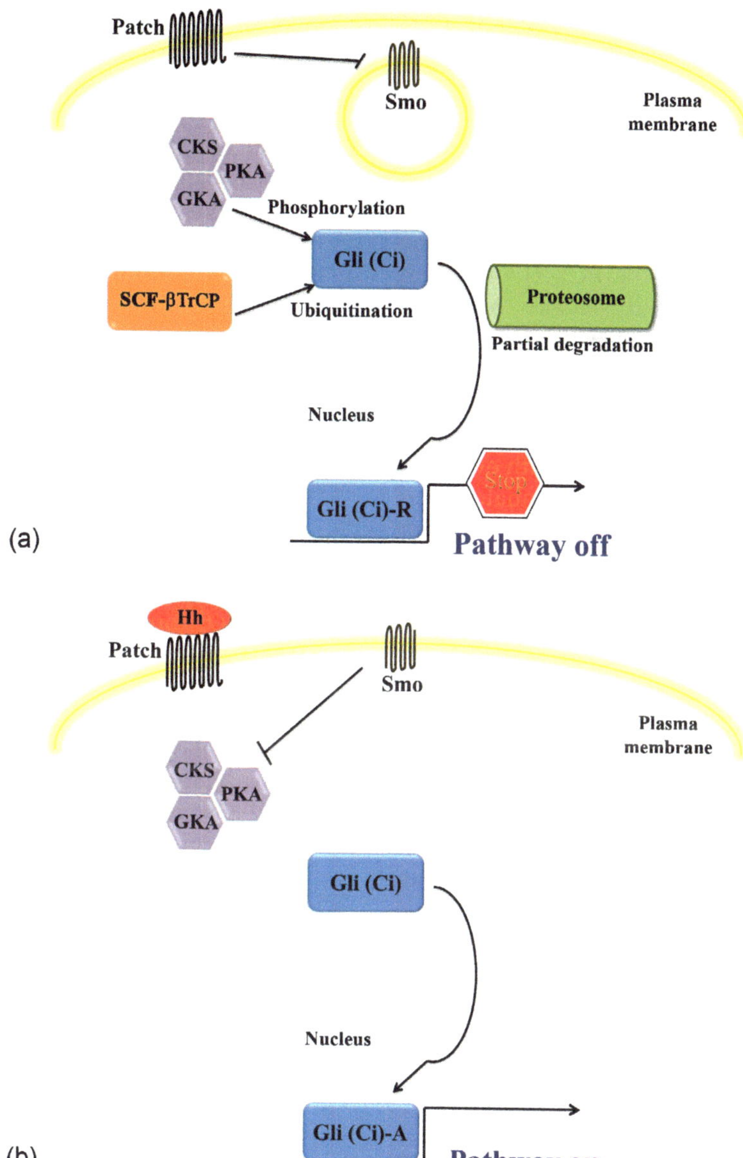

FIGURE 10.1 The Hedgehog-signaling pathway—a simplified representation. (a) In the absence of ligand, Patch represses Smo, that is confined to vesicles, being inactive. That allows the sequential phosphorylation by several kinases: protein kinase A (PKA), glycogen synthase-3β (GSK3β), and casein kinase-1. Thereafter, ubiquitination by Skip-Cullin-F-box (SCF) protein/β-transducing repeat-containing protein primes the phosphorylated Gli to limited proteosomic degradation, exposing the N-terminal repressor domain (Gli-R), which translocates to the nucleus and represses transcription of target genes. (b) Binding of Hedgehog to Patch eliminates the inhibitory effect of Patch in Smo contained in cytoplasmic vesicles. That abrogates phosphorylation and subsequent degradation of Gli. Full-length protein translocates to the nucleus-promoting gene transcription.

partial proteosomal proteolysis to remove the N-terminal repressor domain [19,24].

10.2.3 The Role of Primary Cilium

The canonical Hedgehog pathway (i.e., Hedgehog ligand-Ptc interaction with consequent Smo activation) appears to require the presence of a primary cilium (PC). The PC is an immotile membrane-bound sensory organelle found in most mammalian cells. However, some truly differentiated cell types rarely, if ever, exhibit PC [27,28]. In healthy adult liver, for example, it is believed that hepatocytes do not express PC [27]. This is consistent with evidence that these cells are highly differentiated and rarely divide. Endothelial cells also do not seem to express PC normally but become ciliated under hydrostatic pressure. It is conceivable that cilium expression changes during disease. This is particularly likely to occur when tissue injury triggers regenerative responses because PC are known to emerge during interphase (G1 phase) from the more distal of two centrioles that become the basal body in replicating cells. Cells only possess one PC, which makes it very difficult to detect them. PC have an axoneme built from nine peripheral microtubules doublets that, unlike motile cilia, lack a central pair [29]. The best available approach to detect PC is confocal microscopy. Antibody directed against acetylated tubulin marks PC because microtubules in the PC are acetylated while cytoplasmic microtubules are not [27]. However, this approach can also generate false negative results because PC are not only rare (there is only a single PC/cell), they are dynamic. PC assembly/disassembly changes with the replication status of the cell.

Ptc prevents Hedgehog pathway activation by blocking the entry of Smo into the PC. Binding of Hedgehog ligand to Ptc removes Ptc from the PC, thereby allowing Smo to enter. Activated Smo undergoes anterograde movement along the cilia, directed by kinesin proteins and facilitated by the ciliary Bardet-Biedl syndrome proteins (BBS) and intraflagellar transport proteins (IFP). At the tip of the cilia, active Smo releases Gli from the suppressor complex. Gli then moves retrograde along the cilia, directed by dynein proteins through the binding of BBS and IFPs. The movement of Gli through the cilia is also dependent on the protein Kif7. When in the cytoplasm, unphosphorylated Gli undergoes subsequent processing and translocation for the nucleus (Gli-A), promoting transcription of target genes [19] (Figure 10.2).

10.2.4 Regulation of the Pathway and Noncanonical Pathway

The Hedgehog pathway is tightly regulated. Ptc is a direct transcriptional target gene for Gli-A [30], providing a negative feedback loop for the pathway. Ptc counters Hedgehog signaling not only through inhibition of Smo but also by binding to the ligand and targeting it for lysosome degradation [22]. Another target gene for Gli-A is Hedgehog-interacting protein (Hip), a potent Hedgehog inhibitor that binds and inactivates the ligand. Hip expression is also highly regulated through methylation. In many liver cancers, the Hedgehog pathway becomes constitutively activated through hypermethylation and decreased expression of Hip [31–33]. Gli-1 is also a direct target of Gli-A, functioning in a positive feedback that re-enforces Hedgehog signaling. Gli-1 transcript levels provide an internal measure of Hedgehog pathway activity [30]. In contrast, Hedgehog signaling negatively regulates Gli-3 expression, and Gli-3 levels demonstrate an inverse relationship with Hedgehog ligand gradient [34]. Gli-2 transcript levels fluctuate over time in response to Hedgehog, through complex feedback loops. For example, Gli-2 increases Foxa2 expression, which in turn can repress Gli-2 transcription [30]. Also, Gli-2 transcription/activation can be induced by Hedgehog ligand-independent pathways, including TGF-β, phosphatidylinositol 3-kinase/AKT, Ras, and mitogen-activated

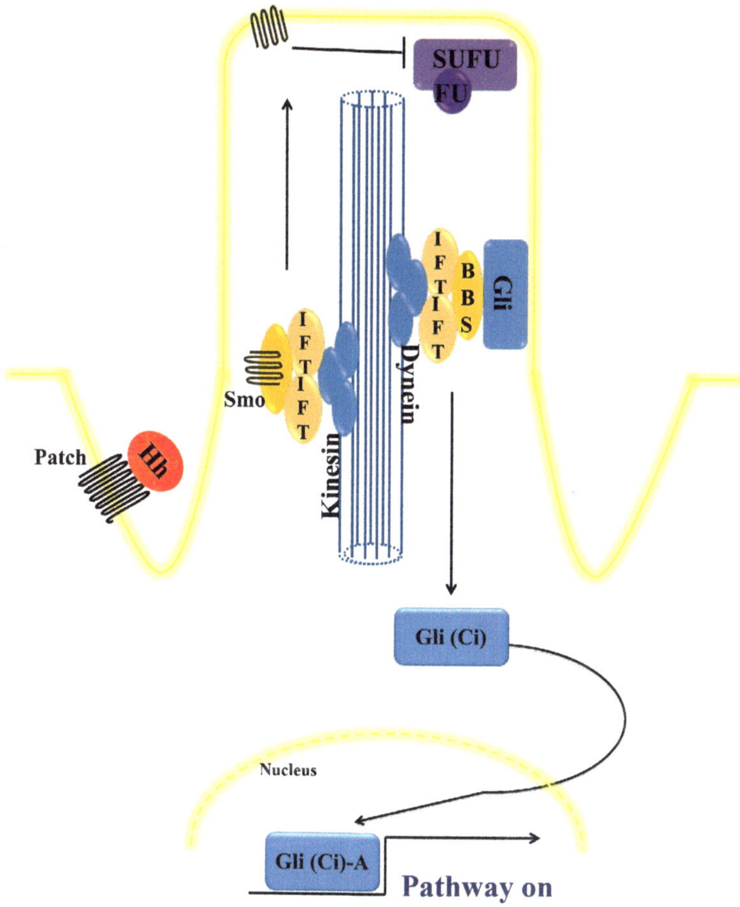

FIGURE 10.2 Hedgehog-signaling pathway and primary cilia. In the absence of ligand, Gli localizes in the cytoplasm as a part of an inhibitory complex with fused kinase (Fu) and suppressor of fused (SuFu), which allows the sequential phosphorylation by several kinases and subsequent processing in its repressor form. When Hedgehog binds to Patch, Smo undergoes anterograde movement along the cilia, directed by kinesin and facilitated by the ciliary proteins Bardet-Biedl syndrome proteins (BBS) and intraflagellar transport proteins (IFP). At the tip of the cilia, Smo releases Gli from the suppressor complex, allowing it to move along the cilia, directed by dynein proteins. Unphosphorylated Gli-2, in the cytoplasm can then be processed in its active form, which translocates to the nucleus-promoting gene transcription.

protein kinases/extracellular signal-regulated kinases [25]. Osteopontin, besides being a target gene of Gli, also inhibits GSK3β, thereby promoting Gli activation [35].

Besides the canonical Hedgehog pathway, there are two types of noncanonical Hedgehog signaling: type I requires Ptc and type II is Smo-dependent [25]. In type I signaling, Hedgehog ligand-free Ptc directly induces apoptosis by physically interacting with, and activating, caspases. It also inhibits proliferation by interacting with cyclin B in the cytoplasm, thereby blocking cyclin B nuclear localization [36]. Type II signaling requires the G-protein-coupled receptor-like function of Smo. Smo, when activated, catalyzes the GDP-GTP exchange of G inhibitory (Gi) proteins, a function that is not essential for canonical Hedgehog signaling [37]. This Gi-activating property may feed a cilium-dependent Smo-calcium-AMP kinase axis that triggers rapid Warburg-like glycolytic metabolic reprogramming (within minutes) in muscle and adipose tissue, promoting robust insulin-independent glucose uptake [38]. It can also induce Rac1 and RhoA activation with cytoskeleton rearrangement, which is known to promote migration in fibroblasts and cholangiocarcinoma cells, and tubulogenesis in endothelial cells [39–41]. This last pathway does not require translocation of Smo to the PC [25].

10.3 HEDGEHOG PATHWAY AFTER PARTIAL HEPATECTOMY—FEEDING PROMETHEUS' LIVER

In the healthy liver, the Hedgehog pathway is dormant, and ligand levels are barely detectable [12]. However, the pathway becomes active in the regenerating liver, and it is crucial for a successful regenerative response [16]. Ochoa et al. were the first to describe the changes in the profile of the Hedgehog pathway over time after 70% partial hepatectomy [16]. Their findings were reproduced in other studies from the same and from different groups [7,42]. After partial hepatectomy, there is an increase in the expression of both Ihh and Shh ligands (Figure 10.3) [16]. Ihh first peaks at 48 h. This corresponds to the initial postpartial hepatectomy peak in hepatocyte proliferation/DNA synthesis. There is also a second Ihh peak at 120 h, which corresponds to a second proliferative peak in hepatocytes. Shh behaves differently, increasing only in the middle to the end of the initial hepatocyte replicative period, but remaining high throughout the post-replicative period. Interestingly, the pattern of Shh expression in whole liver correlates with the expression of α-SMA, a marker of stellate cell-derived myofibroblasts. Post-partial hepatectomy changes in Gli-1 and Gli-2 mRNAs mimic the pattern of Ihh expression. Changes in Gli-2 transcript levels were confirmed at the protein level, by Western blot analysis and immunohistochemistry for Gli-2.

It is important to understand what cells are the sources of Hedgehog ligands after partial hepatectomy because healthy mature hepatocytes do not produce ligands [12]. Stressed hepatocytes, particularly dying ones, release damage-associated molecules such as Shh [43]. Although hepatocytes in the remnant liver are

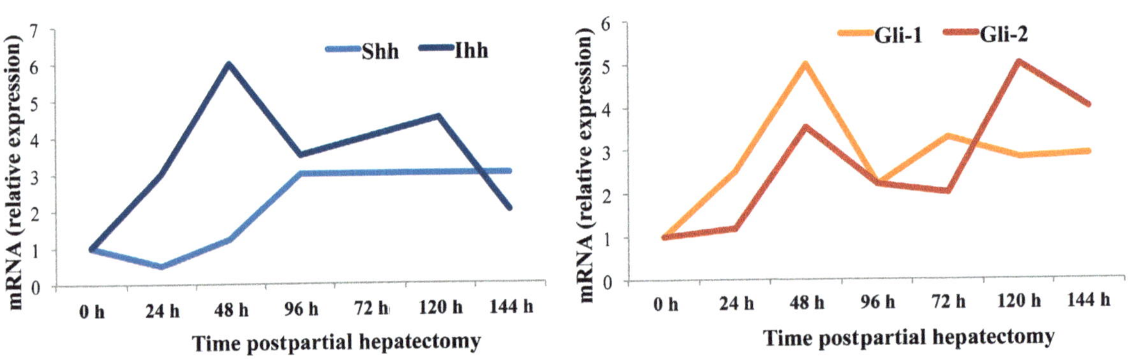

FIGURE 10.3 Expression of components of Hedgehog pathway after partial hepatectomy. Time-course of mRNA expression profile of ligands (Shh, Sonic Hedgehog; Ihh, Indian Hedgehog) and downstream targets (Gli-1 and Gli-2), in whole liver, after partial hepatectomy, in mice.
Adapted from Ochoa et al. [16].

believed to be healthy after partial hepatectomy, they do produce ligands. Ligand production was demonstrated directly in primary hepatocytes isolated after partial hepatectomy. In contrast, ligand production could not be detected in primary hepatocytes isolated after sham surgery. Peak production of Ihh occurred 24 h after surgery, whereas Shh production peaked later (at 48 h) [16]. It is unclear whether the ligand-producing hepatocytes are fully mature. They may be in the process of differentiating from progenitor cells or undergoing a dedifferentiation/EMT-like process. The mechanisms simulating ligand production by the hepatocytic cells in regenerating liver are also unknown. The main accepted source of Hedgehog ligands in injured liver is progenitor cells, i.e., oval cells and ductular-type progenitor cells [12]. Other nonparenchymal cells in injured livers are also able to produce Hedgehog ligands, such as activated HSCs [44,45] and activated liver sinusoidal endothelial cells [46]. The latter observation may be particularly relevant because sinusoidal endothelial cells are subjected to intense shear stress after partial hepatectomy when portal flow has to pass through a much smaller liver mass. Lastly, immune cells, such as NKT cells and Kupffer cells, produce Shh and Ihh, respectively [13,14]. Thus, immune cells might be another source of Hedgehog ligands in regenerating livers after partial hepatectomy.

In Hedgehog-responsive cells, the Hedgehog pathway acts by promoting proliferation, inhibiting apoptosis, and stimulating transition to a more mesenchymal phenotype. After partial hepatectomy, the Hedgehog-responsive cells were identified by their expression of Gli proteins. Gli-2-positive cells included hepatocytic cells, progenitors, ductular, and stromal cells [7,16,42]. Gli-2 protein expression peaked first in hepatocytic cells (48 h after surgery) and 24 h later in ductular cells [16]. Again, current dogma advocates that normal hepatocytes are not responsive to the Hedgehog pathway because they lack PC [27]. However, whereas primary hepatocytes isolated after sham surgery do not have detectable levels of Gli proteins, there is significant Gli expression in hepatocytes isolated after partial hepatectomy [16]. In a hepatocyte preparation isolated in those conditions, 100% of cells expressed albumin protein and 10% coexpressed Gli-2 protein. More interestingly, whereas primary hepatocytes isolated after sham surgery did not respond to cyclopamine (a direct antagonist of Smo), hepatocytes isolated after partial hepatectomy did respond, as evidenced by decreased BrdU labeling/proliferation. Thus, after partial hepatectomy, hepatocytic cells both produce Hedeghog ligands and require Smo activation in order to proliferate. This suggests that regenerative pressure induces at least some hepatocytic cells to acquire Hedgehog signaling capability. Further research is needed to clarify the mechanisms involved. It will be important to determine if the process depends upon PC. Can fully mature hepatocytes develop PC or does PC simply mark immature hepatocytic cells that derived from Hedgehog-responsive progenitors that have not yet lost PC? [16] Alternatively, Gli-2, in hepatocytic cells, could have been activated through a PC-independent pathway. [28]. Such a noncanonical pathway must be Smo-dependent, however, because hepatocytes were demonstrated to respond to cyclopamine, which inhibits Smo by direct physical interaction.

Hepatocytes are not the only cells that proliferate after partial hepatectomy, although regulation of the process in other liver cell types has not been the focus of much research. There is growing evidence that liver regeneration after partial hepatectomy involves an important progenitor proliferative response and epithelial-mesenchymal transitions of HSCs. Some evidence even suggests that HSCs may function as progenitor cells. For example, in recent lineage tracing studies, many progenitors, cholangiocytes, and up to 25% of hepatocytes derived from α-SMA expressing cells after partial hepatectomy in mice [7,16]. α-SMA marks myofibroblasts and HSCs are the predominant source of α-SMA-positive cells during many types of liver injury [47]. Hence, evidence that HSC-derived MF might be facultative progenitors for liver epithelial cells after partial hepatectomy engendered considerable skepticism. However, the issue merits re-examination in light of subsequent evidence that many healthy adult organs, including liver, harbor mesenchymal stem cells (MSCs) that can express α-SMA. These tissue-resident MSC are Hedgehog-responsive (i.e., Gli-1-positive) pericytes [48]. In adult liver, HSCs are Gli-1-positive pericytes, and they exhibit many features of MSC [49–52].

During chronic injury in many tissues, recent fate-mapping approaches demonstrated that many Gli-1(+) pericytes became α-SMA-positive myofibroblasts. Moreover, targeted deletion of the Hedgehog-responsive pericyte population in that context blocked organ fibrosis [48]. These recent findings support an earlier report by Michelotti et al. who used genetic approaches to conditionally delete Smo in α-SMA-positive cells during chronic liver injury. Selectively blocking Hedgehog signaling in the α-SMA-positive population abrogated fibrosis in livers damaged by chronic bile duct ligation or chronic toxin exposure [49]. Thus, there is consistent evidence that Hedgehog-responsive pericytes drive organ fibrosis during chronic injury. After an acute injury, such as partial hepatectomy, blocking Hedgehog signaling by systemic administration of the Smo-specific inhibitor cyclopamine also reduced accumulation of Gli-positive cells. In regenerating livers, decreasing Hedgehog signaling virtually abolished accumulation of α-SMA-positive cells, abrogated matrix remodeling, reduced progenitors, and blocked proliferation of hepatocytes and ductular cells. Those effects profoundly impacted the ability of the previously healthy liver to regenerate, and significantly increased mortality at 48 and 72 h after partial hepatectomy. Moreover, cyclopamine-treated animals that survived exhibited reduced liver weights and liver to body weight ratios compared to nontreated mice [7,16]. Interestingly, a targeted genetic approach that conditionally deleted Smo in α-SMA producing cells after partial hepatectomy had the same dramatic effects that were observed following systemic administration of the pharmacologic Smo inhibitor [7,16]. Thus, the aggregate data from several groups supports the concept that Hedgehog-responsive pericytes in adult livers (a.k.a., HSCs) can exhibit features of both MSCs and myofibroblasts. Although more research is needed to clarify the mechanisms that modulate the plasticity of these cells to control their fate during liver repair, there is no longer any doubt that Hedgehog signaling is critically important in this process.

Glypican-3 is another piece of the puzzle in Hedgehog regulation. Glypican-3 is a heparane sulfate proteoglycan that acts as a negative regulator of hepatocyte growth. It can bind to Hedgehog, inhibiting the pathway by competing with Patch 1. Indeed, in the proliferative phase post-hepatectomy, the binding of Ihh to glypican-3 decreases dramatically, allowing free Ihh to bind to Patch 1. Conversely, at the end of the regeneration process, Ihh-glypican-3 binding returns to baseline [53].

Recently, integrin-β1 signaling has been shown to be crucial for liver regeneration [54]. Integrins are transmembrane receptors that mediate cell-cell and cell-extracellular matrix interactions. They act as sensors of extracellular matrix and neighbor cells, eliciting several signaling pathways involved in cell growth/proliferation, apoptosis, cell adhesion, and migration [55]. A recent study showed that depleting integrin-β1 in hepatocytes (using genetic approaches to knock-out the gene or siRNA approaches to knock-down the gene expression) caused substantial liver necrosis, impaired hepatocyte proliferation, and increased mortality 48 h post-hepatectomy. The authors linked that phenotype to impaired ligand-induced phosphorylation of the epidermal growth factor and hepatocyte growth factor receptors, with resultant attenuation of those signaling pathway responses [54]. Though not addressed in this study, it is conceivable that disrupted Hedgehog signaling might also have contributed to the phenotype described. Integrin-β1 has been shown to induce Shh and Ihh expression in intestinal epithelial cells through increased expression of Forkhead family transcription factor HNF-3β (Foxa2) [56], a known liver-enriched transcriptional activator of the gene encoding Hedgehog ligand [57]. Integrin-β1 can also increase Gli-1 and Gli-2 activity and expression in a Hedgehog ligand-independent manner, through enhancement of insulin-like growth factor and noncanonical Hedgehog signaling [55].

In conclusion, the Hedgehog pathway is crucial for liver regeneration to occur after partial hepatectomy. In this context, there is incredible liver cell plasticity, enabling various cell types to turn on Hedgehog ligand production and modulating which cells respond to them. The process in not only abrupt and dynamic, but it evolves quite rapidly, and the pathway is silenced back to its baseline level of negligible activity within a few days of its activation. The kinetics of Hedgehog pathway activation/deactivation parallels that of liver regeneration itself. More research is needed to clarify the mechanisms by which Hedgehog controls regeneration. To date, it has been clearly shown that Hedgehog-responsive HSCs are pivotal players in the process [7].

10.4 REGENERATING THE SICK LIVER

In chronic liver disease, the injured liver elicits a repair response to confine injury and replace dying cells. Liver repair and liver regeneration are qualitatively similar, although the level and timing of discrete components of the two responses differ. Like liver regeneration after acute partial hepatectomy, liver repair during chronic injury requires appropriate activation and regulation of the Hedgehog pathway. The Hedgehog pathway has been shown to be crucial in the wound-healing response evoked by every cause of chronic liver injury that has been examined to date.

The first evidence for Hedgehog pathway activation in human chronic liver disease was described in biliary diseases. A cDNA microarray analysis performed in patients with primary biliary cirrhosis (PBC) and primary sclerosing cholangitis (PSC) demonstrated eightfold upregulation of *Ptc* mRNA (a target gene of the Hedgehog pathway) in PBC and a trend toward upregulation in PSC [58]. Subsequent immunohistochemistry of PBC livers supported activation in the Hedgehog pathway during chronic biliary injury by demonstrating increased staining for Ptc and nuclear Gli-2 [59]. Reactive-appearing ductular cells and MF cells in portal tracts were Gli-2-positive, and ductular cells expressed Ihh ligand. Recent gene expression profiling studies in a large, international PBC patient cohort validated the concept that Hedgehog signaling is upregulated in this disease [60]. In pediatric cholestatic diseases, such as biliary atresia, progressive familial intra-hepatic cholestasis, and Allagille's syndrome, there is also accumulation of Hedgehog- producing and responding cells [61]. The authors of those studies suggest that excessive Hedgehog activation not only enhances fibrogenesis in biliary atresia but also impairs ductular morphogenesis by inducing EMT responses that arrest immature ductular cell differentiation [61]. This concept was supported by subsequent studies in a zebrafish model of biliary atresia, which proved that treating fish with cyclopamine to inhibit Hedgehog signaling restored normal biliary development [62].

Rodent models of chronic biliary disease also demonstrate Hedgehog pathway activation. Chronic bile duct ligation induces an exuberant fibroductular response that leads to biliary cirrhosis in rodents. The process is accompanied by a dramatic increase in production of Hedgehog ligands by ductular cells and myofibroblasts in portal stroma. Coincidentally, hepatic expression of the inhibitor Hip declines to nearly undetectable levels. Intervening with Roux-en-Y hepaticojejunostomy to reverse biliary obstruction several weeks after bile duct ligation leads to gradual silencing of the Hedgehog pathway, regression of liver fibrosis, and progressive normalization of the hepatic architecture [63–65]. These findings suggest a role for the Hedgehog pathway in repair responses triggered by chronic cholestatic liver injury. This concept was supported by subsequent studies. Treating bile duct ligated rodents with the Smo antagonist, GDC-0449, decreased liver injury (as assessed by plasma levels of aminotransferases and bilirubin), reduced inflammatory cytokines, and substantially improved fibrosis (as assessed by Masson trichrome staining and expression of collagen-α1a and α-SMA) [66]. Hedgehog signaling was also proven to regulate liver repair in genetic models of cholestatic liver damage. In mice, mdr-2 (the murine orthologue of human MDR3) encodes a canalicular phospholipid flippase in hepatocytes. Disruption of mdr2 causes chronic cholestatic liver injury and progressive biliary-type fibrosis, which resembles human PSC [67]. Hepatocellular carcinomas emerge at 50-60 weeks of age. Mdr2-deficient mice exhibit progressive, age-related activation of the Hedgehog pathway. Treating old mice with established liver disease and hepatocellular carcinoma with GDC-0449 for only 9 days had a tremendous impact on liver disease, significantly decreasing various markers of fibrosis, including hepatic hydroxyproline content, and promoting some regression of liver cancer as demonstrated by serial magnetic resonance imaging [67].

The Hedgehog pathway has also been shown to be crucial in alcoholic and nonalcoholic liver disease in humans [68–71] and in rodent models [68,69,72,73]. In two different animal models of nonalcoholic fatty liver disease (methionine-choline-deficient diet supplemented with ethionine, and a Western diet consisting of high-fat diet rich in cholesterol and fructose), treating with different Hedgehog inhibitors (cyclopamine [69] or GDC-0449 (vismodebig) [73]) decreased liver injury and significantly reduced fibrosis. In humans with NASH, posttreatment liver biopsies from responders in a large, multi-center treatment trial demonstrated consistent downregulation of Hedgehog ligands and decreased accumulation of Hedgehog-responsive myofibroblasts and progenitor cells compared to their pretreatment biopsies. In contrast, Hedgehog was not suppressed in subjects who failed to respond to treatment [74].

Finally, whether Hedgehog drives fibrosis progression in other types of human liver disease has not yet been examined directly. However, the pathway has been shown to be active in all forms of chronic liver disease studied to date, including life-threatening liver diseases that are endemic in many parts of the world, such as chronic viral hepatitis B and C [75] and schistosomiasis [14]. Moreover, findings in animal models of other types of human liver injury, including drug induced injury [28], radiotherapy-induced injury [76], and ischemia-reperfusion [77], also demonstrate Hedgehog pathway activation. The aggregate data show that the liver responds to chronic injury in a monotonous way. An insult induces cellular distress and death. Dying cells release damage-associated molecules, including Hedgehog ligands. These ligands act on immune cells, sinusoidal endothelial cells, HSCs, and progenitor cells, eliciting a wound-healing response. Transient pathway activation is essential for regeneration of the injured liver. However, if not promptly shut down, sustained/excessive Hedgehog activity exacerbates injury, promotes fibrogenesis that may result in cirrhosis and its complications, including liver cancer.

10.5 CONCLUSION

The liver has a tremendous capacity for regeneration. Better understanding of the mechanisms that replace liver mass, either after hepatectomy or after injury-induced liver cell death, will enable development of novel approaches to optimize recovery from liver damage. This knowledge will be valuable in various clinical situations that acutely reduce liver mass, such as liver resection to remove metastatic or primary cancer, living donor transplantation, and massive hepatic injury caused by drugs or ischemia. It will also suggest targets that might be manipulated to constrain deregulated liver growth, as occurs in liver cancers. Finally, clarifying mechanisms that control liver pericyte (HSC) plasticity are likely to identify conserved therapeutic and diagnostic targets that control the pathogenesis of liver cirrhosis and its complications. Over the past decade, a body of solid evidence has accrued which consistently identifies a crucial regulatory role for the Hedgehog pathway in adult liver regeneration. Hence, this pathway has now become a potential target for treating liver disease.

References

[1] Mao SA, Glorioso JM, Nyberg SL. Liver regeneration. Transl Res 2014;163:352–62.

[2] Miyagawa S, Kawasaki S, Noike T, Nomura K, Kobayashi A, Shimada R, et al. Liver regeneration after extended right hemihepatectomy in patients with hilar or diffuse bile duct carcinoma. Hepatogastroenterology 1999;46:364–8.

[3] Boulter L, Lu WY, Forbes SJ. Differentiation of progenitors in the liver: a matter of local choice. J Clin Invest 2013;123:1867–73.

[4] Karaca G, Swiderska-Syn M, Xie G, Syn WK, Kruger L, Machado MV, et al. TWEAK/Fn14 signaling is required for liver regeneration after partial hepatectomy in mice. PLoS One 2014;9:e83987.

[5] Katoonizadeh A, Poustchi H, Malekzadeh R. Hepatic progenitor cells in liver regeneration: current advances and clinical perspectives. Liver Int 2014;34:1467–72.

[6] Yang L, Jung Y, Omenetti A, Witek RP, Choi S, Vandongen HM, et al. Fate-mapping evidence that hepatic stellate cells are epithelial progenitors in adult mouse livers. Stem Cells 2008;26:2104–13.

[7] Swiderska-Syn M, Syn WK, Xie G, Kruger L, Machado MV, Karaca G, et al. Myofibroblastic cells function as progenitors to regenerate murine livers after partial hepatectomy. Gut 2014;63:1333–44.

[8] Choi SS, Omenetti A, Witek RP, Moylan CA, Syn WK, Jung Y, et al. Hedgehog pathway activation and epithelial-to-mesenchymal transitions during myofibroblastic transformation of rat hepatic cells in culture and cirrhosis. Am J Physiol Gastrointest Liver Physiol 2009;297:G1093–106.

[9] Munoz-Espin D, Serrano M. Cellular senescence: from physiology to pathology. Nat Rev Mol Cell Biol 2014;15:482–96.

[10] Factor VM, Jensen MR, Thorgeirsson SS. Coexpression of C-myc and transforming growth factor alfa in the liver promotes early replicative senescence and diminishes regenerative capacity after partial hepatectomy in transgenic mice. Hepatology 1997;26:1434–43.

[11] Hubackova S, Krejcikova K, Bartek J, Hodny Z. IL1- and TGFbeta-Nox4 signaling, oxidative stress and DNA damage response are shared features of replicative, oncogene-induced, and drug-induced paracrine 'bystander senescence'. Aging 2012;4:932–51.

[12] Sicklick JK, Li YX, Melhem A, Schmelzer E, Zdanowicz M, Huang J, et al. Hedgehog signaling maintains resident hepatic progenitors throughout life. Am J Physiol Gastrointest Liver Physiol 2006;290: G859–70.

[13] Syn WK, Witek RP, Curbishley SM, Jung Y, Choi SS, Enrich B, et al. Role for Hedgehog pathway in regulating growth and function of invariant NKT cells. Eur J Immunol 2009;39:1879–92.

[14] Pereira TA, Xie G, Choi SS, Syn WK, Voieta I, Lu J, et al. Macrophage-derived Hedgehog ligands promotes fibrogenic and angiogenic responses in human schistosomiasis mansoni. Liver Int 2013;33:149–61.

[15] Syn WK, Agboola KM, Swiderska M, Michelotti GA, Liaskou E, Pang H, et al. NKT-associated Hedgehog and osteopontin drive fibrogenesis in non-alcoholic fatty liver disease. Gut 2012;61:1323–9.

[16] Ochoa B, Syn WK, Delgado I, Karaca GF, Jung Y, Wang J, et al. Hedgehog signaling is critical for normal liver regeneration after partial hepatectomy in mice. Hepatology 2010;51:1712–23.

[17] Nusslein-Volhard C, Wieschaus E. Mutations affecting segment number and polarity in Drosophila. Nature 1980;287:795–801.

[18] Briscoe J, Therond PP. The mechanisms of Hedgehog signalling and its roles in development and disease. Nat Rev Mol Cell Biol 2013;14:416–29.

[19] Merchant JL, Saqui-Salces M. Inhibition of Hedgehog signaling in the gastrointestinal tract: targeting the cancer microenvironment. Cancer Treat Rev 2014;40:12–21.
[20] Farzan SF, Singh S, Schilling NS, Robbins DJ. The adventures of sonic Hedgehog in development and repair. III. Hedgehog processing and biological activity. Am J Physiol Gastrointest Liver Physiol 2008;294: G844–9.
[21] Omenetti A, Choi S, Michelotti G, Diehl AM. Hedgehog signaling in the liver. J Hepatol 2011;54:366–73.
[22] Hooper JE, Scott MP. Communicating with Hedgehogs. Nat Rev Mol Cell Biol 2005;6:306–17.
[23] Zhang Z, Feng J, Pan C, Lv X, Wu W, Zhou Z, et al. Atrophin-Rpd3 complex represses Hedgehog signaling by acting as a corepressor of CiR. J Cell Biol 2013;203:575–83.
[24] Falkenstein KN, Vokes SA. Transcriptional regulation of graded Hedgehog signaling. Semin Cell Dev Biol 2014;33:73–80.
[25] Teperino R, Aberger F, Esterbauer H, Riobo N, Pospisilik JA. Canonical and non-canonical Hedgehog signalling and the control of metabolism. Semin Cell Dev Biol 2014;33:81–92.
[26] Pan Y, Bai CB, Joyner AL, Wang B. Sonic Hedgehog signaling regulates Gli2 transcriptional activity by suppressing its processing and degradation. Mol Cell Biol 2006;26:3365–77.
[27] Wheatley DN, Wang AM, Strugnell GE. Expression of primary cilia in mammalian cells. Cell Biol Int 1996;20:73–81.
[28] Grzelak CA, Martelotto LG, Sigglekow ND, Patkunanathan B, Ajami K, Calabro SR, et al. The intrahepatic signalling niche of Hedgehog is defined by primary cilia positive cells during chronic liver injury. J Hepatol 2014;60:143–51.
[29] Nozawa YI, Lin C, Chuang PT. Hedgehog signaling from the primary cilium to the nucleus: an emerging picture of ciliary localization, trafficking and transduction. Curr Opin Genet Dev 2013;23:429–37.
[30] Peterson KA, Nishi Y, Ma W, Vedenko A, Shokri L, Zhang X, et al. Neural-specific Sox2 input and differential Gli-binding affinity provide context and positional information in Shh-directed neural patterning. Genes Dev 2012;26:2802–16.
[31] Wang F, Anderson PW, Salem N, Kuang Y, Tennant BC, Lee Z. Gene expression studies of hepatitis virus-induced woodchuck hepatocellular carcinoma in correlation with human results. Int J Oncol 2007;30:33–44.
[32] Tada M, Kanai F, Tanaka Y, Tateishi K, Ohta M, Asaoka Y, et al. Down-regulation of Hedgehog-interacting protein through genetic and epigenetic alterations in human hepatocellular carcinoma. Clin Cancer Res 2008;14:3768–76.
[33] Eichenmuller M, Gruner I, Hagl B, Haberle B, Muller-Hocker J, von Schweinitz D, et al. Blocking the Hedgehog pathway inhibits hepatoblastoma growth. Hepatology 2009;49:482–90.
[34] Marigo V, Tabin CJ. Regulation of patched by sonic Hedgehog in the developing neural tube. Proc Natl Acad Sci USA 1996;93:9346–51.
[35] Das S, Samant RS, Shevde LA. Nonclassical activation of Hedgehog signaling enhances multidrug resistance and makes cancer cells refractory to Smoothened-targeting Hedgehog inhibition. J Biol Chem 2013;288:11824–33.
[36] Pasca di Magliano M, Hebrok M. Hedgehog signalling in cancer formation and maintenance. Nat Rev Cancer 2003;3:903–11.
[37] Polizio AH, Chinchilla P, Chen X, Manning DR, Riobo NA. Sonic Hedgehog activates the GTPases Rac1 and RhoA in a Gli-independent manner through coupling of smoothened to Gi proteins. Sci Signal 2011;4(200):pt7.
[38] Teperino R, Amann S, Bayer M, McGee SL, Loipetzberger A, Connor T, et al. Hedgehog partial agonism drives Warburg-like metabolism in muscle and brown fat. Cell 2012;151:414–26.
[39] Polizio AH, Chinchilla P, Chen X, Kim S, Manning DR, Riobo NA. Heterotrimeric Gi proteins link Hedgehog signaling to activation of Rho small GTPases to promote fibroblast migration. J Biol Chem 2011;286:19589–96.
[40] Razumilava N, Gradilone SA, Smoot RL, Mertens JC, Bronk SF, Sirica AE, et al. Non-canonical Hedgehog signaling contributes to chemotaxis in cholangiocarcinoma. J Hepatol 2014;60:599–605.
[41] Chinchilla P, Xiao L, Kazanietz MG, Riobo NA. Hedgehog proteins activate pro-angiogenic responses in endothelial cells through non-canonical signaling pathways. Cell Cycle 2010;9:570–9.
[42] Hanaoka J, Shimada M, Utsunomiya T, Morine Y, Imura S, Ikemoto T, et al. Significance of sonic Hedgehog signaling after massive hepatectomy in a rat. Surg Today 2013;43:300–7.
[43] Jung Y, Witek RP, Syn WK, Choi SS, Omenetti A, Premont R, et al. Signals from dying hepatocytes trigger growth of liver progenitors. Gut 2010;59:655–65.
[44] Sicklick JK, Li YX, Choi SS, Qi Y, Chen W, Bustamante M, et al. Role for Hedgehog signaling in hepatic stellate cell activation and viability. Lab Invest 2005;85:1368–80.
[45] Lin N, Tang Z, Deng M, Zhong Y, Lin J, Yang X, et al. Hedgehog-mediated paracrine interaction between hepatic stellate cells and marrow-derived mesenchymal stem cells. Biochem Biophys Res Commun 2008;372: 260–5.
[46] Xie G, Choi SS, Syn WK, Michelotti GA, Swiderska M, Karaca G, et al. Hedgehog signalling regulates liver

sinusoidal endothelial cell capillarisation. Gut 2013;62: 299–309.
[47] Puche JE, Saiman Y, Friedman SL. Hepatic stellate cells and liver fibrosis. Compr Physiol 2013;3:1473–92.
[48] Kramann R, Schneider RK, DiRocco DP, Machado F, Fleig S, Bondzie PA, et al. Progenitors are key contributors to injury-induced organ fibrosis. Cell Stem Cell 2015;16:51–66.
[49] Michelotti GA, Xie G, Swiderska M, Choi SS, Karaca G, Kruger L, et al. Smoothened is a master regulator of adult liver repair. J Clin Invest 2013;123:2380–94.
[50] Reister S, Kordes C, Sawitza I, Haussinger D. The epigenetic regulation of stem cell factors in hepatic stellate cells. Stem Cells Dev 2011;20:1687–99.
[51] Kordes C, Sawitza I, Gotze S, Haussinger D. Hepatic stellate cells support hematopoiesis and are liver-resident mesenchymal stem cells. Cell Physiol Biochem 2013;31:290–304.
[52] Kordes C, Sawitza I, Gotze S, Herebian D, Haussinger D. Hepatic stellate cells contribute to progenitor cells and liver regeneration. J Clin Invest 2014;124:5503–15.
[53] Bhave VS, Mars W, Donthamsetty S, Zhang X, Tan L, Luo J, et al. Regulation of liver growth by glypican 3, CD81, Hedgehog, and Hhex. Am J Pathol 2013;183:153–9.
[54] Speicher T, Siegenthaler B, Bogorad RL, Ruppert R, Petzold T, Padrissa-Altes S, et al. Knockdown and knockout of beta1-integrin in hepatocytes impairs liver regeneration through inhibition of growth factor signalling. Nat Commun 2014;5:3862.
[55] Goel HL, Underwood JM, Nickerson JA, Hsieh CC, Languino LR. Beta1 integrins mediate cell proliferation in three-dimensional cultures by regulating expression of the sonic Hedgehog effector protein, GLI1. J Cell Physiol 2010;224:210–7.
[56] Jones RG, Li X, Gray PD, Kuang J, Clayton F, Samowitz WS, et al. Conditional deletion of beta1 integrins in the intestinal epithelium causes a loss of Hedgehog expression, intestinal hyperplasia, and early postnatal lethality. J Cell Biol 2006;175:505–14.
[57] Chang BE, Blader P, Fischer N, Ingham PW, Strahle U. Axial (HNF3beta) and retinoic acid receptors are regulators of the zebrafish sonic Hedgehog promoter. EMBO J 1997;16:3955–64.
[58] Shackel NA, McGuinness PH, Abbott CA, Gorrell MD, McCaughan GW. Identification of novel molecules and pathogenic pathways in primary biliary cirrhosis: cDNA array analysis of intrahepatic differential gene expression. Gut 2001;49:565–76.
[59] Jung Y, McCall SJ, Li YX, Diehl AM. Bile ductules and stromal cells express Hedgehog ligands and/or Hedgehog target genes in primary biliary cirrhosis. Hepatology 2007;45:1091–6.
[60] Kar SP, Seldin MF, Chen W, Lu E, Hirschfield GM, Invernizzi P, et al. Pathway-based analysis of primary biliary cirrhosis genome-wide association studies. Genes Immun 2013;14:179–86.
[61] Omenetti A, Bass LM, Anders RA, Clemente MG, Francis H, Guy CD, et al. Hedgehog activity, epithelial-mesenchymal transitions, and biliary dysmorphogenesis in biliary atresia. Hepatology 2011;53:1246–58.
[62] Cui S, Leyva-Vega M, Tsai EA, EauClaire SF, Glessner JT, Hakonarson H, et al. Evidence from human and zebrafish that GPC1 is a biliary atresia susceptibility gene. Gastroenterology 2013;144:1107–15.
[63] Omenetti A, Yang L, Li YX, McCall SJ, Jung Y, Sicklick JK, et al. Hedgehog-mediated mesenchymal-epithelial interactions modulate hepatic response to bile duct ligation. Lab Invest 2007;87:499–514.
[64] Omenetti A, Popov Y, Jung Y, Choi SS, Witek RP, Yang L, et al. The Hedgehog pathway regulates remodelling responses to biliary obstruction in rats. Gut 2008;57:1275–82.
[65] Omenetti A, Porrello A, Jung Y, Yang L, Popov Y, Choi SS, et al. Hedgehog signaling regulates epithelial-mesenchymal transition during biliary fibrosis in rodents and humans. J Clin Invest 2008;118:3331–42.
[66] Kumar V, Mundra V, Mahato RI. Nanomedicines of Hedgehog inhibitor and PPAR-gamma agonist for treating liver fibrosis. Pharm Res 2013;31:1158–69.
[67] Philips GM, Chan IS, Swiderska M, Schroder VT, Guy C, Karaca GF, et al. Hedgehog signaling antagonist promotes regression of both liver fibrosis and hepatocellular carcinoma in a murine model of primary liver cancer. PLoS One 2011;6:e23943.
[68] Jung Y, Brown KD, Witek RP, Omenetti A, Yang L, Vandongen M, et al. Accumulation of Hedgehog-responsive progenitors parallels alcoholic liver disease severity in mice and humans. Gastroenterology 2008;134:1532–43.
[69] Syn WK, Jung Y, Omenetti A, Abdelmalek M, Guy CD, Yang L, et al. Hedgehog-mediated epithelial-to-mesenchymal transition and fibrogenic repair in nonalcoholic fatty liver disease. Gastroenterology 2009;137: 1478–88.
[70] Guy CD, Suzuki A, Zdanowicz M, Abdelmalek MF, Burchette J, Unalp A, et al. Hedgehog pathway activation parallels histologic severity of injury and fibrosis in human nonalcoholic fatty liver disease. Hepatology 2012;55:1711–21.
[71] Swiderska-Syn M, Suzuki A, Guy CD, Schwimmer JB, Abdelmalek MF, Lavine JE, et al. Hedgehog pathway and pediatric nonalcoholic fatty liver disease. Hepatology 2013;57:1814–25.
[72] Fleig SV, Choi SS, Yang L, Jung Y, Omenetti A, VanDongen HM, et al. Hepatic accumulation of Hedgehog-reactive progenitors increases with severity of fatty liver damage in mice. Lab Invest 2007;87:1227–39.

[73] Hirsova P, Ibrahim SH, Bronk SF, Yagita H, Gores GJ. Vismodegib suppresses TRAIL-mediated liver injury in a mouse model of nonalcoholic steatohepatitis. PLoS One 2013;8:e70599.

[74] Guy CD, Suzuki A, Abdelmalek MF, Burchette JL, Diehl AM, for the NC. Treatment response in the PIVENS trial is associated with decreased Hedgehog pathway activity. Hepatology 2014;61:98–107.

[75] Pereira Tde A, Witek RP, Syn WK, Choi SS, Bradrick S, Karaca GF, et al. Viral factors induce Hedgehog pathway activation in humans with viral hepatitis, cirrhosis, and hepatocellular carcinoma. Lab Invest 2010;90: 1690–703.

[76] Wang S, Hyun J, Youn B, Jung Y. Hedgehog signaling regulates the repair response in mouse liver damaged by irradiation. Radiat Res 2013;179:69–75.

[77] Pratap A, Panakanti R, Yang N, Eason JD, Mahato RI. Inhibition of endogenous Hedgehog signaling protects against acute liver injury after ischemia reperfusion. Pharm Res 2010;27:2492–504.

CHAPTER

11

Regulation of Cell Cycle During Liver Regeneration

Yulia A. Nevzorova, Christian Trautwein

Department of Medicine III, University Hospital, Aachen, Germany

ABBREVIATIONS

Cdc	cell division control protein
Cdk	cyclin-dependent kinases
Cip/Kip	cyclin-dependent kinase inhibitor
DEN	diethylnitrosamine
G0-phase	Gap0 phase (quiescence)
G1-phase	Gap1 phase
G2-phase	Gap2 phase
HCC	hepatocellular carcinoma
MCM	mini chromosome maintenance complex
MEFs	mouse embryonic fibroblasts
PCNA	proliferating cell nuclear antigen
PH	partial hepatectomy
pre-RC	pre-replication complex
Rb	retinoblastoma protein
Skp2	S-phase kinase-associated protein 2

11.1 INTRODUCTION

The cell cycle is an ordered set of events eventually culminating in cell growth and division into two daughter cells. In eukaryotic cells, it is composed of four distinct phases: Gap1 phase (G1-phase), S-phase (DNA synthesis), Gap2 phase (G2-phase), and mitosis. The first three phases are collectively known as interphase (Figures 11.1 and 11.2). In addition, quiescent cells such as mature hepatocytes reside in a reversible nonreplicative phase termed Gap0 phase (G0-phase) [1].

The transition from one cell-cycle phase to another occurs in a well-ordered manner and is regulated by different cellular proteins. Key regulatory proteins are the cyclin-dependent kinases (Cdk), a family of serine/threonine protein kinases that are activated at specific points of the cell cycle. Cdk protein levels remain stable during the cell cycle, in contrast to their activating proteins, the cyclins. Cyclin protein levels rise and fall during the cell cycle and in this way they periodically activate Cdk [2].

In primitive eukaryotes such as yeast, a single Cdk (Cdc28 in *Saccharomyces cerevisiae* and Cdc2 in *Schizosaccharomyces pombe*) drives all phases of cell cycle by interaction with different cyclins. In mammalian cells, the sequential activation of at least four different Cdks (Cdk1, equivalent to Cdc2, Cdk2, Cdk4, and Cdk6) is responsible for the whole process. Accordingly, four cyclin

Liver Regeneration
http://dx.doi.org/10.1016/B978-0-12-420128-6.00011-7

FIGURE 11.1 Cell cycle is the central cross road of the liver regeneration signaling pathways. The proper activation is important for adequate hepatocytes proliferation, restoration of liver mass, and liver function. Whereas the dysregulation can drivehepatic cell straight to tumorigenesis.

families specifically activate corresponding Cdks via complex formation in different phases of cell cycle: cyclin D (D1, D2, and D3) activates Cdk4 and Cdk6; cyclin E (E1 and E2) activates Cdk2; cyclin A (A1 and A2) activates Cdk2 and Cdk1; and cyclin B (B1, B2 and B3) activates Cdk1 [3–5].

During the last decades, tremendous progress has been made in understanding the molecular model of the cell-cycle regulation. According to this model, the mammalian cell cycle starts with expression of D-type cyclins, which are induced by extracellular mitogens. In early G1-phase, D-type cyclins activate Cdk4 and Cdk6 and initiate the phosphorylation of the retinoblastoma protein (Rb) family [6]. The unphosphorylated Rb proteins repress gene transcription by binding to and thereby inhibiting E2F transcription factors. Phosphorylation of Rb leads to the release of E2F transcription factors and results in the initiation of E2F early responsive genes required for cell-cycle progression such as cyclin E. Cyclin E activates Cdk2 in the later G1-phase and completes the phosphorylation of Rb leading to further activation of E2F-mediated transcription. Cells are then prepared to move through the restriction point at boundary of G1/S-phase and to S-phase initiation [7–9].

S-phase of the cell cycle is defined as the time interval necessary for DNA synthesis. A-type cyclins are synthesized at the onset of the S-phase and orchestrate the process of DNA replication by forming a complex with Cdk2. Once DNA replication is finished, cells enter G2-phase, where Cdk1/cyclin A2 kinase complexes prepare the cells for mitosis. Finally, Cdk1/cyclin B complexes actively participate and complete mitosis [10–13].

The activities and functions of Cdk/cyclin complexes are regulated by two families of Cdk inhibitors; the INK4 family (p16, p15, p18, p19) specifically binds to Cdk4 and Cdk6 and prevents D-type cyclin activity and the cyclin-dependent kinase inhibitor (Cip/Kip) family (p21, p27, p57) inhibits Cdk2/cyclin E, Cdk2/cyclin A, Cdk1/cyclin A, as well as Cdk1/cyclin B activity [14].

The recent generation of knockout mouse models for individual cyclins and Cdks, or different combined deletions, has provided information about widespread redundancy in the physiological function of these proteins in development and cell homeostasis (Table 11.1). In this chapter, we will summarize the most important findings related to the function of Cdks and cyclins in the control of liver regeneration based on the information obtained from the phenotypical characterization of these gene-targeted mouse strains.

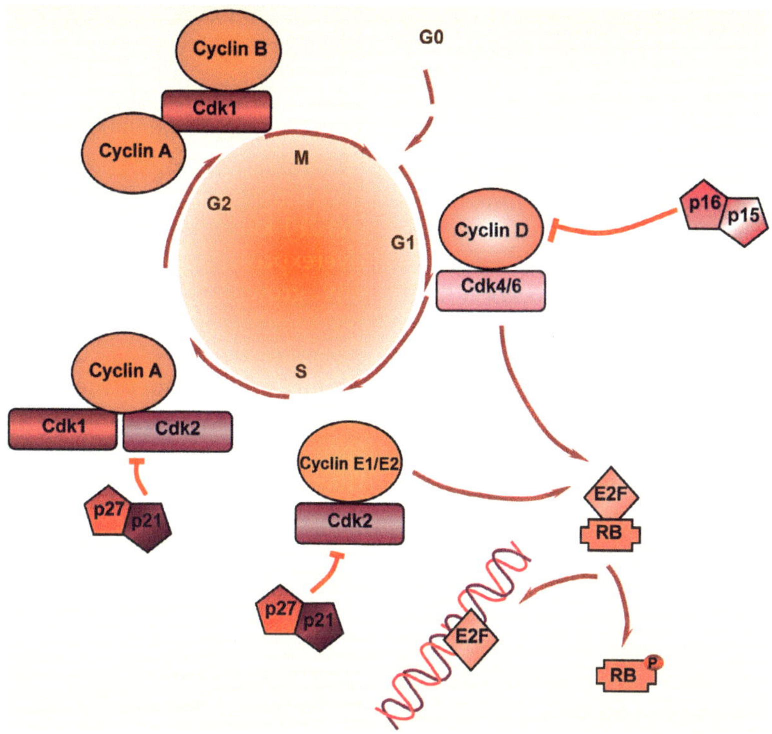

FIGURE 11.2 Cell-cycle machinery: specific Cdk/cyclin complexes regulate cell-cycle progression.

TABLE 11.1 Phenotype of Mice Deficient in Specific Genes Involved in Cell-Cycle Regulation After PH

Disrupted Gene	Phenotype After Partial Hepatectomy
E2F1	No detectable abnormalities [33]
Cyclin D1	Transiently delay the into S-phase entry [41]
Cyclin E1	Minor delay in S-phase entry [62]
Cyclin E2	Accelerated and extended DNA synthesis, hepatomegaly, and rampant endoreduplication [62]
CyclinE1/Cyclin E2	Impaired DNA replication and liver regeneration [64]
Cdk2	No detectable abnormalities [18,64]
Cdk2/Cdk4	No detectable abnormalities [67]
Cdk2/CyclinE1	Reduced hepatocyte proliferation and liver regeneration [64]
Cdk1	Liver regeneration is not impaired. No cell divisions, enlarged hepatocytes [92]
p21	Markedly accelerated hepatocytes progression through G1-phase [98], the degree of liver injury determines the effect [100]
p27	The timing of DNA synthesis is significantly accelerated [104]
SKP2	Restoration of liver mass and function achieves not by proliferation but by enlargement of cells [104]
p27/SKP2	Normal kinetic of cell-cycle entry and progression [105]
p18	No obvious phenotype [106]
p18/p21	Shortened G1-phase, early onset of cyclin D and PCNA expression and Cdk2 activation [106]
p18/p27	G1 was unchanged, but amount of proliferating hepatocytes 48 h after PH was elevated [106]

11.2 Rb AND E2F1

The **retinoblastoma tumor suppressor (Rb)** plays a critical role in cellular proliferation control and inhibition of oncogenic transformation. Inactivation of both Rb alleles in mice results in unscheduled cell proliferation, apoptosis, and widespread development defects, leading to embryonic death by day 14.5 [25].

Studies using liver-specific Rb knockout mice showed that pRB loss is sufficient to deregulate E2F target gene expression and cell-cycle control in postnatal hepatocytes. However, despite this deregulated cell-cycle progression, pRB loss is compensated for such that in adult mice pRB-deficient hepatocytes are biochemically normal and do not exhibit hyperplasia. Consistent with compensation within the animal, the acute elimination of pRB in adult liver is sufficient to deregulate E2F target gene expression and elicit cell-cycle entry. As a consequence of pRB loss, both chronic and acute, there is a significant development of aberrant hepatocyte ploidy [26].

The Rb protein has been shown to bind and regulate a large number of cellular proteins including members of the E2F family. The transcription factors of the E2F family are critical regulators of numerous genes including those important for cell-cycle progression (e.g., cyclin E and cyclin A) and DNA replication. Inhibition or lack of E2F activity will block G1 to S-phase progression in mammalian cells. Consistently, ectopic expression of several E2F proteins is sufficient to induce S-phase in quiescent cells. Various properties of the individual E2F family members suggest distinct functional role for these proteins: the E2F1, E2F2, and E2F3 activities appear to play a positive role in cell-cycle progression, while E2F4, E2F5, and E2F6 most likely contribute to the repression of cell growth [27–29].

E2F1 is the best-characterized member of the E2F family. Fundamental *in vivo* studies demonstrated that the quiescent embryonic myofibroblasts derived from E2F1-nullizygous mice have a defect in cell-cycle reentry [30]. Indeed, E2F1 deficiency in liver results only in minor changes in gene expression. The overall rate and timing of liver regeneration is not altered in E2F1 knockout mice. It is also noteworthy that overexpression of E2F1 in liver does not provide any growth advantage during compensatory regeneration after partial hepatectomy (PH). Furthermore, the loss of E2F1 in liver has no effect on tumor growth after diethylnitrosamine treatment. In sharp contrast, loss of E2F1 results in increased tumor development in the lung, reproductive tract, and lymphatic system. Evidence is mounting that the role of E2F1 in the transition of resting hepatocytes to a proliferating state is probably diverse in different tissues [31–33].

Indeed the overexpression of E2F-1 in the liver leads to the development of hepatocellular adenoma and, ultimately, hepatocellular carcinoma (HCC) in transgenic mice with 100% incidence by 10 months [32].

11.3 CYCLIN D-Cdk4/Cdk6

D-type cyclins represent very unique components of cell-cycle control. As main target of - several signal transduction pathways they represent a crucial crossroad and serve as "links" between the extracellular environment and the core of cell-cycle machinery [34,35]. The D-type cyclin family consist of cyclins D1, D2, and D3 [4]. It has been shown that knockout mice lacking individual D-cyclins are viable, and display only minor phenotypes, revealing that these proteins are dispensable for development of the overwhelming majority of organs [36].

All three D-cyclin family members show significant similarity on the protein level and in their biochemical functions during G1-phase. However, their role appears to vary significantly between different cell types [15]. Hence, the requirement for a particular cyclin in a given tissue is likely caused by specific transcription factors, rather than by unique properties of cyclins [37].

For example, cyclin D1 is a pivotal mediator of G1 progression in many cell types, including hepatocytes. It is undetectable in quiescent mouse liver and is markedly upregulated during G1-phase after PH (approximately 12 h in rat) [38,39]. Furthermore, the transfection of hepatocytes with cyclin D1 is sufficient to trigger hepatocyte proliferation *in vivo* and results in liver enlargement [40]. The role of cyclin D1 for liver regeneration has been studied in respective knockout mice in combination with mitogens such as TCPOBOP. Although lack of cyclin D1 expression transiently delays S-phase entry for a relatively short time, this cyclin seems not to be essential for liver growth and hepatocyte proliferation [41].

While the role of cyclin D1 in liver regeneration has been well documented, the involvement of other D-type cyclins is less clear. For instant, cyclin D2 is modestly upregulated and barely detectable during cell-cycle progression [42]. In contrast, cyclin D3 is readily detectable in quiescent liver but shows only minor changes after PH and may not be directly involved in the mitogenic response. Possibly, it supports some unidentified differentiated functions of hepatocytes [43].

Recent studies using transgenic mice demonstrate that overexpression of cyclin D1 specifically in hepatocytes eventually leads to the development of multiple adenomatous lesions and ultimately to HCC with a high degree of penetrance [44]. In addition to the roles of overexpressed D-cyclins in tumor initiation, recent studies clearly showed that the presence of D-cyclins is essential for tumor maintenance. Moreover, higher cyclin D1 expression is associated with aggressive forms of HCC [45–47].

The catalytic partners of the D-type cyclins, Cdk4 and Cdk6, are closely related proteins with basically indistinguishable biochemical properties. Mice lacking individual Cdk4 or Cdk6 are viable and display very minor tissue-specific defects indicating that Cdk4 and Cdk6 can compensate for each other in most cells. However, specific roles of Cdk4 or Cdk6 were also observed in several cell types, which might be attributed to distinct substrate specificity and subcellular localization [16,48,49]. It is interesting to note that mice lacking cyclin D, or their catalytic partners Cdk4 and Cdk6, displayed very similar phenotype. Embryos with combined depletion of Cdk4 and Cdk6 die during the late stages of embryonic development due to severe anemia. However, these embryos display normal organogenesis and most cell types proliferate normally. Moreover, quiescent Cdk4/Cdk6-null cells enter S-phase with normal kinetics although with lower efficiency. These results indicate that D-type Cdk are not essential for cell-cycle reentry. The molecule possibly responsible for this compensatory activity is probably Cdk2 [34,50].

Cdk4 complexes were also shown to play a key role in the initiation of malignant growth, as evidenced by the reduced susceptibility of mice lacking Cdk4 or Cdk6 to several types of tumors. However, it remains to be investigated if conditional ablation of Cdk4 or Cdk6 after tumor initiation blocks tumor progression. These studies are of critical importance as cyclin D/Cdk4-Cdk6 complexes are potential molecular targets to treat cancer also in humans [34,51–54].

11.4 CYCLIN E-Cdk2

In contrast to the growth factor inducible D-type cyclins, the expression of E-type cyclins is controlled by an autonomous mechanism and peaks sharply at G1/S border [55]. After passing the restriction point, the cell cycle becomes substantially less responsive to extracellular factors, which can delay entry into the S-phase or even arrest the cell cycle. Regulation of cell-cycle progression through the restriction point is believed to be the main function of the E-cyclins [8].

Two members of cyclin E family have been described: cyclin E1 and more recently discovered cyclin E2. Both show significant amino acid similarity (75% within the cyclin box, 47% throughout the entire coding sequence); nevertheless, there is increasing evidence for their differential and nonredundant functions [56–59].

Single genetic deletion of either E-type cyclin does not affect viability or embryonal development in mice. However, E-cyclins were shown to be essential for the transition from the quiescent state into the active cell cycle, because cyclin E1/E2 double-deficient fibroblasts are unable to reenter the cell cycle from starvation-induced quiescence [60]. After PH, normally quiescent hepatocytes reenter the cell cycle; therefore, it has been suggested that E-cyclins must be important for the regenerating liver [61].

The first attempts to untangle the role of E-type cyclins for liver regeneration demonstrated that cyclin E1 and cyclin E2 have unique functions after PH and even play antagonistic roles. Single gene deletions showed that cyclin E1 knockout mice display a relatively normal regenerative response with a minor delay in S-phase entry. In contrast, cyclin E2 knockout mice overexpress cyclin E1 and "hyperactivate" Cdk2, resulting in accelerated and extended DNA synthesis, hepatomegaly, and enhanced endoreduplication [62]. Likewise, complete loss of cyclin E in the liver results in impaired DNA replication and liver regeneration and clearly demonstrates that cyclin E2 shares at least some overlapping functions with cyclin E1 [63,64].

Cdk2 is the canonical binding partner of E-type cyclins, once believed to be essential for G1/S transition in all mammalian cells [65]. Surprisingly, genetic knockout experiments in mice demonstrated that Cdk2 is largely dispensable for cell proliferation or development *in vivo* [65–67]. Moreover, the deletion of Cdk2 does not reproduce any of the phenotypes observed in cyclin E1/E2 double knockout mice indicating that at least some of the functions of E-type cyclins are Cdk2 independent [60,65]. These observations have been attributed, at least in part to compensatory activities between Cdk2 and Cdk4. Indeed, Cdk2/Cdk4 double knockout mice die during embryogenesis around E15 as results of heart defect [68]. Nevertheless, there are specific requirements for Cdk2 in some cell types [17].

Two independent studies using knockout mice lacking Cdk2 in the liver clearly demonstrated that Cdk2 is also fully dispensable for cell-cycle progression in hepatocytes [18,64]. Moreover, Cdk2/Cdk4 double knockout mice recover normally after PH [67].

Importantly, the conditional deletion of Cdk2 alone in hepatocytes resulted in normal initiation of S-phase *in vivo* and *in vitro*, but otherwise accelerated induction of cyclin E1. The excessive cyclin E1 does not contribute to a canonical kinase activity, but is found to be associated with the chromatin together with components of the pre-replication complex (pre-RC), such as the mini chromosome maintenance helicase. This kinase-independent, enhanced association of cyclin E1 with chromatin is absolutely essential for S-phase reentry in Cdk2-deficient hepatocytes, because combined ablation of Cdk2 and cyclin E1 abolished mini chromosome maintenance complex loading and induction of the S-phase. This result provides confirmation that combined genetic ablation of complete cyclin E1/Cdk2 complex leads to dramatically impaired S-phase progression and cell death *in vitro* and substantially reduced hepatocyte proliferation and liver regeneration after PH *in vivo* (Figure 11.3). The loss of Cdk2, in mice lacking both cyclin E1 and E2 causes no additional deficiencies, supporting cyclin E1 as key player and thus the most important kinase subunit for efficient hepatocyte proliferation [63,64].

Apart from the contribution to normal cell proliferation and development, aberrant cyclin E expression became a subject of particular interest in the field of liver disease. For instance, cyclin E1 has been recently detected as novel key mediator of hepatic fibrosis due to its essential role for the proliferation and survival of hepatic stellate cells [19].

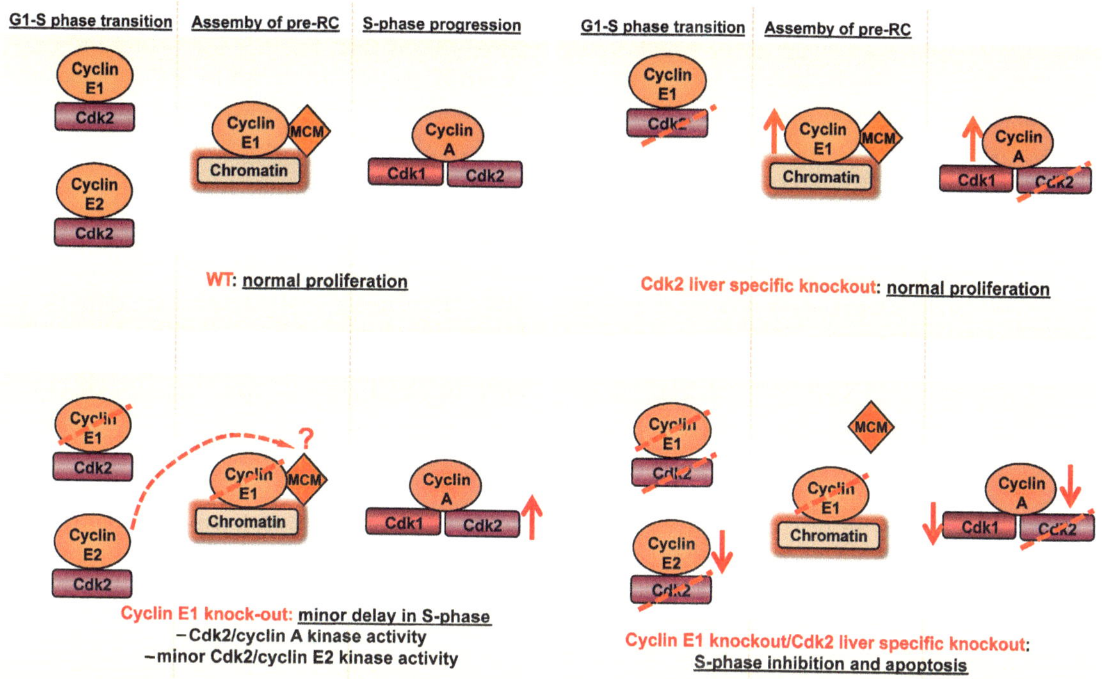

FIGURE 11.3 Model explaining the hepatic phenotype of mice deficient for cyclin E1 or/and Cdk2 in liver. *Modified from Ref. [18].*

Several reports have shown that overexpression of cyclin E1 accelerates G1-phase, decreases size of mitotic cells, promotes cell division independent from growth factors, and causes chromosomal instability in cancer cells [21,69,70]. Hence, deregulated cyclin E1 has been shown to act as a potent oncogene in some types of human cancer, including HCC and often is associated with poor survival of patients [71,72].

In this regard, recent studies provide a novel therapeutic view to inhibit proliferation of cancer cells by inhibiting cyclin E1 together with Cdk2 or both E-type cyclins. Indeed, the inhibition of Cdk2 alone might not be sufficient to block cell proliferation of certain cancer cells. Hence, in the future, cyclin E may become potentially a powerful tool in redefining prognosis and treatment [20,73,74].

11.5 CYCLIN A

Cyclin A is an essential component of the cell-cycle machinery. Cyclin A is especially interesting as it can activate two different cyclin-dependent kinases (Cdk2 and Cdk1) and function in both S-phase and mitosis [10]. Mammalian cells encode two A-type cyclins: cyclin A1 and cyclin A2. Cyclin A1 is expressed almost exclusively in the testes and during meiosis in the male germline. Consistently, male knockout mice lacking cyclin A1 are sterile [75,76]. The second mammalian A-type cyclin, cyclin A2, is ubiquitously expressed in all proliferating cells and has functions in both S-phase and mitosis [11,46,77].

During cell-cycle progression, cyclin A level increases at the onset of S-phase and contributes to the stimulation of DNA synthesis. In early mitosis, it drives the initiation of chromosome condensation and nuclear envelope breakdown [11,78,79].

Constitutive cyclin A2 knockout mice die at embryonic stage E5.5, indicating an essential role for cyclin A2 in embryonic development [80]. The experiments with conditional knockout mice revealed that cyclin A2 is dispensable in mouse embryonic fibroblasts (MEFs) but essential for cell-cycle progression of hematopoietic and embryonic stem cells. This was attributed to the role of E-type cyclins, which are expressed at sufficient levels in MEFs to compensate for absence of cyclin A2, but not in the two studied stem cell types [79].

In regenerating liver, proliferation of hepatocytes is associated with upregulation and intracellular localization of cyclin A. PH in rats reveals that the level of cyclin A increasing during G1 (at 6 h after hepatectomy) and reaching maximal levels during S-phase (at 18 h) [81,82]. Additionally, recent experiments with hepatocyte-specific knockout mice clearly demonstrated that loss of cyclin A2 resulted in a significant delay of hepatic tumor formation [83].

11.6 CYCLIN B

The entry into mitosis is under the control of B-type cyclins [84]. Two B-type cyclins, B1 and B2, have been identified in mammals. Proliferating cells express both cyclins, which bind to and activate Cdk1. Experiments with knockout mice demonstrated that the two B-type cyclins have distinct roles. Cyclin B2 knockout mice develop normally and did not display any obvious abnormalities, except reduced body size. In sharp contrast, cyclin B1 proved to be an essential gene and its deletion in mice leads to embryonic lethality. This observation needs to be seen in the context that most proliferating cells show higher levels of cyclin B1 expression [85]. For example, in the liver, quiescent hepatocytes sustain a basal rate of cyclin B1 transcription. After PH in rats, there is a gradual increase in cyclin B1 expression with biphasic peaks at 42-48 h and at 72 h correlating with phosphorylation, and nuclear translocation of cyclin B1 [86,87].

A recent study demonstrated that cyclin B1 and B2 transgenic mice are highly prone to tumors, due to chromosome missegregation and improper centrosome separation [88]. Deregulated expression of cyclin B1 has been reported in different types of tumors including HCC. Interestingly, there is evidence that abnormal expression of cyclin B1 is a result of p53 inactivation. Therefore, cyclin B1 can be proposed as an appropriate tumor antigen applicable to all tumor types for which p53 mutation is the common denominator [89–91].

11.7 Cdk1

Cdk1 can interact with either A-type cyclins or B-type cyclins. Due to the essential role of Cdk1 for mitosis, deletion of Cdk1 leads to early embryonic lethality [92]. No evidence so far indicated that other Cdks could compensate for the loss of Cdk1. Even when Cdk2 is directly expressed from the Cdk1 locus, it still leads to early embryonic lethality [5,14]. Moreover, it has been shown that Cdk1 activity alone can drive the mammalian cell cycle through cell division, as in case in unicellular organisms such as yeast. The mouse embryos lacking all interphase Cdks (Cdk 2, 3, 4, and 6) undergo organogenesis and develop to midgestation. In these embryos, Cdk1 binds to all cyclins, resulting in the phosphorylation of the Rb and the expression of genes that are regulated by E2F transcriptional factors [17].

Interestingly, liver-specific deletion of Cdk1 is well tolerated, and liver regeneration after PH is not impaired. However, this occurs in the absence of cell divisions, resulting in enlarged hepatocytes. The large increase in cell size is accompanied by downregulation of mitochondrial gene expression, similar to that observed in diabetic individuals [92,93]. More importantly, loss of Cdk1 in the liver confers complete resistance against tumorigenesis induced by activated Ras and silencing of p53. Consistently, another *in vitro* study clearly demonstrates that targeting Cdk1 rather than Cdk4/6 or Cdk2 selectively induced apoptosis and reduced viability of myc-dependent breast cancer cells [94].

These discoveries are in sharp contrast to other Cdks, whose inactivation had little effect on tumor formation. Altogether, it indicates the potential of Cdk1 inhibitors in cancer therapy if we can prevent the detrimental side effects resulting from unintentionally interfering with the essential functions of Cdk1 in proliferative tissues [92].

11.8 CELL-CYCLE INHIBITORS

The activities of cyclin/Cdks complexes are in turn constrained by Cdk inhibitors. Cdk-inhibitory proteins that govern these events have been subdivided into two families based on their structures and Cdk targets: the Cip/Kip family (p21, p27) and INK4 family (p15, p16, p18, and p19). *In vitro,* INK4 family members specifically form inactive stable binary complexes with Cdk4 and Cdk6, whereas the CIP/KIP inhibitors have broader substrate specificity and inhibit kinase function through the formation of ternary cyclin/Cdk complexes [6,95].

p21 (Cip1) mediates its various biological activities primarily by binding to and inhibiting the kinase activity of the cyclin-dependent kinases Cdk2 and Cdk1 leading to growth arrest at specific stages in the cell cycle. In addition, by binding to proliferating cell nuclear antigen (PCNA), p21 interferes with PCNA-dependent DNA polymerase activity, thereby inhibiting DNA replication and modulating various PCNA-dependent DNA repair processes [96]. Mice homozygous for a null mutation in the *p21* gene, however, developed normally, suggesting that p21 is not required for normal differentiation [97].

In regenerating liver, p21 is induced during the prereplicative phase and is maximally expressed after peak hepatocyte DNA synthesis [98]. Inappropriate expression of p21 leads to severe abnormalities due to improper control of the hepatocyte cell cycle. Overexpression of p21 impairs hepatocyte proliferation during liver development and after PH. Hepatocytes from hepatectomized transgenic mice display almost complete inhibition of induced cell division [99]. Conversely, p21 knockout mice after PH demonstrate evidence of markedly accelerated hepatocytes progression through G1-phase. DNA synthesis, upregulation of cyclin A and PCNA, induction of cyclin D1- and Cdk2-associated kinase activity, and appearance of Rb phosphorylation occurred earlier in the p21 knockout mice [98]. However, the degree of overall (acute and chronic) liver injury determines the strength of p21 induction in the liver and, subsequently, its effect on hepatocyte proliferation. As it has been shown, that hepatocyte proliferation is significantly reduced in p21-deficient mice with moderate liver injury [100]. Despite its profound role in halting cellular proliferation and its ability to promote differentiation and cellular senescence, recent studies suggest that, under certain conditions, p21 can promote cellular proliferation and oncogenicity.

p27 (also known as KIP1) is an atypical tumor suppressor that regulates G0- to S-phase transition by binding to and regulating the activity of Cdks [22]. Mice in which the expression of p27 or of its regulators is deficient exhibit an imbalance between cell growth and proliferation. For example, p27-deficient mice manifest multiple organ hyperplasia and a decrease in the size of individual cells [101]. In contrast, mice lacking S-phase kinase-associated protein 2 (Skp2), a F-box protein of the SCF ubiquitin ligase complex that targets p27 for degradation, exhibit cellular accumulation of p27, a reduction in both body size and the size of individual organs, an increase in the mass of individual cells, and polyploidy [102]. These observations suggest that p27 is a determinant of organ size [103].

After PH in p27-deficient mice, the timing of DNA synthesis is significantly accelerated, but timing of S-phase entry most strongly affects the cells around the pericentral region, without affecting the total cell population [104]. Degradation of p27 is required for the transition of hepatic cells from quiescence to the proliferative state. Skp2 is a rate-limiting component of the machinery that ubiquitinates and degrades phosphorylated p27. Restoration of liver mass and function in the Skp2 knockout mice after PH is achieved not by cellular proliferation but by the enlargement of individual cells. Skp2 knockout cells are able to enter S-phase but not mitosis, a characteristic of endoreplication. The enlargement and polyploidy of Skp2 knockout hepatocytes are consistent with the repeated occurrence of S-phase without mitosis [103]. Indeed, double Skp2/p27 knockout mice subjected to PH showed normal kinetics of cell-cycle entry and progression [105].

p18, a member of the INK4 family of Cdk inhibitors, suppresses cell growth when ectopically expressed in an Rb-dependent manner. Moreover, it is expressed during mouse embryogenesis, and mice lacking functional p18 protein exhibit a phenotype characterized by increased body size and organomegaly [23]. p18 protein is expressed in the quiescent liver and is of functional relevance in regulating cell-cycle progression of the liver in adult mice. p18 knockout animals have no obvious phenotype in the replicative response of hepatocytes after PH, mainly due to functional redundancy of p18 with other cell-cycle inhibitors. However, combining p18 knockout mice with mice lacking genes from the Cip/Kip family of Cdk inhibitors revealed a role of p18 in the control of liver growth. Such as in p21/p18 knockout animals, the G1-phase was shortened as evidence by an earlier onset of cyclin D and PCNA expression and Cdk2 activation after surgery. After PH in double p18/p21 knockout mice the G1 was unchanged, but 48 h after PH the amount of proliferating hepatocytes was elevated. This means that the two different Cdk inhibitors normally collaborate in the regulation of hepatocytes entering S-phase [106]. Treatment of p18 null and heterozygous mice with a chemical carcinogen resulted in accelerated tumor development in many organs, including the liver. Interestingly, the remaining p18 wild-type allele is neither mutated nor silenced in tumors derived from heterozygote animals. Indicating that, p18 is a haploinsufficient tumor suppressor in mice [24].

p15 expression in mouse liver increases during and beyond the first postnatal week to levels that is highest in adult animals. Therefore, p15 may contribute to the maintenance of adult hepatocytes in a quiescent state. The functional significance of high p15 expression in adult liver is further supported by the finding that PH is associated with a rapid loss of nuclear p15. Consistently, it has been shown that c-myc-induced hepatocarcinogenesis in transgenic mice is associated with disrupted induction of p15 [107]. Finally, deletion of p15 has been observed in human HCC [108].

11.9 CONCLUDING REMARKS

Liver regeneration after PH provides an excellent system to tackle basic biological questions. Through the characterization of knockout mice of members of the cell-cycle machinery, our knowledge in its control during liver regeneration has significantly improved. Extensive analyses of these mouse models revealed that most of the Cdks and cyclins, originally thought essential for cell-cycle control, are in fact largely dispensable. However, individual cyclins and Cdks have been shown to have a cell-type specific role during liver regeneration.

Understanding the mechanisms underlying liver regeneration is clinically important because hepatectomy is a practical treatment for liver tumors, and liver transplantation is an important therapeutic option in patients with severe liver diseases.

Additionally, it is essential to better understand why normal hepatocytes and hepatic tumor cells have specific requirements for individual Cdks or cyclins. However, the therapeutic value of inhibiting the activity of Cdks and cyclins deserves more in detail evaluation. The implication of individual cyclins and Cdks in hepatocyte proliferation and their therapeutic potential as targets for anti-tumor drugs needs to be revised according to the recent findings in transgenic and gene-targeted mice. Synthesis of more selective and potent inhibitors would be required [16,50,109].

We believe that further understanding of the mechanism of cell-cycle progression during liver regeneration will lead to the development of promising therapeutic strategies [110].

References

[1] Boonstra J. Progression through the G1-phase of the on-going cell cycle. J Cell Biochem 2003;90:244–52.

[2] Vermeulen K, Van Bockstaele DR, Berneman ZN. The cell cycle: a review of regulation, deregulation and therapeutic targets in cancer. Cell Prolif 2003;36:131–49.

[3] Nigg EA. Cyclin-dependent protein kinases: key regulators of the eukaryotic cell cycle. Bioessays 1995; 17:471–80.

[4] Sherr CJ. D-type cyclins. Trends Biochem Sci 1995;20: 187–90.

[5] Satyanarayana A, Kaldis P. Mammalian cell-cycle regulation: several Cdks, numerous cyclins and diverse compensatory mechanisms. Oncogene 2009;28:2925–39.

[6] Sherr CJ, Roberts JM. CDK inhibitors: positive and negative regulators of G1-phase progression. Genes Dev 1999;13:1501–12.

[7] Giacinti C, Giordano A. RB and cell cycle progression. Oncogene 2006;25:5220–7.

[8] Moroy T, Geisen C. Cyclin E. Int J Biochem Cell Biol 2004;36:1424–39.

[9] Grana X, Garriga J, Mayol X. Role of the retinoblastoma protein family, pRB, p107 and p130 in the negative control of cell growth. Oncogene 1998;17:3365–83.

[10] Pagano M, Pepperkok R, Verde F, Ansorge W, Draetta G. Cyclin A is required at two points in the human cell cycle. Embo J 1992;11:961–71.

[11] Yam CH, Fung TK, Poon RY. Cyclin A in cell cycle control and cancer. Cell Mol Life Sci 2002;59:1317–26.

[12] Porter LA, Donoghue DJ. Cyclin B1 and CDK1: nuclear localization and upstream regulators. Prog Cell Cycle Res 2003;5:335–47.

[13] Yu H, Yao X. Cyclin B1: conductor of mitotic symphony orchestra. Cell Res 2008;18:218–20.

[14] Satyanarayana A, Berthet C, Lopez-Molina J, Coppola V, Tessarollo L, Kaldis P. Genetic substitution of Cdk1 by Cdk2 leads to embryonic lethality and loss of meiotic function of Cdk2. Development 2008;135:3389–400.

[15] Musgrove EA, Caldon CE, Barraclough J, Stone A, Sutherland RL. Cyclin D as a therapeutic target in cancer. Nat Rev Cancer 2011;11:558–72.

[16] Tsutsui T, Hesabi B, Moons DS, Pandolfi PP, Hansel KS, Koff A, et al. Targeted disruption of CDK4 delays cell cycle entry with enhanced p27(Kip1) activity. Mol Cell Biol 1999;19:7011–9.

[17] Santamaria D, Barriere C, Cerqueira A, Hunt S, Tardy C, Newton K, et al. Cdk1 is sufficient to drive the mammalian cell cycle. Nature 2007;448:811–5.

[18] Satyanarayana A, Hilton MB, Kaldis P. p21 Inhibits Cdk1 in the absence of Cdk2 to maintain the G1/S phase DNA damage checkpoint. Mol Biol Cell 2008;19:65–77.

[19] Nevzorova YA, Bangen JM, Hu W, Haas U, Weiskirchen R, Gassler N, et al. Cyclin E1 controls proliferation of hepatic stellate cells and is essential for liver fibrogenesis in mice. Hepatology 2012;56:1140–9.

[20] Donnellan R, Chetty R. Cyclin E in human cancers. FASEB J 1999;13:773–80.

[21] Spruck CH, Won KA, Reed SI. Deregulated cyclin E induces chromosome instability. Nature 1999;401: 297–300.

[22] Chu IM, Hengst L, Slingerland JM. The Cdk inhibitor p27 in human cancer: prognostic potential and relevance to anticancer therapy. Nat Rev Cancer 2008;8:253–67.

[23] Franklin DS, Godfrey VL, Lee H, Kovalev GI, Schoonhoven R, Chen-Kiang S, et al. CDK inhibitors p18(INK4c) and p27(Kip1) mediate two separate pathways to collaboratively suppress pituitary tumorigenesis. Genes Dev 1998;12:2899–911.

[24] Bai F, Pei XH, Godfrey VL, Xiong Y. Haploinsufficiency of p18(INK4c) sensitizes mice to carcinogen-induced tumorigenesis. Mol Cell Biol 2003;23:1269–77.

[25] Jacks T, Fazeli A, Schmitt EM, Bronson RT, Goodell MA, Weinberg RA. Effects of an Rb mutation in the mouse. Nature 1992;359:295–300.

[26] Mayhew CN, Bosco EE, Fox SR, Okaya T, Tarapore P, Schwemberger SJ, et al. Liver-specific pRB loss results

in ectopic cell cycle entry and aberrant ploidy. Cancer Res 2005;65:4568–77.

[27] Leone G, Sears R, Huang E, Rempel R, Nuckolls F, Park CH, et al. Myc requires distinct E2F activities to induce S phase and apoptosis. Mol Cell 2001;8:105–13.

[28] Kowalik TF, DeGregori J, Schwarz JK, Nevins JR. E2F1 overexpression in quiescent fibroblasts leads to induction of cellular DNA synthesis and apoptosis. J Virol 1995;69:2491–500.

[29] Shan B, Lee WH. Deregulated expression of E2F-1 induces S-phase entry and leads to apoptosis. Mol Cell Biol 1994;14:8166–73.

[30] Wang ZM, Yang H, Livingston DM. Endogenous E2F-1 promotes timely G0 exit of resting mouse embryo fibroblasts. Proc Natl Acad Sci USA 1998;95:15583–6.

[31] Yamasaki L, Jacks T, Bronson R, Goillot E, Harlow E, Dyson NJ. Tumor induction and tissue atrophy in mice lacking E2F-1. Cell 1996;85:537–48.

[32] Conner EA, Lemmer ER, Omori M, Wirth PJ, Factor VM, Thorgeirsson SS. Dual functions of E2F-1 in a transgenic mouse model of liver carcinogenesis. Oncogene 2000;19:5054–62.

[33] Lukas ER, Bartley SM, Graveel CR, Diaz Z,M, Dyson N, Harlow E. No effect of loss of E2F1 on liver regeneration or hepatocarcinogenesis in C57BL/6J or C3H/HeJ mice. Mol Carcinog 1999;25:295–303.

[34] Kozar K, Sicinski P. Cell cycle progression without cyclin D-CDK4 and cyclin D-CDK6 complexes. Cell Cycle 2005;4:388–91.

[35] Crescenzi M. Levels matter: miR-206 and cyclin D1. Cell Cycle 2013;12:3715.

[36] Sherr CJ, Roberts JM. Living with or without cyclins and cyclin-dependent kinases. Genes Dev 2004;18:2699–711.

[37] Ciemerych MA, Kenney AM, Sicinska E, Kalaszczynska I, Bronson RT, Rowitch DH, et al. Development of mice expressing a single D-type cyclin. Genes Dev 2002;16:3277–89.

[38] Albrecht JH, Hu MY, Cerra FB. Distinct patterns of cyclin D1 regulation in models of liver regeneration and human liver. Biochem Biophys Res Commun 1995;209:648–55.

[39] Corlu A, Loyer P. Regulation of the g1/s transition in hepatocytes: involvement of the cyclin-dependent kinase cdk1 in the DNA replication. Int J Hepatol 2012;2012:689324.

[40] Nelsen CJ, Rickheim DG, Timchenko NA, Stanley MW, Albrecht JH. Transient expression of cyclin D1 is sufficient to promote hepatocyte replication and liver growth in vivo. Cancer Res 2001;61:8564–8.

[41] Ledda-Columbano GM, Pibiri M, Concas D, Cossu C, Tripodi M, Columbano A. Loss of cyclin D1 does not inhibit the proliferative response of mouse liver to mitogenic stimuli. Hepatology 2002;36:1098–105.

[42] Loyer P, Cariou S, Glaise D, Bilodeau M, Baffet G, Guguen-Guillouzo C. Growth factor dependence of progression through G1 and S phases of adult rat hepatocytes in vitro. Evidence of a mitogen restriction point in mid-late G1. J Biol Chem 1996;271:11484–92.

[43] Teixeira A, Chaverot N, Strosberg AD, Cazaubon S. Differential regulation of cyclin D1 and D3 expression in the control of astrocyte proliferation induced by endothelin-1. J Neurochem 2000;74:1034–40.

[44] Deane NG, Parker MA, Aramandla R, Diehl L, Lee WJ, Washington MK, et al. Hepatocellular carcinoma results from chronic cyclin D1 overexpression in transgenic mice. Cancer Res 2001;61:5389–95.

[45] Choi YJ, Li X, Hydbring P, Sanda T, Stefano J, Christie AL, et al. The requirement for cyclin D function in tumor maintenance. Cancer Cell 2012;22:438–51.

[46] Nishida N, Fukuda Y, Komeda T, Kita R, Sando T, Furukawa M, et al. Amplification and overexpression of the cyclin D1 gene in aggressive human hepatocellular carcinoma. Cancer Res 1994;54:3107–10.

[47] Nishida N, Fukuda Y, Ishizaki K, Nakao K. Alteration of cell cycle-related genes in hepatocarcinogenesis. Histol Histopathol 1997;12:1019–25.

[48] Rane SG, Dubus P, Mettus RV, Galbreath EJ, Boden G, Reddy EP, et al. Loss of Cdk4 expression causes insulin-deficient diabetes and Cdk4 activation results in beta-islet cell hyperplasia. Nat Genet 1999;22:44–52.

[49] Martin J, Hunt SL, Dubus P, Sotillo R,, Nehme-Pelluard F, Magnuson MA, et al. Genetic rescue of Cdk4 null mice restores pancreatic beta-cell proliferation but not homeostatic cell number. Oncogene 2003;22:5261–9.

[50] Malumbres M, Sotillo R, Santamaria D, Galan J, Cerezo A, Ortega S, et al. Mammalian cells cycle without the D-type cyclin-dependent kinases Cdk4 and Cdk6. Cell 2004;118:493–504.

[51] Miliani de Marval PL, Macias E, Rounbehler R, Sicinski P, Kiyokawa H, Johnson DG, et al. Lack of cyclin-dependent kinase 4 inhibits c-myc tumorigenic activities in epithelial tissues. Mol Cell Biol 2004;24:7538–47.

[52] Rodriguez-Puebla ML, Miliani de Marval PL, LaCava M, Moons DS, Kiyokawa H, Conti CJ. Cdk4 deficiency inhibits skin tumor development but does not affect normal keratinocyte proliferation. Am J Pathol 2002;161:405–11.

[53] Bullrich F, MacLachlan TK, Sang N, Druck T, Veronese ML, Allen SL, et al. Chromosomal mapping of members of the Cdc2 family of protein kinases, cdk3, cdk6, PISSLRE, and PITALRE, and a cdk inhibitor, p27Kip1, to regions involved in human cancer. Cancer Res 1995;55:1199–205.

[54] Cram EJ, Liu BD, Bjeldanes LF, Firestone GL. Indole-3-carbinol inhibits CDK6 expression in human MCF-7

breast cancer cells by disrupting Sp1 transcription factor interactions with a composite element in the CDK6 gene promoter. J Biol Chem 2001;276:22332–40.
[55] Koff A, Giordano A, Desai D, Yamashita K, Harper JW, Elledge S, et al. Formation and activation of a cyclin E-cdk2 complex during the G1 phase of the human cell cycle. Science 1992;257:1689–94.
[56] Zariwala M, Liu J, Xiong Y. Cyclin E2, a novel human G1 cyclin and activating partner of CDK2 and CDK3, is induced by viral oncoproteins. Oncogene 1998;17: 2787–98.
[57] Gudas JM, Payton M, Thukral S, Chen E, Bass M, Robinson MO, et al. Cyclin E2, a novel G1 cyclin that binds Cdk2 and is aberrantly expressed in human cancers. Mol Cell Biol 1999;19:612–22.
[58] Caldon CE, Musgrove EA. Distinct and redundant functions of cyclin E1 and cyclin E2 in development and cancer. Cell Div 2010;5:2.
[59] Lauper N, Beck AR, Cariou S, Richman L, Hofmann K, Reith W, et al. Cyclin E2: a novel CDK2 partner in the late G1 and S phases of the mammalian cell cycle. Oncogene 1998;17:2637–43.
[60] Geng Y, Yu Q, Sicinska E, Das M, Schneider JE, Bhattacharya S, et al. Cyclin E ablation in the mouse. Cell 2003;114:431–43.
[61] Ehrenfried JA, Ko TC, Thompson EA, Evers BM. Cell cycle-mediated regulation of hepatic regeneration. Surgery 1997;122:927–35.
[62] Nevzorova YA, Tschaharganeh D, Gassler N, Geng Y, Weiskirchen R, Sicinski P, et al. Aberrant cell cycle progression and endoreplication in regenerating livers of mice that lack a single E-type cyclin. Gastroenterology 2009;137:691–703, e1-6.
[63] Stratton SA, Barton MC. Hierarchy of a regenerative cell cycle: cyclin E1 multitasks. Hepatology 2014;59: 370–1.
[64] Hu W, Nevzorova YA, Haas U, Moro N, Sicinski P, Geng Y, et al. Concurrent deletion of cyclin E1 and cyclin-dependent kinase 2 in hepatocytes inhibits DNA replication and liver regeneration in mice. Hepatology 2014;59:651–60.
[65] Ortega S, Prieto I, Odajima J, Martin A, Dubus P, Sotillo R, et al. Cyclin-dependent kinase 2 is essential for meiosis but not for mitotic cell division in mice. Nat Genet 2003;35:25–31.
[66] Berthet C, Aleem E, Coppola V, Tessarollo L, Kaldis P. Cdk2 knockout mice are viable. Curr Biol 2003;13:1775–85.
[67] Barriere C, Santamaria D, Cerqueira A, Galan J, Martin A, Ortega S, et al. Mice thrive without Cdk4 and Cdk2. Mol Oncol 2007;1:72–83.
[68] Berthet C, Klarmann KD, Hilton MB, Suh HC, Keller JR, Kiyokawa H, et al. Combined loss of Cdk2 and Cdk4 results in embryonic lethality and Rb hypophosphorylation. Dev Cell 2006;10:563–73.
[69] Resnitzky D, Gossen M, Bujard H, Reed SI. Acceleration of the G1/S phase transition by expression of cyclins D1 and E with an inducible system. Mol Cell Biol 1994;14:1669–79.
[70] Wimmel A, Lucibello FC, Sewing A, Adolph S, Muller R. Inducible acceleration of G1 progression through tetracycline-regulated expression of human cyclin E. Oncogene 1994;9:995–7.
[71] Ohashi R, Gao C, Miyazaki M, Hamazaki K, Tsuji T, Inoue Y, et al. Enhanced expression of cyclin E and cyclin A in human hepatocellular carcinomas. Anticancer Res 2001;21:657–62.
[72] Jung YJ, Lee KH, Choi DW, Han CJ, Jeong SH, Kim KC, et al. Reciprocal expressions of cyclin E and cyclin D1 in hepatocellular carcinoma. Cancer Lett 2001;168:57–63.
[73] Zhou Q, He Q, Liang LJ. Expression of p27, cyclin E and cyclin A in hepatocellular carcinoma and its clinical significance. World J Gastroenterol 2003;9:2450–4.
[74] Hwang HC, Clurman BE. Cyclin E in normal and neoplastic cell cycles. Oncogene 2005;24:2776–86.
[75] Sweeney C, Murphy M, Kubelka M, Ravnik SE, Hawkins CF, Wolgemuth DJ, et al. A distinct cyclin A is expressed in germ cells in the mouse. Development 1996;122:53–64.
[76] Hochegger H, Takeda S, Hunt T. Cyclin-dependent kinases and cell-cycle transitions: does one fit all? Nat Rev Mol Cell Biol 2008;9:910–6.
[77] Pines J, Hunter T, Bachs O. Human cyclin A is adenovirus E1A-associated protein p60 and behaves differently from cyclin B. Nature 1990;346:760–3.
[78] Mateo F, Vidal-Laliena M, Pujol MJ, Bachs O. Acetylation of cyclin A: a new cell cycle regulatory mechanism. Biochem Soc Trans 2010;38:83–6.
[79] Kalaszczynska I, Geng Y, Iino T, Mizuno S, Choi Y, Kondratiuk I, et al. Cyclin A is redundant in fibroblasts but essential in hematopoietic and embryonic stem cells. Cell 2009;138:352–65.
[80] Murphy M, Stinnakre MG, Senamaud-Beaufort C, Winston NJ, Sweeney C, Kubelka M, et al. Delayed early embryonic lethality following disruption of the murine cyclin A2 gene. Nat Genet 1997;15:83–6.
[81] Castro A, Jaumot M, Verges M, Agell N, Bachs O. Microsomal localization of cyclin A and cdk2 in proliferating rat liver cells. Biochem Biophys Res Commun 1994;201: 1072–8.
[82] Verges M, Castro A, Jaumot M, Bachs O, Enrich C. Cyclin A is present in the endocytic compartment of rat liver cells and increases during liver regeneration. Biochem Biophys Res Commun 1997;230:49–53.
[83] Gopinathan L, Tan SL, Padmakumar VC, Coppola V, Tessarollo L, Kaldis P. Loss of Cdk2 and cyclin A2

impairs cell proliferation and tumorigenesis. Cancer Res 2014;74:3870–9.
[84] Smits VA, Medema RH. Checking out the G(2)/M transition. Biochim Biophys Acta 2001;1519:1–12.
[85] Brandeis M, Rosewell I, Carrington M, Crompton T, Jacobs MA, Kirk J, et al. Cyclin B2-null mice develop normally and are fertile whereas cyclin B1-null mice die in utero. Proc Natl Acad Sci USA 1998;95:4344–9.
[86] Trembley JH, Kren BT, Steer CJ. Posttranscriptional regulation of cyclin B messenger RNA expression in the regenerating rat liver. Cell Growth Differ 1994;5:99–108.
[87] Freimuth J, Bangen JM, Lambertz D, Hu W, Nevzorova YA, Sonntag R, et al. Loss of caspase-8 in hepatocytes accelerates the onset of liver regeneration in mice through premature nuclear factor kappa B activation. Hepatology 2013;58:1779–89.
[88] Nam HJ, van Deursen JM. Cyclin B2 and p53 control proper timing of centrosome separation. Nat Cell Biol 2014;16:538–49.
[89] Yu M, Zhan Q, Finn OJ. Immune recognition of cyclin B1 as a tumor antigen is a result of its overexpression in human tumors that is caused by non-functional p53. Mol Immunol 2002;38:981–7.
[90] Warenius HM, Seabra L, Kyritsi L, White R, Dormer R, Anandappa S, et al. Theranostic proteomic profiling of cyclins, cyclin dependent kinases and Ras in human cancer cell lines is dependent on p53 mutational status. Int J Oncol 2008;32:895–907.
[91] Wang A, Yoshimi N, Ino N, Tanaka T, Mor H. Overexpression of cyclin B1 in human colorectal cancers. J Cancer Res Clin Oncol 1997;123:124–7.
[92] Diril MK, Ratnacaram CK, Padmakumar VC, Du T, Wasser M, Coppola V, et al. Cyclin-dependent kinase 1 (Cdk1) is essential for cell division and suppression of DNA re-replication but not for liver regeneration. Proc Natl Acad Sci USA 2012;109:3826–31.
[93] Miettinen TP, Pessa HK, Caldez MJ, Fuhrer T, Diril MK, Sauer U, et al. Identification of transcriptional and metabolic programs related to mammalian cell size. Curr Biol 2014;24:598–608.
[94] Kang J, Sergio CM, Sutherland RL, Musgrove EA. Targeting cyclin-dependent kinase 1 (CDK1) but not CDK4/6 or CDK2 is selectively lethal to MYC-dependent human breast cancer cells. BMC Cancer 2014;14:32.
[95] Sherr CJ, Roberts JM. Inhibitors of mammalian G1 cyclin-dependent kinases. Genes Dev 1995;9: 1149–63.
[96] Abbas T, Dutta A. p21 in cancer: intricate networks and multiple activities. Nat Rev Cancer 2009;9:400–14.
[97] Deng C, Zhang P, Harper JW, Elledge SJ, Leder P. Mice lacking p21CIP1/WAF1 undergo normal development, but are defective in G1 checkpoint control. Cell 1995;82:675–84.
[98] Albrecht JH, Poon RY, Ahonen CL, Rieland BM, Deng C, Crary GS, et al. Involvement of p21 and p27 in the regulation of CDK activity and cell cycle progression in the regenerating liver. Oncogene 1998;16:2141–50.
[99] Wu H, Wade M, Krall L, Grisham J, Xiong Y, Van Dyke T. Targeted in vivo expression of the cyclin-dependent kinase inhibitor p21 halts hepatocyte cell-cycle progression, postnatal liver development and regeneration. Genes Dev 1996;10:245–60.
[100] Buitrago-Molina LE, Marhenke S, Longerich T, Sharma AD, Boukouris AE, Geffers R, et al. The degree of liver injury determines the role of p21 in liver regeneration and hepatocarcinogenesis in mice. Hepatology 2013;58:1143–52.
[101] Nakayama K, Ishida N, Shirane M, Inomata A, Inoue T, Shishido N, et al. Mice lacking p27(Kip1) display increased body size, multiple organ hyperplasia, retinal dysplasia, and pituitary tumors. Cell 1996;85:707–20.
[102] Nakayama K, Nagahama H, Minamishima YA, Matsumoto M, Nakamich I, Kitagawa K, et al. Targeted disruption of Skp2 results in accumulation of cyclin E and p27(Kip1), polyploidy and centrosome overduplication. EMBO J 2000;19:2069–81.
[103] Minamishima YA, Nakayama K. Recovery of liver mass without proliferation of hepatocytes after partial hepatectomy in Skp2-deficient mice. Cancer Res 2002;62:995–9.
[104] Hayashi E, Yasui A, Oda K, Nagino M, Nimura Y, Nakanishi M, et al. Loss of p27(Kip1) accelerates DNA replication after partial hepatectomy in mice. J Surg Res 2003;111:196–202.
[105] Kossatz U, Dietrich N, Zender L, Buer J, Manns MP, Malek NP. Skp2-dependent degradation of p27kip1 is essential for cell cycle progression. Genes Dev 2004;18:2602–7.
[106] Luedde T, Rodriguez ME, Tacke F, Xiong Y, Brenner DA, Trautwein C. p18(INK4c) collaborates with other CDK-inhibitory proteins in the regenerating liver. Hepatology 2003;37:833–41.
[107] Santoni-Rugiu E, Jensen MR, Factor VM, Thorgeirsson SS. Acceleration of c-myc-induced hepatocarcinogenesis by co-expression of transforming growth factor (TGF)-alpha in transgenic mice is associated with TGF-beta1 signaling disruption. Am J Pathol 1999;154:1693–700.
[108] Awad MM, Sanders JA, Gruppuso PA. A potential role for p15(Ink4b) and p57(Kip2) in liver development. FEBS Lett 2000;483:160–4.
[109] Lapenna S, Giordano A. Cell cycle kinases as therapeutic targets for cancer. Nat Rev Drug Discov 2009;8: 547–66.
[110] Miyaoka Y, Miyajima A. To divide or not to divide: revisiting liver regeneration. Cell Div 2013;8:8.

CHAPTER

12

Changes in Hepatocyte Ploidy During Liver Regeneration

Andrew W. Duncan

Department of Pathology, McGowan Institute for Regenerative Medicine, University of Pittsburgh, Pittsburgh, Pennsylvania, USA

ABBREVIATIONS

2n	diploid cell
4n	tetraploid cell
8n	octaploid cell
Ch	chromosome
FACS	fluorescence-activated cell sorting
FISH	fluorescence *in situ* hybridization
FAH	fumarylacetoacetate hydrolase
HGD	homegentisic acid dioxygenase
NTBC	2-(2-nitro-4-trifluoro- methylbenzoyl)-1,3-cyclo-hexanedione

12.1 INTRODUCTION

The liver represents 2% of total body weight in humans and 5% in mice and is the largest solid organ in the body [1,2]. The liver is responsible for a myriad of functions, including synthesis of serum proteins, bile production/secretion, metabolism of carbohydrates, lipids, and amino acids, and detoxification of xenobiotic compounds. These roles are performed primarily by hepatocytes that comprise 70-80% of the liver mass. The liver is unique among other organs in its regenerative capacity. Hepatocyte turnover occurs slowly in adults with average hepatic lifespan ranging from 200 to 300 days [3]. It has been estimated that fewer than 0.1% of hepatocytes are cycling at any given time [2]. Although hepatocytes are mostly quiescent in noninjury situations, these cells harbor tremendous regenerative potential in response to liver injury. For example, rats survived and regenerated their livers in response to seven sequential 50% partial hepatectomy surgeries [4]. Similarly, mature mouse and human hepatocytes transplanted into mice undergoing liver failure proliferated 50-100-fold to restore the entire liver mass [5,6]. Moreover, serial transplantation revealed even greater proliferative capacity with 10^{20}-fold hepatocyte expansion. Mature hepatocytes are the primary drivers of liver regeneration in the adults, but hepatic stem/progenitor cells can also contribute to the regenerative response, especially when mature hepatocytes are inhibited [7–9]. Processes such as cell fusion [10–12] and transdifferentiation [13,14] have also been shown to promote liver regeneration.

In addition to the diverse functions required for homeostasis and regeneration, the liver is

Liver Regeneration
http://dx.doi.org/10.1016/B978-0-12-420128-6.00012-9

also characterized by variations in nuclear content. The most extensively described type of nuclear alteration in the liver is polyploidy, an increase in the number of chromosome sets per cell [15]. Although polyploid hepatocytes have been well described [16,17], the function of these cells is not well understood. A second and more recently characterized type of hepatic nuclear alteration is aneuploidy. Aneuploidy refers to the gain or loss of individual chromosomes. While aneuploidy is frequently associated with cancer, this does not seem to be the case in the liver. This chapter will focus on hepatic polyploidy and aneuploidy and explore how these processes cooperate to regulate liver function.

12.2 POLYPLOIDY IN THE LIVER

Polyploidy in the liver has been described in the literature for well over 100 years [18]. Initially, analysis of liver sections revealed a great deal of heterogeneity among hepatocytes. Cell and nuclear size varied between hepatocytes, as well as the number of nuclei per cell. We now know that hepatocyte ploidy depends on the DNA content of each nucleus (e.g., diploid, tetraploid, octaploid, etc.) plus the number of nuclei per cell [19]. The vast majority of hepatocytes are either mononucleate or binucleate, but rare trinucleate and tetranucleate hepatocytes are occasionally seen. Hepatic polyploidization is an age-dependent process. Most hepatocytes are diploid in young individuals, and in humans approximately half of adult hepatocytes are polyploid [20]. The degree of polyploidy is even more striking in rodents. Up to 90% of hepatocytes are polyploid in adult C57BL mice [21].

12.2.1 Mechanisms for Hepatic Polyploidization

Polyploid cells are born in a tissue-type specific manner [22]. For instance, cell fusion occurs between myoblasts to generate myofibrils [23] and macrophages to produce osteoclasts [24]; endoreplication involves DNA replication without nuclear division and occurs in megakaryocytes [25]; and cytokinesis failure can generate tetraploid cardiac cells [26] and cancer cells [27]. In the liver, the predominant mechanism leading to polyploidy is failed cytokinesis. Elegant studies in the rat by Desdouets and colleagues clearly elucidated the process (Figure 12.1). Prior

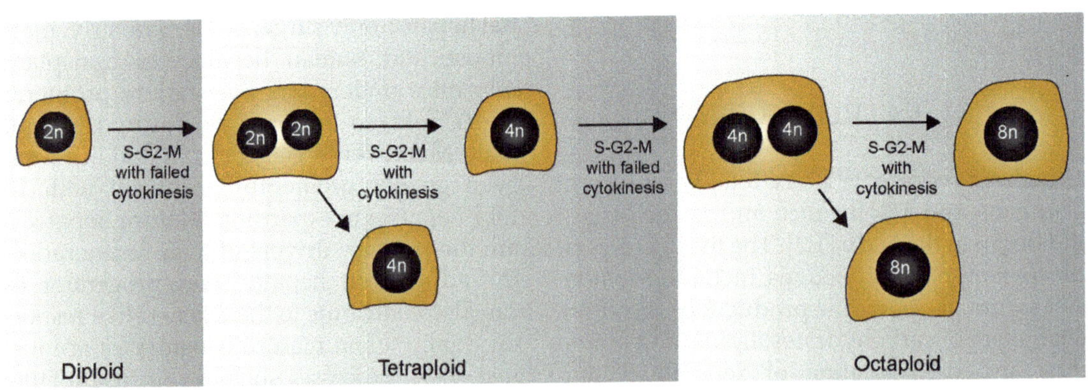

FIGURE 12.1 Polyploidization by failed cytokinesis. Diploid hepatocytes undergo a complete cell cycle with failed cytokinesis to form a binucleate tetraploid hepatocyte (where nuclei are 2n). The binucleate tetraploid completes a successful cell cycle plus mitosis, generating two mononucleate tetraploids (where each nucleus is 4n). A subsequent cell cycle plus mitosis with failed cytokinesis produces a binucleate octaploid (where each nucleus is 4n). The cycle continues, producing mononucleate octaploids, hexadecaploids, etc.

to weaning, rat hepatocytes are nearly exclusively diploid. Changes in insulin signaling occur at weaning in an AKT-dependent manner that induces cytokinesis failure [28]. First, a subset of diploid hepatocytes undergoes failed cytokinesis to generate tetraploid daughter cells, each with two diploid nuclei [19,29]. Second, binucleate tetraploid hepatocytes go through DNA replication with successful cytokinesis, generating pairs of mononucleate tetraploid hepatocytes (with a single tetraploid nucleus per cell). Next, a subset of mononucleate tetraploid hepatocytes undergoes cytokinesis failure during mitosis, leading to the formation of binucleate octaploid daughters. Finally, the process continues, generating mononucleate octaploids, binucleate hexadecaploids, and so on. In mice and humans, polyploidization is believed to also occur primarily by failed cytokinesis. However, the kinetics of polyploidization differs somewhat between mice and rats. Whereas polyploidization begins at weaning in rats, nearly half of mouse hepatocytes are already polyploid at this time [21].

In addition to insulin, many signals have been implicated in the regulation of hepatic polyploidy. Table 12.1 summarizes key genes that either promote or inhibit polyploidy in the liver, as observed in genetically modified mouse models. The table also summarizes stimuli, including partial hepatectomy [42,43], oxidative stress [44], and iron overload [45], that increase the degree of ploidy in the liver. Pathological

TABLE 12.1 Select Signals that Alter Polyploidy in the Liver

Signal	Manipulation	Effect on Hepatic Polyploidy	References
Genes			
Birc5 (Survivin)	Deletion via knockout mouse	Increased	[30]
Cdkn1a (p21)	Overexpression via transgenic mouse	Increased	[31]
c-Myc	Overexpression via transgenic mouse	Increased	[32]
c-Myc	Deletion via knockout mouse	Decreased	[33]
E2f1	Deletion via knockout mouse	Increased	[34]
E2f1	Overexpression via transgenic mouse	Decreased	[32]
E2f7/E2f8	Deletion via knockout mouse	Decreased	[34,35]
Ercc1	Deletion via knockout mouse	Increased	[36,37]
Mapk14 (p38alpha)	Deletion via knockout mouse	Increased	[38]
Rb1	Deletion via knockout mouse	Increased	[39]
Skp2	Deletion via knockout mouse	Increased	[40]
Trp53	Deletion via knockout mouse	Increased	[41]
Conditions			
Liver regeneration	Partial hepatectomy	Increased	[42,43]
Oxidative stress	Partial hepatectomy	Increased	[44]
Iron overload	Iron diet	Increased	[45]
Viral hepatitis	Infection with hepatitis B or C virus	Increased	[46,47]

conditions, such as viral hepatitis, are also associated with enrichment for polyploid hepatocytes [46,47]. Although numerous signals have been identified that regulate polyploidy, it remains to be seen which signals are the most important—in terms of liver development/maturation and in response to regeneration/injury. To date, transcription factors E2F1, E2F7, and E2F8 are the most promising upstream regulators of polyploidy during postnatal liver development. The E2F family of transcription factors is divided into an activator class (E2F1-3) and repressor class (E2F4-8) [48]. E2F7 and E2F8 are expressed maximally between weeks 1 and 7 of postnatal development, a period that correlates with hepatocyte polyploidization [34]. The role of E2F7 and E2F8 was studied using liver-specific *E2f7/E2f8* double knockout mice [34] where Cre recombinase expression was driven by the liver-specific albumin promoter [49]. Livers in these mice were reported to develop and function normally, but they were significantly depleted of polyploid hepatocytes [34,35]. Binucleate hepatocytes were reduced by 75% in double knockout livers by 4 weeks of age, and there was a corresponding reduction in nuclear ploidy [34]. Thus, the data indicate that E2F7/E2F8 deficiency in the liver specifically affects hepatic ploidy. E2F7 and E2F8 are believed to regulate ploidy by antagonizing *E2f1* expression and inhibiting pro-cytokinesis genes, including *Ect2*, *Mklp1*, and *Racgap1* [34].

Although cytokinesis failure is the major pathway leading to hepatic polyploidy, cell fusion has also been shown to occur in the liver. Homotypic cell fusion between hepatocytes is controversial. One report indicates that hepatocyte-hepatocyte fusion occurs readily during development [12], whereas another report concludes that hepatocyte-hepatocyte fusion in adults does not occur at all [11]. It remains to be seen if, when and to what extent homotypic hepatocyte fusion contributes to liver polyploidy. It is clear, however, that heterotypic cell fusion occurs in the liver. Cell fusion between hepatocytes and myelomonocytic cells, presumably liver-resident macrophages known as Kupffer cells, happens with a frequency of ~1/100,000 hepatocytes [50]. Despite the paucity of fusion events, these fusion-derived hepatocytes are functionally competent and capable of extensive *in vivo* proliferation under experimental conditions [10,11,51,52].

12.2.2 Function of Polyploid Hepatocytes

A number of ideas have been proposed to explain the functional significance of polyploidy in the liver. The first hypothesis posits that hepatic polyploidy represents a stage of terminal differentiation. In this thinking, polyploidy is associated with increased quiescence and decreased proliferative potential. Thus, polyploidy characterizes the final stage in the life of a hepatocyte. In support of this hypothesis, aging has been associated with shortened telomeres and replicative senescence in human livers [53,54]. Since polyploidy increases with age, it is assumed that polyploid hepatocytes have short telomeres and are replication compromised. The geographical location of hepatocytes within the hepatic lobule has also been used to support the association between polyploidy and terminal differentiation. Numerous publications suggest that diploid hepatocytes are preferentially located in the periportal region and polyploid hepatocytes in the pericentral zone [55,56]. Hepatocytes are believed to "mature" as they transition from the periportal to pericentral region of the liver where they are ultimately eliminated [57]. In contrast, others have suggested that diploid and polyploid hepatocytes are distributed randomly throughout the hepatic lobule [29]. The most compelling evidence against terminal differentiation, as a defining feature of hepatic polyploidy, is that polyploid hepatocytes are actually highly proliferative. In response to a strong regenerative stimulus such as partial hepatectomy, polyploid hepatocytes proliferate to restore liver mass [43,51]. Moreover, polyploid hepatocytes, prospectively isolated by fluorescence-activated cell sorting (FACS) or centrifugal elutriation, proliferated extensively in transplantation experiments and in cell culture [51,58,59]. Cell mixing experiments in which genetically marked octaploid and diploid hepatocytes were cotransplanted in defined ratios into *Fumarylacetoacetate hydrolase* (*Fah*)-deficient mice proved that proliferative capacity is nearly identical between ploidy classes [21]. Thus, terminal differentiation does not accurately characterize hepatocytes with polyploid nuclear content.

Second, it has been suggested that hepatic polyploidy may serve to enhance the functional capacity of the liver. The liver participates in a wide array of activities related to detoxification, protein synthesis/secretion, metabolism, etc. Since polyploid hepatocytes have an increased complement of DNA compared to diploids, it is conceivable that they have altered gene/protein synthesis, leading to robust cellular function. Similar to observations in plants [60], hepatic genes could be expressed differentially by diploids, tetraploids, octaploids, etc. The expression of unique subsets of genes would, therefore, endow each ploidy class with specialized functions. To test this idea directly, microarray analysis was performed on equivalent amounts of RNA from FACS-isolated diploid, tetraploid, and octaploid hepatocytes from mice [61]. Interestingly, the study found very few differences in gene expression among the different ploidy classes. In contrast to differential gene expression, the total amount of gene/protein expression could simply increase with nuclear content. For example, gene expression and functional capacity could be elevated fourfold in octaploid hepatocytes compared to diploids. Additional studies are necessary to determine whether the overall transcriptional or translational output per cell varies between diploid and polyploid hepatocytes.

A third hypothesis to explain the functional significance of hepatic polyploidy concerns genomic damage. The liver protects the body from xenobiotic-induced injury, but it remains susceptible to many forms of damage, including mutagenesis. For example, chronic exposure to alcohol [62] or naturally occurring catabolic by-products like fumarylacetoacetate in patients suffering from Hereditary Tyrosinemia type I [63] can lead to point mutations within the genome or other forms of chromosomal damage. Dysregulated gene function can contribute to liver cancer. It has been suggested that hepatic polyploidy serves to "buffer" the genome against genetic damage. Tetraploid and octaploid hepatocytes have two- and four-times, respectively, the amount of genetic material compared to diploid cells. Loss of a tumor suppressor on one or both homologs in a diploid cell could have disastrous consequences. In contrast, the same type of mutation affecting one or two homologs in tetraploid and octaploid hepatocytes should have minimal impact. The remaining homologs containing functional gene products potentially maintain normal cellular function.

12.3 GENETIC DIVERSITY IN THE LIVER

Polyploidization is not the only atypical process in the liver. Recent studies have shown that polyploid hepatocytes undergo aberrant cell divisions leading to reductive divisions (i.e., the formation of daughter cells with one-half ploidy of the parental cell) and aneuploidy.

12.3.1 Ploidy Reversal and Aneuploidy in the Liver

The initial hint of karyotypic alterations (beyond polyploidy) came from studies of heterotypic cell fusion in the liver. Bone marrow from wild-type males was transplanted into lethally irradiated female $Fah^{-/-}$ mice. In this system, donor hematopoietic stem cells regenerate the entire immune system in the host [10]. Donor-derived myeloid cells are then available for fusion with endogenous $Fah^{-/-}$ hepatocytes. Fusion-derived hepatocytes formed by this process are reprogrammed to express *Fah*, and they acquire the capacity to regenerate the host liver. Cell fusion between a donor-derived diploid myeloid cell and a diploid $Fah^{-/-}$ hepatocyte produces a tetraploid fusion product. Fusion with polyploid hepatocytes produces fusion products with even higher ploidies. Karyotypic analysis of livers repopulated with fusion-derived hepatocytes revealed two very striking observations [10,51]. First, diploid hepatocytes were present with characteristics of both fusion partners, indicating that fusion-derived hepatocytes could somehow generate diploid daughters. Second, gains and losses of whole chromosomes were seen in fusion-derived hepatocytes of all ploidy classes. These results suggested that fusion-derived hepatocytes could tolerate sweeping karyotypic changes. Although very intriguing, it was unclear if these changes were simply artifacts of rare fusion-derived hepatocytes subjected to forced proliferation.

To determine whether aneuploidy and ploidy reversal were features of the normal liver, hepatocytes from wild-type mice were examined. Karyotype analysis of metaphase chromosomes of hepatocytes from wild-type mice showed high levels of aneuploidy. Nearly 25% of hepatocytes from 3-week old mice were aneuploid, and aneuploidy increased to 70% in mice 4-15 months old [21]. Entire chromosomes were gained and/or lost, and structural rearrangements were rarely seen. Chromosome counts clustered around the expected number for the different ploidy classes: diploid (~40 chromosomes), tetraploid (~80 chromosomes), and octaploid (~160 chromosomes). Aneuploidy in wild-type mice occurred randomly as all chromosomes were affected equally [64]. One or more chromosomes were gained or lost by each aneuploid hepatocyte, and occasionally chromosome gains balanced losses, resulting in aneuploid cells with the expected number of chromosomes (Figure 12.2a) [21]. Aneuploidy was also demonstrated by fluorescence *in situ* hybridization (FISH) on interphase nuclei that were isolated without any *in vitro* expansion [65]. Furthermore, hepatic aneuploidy was identified in human hepatocytes. Karyotype and FISH analysis revealed random aneuploidy in 30-90% of hepatocytes from patients with normal liver tissue [20]. Importantly, aneuploidy was present in healthy tissue lacking any signs of tumorigenesis in mice and humans. These results suggest that hepatic aneuploidy is a normal feature of the healthy liver.

Ploidy reversal was examined using transplantation and *in vitro* approaches. Octaploid hepatocytes isolated from wild-type or genetically marked *Fah*-positive livers were transplanted into $Fah^{-/-}$ mice [21]. Ploidy levels were examined upon complete liver repopulation by donor hepatocytes. As expected, livers were repopulated with donor-derived octaploid hepatocytes but also tetraploid and diploid hepatocytes (Figure 12.2b). The degree of aneuploidy found in donor-derived hepatocytes (~90%) was elevated compared to hepatocytes from normal adult donors, suggesting that aneuploidy increases as hepatocytes are forced to divide extensively. Cell culture experiments also illustrated ploidy reversal [21]. Expansion of FACS-isolated octaploid hepatocytes for 3-5 days generated daughter cells representing all the ploidy classes. Thus, the data show that chromosome alterations in normal hepatocytes are highly dynamic. Polyploidization, ploidy reversal, and aneuploidy occur frequently in the liver.

12.3.2 Hepatic Cell Divisions with Multipolar Spindles

Aneuploidy and ploidy reversal in the liver could potentially arise by a number of different mechanisms [51]. First, one or more

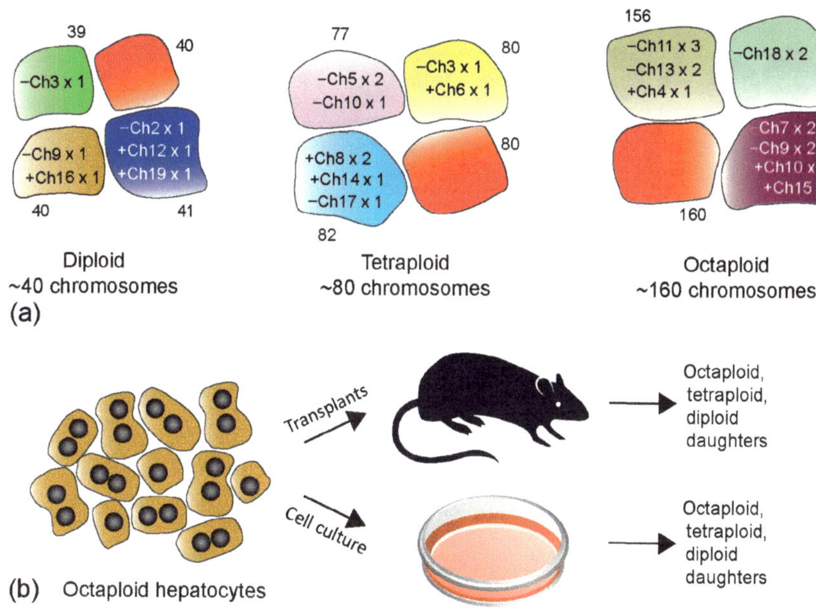

FIGURE 12.2 Aneuploidy and ploidy reversal in the liver. (a) Most hepatocytes in the adult liver are aneuploid. Hepatocytes with unique karyotypes are shown with different colors. Gains (+) and losses (−) are illustrated for different chromosomes (Ch), and the number of affected homologs is indicated. The total chromosome count is listed adjacent to each cell. For example, the blue tetraploid hepatocyte has 82 chromosomes, including gain of two Ch8, gain of one Ch14, and loss of one Ch17. The red cells are euploid. (b) Proliferating octaploid hepatocytes (*in vivo* and *in vitro*) generate octaploid daughters, as well as reduced ploidy tetraploid and diploid daughters.

chromosomes could be lost with every cell division with ploidy levels equilibrating near diploidy. Hepatocytes undergoing this process would appear, for example, as tetraploids with chromosome losses or diploids with chromosome gains. Second, polyploid hepatocytes could undergo a specialized form of mitosis without DNA synthesis. Successful cell division by a tetraploid hepatocyte would therefore generate a pair of diploid daughters. Depending on whether pairing between homologous chromosomes occurred, such divisions could generate either euploid or aneuploid progeny. Finally, since the majority of polyploid hepatocytes arise by a failure of cytokinesis, it is possible that binucleate hepatocytes could simply resume cytokinesis. Initiation of cytokinesis by a binucleate tetraploid hepatocyte would generate two diploid progeny. Aneuploidy would not be associated with this method of ploidy reversal.

To elucidate the mechanism(s) leading to hepatic aneuploidy and ploidy reversal, imaging studies were performed. FACS-isolated tetraploid hepatocytes were induced to divide in culture and mitoses monitored in both fixed and living cells. Centrosome number significantly affects the outcome of cell division. Whereas dividing diploid cells have two centrosomes, polyploid hepatocytes have supernumerary centrosomes (e.g., dividing tetraploid hepatocytes typically have four centrosomes) [65]. Following S phase and G2 of the cell cycle, tetraploid hepatocytes enter mitosis (Figure 12.3a). The nuclear membrane breaks down, and chromosomes start to condense. Chromosomes from each nucleus in a binucleate tetraploid appear to intermingle. At this point,

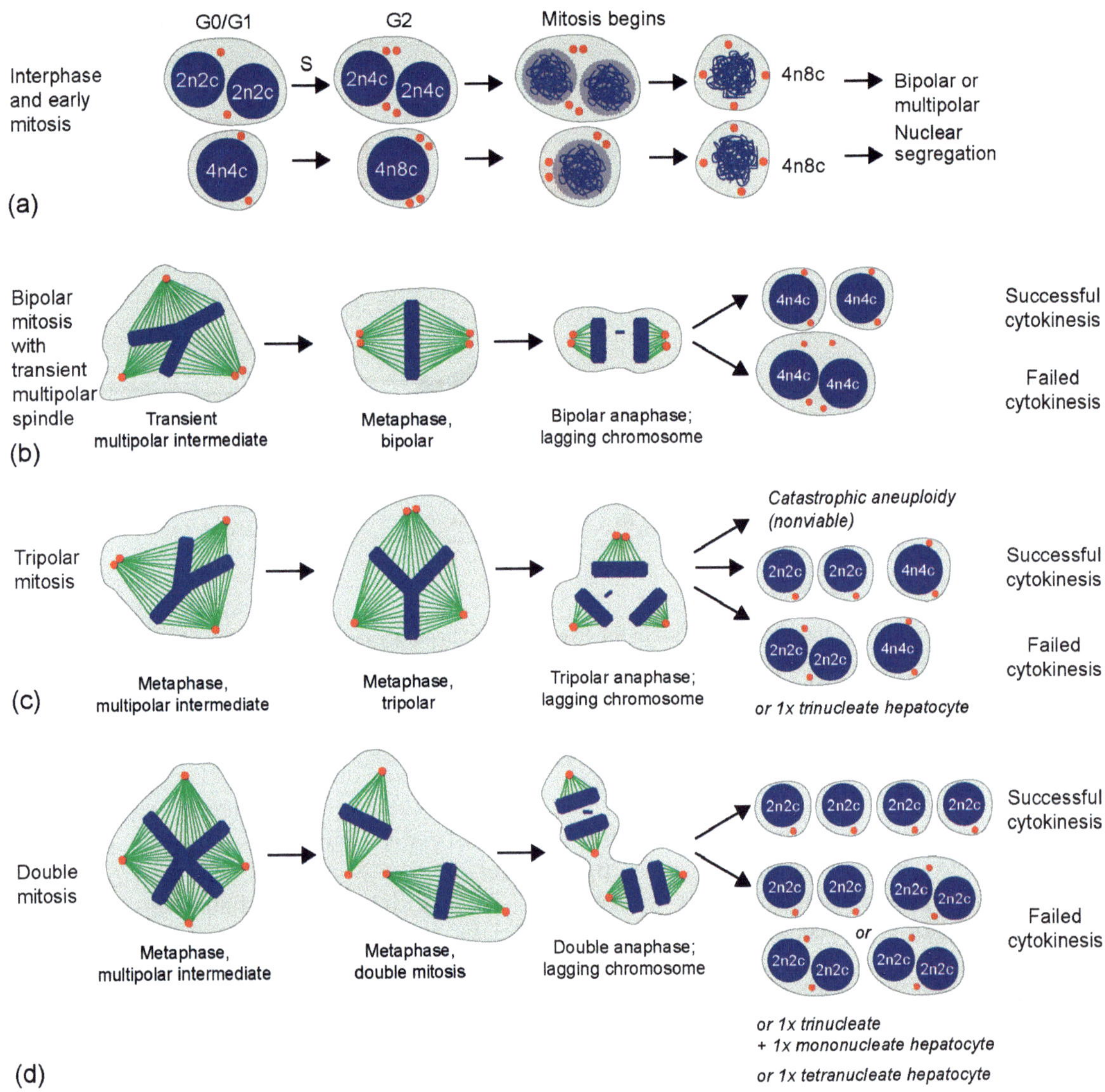

FIGURE 12.3 Cell divisions by tetraploid hepatocytes. (a) Tetraploid hepatocytes go through S phase of the cell cycle. Mononucleate tetraploids (4n4c) become 4n8c, and binucleate tetraploids (2n2c/nucleus) become 2n4c/nucleus. In this convention, "n" refers to chromosome number, whereas "c" refers to chromatid number; thus "4n8c" indicates a tetraploid cell with replicated DNA. Centrosomes (red) are duplicated during S phase. The beginning of mitosis is marked by nuclear breakdown and chromatin condensation (blue). (b) In most cases, tetraploid hepatocytes form transient multipolar spindles (green, microtubules) that reorient into a bipolar fashion. Nuclear segregation proceeds, but lagging chromosomes can occur. Successful cytokinesis generates two mononucleate tetraploid daughters, whereas failed cytokinesis generates a binucleate octaploid daughter. Multipolar cell division can also occur. Tripolar mitosis (c) can generate three daughter nuclei and double mitosis (d) can generate four daughter nuclei. The approximate ploidy of daughter nuclei is shown. Completion of cytokinesis determines the number of nuclei/cell.

centrosome orientation directs mitosis to proceed along either a bipolar or multipolar direction. It should be noted that mononucleate and binucleate tetraploids perform the same type of mitoses [21]. Thus, the nucleation status of the dividing cell affects neither centrosome orientation nor the outcome of cell division.

Approximately 95% of tetraploid hepatocytes undergo bipolar cell division (Figure 12.3b) [21]. In prometaphase or metaphase, the four centrosomes orient along three or four distinct poles, forming a multipolar spindle. The multipolar spindle, however, is transient. Pairs of centrosomes rapidly cluster, leading to the formation of two discrete poles. Nuclear segregation then proceeds in a bipolar manner to generate two tetraploid nuclei. Successful cytokinesis yields two mononucleate tetraploids; failed cytokinesis generates a single binucleate octaploid. Nuclear segregation errors are very common in cell divisions by tetraploid hepatocytes. Early in mitosis when the transient multipolar spindle is setup, microtubules emanating from multiple poles can attach to the same kinetochore. This type of error, termed a merotelic attachment, is poorly recognized by the spindle assembly complex and is frequently left unrepaired [66,67]. Merotelic chromosomes fail to migrate to distinct poles during anaphase and appear as "lagging" chromosomes. Lagging chromosomes are ultimately excluded from the newly formed nuclei, which leads to aneuploidy and other consequences [68]. It is estimated that up to 50% of bipolar cell divisions by tetraploid hepatocytes have lagging chromosomes [21].

The remaining 5% of dividing tetraploid hepatocytes undergo full-fledged multipolar division (Figure 12.3c and d). A multipolar spindle is formed in prometaphase/metaphase, but centrosomes never reorient to produce two discrete poles. Incomplete centrosome clustering can lead to a tripolar metaphase spindle in which chromosomes segregate to distinct poles, forming three independent nuclei (Figure 12.3c). Lagging chromosomes can occur. The mechanism by which chromosomes segregate along three axes is not well understood. Equal chromosome distribution would generate highly aneuploid daughters that are likely nonviable. However, live-cell imaging shows that progeny resulting from tripolar divisions are, indeed, viable and mitotically active. Examination of nuclear content following cell division suggests that many tripolar mitoses progress in an orderly fashion, producing two near-diploid daughters and one near-tetraploid daughter. Thus, in many cases, successful cytokinesis leads to three mononucleate daughters: two near-diploids and a near-tetraploid. Partial failed cytokinesis can produce two cells: a binucleate near-tetraploid and a mononucleate near-tetraploid. Complete failure of cytokinesis potentially generates a rare trinucleate hepatocyte. When centrosomes never cluster, two independent metaphase plates are established (Figure 12.3d). This type of cell division has been termed "double mitosis." Four-way nuclear segregation occurs (with or without lagging chromosomes) to generate four diploid nuclei. Successful cytokinesis yields diploid daughters, and failed cytokinesis generates a variety of daughters as shown in Figure 12.3d.

The formation of multipolar spindles by dividing polyploid hepatocytes provides a unifying mechanism for aneuploidy and ploidy reversal in the liver. Aneuploidy is a consequence of bipolar cell divisions with lagging chromosomes. Aneuploidy also potentially arises from uneven nuclear segregation during multipolar cell divisions. In a multipolar division by a tetraploid hepatocyte, chromosomes segregate along three-four distinct poles, and the extent to which every chromosome is properly segregated is unknown. Ploidy reversal occurs via multipolar cell divisions when a single cell produces three or more daughter cells. Tripolar divisions and double mitoses, for instance, can produce daughters with diploid or near-diploid nuclear content. Although multipolar hepatic divisions have been extensively described in tetraploids, much less is known about cell divisions by octaploid hepatocytes. Octaploids are predicted to generate as many as eight diploid daughter cells.

12.3.3 Control of Liver Diversity and the Link to Cancer

The liver is not the only healthy organ with aneuploid cells. Low-level aneuploidy has been described in diverse tissues, including blood [69], brain [70], kidney [71], cytotrophoblast [72], etc. The cell cycle is typically a tightly controlled process where checkpoints are utilized to arrest/eliminate cells with chromosomal abnormalities. How aneuploid cells persist and thrive in healthy liver tissue and elsewhere is unknown.

Very little is also known about the unique molecular processes that regulate mitosis by polyploid hepatocytes. It is clear that cell division is critically influenced by centrosome clustering. When supernumerary centrosomes cluster and produce two independent poles, mitosis proceeds in a typical manner to generate a pair of daughter nuclei equal in ploidy to the parental cell. Failure of supernumerary centrosomes to cluster can lead to multipolar cell divisions that generate daughters of reduced ploidy and aneuploidy. It is unclear why centrosome clustering occurs in most, but not all, polyploid hepatocytes. In cases when centrosomes fail to cluster, is this simply a stochastic event, or are the cells somehow programmed for a multipolar division? Answers to these questions may come from investigation of signals known to induce ploidy changes *in vivo*. Rodents treated with hepatotoxic agents carbon tetrachloride [73] and thioacetamide [74] dramatically increased the frequency of diploid hepatocytes and simultaneously decreased the frequency of polyploid hepatocytes. Similarly, overexpression of Cu, Zn-superoxide dismutase or intracellular glutathione peroxidase induced differential expansion of diploid hepatocytes following partial hepatectomy [75]. Additionally, a better understanding of the mechanisms that control hepatic polyploidization (discussed in Section 12.2.1) may also shed light on ploidy reversal and aneuploidy.

There are many similarities between the liver and cancer in terms of ploidy diversity. Aneuploidy occurs in >75% of human cancers and tetraploidy is common [27,76,77]. In the liver, aneuploidy affects 30-90% of hepatocytes in humans, and approximately half of human hepatocytes are polyploid. Supernumerary centrosomes, multipolar spindles, and lagging chromosomes are features in both tumor cells and proliferating hepatocytes [66]. Whether aneuploidy is the stimulus for cancer or merely the consequence of malignant transformation remains to be elucidated. Interestingly, spontaneous liver cancer is rare, <4% [78], especially considering high degree of aneuploidy in the normal liver. Thus, hepatic chromosomal instability *per se* is not associated with malignant growth. Understanding how hepatocytes maintain normal function, even in the face of extreme chromosomal instability, may reveal new insights in tumor pathogenesis and targets for cancer therapy [79].

12.4 AN INTEGRATED MODEL FOR POLYPLOIDY, PLOIDY REVERSAL, AND ANEUPLOIDY IN THE LIVER

New evidence indicates that polyploidy, ploidy reversal, and aneuploidy are interconnected and may collectively protect the liver from injury.

12.4.1 The Ploidy Conveyor

The "ploidy conveyor" is a model describing the interconnectedness between polyploidy, ploidy reversal, and aneuploidy in the liver [21] (Figure 12.4). In terms of overall ploidy changes,

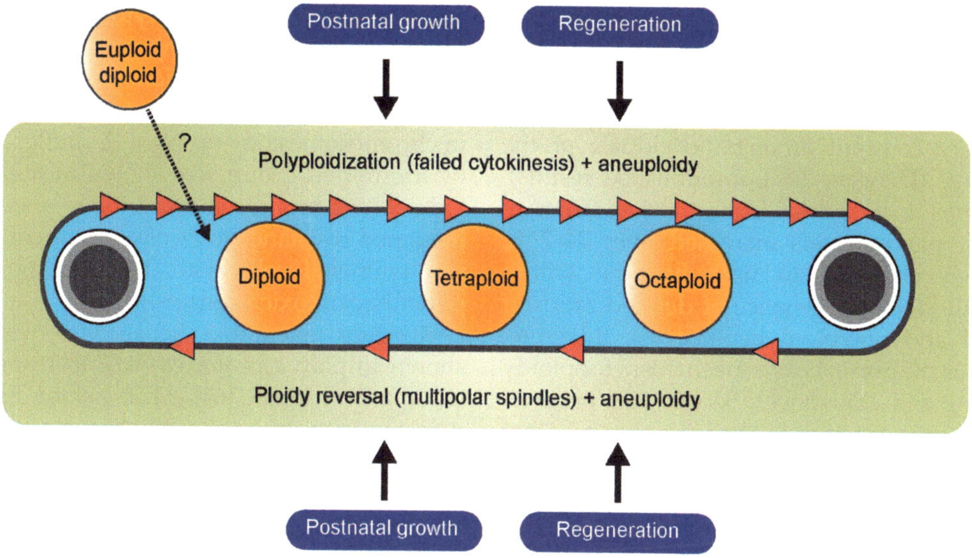

FIGURE 12.4 Ploidy conveyor model. The ploidy conveyor model describes the dynamic relationship between polyploidization, ploidy reversal, and aneuploidy. Compared to ploidy reversal, polyploidization is the dominant event, as emphasized with red arrows. The conveyor is controlled by signals for growth and regeneration. The origin of the euploid diploid cell is unclear.

hepatic polyploidization is the dominant event during liver growth and regeneration, whereas ploidy reversal occurs much less frequently. Hepatocytes polyploidize during postnatal growth and regeneration. Proliferating diploid hepatocytes give rise to identical diploid daughters, or they undergo failed cytokinesis to generate tetraploid hepatocytes. In turn, cell divisions by tetraploid hepatocytes generate tetraploid daughters upon the completion of cytokinesis, or cytokinesis failure produces an octaploid cell. The process repeats to produce hepatocytes of even greater ploidies. Cell divisions by polyploid hepatocytes usually involve transient multipolar spindles that reorient into a traditional bipolar mitotic spindle. Chromosome segregation errors, like merotelic attachments and lagging chromosomes, are one of the consequences of spindle reorientation, resulting in the formation of aneuploid daughter nuclei.

Ploidy reversal occurs throughout postnatal development and during liver regeneration (Figure 12.4). Polyploid hepatocytes undergo multipolar mitosis to produce three or more daughter nuclei. For example, a single tetraploid hepatocyte generates four diploid daughters, or an octaploid produces up to eight diploid daughters. Aneuploidy arises from either the parental cell (which can, itself, have chromosome gains/losses), uneven nuclear segregation to daughter nuclei during the multipolar division or a combination of both. The cycle begins again as newly formed diploid (and most likely aneuploid) hepatocytes start to polyploidize.

The ploidy conveyor model explains the dramatic and dynamic chromosomal changes associated with liver growth and regeneration. In addition to the molecular cues that regulate polyploidization and ploidy reversal, proper function of the ploidy conveyor is predicted to require additional layers of control. First, hepatocytes belonging to each ploidy category must be present in the adult liver. Studies have shown that the ratio of cells within each ploidy class is tightly regulated. For example, analysis of ploidy levels in livers from different aged wild-type mice showed that the frequency of cells within each ploidy population remained relatively constant among individuals of the same age (Duncan, 2012, unpublished results). Moreover, in response to partial hepatectomy in rats, ploidy levels increased over 24-72 h, but they returned to near baseline within 1 month [43]. Maintenance of defined ratios of diploid and polyploid hepatocytes, therefore, indicates the presence of a mechanism for ploidy counting and adjustment. Second, the progressive aneuploidization of the liver suggests that nearly all hepatocytes could become aneuploid. This is not the case, however, because the degree of aneuploidy in 4-5-month-old mice and 10 to 15-month-old mice is ~70% [21]. Similarly, in humans the frequency of aneuploid hepatocytes is constant in individuals <10 years old and >60 years old [20]. In every normal liver examined, there is always a balance between aneuploidy and euploidy. How does the liver regulate the ratio? A potential mechanism is a steady supply of fresh diploid/euploid hepatocytes. Entry of these cells into the ploidy conveyor would provide a continuous and lifetime supply of euploid hepatocytes. Whether such cells are derived from a hepatic progenitor cell or simply a subset of karyotypically normal diploid hepatocytes is not known.

12.4.2 The Ploidy Conveyor as a Mechanism for Liver Adaptation

The idea that aneuploidy could perform a cytoprotective function was demonstrated in the budding yeast. In response to extreme cellular stress induced by loss of proteins required for cell division, chemotherapeutic agents, antifungals, etc., aneuploid yeast cells with specific karyotypes adapted to the injury [80–83]. For instance, yeast cells deficient for myosin II (which is required for cytokinesis) ordinarily divided poorly. Left alone, these cells acquired the capacity for proper cytokinesis through aneuploidization [81]. Injury resistance was mediated by acute alterations in gene and protein expression commensurate with variations in chromosome copy number. A similar finding by a different group was shown in the mouse kidney. Chromosome-specific aneuploidy mediated resistance to *ex vivo* treatment of kidney epithelial cells with 2,6-diaminopurin, a potently cytotoxic agent [84].

Recently, aneuploid hepatocytes were also shown to play a protective role in response to chronic liver injury [64]. Mice lacking FAH are defective in the tyrosine catabolic pathway, a condition called tyrosinemia. FAH loss leads to the accumulation of toxic metabolites that damage the liver, induce liver failure, and cause death. Blocking the pathway upstream of FAH using the drug 2-(2-nitro-4-trifluoro-methylbenzoyl)-1,3-cyclo-hexanedione (NTBC) or knockout of homegentisic acid dioxygenase (HGD) prevents tyrosinemia. Thus, *Fah*$^{-/-}$ mice maintained on NTBC are protected from tyrosinemia and so are *Fah*$^{-/-}$ *Hgd*$^{-/-}$ mice without NTBC therapy.

Aneuploid hepatocytes were studied using *Fah*$^{-/-}$ *Hgd*$^{+/-}$ mice [64]. These mice were maintained on NTBC until adulthood, and then the drug was removed. NTBC withdrawal leads to liver failure and death within 2-4 weeks in *Fah*$^{-/-}$ mice, and the same was expected for *Fah*$^{-/-}$ *Hgd*$^{+/-}$ mice since they have a functional allele of *Hgd*. Several weeks after NTBC withdrawal, livers were mostly pale and sickly, but they were also decorated with hundreds of small random, healthy nodules. Over the ensuing weeks, the healthy hepatic nodules expanded to $>75\%$ of the liver mass, curing the mice of tyrosinemia and restoring normal liver function. Analysis of healthy liver tissue in rescued mice showed a complete loss of *Hgd*. Approximately 30% of the nodules contained inactivating mutations in the wild-type *Hgd* allele [85]. Most strikingly, the remaining nodules were completely deficient for functional *Hgd*. Karyotype and array comparative genomic hybridization studies proved that the livers were repopulated with a chromosome-specific aneuploid population: monosomy of Chromosome 16, the home of *Hgd* [85]. It is hypothesized that naturally occurring aneuploid hepatocytes (monosomic for Chromosome 16) were present in livers of $Fah^{-/-}$ $Hgd^{+/-}$ mice prior to liver stress. Liver stress, induced by removal of NTBC, was toxic to most of the liver except those aneuploid hepatocytes that had lost Chromosome 16 with the wild-type *Hgd*. In response to massive liver failure, these injury-resistant cells expanded, repopulated with liver with healthy tissue, and restored liver function.

The ploidy conveyor is believed to operate in a multistep process to guard against many forms of chronic liver injury, including tyrosinemia, alcohol liver disease, high fat, etc. (Figure 12.5). First, the adult liver is highly genetically diverse. All of the hepatic ploidy classes are present, and a large percentage of hepatocytes are aneuploid. Second, chronic liver

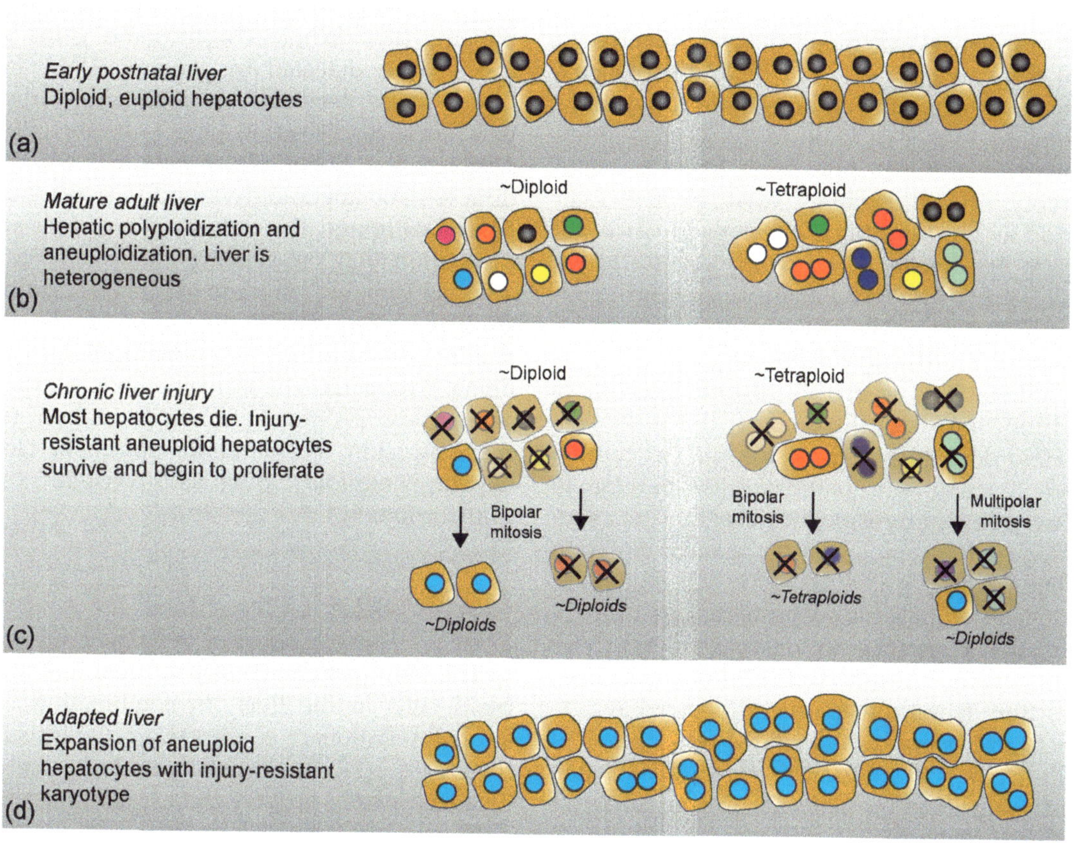

FIGURE 12.5 Liver adaptation by aneuploid hepatocytes. (a) Early in life, hepatocytes are diploid and euploid. (b) Hepatocytes polyploidize and become aneuploid during aging. Different colored nuclei represent aneuploid karyotypes. Tetraploids are the only polyploid population shown. To account for alterations in chromosome number resulting from aneuploidy, populations are marked as "~diploid" (near-diploid) and "~tetraploid" (near-tetraploid). (c) In response to chronic liver injury, most of the hepatocytes die (marked with an "X"). However, diploid hepatocytes with the resistant aneuploid karyotype (blue nuclei) survive. (d) These injury-resistant aneuploid hepatocytes proliferate to restore liver mass and function.

damage places a tremendous amount of stress on the liver. Most of the hepatocytes are sensitive to the damage and ultimately die. However, a subpopulation of aneuploid hepatocytes with chromosome-specific alterations has the capacity to resist the injury. The injury-resistant cells are most likely near-diploid because copy number variation in near-tetraploid or near-octaploid hepatocytes is expected to minimally affect cellular function. The loss or gain of even a single chromosome in a near-diploid hepatocyte could have profound effects on gene expression, especially considering that 10% of genes are monoallelically expressed [86]. This means that a near-diploid hepatocyte with a monosomic chromosome could have complete loss-of-function of multiple genes. Near-diploid hepatocytes with injury-resistant karyotypes could either exist in the liver (prior to the injury), or they could arise via multipolar divisions (i.e., ploidy reversal) by a polyploid hepatocyte. Finally, aneuploid hepatocytes that are protected from the injury have a potent proliferative advantage. These cells proliferate extensively to restore liver mass and function.

How does hepatic aneuploidy affect human liver function and pathogenesis? Similar to observations in tyrosinemic mice, hepatic aneuploidy could promote resistance and adaptation to injury. Indeed, analysis of livers from hepatitis C-infected patients suggests that aneuploidy promotes resistance to disease. Hepatocytes from healthy, noncancerous cirrhotic liver nodules from females were microdissected and X-inactivation status examined [87–89]. Up to half of the cirrhotic nodules were shown to be clonal, containing either the maternal or paternal active X chromosome. This result suggests that each clone was derived from a single hepatocyte with a regenerative advantage. It is hypothesized that the clones were derived from hepatocytes with aneuploid karyotypes specifically resistant to hepatitis C infection. Further studies are required to determine whether clonal nodules comprised a dominant aneuploid karyotype. Aneuploidy should also be examined in cirrhotic nodules in other forms of liver disease such as nonalcoholic steatohepatitis and alcoholic steatohepatitis.

In addition to performing a positive role in the liver, hepatic aneuploidy could contribute to pathogenesis. Spontaneous hepatocellular carcinoma is rare, but it is possible that pre-existing aneuploid karyotypes promote neoplastic transformation. In a very interesting study, chromosome copy number variation was interrogated in hepatocellular carcinomas from 76 patients [90]. Whole chromosome gains and losses were detected in genomic DNA from tumor tissue derived from dozens of individuals. These results are striking as the aneuploidy detected was clonal, which contrasts with random hepatic aneuploidy found in normal liver tissue. Although the data are consistent with the notion that pre-existing aneuploidy could serve as "first hit" en route to cancer, additional studies are required to determine exactly how aneuploidy contributes to hepatocellular carcinoma. It remains to be seen whether pre-existing aneuploidy promotes tumor formation (and the expansion of aneuploid clones) or clonal hepatic aneuploidy arises as a consequence of transformation.

12.5 CONCLUSION

The high percentage of polyploid and aneuploid cells in the liver, in combination with strong evolutionary conservation, suggests that these cells perform a functional role. Numerous ideas have been proposed to explain the role of polyploid hepatocytes, but none have provided a completely satisfactory answer. Considering the link between polyploidy and aneuploidy, it is now hypothesized that the primary function of polyploid hepatocytes is to generate aneuploidy. The idea that naturally occurring aneuploid hepatocytes could influence liver fate is exciting. It is clear that hepatic aneuploidy facilitates adaptation to chronic liver damage in a mouse model of tyrosinemia. How aneuploidy contributes to liver fate in other contexts remains to be seen; it is possible that injury-resistant aneuploid karyotypes exist for many different forms of liver damage. Future work should unravel (a) the molecular cues that control the ploidy conveyor and (b) how hepatic aneuploidy, whether beneficial or pathological, affects human health. Aneuploid cells have been identified in many tissues beyond the liver, albeit at low frequencies, but these observations are largely underappreciated. It is time to seriously explore the potential role of aneuploidy in other tissues. The hope, of course, is to translate our knowledge of inherent aneuploidy into new diagnostics and therapeutic strategies.

Acknowledgments

This work was supported by grants to A. W. D. from the NIH (R01 DK103645) and the Commonwealth of Pennsylvania. Thanks to Shu-hao Hsu and Anthony Otero for insightful discussions and helpful comments. A. W. D. apologizes to authors whose papers were not cited due to space restrictions.

References

[1] Duncan AW, Dorrell C, Grompe M. Stem cells and liver regeneration. Gastroenterology 2009;137:466–81.

[2] Fausto N, Campbell JS. The role of hepatocytes and oval cells in liver regeneration and repopulation. Mech Dev 2003;120:117–30.

[3] Bucher NLR, Malt RA. Regeneration of liver and kidney. 1st ed. Boston: Little Brown; 1971.

[4] Stocker E, Wullstein HK, Brau G. Capacity of regeneration in liver epithelia of juvenile, repeated partially hepatectomized rats. Autoradiographic studies after continuous infusion of 3H-thymidine. Virchows Arch B Cell Pathol 1973;14:93–103.

[5] Azuma H, Paulk N, Ranade A, Dorrell C, Al-Dhalimy M, Ellis E, et al. Robust expansion of human hepatocytes in $Fah^{-/-}/Rag2^{-/-}/Il2rg^{-/-}$ mice. Nat Biotechnol 2007;25:903–10.

[6] Overturf K, Al-Dhalimy M, Ou CN, Finegold M, Grompe M. Serial transplantation reveals the stem-cell-like regenerative potential of adult mouse hepatocytes. Am J Pathol 1997;151:1273–80.

[7] Dorrell C, Erker L, Schug J, Kopp JL, Canaday PS, Fox AJ, et al. Prospective isolation of a bipotential clonogenic liver progenitor cell in adult mice. Genes Dev 2011;25:1193–203.

[8] Shin S, Walton G, Aoki R, Brondell K, Schug J, Fox A, et al. Foxl1-Cre-marked adult hepatic progenitors have clonogenic and bilineage differentiation potential. Genes Dev 2011;25:1185–92.

[9] Zhang L, Theise N, Chua M, Reid LM. The stem cell niche of human livers: symmetry between development and regeneration. Hepatology 2008;48:1598–607.

[10] Wang X, Willenbring H, Akkari Y, Torimaru Y, Foster M, Al-Dhalimy M, et al. Cell fusion is the principal source of bone-marrow-derived hepatocytes. Nature 2003;422:897–901.

[11] Willenbring H, Bailey AS, Foster M, Akkari Y, Dorrell C, Olson S, et al. Myelomonocytic cells are sufficient for therapeutic cell fusion in liver. Nat Med 2004;10:744–8.

[12] Faggioli F, Sacco MG, Susani L, Montagna C, Vezzoni P. Cell fusion is a physiological process in mouse liver. Hepatology 2008;48:1655–64.

[13] Michalopoulos GK. Liver regeneration: alternative epithelial pathways. Int J Biochem Cell Biol 2011;43: 173–9.

[14] Michalopoulos GK, Barua L, Bowen WC. Transdifferentiation of rat hepatocytes into biliary cells after bile duct ligation and toxic biliary injury. Hepatology 2005;41: 535–44.

[15] Kudryavtsev BN, Kudryavtseva MV, Sakuta GA, Stein GI. Human hepatocyte polyploidization kinetics in the course of life cycle. Virchows Arch B Cell Pathol Incl Mol Pathol 1993;64:387–93.

[16] Gentric G, Celton-Morizur S, Desdouets C. Polyploidy and liver proliferation. Clin Res Hepatol Gastroenterol 2012;36:29–34.

[17] Gupta S. Hepatic polyploidy and liver growth control. Semin Cancer Biol 2000;10:161–71.

[18] Milne LS. The histology of liver tissue regeneration. J Pathol Bacteriol 1909;13:127–58.

[19] Guidotti JE, Bregerie O, Robert A, Debey P, Brechot C, Desdouets C. Liver cell polyploidization: a pivotal role for binuclear hepatocytes. J Biol Chem 2003;278: 19095–101.

[20] Duncan AW, Hanlon Newell AE, Smith L, Wilson EM, Olson SB, Thayer MJ, et al. Frequent aneuploidy among normal human hepatocytes. Gastroenterology 2012;142: 25–8.

[21] Duncan AW, Taylor MH, Hickey RD, Hanlon Newell AE, Lenzi ML, Olson SB, et al. The ploidy conveyor of mature hepatocytes as a source of genetic variation. Nature 2010;467:707–10.

[22] Lacroix B, Maddox AS. Cytokinesis, ploidy and aneuploidy. J Pathol 2012;226:338–51.

[23] Abmayr SM, Pavlath GK. Myoblast fusion: lessons from flies and mice. Development 2012;139:641–56.

[24] Helming L, Gordon S. Molecular mediators of macrophage fusion. Trends Cell Biol 2009;19:514–22.

[25] Vitrat N, Cohen-Solal K, Pique C, Le Couedic JP, Norol F, Larsen AK, et al. Endomitosis of human megakaryocytes are due to abortive mitosis. Blood 1998;91: 3711–23.

[26] Engel FB, Schebesta M, Keating MT. Anillin localization defect in cardiomyocyte binucleation. J Mol Cell Cardiol 2006;41:601–12.

[27] Ganem NJ, Storchova Z, Pellman D. Tetraploidy, aneuploidy and cancer. Curr Opin Genet Dev 2007;17: 157–62.

[28] Celton-Morizur S, Merlen G, Couton D, Margall-Ducos G, Desdouets C. The insulin/Akt pathway controls a specific cell division program that leads to generation of binucleated tetraploid liver cells in rodents. J Clin Invest 2009;119:1880–7.

[29] Margall-Ducos G, Celton-Morizur S, Couton D, Bregerie O, Desdouets C. Liver tetraploidization is controlled by a new process of incomplete cytokinesis. J Cell Sci 2007;120:3633–9.

[30] Hagemann S, Wohlschlaeger J, Bertram S, Levkau B, Musacchio A, Conway EM, et al. Loss of Survivin influences liver regeneration and is associated with impaired Aurora B function. Cell Death Differ 2013;20:834–44.

[31] Wu H, Wade M, Krall L, Grisham J, Xiong Y, Van Dyke T. Targeted in vivo expression of the cyclin-dependent kinase inhibitor p21 halts hepatocyte cell-cycle progression, postnatal liver development and regeneration. Genes Dev 1996;10:245–60.

[32] Conner EA, Lemmer ER, Sanchez A, Factor VM, Thorgeirsson SS. E2F1 blocks and c-Myc accelerates hepatic ploidy in transgenic mouse models. Biochem Biophys Res Commun 2003;302:114–20.

[33] Baena E, Gandarillas A, Vallespinos M, Zanet J, Bachs O, Redondo C, et al. c-Myc regulates cell size and ploidy but is not essential for postnatal proliferation in liver. Proc Natl Acad Sci USA 2005;102: 7286–91.

[34] Pandit SK, Westendorp B, Nantasanti S, van Liere E, Tooten PC, Cornelissen PW, et al. E2F8 is essential for polyploidization in mammalian cells. Nat Cell Biol 2012;14:1181–91.

[35] Chen HZ, Ouseph MM, Li J, Pecot T, Chokshi V, Kent L, et al. Canonical and atypical E2Fs regulate the mammalian endocycle. Nat Cell Biol 2012;14:1192–202.

[36] Chipchase MD, O'Neill M, Melton DW. Characterization of premature liver polyploidy in DNA repair (Ercc1)-deficient mice. Hepatology 2003;38:958–66.

[37] Nunez F, Chipchase MD, Clarke AR, Melton DW. Nucleotide excision repair gene (ERCC1) deficiency causes G(2) arrest in hepatocytes and a reduction in liver binucleation: the role of p53 and p21. FASEB J 2000;14:1073–82.

[38] Tormos AM, Talens-Visconti R, Nebreda AR, Sastre J. p38 MAPK: a dual role in hepatocyte proliferation through reactive oxygen species. Free Radic Res 2013;47:905–16.

[39] Mayhew CN, Bosco EE, Fox SR, Okaya T, Tarapore P, Schwemberger SJ, et al. Liver-specific pRB loss results in ectopic cell cycle entry and aberrant ploidy. Cancer Res 2005;65:4568–77.

[40] Minamishima YA, Nakayama K, Nakayama K. Recovery of liver mass without proliferation of hepatocytes after partial hepatectomy in Skp2-deficient mice. Cancer Res 2002;62:995–9.

[41] Kurinna S, Stratton SA, Coban Z, Schumacher JM, Grompe M, Duncan AW, et al. p53 regulates a mitotic transcription program and determines ploidy in normal mouse liver. Hepatology 2013;57:2004–13.

[42] Faktor VM, Uryvaeva IV. Progressive polyploidy in mouse liver following repeated hepatectomy. Tsitologiia 1975;17:909–16.

[43] Sigal SH, Rajvanshi P, Gorla GR, Sokhi RP, Saxena R, Gebhard Jr DR, et al. Partial hepatectomy-induced polyploidy attenuates hepatocyte replication and activates cell aging events. Am J Physiol 1999;276:G1260–72.

[44] Gorla GR, Malhi H, Gupta S. Polyploidy associated with oxidative injury attenuates proliferative potential of cells. J Cell Sci 2001;114:2943–51.

[45] Troadec MB, Courselaud B, Detivaud L, Haziza-Pigeon C, Leroyer P, Brissot P, et al. Iron overload promotes Cyclin D1 expression and alters cell cycle in mouse hepatocytes. J Hepatol 2006;44:391–9.

[46] Toyoda H, Bregerie O, Vallet A, Nalpas B, Pivert G, Brechot C, et al. Changes to hepatocyte ploidy and binuclearity profiles during human chronic viral hepatitis. Gut 2005;54:297–302.

[47] Toyoda H, Kumada T, Bregerie O, Brechot C, Desdouets C. Conserved balance of hepatocyte nuclear DNA content in mononuclear and binuclear hepatocyte populations during the course of chronic viral hepatitis. World J Gastroenterol 2006;12: 4546–8.

[48] Xanthoulis A, Tiniakos DG. E2F transcription factors and digestive system malignancies: how much do we know? World J Gastroenterol 2013;19:3189–98.

[49] Postic C, Magnuson MA. DNA excision in liver by an albumin-Cre transgene occurs progressively with age. Genesis 2000;26:149–50.

[50] Wang X, Montini E, Al-Dhalimy M, Lagasse E, Finegold M, Grompe M. Kinetics of liver repopulation after bone marrow transplantation. Am J Pathol 2002;161:565–74.

[51] Duncan AW, Hickey RD, Paulk NK, Culberson AJ, Olson SB, Finegold MJ, et al. Ploidy reductions in murine fusion-derived hepatocytes. PLoS Genet 2009;5:e1000385.

[52] Vassilopoulos G, Wang PR, Russell DW. Transplanted bone marrow regenerates liver by cell fusion. Nature 2003;422:901–4.

[53] Aini W, Miyagawa-Hayashino A, Tsuruyama T, Hashimoto S, Sumiyoshi S, Ozeki M, et al. Telomere shortening and karyotypic alterations in hepatocytes in long-term transplanted human liver allografts. Transpl Int 2012;25:956–66.

[54] Hoare M, Das T, Alexander G. Ageing, telomeres, senescence, and liver injury. J Hepatol 2010;53:950–61.

[55] Gandillet A, Alexandre E, Royer C, Cinqualbre J, Jaeck D, Richert L. Hepatocyte ploidy in regenerating livers after partial hepatectomy, drug-induced necrosis, and cirrhosis. Eur Surg Res 2003;35:148–60.

[56] Asahina K, Shiokawa M, Ueki T, Yamasaki C, Aratani A, Tateno C, et al. Multiplicative mononuclear small hepatocytes in adult rat liver: their isolation as a homogeneous population and localization to periportal zone. Biochem Biophys Res Commun 2006;342: 1160–7.

[57] Turner R, Lozoya O, Wang Y, Cardinale V, Gaudio E, Alpini G, et al. Human hepatic stem cell and maturational liver lineage biology. Hepatology 2011;53: 1035–45.

[58] Overturf K, Al-Dhalimy M, Finegold M, Grompe M. The repopulation potential of hepatocyte populations differing in size and prior mitotic expansion. Am J Pathol 1999;155:2135–43.

[59] Weglarz TC, Degen JL, Sandgren EP. Hepatocyte transplantation into diseased mouse liver. Kinetics of parenchymal repopulation and identification of the proliferative capacity of tetraploid and octaploid hepatocytes. Am J Pathol 2000;157:1963–74.

[60] Otto SP. The evolutionary consequences of polyploidy. Cell 2007;131:452–62.

[61] Lu P, Prost S, Caldwell H, Tugwood JD, Betton GR, Harrison DJ. Microarray analysis of gene expression of mouse hepatocytes of different ploidy. Mamm Genome 2007;18:617–26.

[62] Seitz HK, Stickel F. Molecular mechanisms of alcohol-mediated carcinogenesis. Nat Rev Cancer 2007;7: 599–612.

[63] Jorquera R, Tanguay RM. The mutagenicity of the tyrosine metabolite, fumarylacetoacetate, is enhanced by glutathione depletion. Biochem Biophys Res Commun 1997;232:42–8.

[64] Duncan AW, Hanlon Newell AE, Bi W, Finegold MJ, Olson SB, Beaudet AL, et al. Aneuploidy as a mechanism for stress-induced liver adaptation. J Clin Invest 2012;122:3307–15.

[65] Faggioli F, Vezzoni P, Montagna C. Single-cell analysis of ploidy and centrosomes underscores the peculiarity of normal hepatocytes. PLoS One 2011;6:e26080.

[66] Ganem NJ, Godinho SA, Pellman D. A mechanism linking extra centrosomes to chromosomal instability. Nature 2009;460:278–82.

[67] Cimini D, Moree B, Canman JC, Salmon ED. Merotelic kinetochore orientation occurs frequently during early mitosis in mammalian tissue cells and error correction is achieved by two different mechanisms. J Cell Sci 2003;116:4213–25.

[68] Crasta K, Ganem NJ, Dagher R, Lantermann AB, Ivanova EV, Pan Y, et al. DNA breaks and chromosome pulverization from errors in mitosis. Nature 2012;482: 53–8.

[69] Conlin LK, Thiel BD, Bonnemann CG, Medne L, Ernst LM, Zackai EH, et al. Mechanisms of mosaicism, chimerism and uniparental disomy identified by single nucleotide polymorphism array analysis. Hum Mol Genet 2010;19:1263–75.

[70] Westra JW, Peterson SE, Yung YC, Mutoh T, Barral S, Chun J. Aneuploid mosaicism in the developing and adult cerebellar cortex. J Comp Neurol 2008;507:1944–51.

[71] Turker MS, Schwartz JL, Jordan R, Ponomareva ON, Connolly L, Kasameyer E, et al. Persistence of chromatid aberrations in the cells of solid mouse tissues exposed to ^{137}Cs gamma radiation. Radiat Res 2004;162:357–64.

[72] Weier JF, Weier HU, Jung CJ, Gormley M, Zhou Y, Chu LW, et al. Human cytotrophoblasts acquire aneuploidies as they differentiate to an invasive phenotype. Dev Biol 2005;279:420–32.

[73] Steele PR, Yim AP, Herbertson BM, Watson J. Some flow cytofluorimetric studies of the nuclear ploidy of mouse hepatocytes. II. Early changes in nuclear ploidy of mouse hepatocytes following carbon tetrachloride administration: evidence for polyploid nuclei arrested in telophase. Br J Exp Pathol 1981;62:474–9.

[74] Diez-Fernandez C, Bosca L, Fernandez-Simon L, Alvarez A, Cascales M. Relationship between genomic DNA ploidy and parameters of liver damage during necrosis and regeneration induced by thioacetamide. Hepatology 1993;18:912–8.

[75] Nakatani T, Inouye M, Mirochnitchenko O. Overexpression of antioxidant enzymes in transgenic mice decreases cellular ploidy during liver regeneration. Exp Cell Res 1997;236:137–46.

[76] Weaver BA, Cleveland DW. The aneuploidy paradox in cell growth and tumorigenesis. Cancer cell 2008;14:431–3.

[77] Thompson SL, Bakhoum SF, Compton DA. Mechanisms of chromosomal instability. Curr Biol 2010;20: R285–95.

[78] Leenders MW, Nijkamp MW, Borel Rinkes IH. Mouse models in liver cancer research: a review of current literature. World J Gastroenterol 2008;14:6915–23.

[79] Speidel D. The magic trick of hepatocytes: uncoupling genomic instability from cancer. Trends Mol Med 2011;17:117–8.

[80] Pavelka N, Rancati G, Zhu J, Bradford WD, Saraf A, Florens L, et al. Aneuploidy confers quantitative proteome changes and phenotypic variation in budding yeast. Nature 2010;468:321–5.

[81] Rancati G, Pavelka N, Fleharty B, Noll A, Trimble R, Walton K, et al. Aneuploidy underlies rapid adaptive evolution of yeast cells deprived of a conserved cytokinesis motor. Cell 2008;135:879–93.

[82] Chen G, Bradford WD, Seidel CW, Li R. Hsp90 stress potentiates rapid cellular adaptation through induction of aneuploidy. Nature 2012;482:246–50.

[83] Zhu J, Pavelka N, Bradford WD, Rancati G, Li R. Karyotypic determinants of chromosome instability in aneuploid budding yeast. PLoS Genet 2012;8:e1002719.

[84] Dan C, Grygoryev D, Sandfort K, Connolly M, Cross B, Lasarev M, et al. Marked aneuploidy and loss of multiple chromosomes are common in autosomal mutants isolated from normal mouse kidney epithelium. Genes Chromosomes Cancer 2011;50:239–49.

[85] Manning K, Al-Dhalimy M, Finegold M, Grompe M. In vivo suppressor mutations correct a murine model of hereditary tyrosinemia type I. Proc Natl Acad Sci USA 1999;96:11928–33.

[86] Gimelbrant A, Hutchinson JN, Thompson BR, Chess A. Widespread monoallelic expression on human autosomes. Science 2007;318:1136–40.

[87] Ochiai T, Urata Y, Yamano T, Yamagishi H, Ashihara T. Clonal expansion in evolution of chronic hepatitis to hepatocellular carcinoma as seen at an X-chromosome locus. Hepatology 2000;31:615–21.

[88] Paradis V, Dargere D, Bonvoust F, Rubbia-Brandt L, Ba N, Bioulac-Sage P, et al. Clonal analysis of micronodules in virus C-induced liver cirrhosis using laser capture microdissection (LCM) and HUMARA assay. Lab Invest 2000;80:1553–9.

[89] Paradis V, Laurendeau I, Vidaud M, Bedossa P. Clonal analysis of macronodules in cirrhosis. Hepatology 1998;28:953–8.

[90] Kim TM, Yim SH, Shin SH, Xu HD, Jung YC, Park CK, et al. Clinical implication of recurrent copy number alterations in hepatocellular carcinoma and putative oncogenes in recurrent gains on 1q. Int J Cancer 2008;123:2808–15.

CHAPTER

13

Computational Modeling as an Approach to Study the Cellular and Molecular Regulatory Networks Driving Liver Regeneration

Daniel Cook,†, Rajanikanth Vadigepalli*,†*

*Daniel Baugh Institute for Functional Genomics and Computational Biology, Department of Pathology, Anatomy and Cell Biology, Sidney Kimmel Medical College, Thomas Jefferson University, Philadelphia, Pennsylvania, USA

†Department of Chemical and Biomolecular Engineering, University of Delaware, Newark, Delaware, USA

13.1 INTRODUCTION

Liver regeneration involves a highly coordinated response to tissue damage involving multiple cell types (parenchymal and nonparenchymal), extra-hepatic tissues, multiple size scales (from molecular signaling at the atomic level to overall tissue function at the macroscale), and multiple time scales (from nearly instantaneous release of ATP from hepatocytes following a partial hepatectomy to the weeks, or months, it may take for humans to regrow lost liver mass following a resection). Understanding the complexity of the liver's regenerative response to tissue damage and being able to predict the outcome of resection would provide an invaluable tool to patients undergoing partial liver resection for hepatocellular carcinoma or live liver transplant. Computational modeling provides a framework to help understand the complex, multi-scale interactions (molecular, cellular, and organ to organ) governing the liver's innate regenerative ability. Computational modeling enables researchers to collect and organize the available phenomenological as well as mechanistic knowledge, evaluate a wide range of scenarios in simulations that are experimentally intractable, predict the impact of altering particular underlying mechanisms on the cellular as well as the whole-tissue response, and design experiments to test the model-predicted hypotheses on the key control mechanisms governing the overall physiological response.

Liver Regeneration
http://dx.doi.org/10.1016/B978-0-12-420128-6.00013-0

Researchers in recent decades have made significant progress toward building computational representations of the regulatory events that are known to occur during liver regeneration. These models span a wide spectrum of detail incorporated into the model including spatial aspects, molecular processes, and whole-tissue scale phenomena (Figure 13.1a). At one end of the spectrum, agent-based models allow for high-spatial resolution, but often for only a section of tissue. At the other end, physiological models account for the regeneration of the whole liver but offer limited molecular mechanistic insight into how these responses are orchestrated. Models based on the so-called "omics" data provide a middle ground with high resolution of molecular events (e.g., transcriptome, proteome, metabolome) while empirically accounting for whole-tissue response. A broader challenge of connecting the "omics"-based modeling to tissue-scale and organ-scale physiology in a multi-scale framework remains unsolved.

A computational model recently published by the Drasdo lab took an agent-based modeling approach to simulating liver regeneration following carbon tetrachloride administration [1]. This model integrated three disparate scales of data: liver lobule architecture, cell processes, and physical forces (Figure 13.1b). Liver architecture was reconstructed in 3D from immunohistochemically stained confocal microscope images of livers. Liver architecture was considered as a fixed parameter as carbon tetrachloride does not drastically alter this aspect. Several cellular processes considered in the model were spatially distributed cell replication, necrosis following toxin administration, and hepatocyte-sinusoid contact area and orientation. The cell replication and necrosis were modeled as stochastic processes for each cell in the model, while hepatocyte-sinusoid orientation was simulated as a directed stochastic process with orientation biased toward an alignment along sinusoids. In addition, this model considered individual forces acting between cells. Each cell was modeled as a polarized, deformable sphere (with the bile canalicular structure oriented toward other hepatocytes and the basolateral membrane oriented toward the sinusoid). Physical forces, governed by cell adhesion, compression, and deformation were considered between cells. Model simulations predicted that hepatocytes must have a strong propensity for alignment along the sinusoid in order to capture the observed regeneration phenotype following chemical injury. Hoehme et al. [1] speculated that endothelial cells that survive the toxic liver damage contribute to such an alignment propensity.

"Omics"-based modeling involves a more detailed account of molecular aspects. In this approach, high-dimensional gene or protein expression data or metabolomics data are used to train or inform a model structure and parameters. Often, these models are algebraic because they necessarily contain a large number of terms, not amenable for typical ordinary differential equation-based dynamic modeling. An example of an "omics" model of liver function is the HepatoNet model constructed by the Holzhutter lab [2]. This model contains 2539 individual reactions and 777 metabolites to simulate the metabolic function of the liver (Figure 13.1c). The reactions were formulated into a stoichiometric matrix of fluxes and flux balance analysis was used to match the externally measured metabolites under an optimization objective of minimizing internal fluxes. The authors employed this model to investigate the robustness of the metabolic network active in hepatocytes by simulating single-enzyme knockouts. This model, though molecularly detailed, did not connect molecular regulation to the whole-tissue function in the context of regeneration.

As an example at the other end of the modeling spectrum, the Periwal group developed a physiologically based model that simulates tissue-scale liver regeneration following partial hepatectomy [3]. This model includes the JAK-STAT signaling

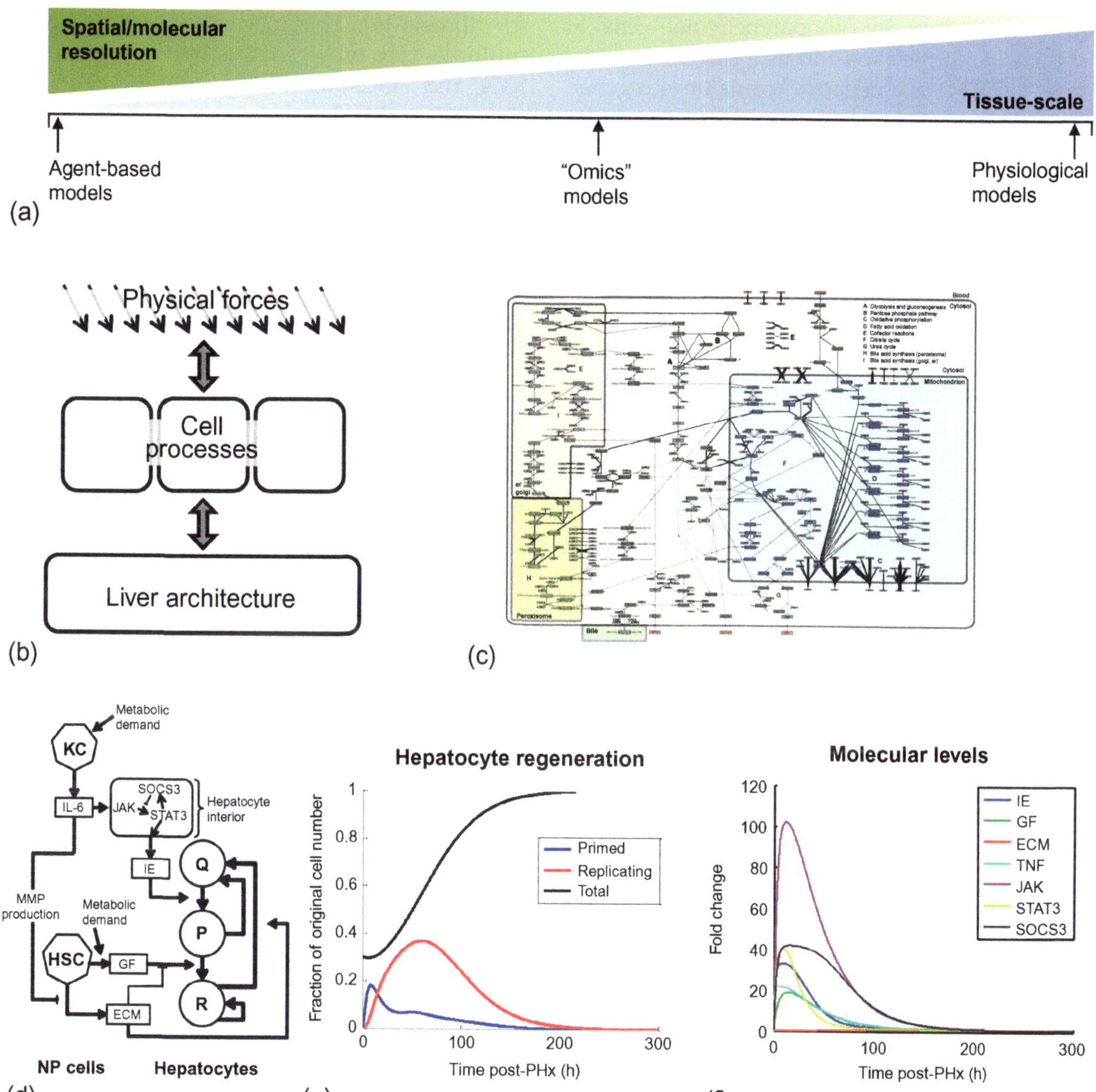

FIGURE 13.1 Computational models of liver regeneration. (a) Computational limitations require a trade-off between high resolution (molecular or spatial) and the ability to predict whole-tissue function. (b) Agent-based models, such as that developed by Hoehme et al. [1], based on CellSys, allow for high-spatial resolution, but often for only a section of tissue. (c) Models based on "omics" data, such as HepatoNet1 developed by Gille et al. [2], provide a middle ground where there is high resolution of molecular events at the desired scale (transcriptome, proteome, metabolome) and the molecular regulation can be extrapolated to whole tissue. (d) Physiological models, such as that developed by Furchtgott et al. [3], predict the regeneration of the whole liver but offer limited mechanistic insight into the molecular events. (e) Outputs from the physiological model can show cells in discrete physiological states as well as (f) molecular regulation governing those cell-state transitions.
Panel (d) was reproduced from Gille et al. [2].

pathway induced by IL-6 and growth factors (GFs) produced by nonparenchymal cells as drivers of regeneration and includes extracellular matrix (ECM) as a negative regulator of regeneration (Figure 13.1d). Hepatocytes were considered as existing in one of the three physiological states: quiescent (*Q*), primed (*P*), or replicating (*R*), with shifts between these states governed by factors produced by nonparenchymal cells, specifically Kupffer cells (KCs) and hepatic stellate cells (HSCs). This physiologically based model allowed the authors to connect the dynamic hepatocyte regeneration state (Figure 13.1e) to molecular regulation (Figure 13.1f). Using this model, Furchgott et al. predicted that two smaller resections can result in faster recovery than a single large resection and that decreasing metabolic load in the liver through fasting could lead to recovery following massive hepatectomy.

Other studies have also used computational modeling to explore liver function beyond the context of regeneration. For example, the Hunt lab used a game-theory model to explore the zonated liver metabolism of toxin clearance [4]. Their model consisted of multiple agents that can either eliminate a toxin (using nutrients) or ignore the toxin. Each hepatocyte (agent) in the model learned rules for eliminating toxins through a reinforcement learning rule to optimize long-term toxin elimination. They found that, for certain classes of toxins with specific harm to nutrient cost ratios, the most efficient elimination strategy is a zonated approach. In contrast to utilizing an agent-based approach, the Kuepfer lab developed a spatially resolved model of the liver architecture coupled with a physiologically based pharmacokinetic model to simulate blood flow and drug perfusion in the liver using partial differential equations [5]. This model was employed to assess and make predictions on how increasingly severe chronic liver disease would affect liver drug metabolism. In the context of liver fibrosis, the Vodovotz lab used an agent-based model to investigate the progression of and treatment for fibrosis [6]. This model takes into account multiple cell types in the liver (hepatocytes, KCs, and HSCs) as well as multiple cell states (i.e., M1 and M2 polarization of KCs). Model simulations indicated that treating fibrotic livers with anti-TNF-α could lead to decreased collagen deposition. However, clinical treatment of patients with alcoholic steatohepatitis using anti-TNF-α has been poorly received due to a suppressed immune system and increased likelihood of severe infections [7].

In addition to the classes of models and examples summarized above, there exist several opportunities where the disparate models with varying level of detail can be combined or merged to more accurately represent liver function during regeneration (Figure 13.1a). For example, it is possible to combine the metabolomic modeling methods with the agent-based models by incorporating the detailed metabolic model into each hepatocyte component. This formulation may be able to capture the pericentral to periportal gradient of hepatocyte metabolism seen in the homeostatic liver and predict how this gradient may change in response to tissue resection. Alternatively, physiologically based models could be expanded to include data from transcriptomic or proteomic studies to account for a larger context of molecular mechanisms governing physiological changes. In such an extended multiscale framework, the combined physiological-molecular modeling could be used to inform and prioritize where additional molecular details are required to iteratively expand the molecular detail, connecting the physiological-scale phenomenology with molecular-scale mechanisms.

In this chapter, we employ a physiologically based model to explore and understand the control principles governing liver regeneration. In contrast to "omics" data analysis studies aimed at identifying global molecular changes in the liver during regeneration, the present approach employs a targeted approach to identify classes of molecular regulatory events that strongly contribute to liver regeneration or deficiencies in liver repair.

13.2 EXTENDED COMPUTATIONAL MODEL OF LIVER REGENERATION INCLUDING CELL GROWTH

Liver regeneration following partial resection typically follows a well-documented set of steps involving hepatocytes and nonparenchymal cells, summarized in Table 13.1 [8–13]. Immediately following resection, hepatocytes sense the injury due to some combination of increased portal pressure, increased blood flow and shear rate, and increased metabolic demand per hepatocyte or increased toxin load per hepatocyte. These immediate changes cause hepatocytes to release ATP, initiate calcium signaling, and begin WNT signaling. During the priming phase, a proinflammatory response mediated predominantly through the KCs primes hepatocytes to enter the cell cycle as a response to the immediate PHx response. Following the priming phase, the replicating phase begins when hepatocytes enter the cell cycle. This phase is characterized by matrix remodeling, high levels of GFs, and an increase in hepatocyte number. Once hepatocytes have replaced much of the lost liver mass, nonparenchymal cells begin to regenerate. This regeneration phase is characterized by increased levels of VEGF, nonparenchymal cell replication, and increased angiogenesis. As recovery of lost liver mass completes, the termination phase begins. This phase is less well understood but is thought to occur through a

TABLE 13.1 Current Understanding of the Timing of Regulatory Events Occurring Following Liver Resection

Timing of Response in Rats	Regulatory Events
Immediate response (<30 min)	Hepatocytes release ATP Calcium signaling occurs WNT signaling occurs
Priming phase (0-6 h)	LPS levels increase in the liver Nonparenchymal cells produce cytokines and chemokines NF-κB activated in hepatocytes and nonparenchymal cells C/EBP-β to C/EBP-α ratio increases JAK-STAT signaling pathway activated in hepatocytes Hepatocytes begin to show signs of G0 to G1 transition
Hepatocyte Replicating phase (12-48 h)	Nonparenchymal cells produce MMPs, cleaving matrix Growth factors (HGF, EGF, FGF, etc.) are liberated from cleaved matrix Additional growth factors are synthesized by nonparenchymal cells New, basement-membrane matrix is deposited to facilitate proliferation Hepatocytes synchronously enter the cell cycle
Nonparenchymal cell Replicating phase (48-72 h)	VEGF signaling increases Angiogenesis occurs Nonparenchymal cells enter the cell cycle
Termination phase (approximately 1 week)	Matrix accumulates back to baseline levels Nonparenchymal cell activation decreases C/EBP-β to C/EBP-α ratio decreases to baseline levels Hepatocytes exit the cell cycle and return to G0 Liver mass may overshoot baseline, followed by a wave of hepatocyte apoptosis

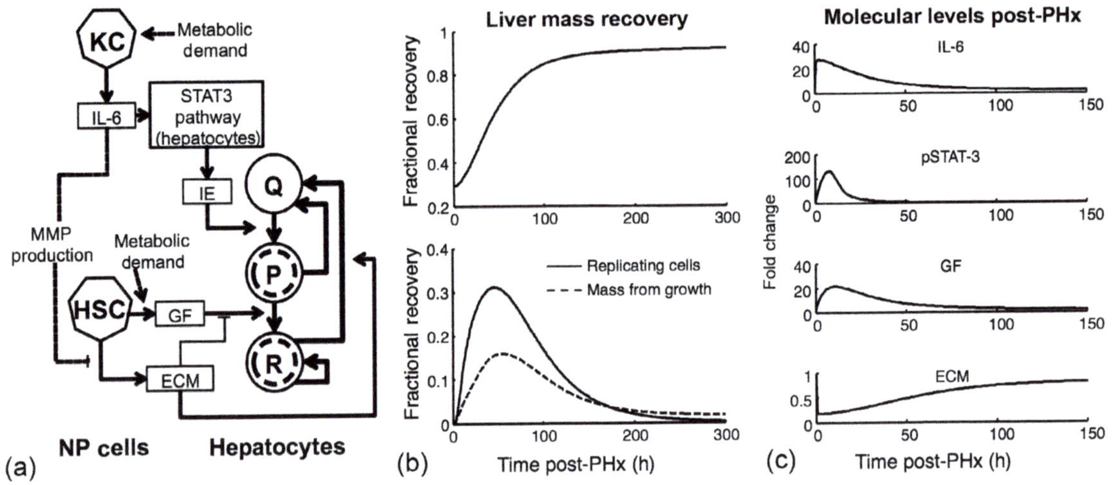

FIGURE 13.2 Extended computational model of liver regeneration including cell growth. (a) Schematic representation of the model. Hepatocytes in the primed and replicating states may increase their functional mass to compensate for increased metabolic load. (b) Simulated liver mass recovery in response to 70% PHx in rats showed that most of the mass is recovered by 1 week following resection. Mass from growth compensated for reduced mass early following resection but at later times replication was the predominant method of mass recovery. (c) Molecular levels of factors underlying the regeneration profile.

combination of physical size constraints, metabolic constraints, hepatic signaling, and an accumulation of ECM [14].

Our computational model of liver regeneration (represented schematically in Figure 13.2a) was used to investigate quantitatively how altering the molecular regulation of hepatocytes affects the liver's innate repair ability. Our model extends the model originally developed by the Periwal group to include cell growth in addition to hepatocyte replication; a detailed explanation of initial model derivation and parameter estimation is available in Ref. [3]. Our extended model with cell growth maintains the framework of the previously published initial model by allowing hepatocytes to exist in one of three states: quiescent (Q), primed (P), or replicating (R). Factors produced by nonparenchymal cells in response to liver metabolic load (metabolic demand per cell or M/N) shift hepatocytes between states, according to the following equations:

$$\frac{\mathrm{d}}{\mathrm{d}t}Q = -k_Q([\mathrm{IE}]-[\mathrm{IE}_0])Q + k_R[\mathrm{ECM}]R + k_{req}\sigma_{req}P - k_{ap}\sigma_{ap}Q \quad (13.1)$$

$$\frac{\mathrm{d}}{\mathrm{d}t}P = k_Q([\mathrm{IE}]-[\mathrm{IE}_0])Q - k_P([\mathrm{GF}] - [\mathrm{GF}_0])P - k_{req}\sigma_{req}P - k_{ap}\sigma_{ap}Q \quad (13.2)$$

$$\frac{\mathrm{d}}{\mathrm{d}t}R = k_P([\mathrm{GF}]-[\mathrm{GF}_0])P - k_R[\mathrm{ECM}]R + k_{prol}R - k_{ap}\sigma_{ap}R \quad (13.3)$$

where [IE] represents the concentration of immediate early genes expressed in response to STAT-3 transcriptional regulation, and [ECM] represents the amount of extracellular matrix. σ_{ap} and σ_{req} are sigmoidal functions defined as:

$$\sigma_{ap} = 0.5^* \left(1 + \tanh\left(\frac{\left(\theta_{ap} - \frac{M}{N}\right)}{\beta_{ap}}\right)\right) \quad (13.4)$$

$$\sigma_{req} = 0.5^* \left(1 + \tanh\left(\frac{(\theta_{req} - [\mathrm{GF}])}{\beta_{req}}\right)\right) \quad (13.5)$$

The parameters β and θ in each of these equations are tuned so that when metabolic load is high, σ_{ap} is high; conversely, when [GF] is high, σ_{req} is low. Therefore, when cells are highly stressed (high metabolic load), apoptosis occurs at a high rate; when GFs are available, cells remain in the "replicating" state.

The JAK-STAT signaling pathway, GF production, and ECM production are modeled as a combination of first-order and Michealis-Menton kinetics, as shown in the following equations. For a schematic of the JAK-STAT signaling pathway modeled, see Figure 13.1d.

$$\frac{d}{dt}[IL6] = k_{IL6}\frac{M}{N} - \frac{V_{JAK}[IL6]}{[IL6]+k_M^{JAK}} - \kappa_{IL6}[IL6] + k_1 \tag{13.6}$$

$$\frac{d}{dt}[JAK] = \frac{V_{JAK}[IL6]}{[IL6]+k_M^{JAK}} - \kappa_{JAK}[JAK] + k_2 \tag{13.7}$$

$$\frac{d}{dt}[STAT3] = \frac{V_{ST3}[JAK][proSTAT3]^2}{[proSTAT3]^2 + k_M^{ST3}\left(1+[SOCS3]/k_I^{SOCS3}\right)} - \frac{V_{IE}[STAT3]}{[STAT3]+k_M^{IE}} - \frac{V_{SOCS3}[STAT3]}{[STAT3]+k_M^{SOCS3}} - \kappa_{ST3}[STAT3] + k_3 \tag{13.8}$$

$$\frac{d}{dt}[SOCS3] = \frac{V_{SOCS3}[STAT3]}{[STAT3]+k_M^{SOCS3}} - \kappa_{SOCS3}[SOCS3] + k_4 \tag{13.9}$$

$$\frac{d}{dt}[IE] = \frac{V_{IE}[STAT3]}{[STAT3]+k_M^{IE}} - \kappa_{IE}[IE] + k_5 \tag{13.10}$$

$$\frac{d}{dt}[GF] = k_{GF}\frac{M}{N} - k_{up}[GF][ECM] - \kappa_{GF}[GF] + k_7 \tag{13.11}$$

$$\frac{d}{dt}[ECM] = -k_{deg}[IL6][ECM] - \kappa_{ECM}[ECM] + k_6 \tag{13.12}$$

where [proSTAT3] represents the concentration of monomeric STAT-3 available to dimerize following IL-6 signaling.

The overall cell mass, N, was modified from the initial model to include cell growth of primed and replicating cells in response to metabolic load as follows:

$$N = Q + G(P+R) \tag{13.13}$$

where G represents the relative cell mass, which is initially set to 1.

Explicitly considering cell mass and number of hepatocytes allowed us to identify fractional recovery of liver mass due to hepatocyte replication and hepatocyte growth (Figure 13.2b). Furthermore, our model allowed us to understand the molecular regulation underlying liver regeneration (Figure 13.2c). All mass increase in this model is assumed to be functional. It is conceivable that hepatocytes may increase in mass but maintain a specific functional capacity; however, such a scenario is not considered in our model.

All simulations were performed in Matlab (Mathworks). Model equations were set up to prevent molecular levels from becoming negative; however, a small fraction of parameter sets in combination with the integration tolerances of ode15s led to GF levels becoming negative at longer simulation times (greater than 150 h). These physically impossible GF levels did not significantly impact the regeneration profile because most of the growth had concluded by the time GF became negative. Because of such numerical instabilities of model simulations, however, GF levels were constrained to a minimum of 1.

13.3 MODES OF REGENERATION IDENTIFIED BY SAMPLING COMPUTATIONAL MODEL PARAMETER SPACE

We explored the model behavior across a range of biologically reasonable parameter values to map the spectrum of likely liver regeneration phenotypes. We varied the parameters in the model by ±50% of their nominal values using a Latin hypercube strategy to sample from a uniform distribution for each parameter. This technique allowed us to efficiently and uniformly sample throughout the parameter space, as shown using a three-dimensional projection of a principal component analysis decomposition of the sample space (Figure 13.3a).

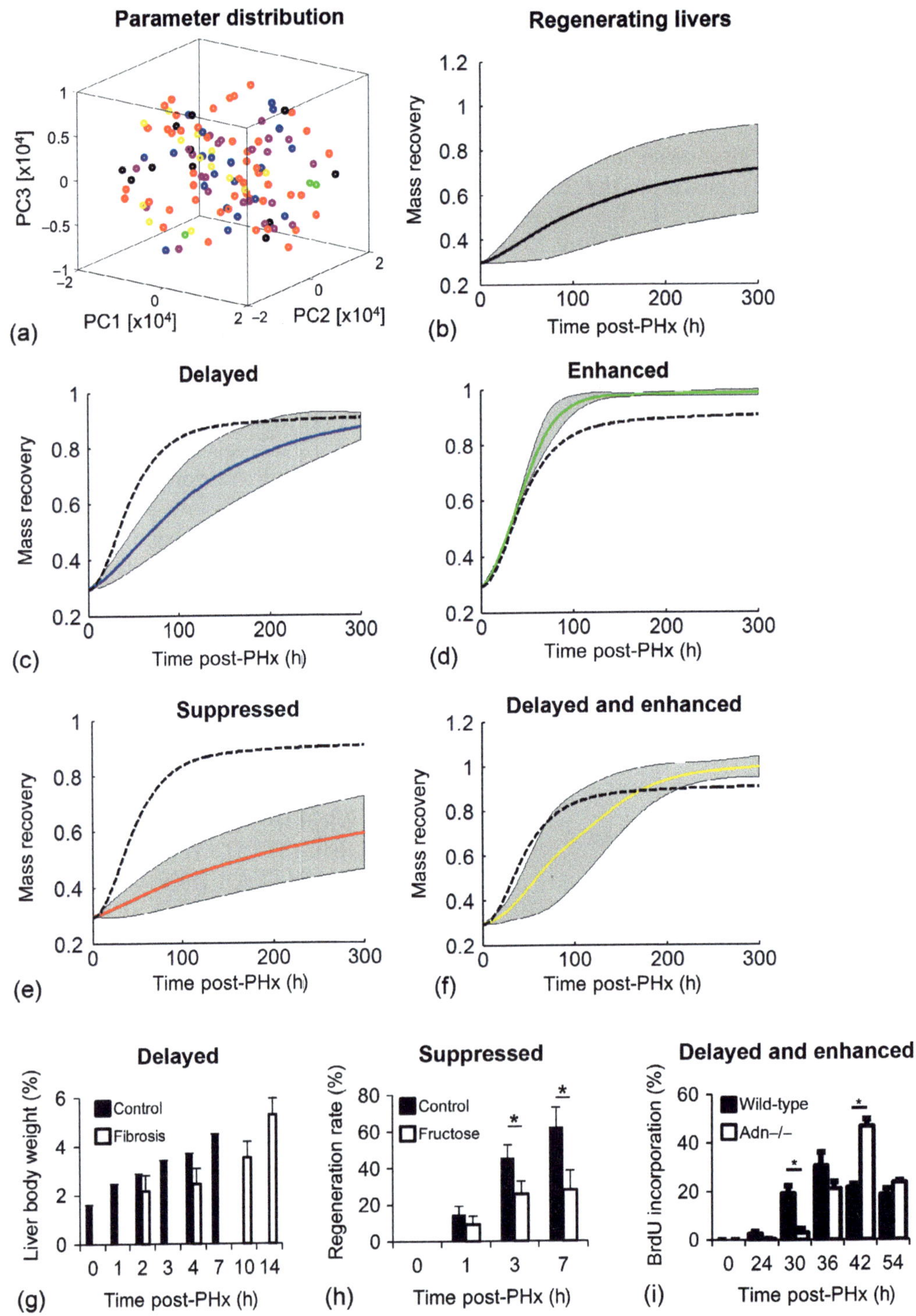
Parameter distribution
PC3 [x10⁴]
PC1 [x10⁴]
PC2 [x10⁴]
(a)
Regenerating livers
Mass recovery
Time post-PHx (h)
(b)
Delayed
(c)
Enhanced
(d)
Suppressed
(e)
Delayed and enhanced
(f)
Delayed
Liver body weight (%)
Control
Fibrosis
(g)
Suppressed
Regeneration rate (%)
Control
Fructose
(h)
Delayed and enhanced
BrdU incorporation (%)
Wild-type
Adn–/–
(i)

FIGURE 13.3 Multiple modes of regeneration revealed through simulation. (a) The parameter space for all parameters was sampled using a Latin hypercube method to efficiently sample the biologically feasible sample space (±50% of nominal value). A projection of the sampling results on the first three Principal Component space showing the uniform coverage of the parameter range. (b) There was wide variability in regeneration dynamics following resection. This variability was then categorized into altered modes of regeneration: (c) delayed regeneration, (d) enhanced regeneration, (e) suppressed regeneration, and (f) delayed and enhanced regeneration. Solid line = mean value, gray area = mean ± 1 standard deviation, dashed black line = nominal regeneration profile. Blue = delayed, red = suppressed, green = enhanced, yellow = delayed and enhanced, magenta = no regeneration, black = liver failure. Experimental studies suggest that chronic liver conditions can shift the liver into altered modes of regeneration. (g) Livers of mice subjected to CCl_4-induced fibrosis and 70% PHx showed delayed regeneration when compared to wild-type mice subjected to 70% PHx in another experiment [15,16]. (h) Rats with nonalcoholic steatohepatitis caused by a diet high in fructose had a suppressed regenerative response compared to controls [17]. (i) Mice with a genetic deficiency in the gene adiponectin had a delay in the initiation of regeneration but a later acceleration of hepatocyte replication [18].

The dynamic mass recovery profiles calculated using these parameters revealed a wide variation following PHx (Figure 13.3b). The mass recovery profiles were manually categorized into separate modes of regeneration: delayed (Figure 13.3c), enhanced (Figure 13.3d), suppressed (Figure 13.3e), and delayed and enhanced (Figure 13.3f). These phenotypes are associated with differential balance of the underlying molecular processes that affect the hepatocyte transitions between the three states.

We sought to relate the model-predicted regeneration modes to the phenotypes observed in multiple liver disease and injury contexts. For example, CCl_4-induced fibrosis appears to delay regeneration in mice compared to wild-type mice from another study, extending the time to complete recovery from approximately 1 to 2 weeks (Figure 13.3g) [15,16]. Rats with nonalcoholic steatohepatitis caused by a diet high in fructose had a suppressed regenerative response compared to controls (Figure 13.3h) [17]. Mice with a genetic deficiency in the gene adiponectin, in contrast, had a delay in the initiation of regeneration but a later acceleration of hepatocyte replication (Figure 13.3i) [18]. These biological correlates demonstrate that the physiologically based model can exhibit the range of mass recovery behavior as observed in the experimental studies. The model predictions on the molecular balances underlying the mass recovery phenotypes need to be tested for confirming the insights and for further model improvements, closing the loop between experiments and modeling pursuits. We have recently demonstrated such an iterative model improvement and experimental validation in the case of adiponectin fine-tuning of liver regeneration where a physiologically based model predictions on GF dynamics during regeneration were experimentally validated [18].

In addition to the regenerating modes, our model found additional modes of response to PHx that resulted in high variability in mass recovery (Figure 13.4a). Model simulations indicated that certain sets of parameters drive the system into a nonregenerative mode (Figure 13.4b) or a liver failure mode (Figure 13.4c). We filtered these parameter sets further to identify the parameter combinations that can shift regeneration between modes after crossing a certain threshold. For instance, increasing the apoptosis rate over 0.23 caused an otherwise regenerating liver phenotype to shift into a liver failure mode (Figure 13.4d).

13.4 DYNAMIC REGULATION OF THE REGENERATING LIVER REVEALED USING SENSITIVITY ANALYSES

We found that certain combinations of parameters resulted in profiles characterized

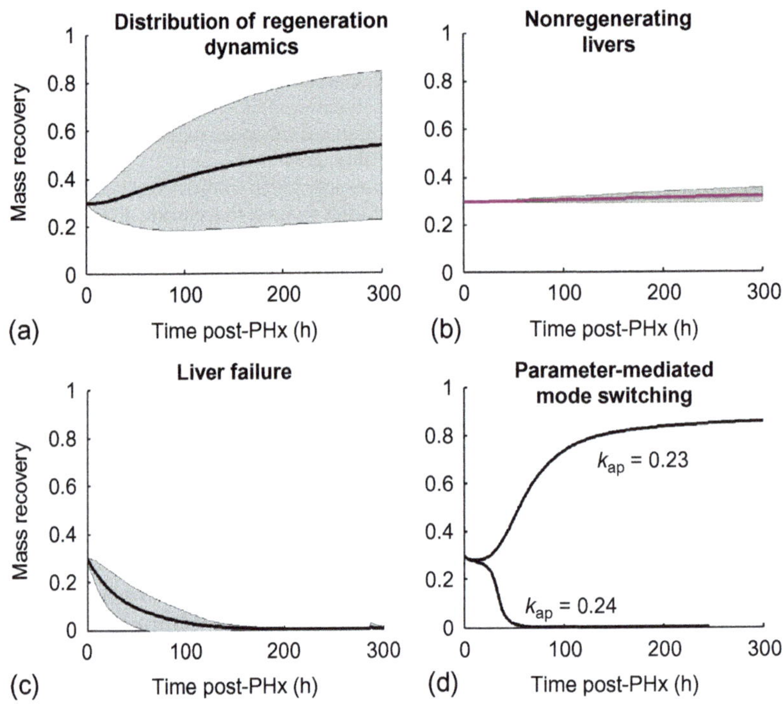

FIGURE 13.4 Identification of nonregenerating and liver failure modes. Decomposition of (a) model output also included response modes that did not regenerate, including (b) lack of regeneration and (c) liver failure. (d) Increasing the apoptosis rate of hepatocytes above a certain boundary limit caused delayed regeneration to lead to a complete liver failure post-PHx. Gray lines indicate ±1 standard deviation.

by mass recovery from predominantly hepatocyte replication and from predominantly mass increase in existing hepatocytes (Figure 13.5a). The replication-dominant regeneration mode was characterized by a relatively short delay prior to replication and a short duration of cell growth; whereas the growth-dominant mode was characterized by a relatively long delay in cell replication and a sustained cell growth (Figure 13.5b). This suggests that there is a balance between hepatocyte mass increase and replication. Hepatocytes may become larger to compensate for modes of regeneration with limited replication. Alternatively, livers containing hepatocytes that can increase their functional mass may not need as large of a replication response to tissue damage. The relative contributions of hepatocyte growth rate and metabolic demand leading to replication could inform whether these alternative but not mutually exclusive mechanisms are balanced in a dynamic fashion.

We pursued a sensitivity analysis-based approach that enables us to identify dynamic contributions of all the model parameters to the recovery profile. The computed sensitivity values (S_i) represent the extent to which the model output (mass recovery, in this case) would change due to a defined level of perturbation in each of the parameters. Normalized sensitivity coefficients were estimated by changing each parameter (p_i) by ±10% of its nominal value and calculating sensitivity at each simulation time point according to the following equation:

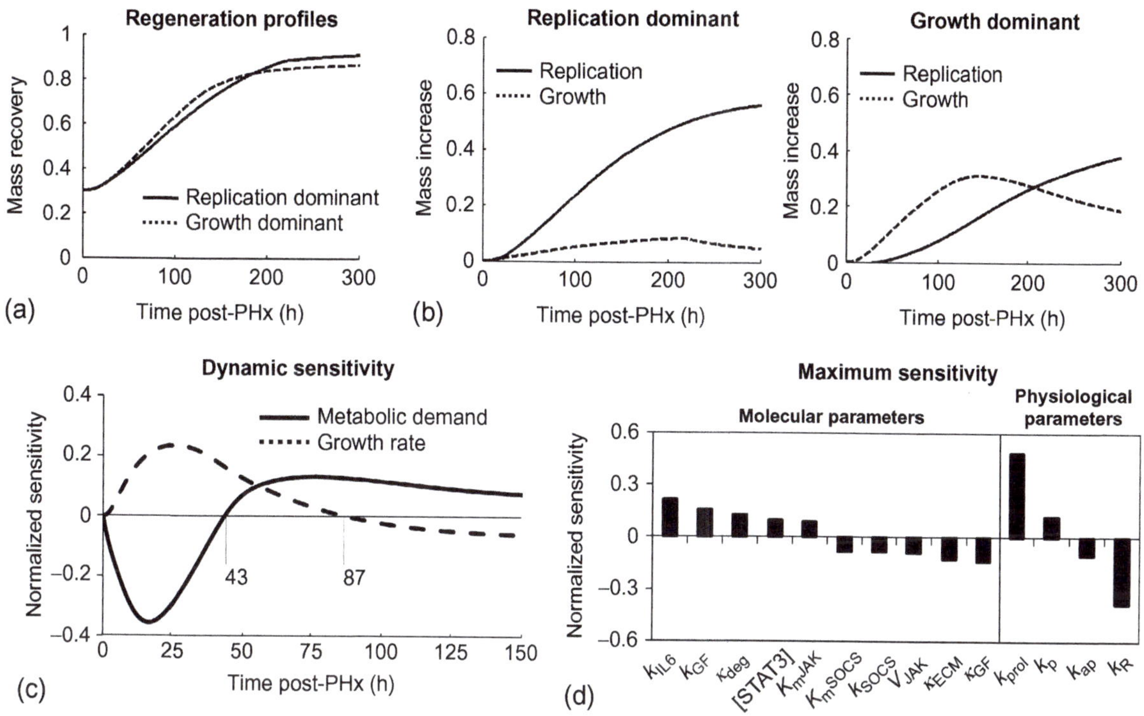

FIGURE 13.5 Local sensitivity analysis of the extended model. (a) Combinations of parameters resulted in profiles categorized as replication-dominant (solid line) and growth-dominant (dashed line) modes. (b) The replication-dominant regeneration mode is characterized by a relatively short delay prior to replication and a short duration of cell growth; whereas the growth-dominant mode is characterized by a relatively long delay in cell replication and a sustained cell growth. (c) Normalized local sensitivities of metabolic load and growth rate reveal a dynamic competition between replication and growth. During the replication phase (from 12 to 80 h), growth and metabolic demand both drive regeneration. During initiation and termination, however, imbalances between growth and metabolic load can inhibit regeneration. (d) Normalized sensitivities corresponding to the molecular parameters that can dynamically affect liver regeneration.

$$S_i(t) = \frac{\Delta \mathrm{Mass}(t)/\mathrm{Mass}(t)}{\Delta p_i/p_i} \tag{13.14}$$

where Mass(t) represents the nominal mass fraction of hepatocytes at any given time, t, ΔMass(t) is the deviation from nominal caused by the parameter change, and Δp_i is the change in the parameter value. This calculation yields estimates of dynamic parametric sensitivities, showing how the profile of liver regeneration responds to changes in parameters as a function of time. The sensitivity analysis predicts not only which parameters (and therefore biological functions) are likely to be critical for complete regeneration but also which parameters may be beneficial or detrimental to mass recovery.

We performed a parametric local sensitivity analysis to identify the model parameters that contributed most to the shifting regeneration modes. This analysis also allowed us to identify additional factors and network balances significantly affecting the liver regeneration dynamics. Our analysis revealed a dynamic balance between hepatocyte growth and regeneration induced by increased metabolic load

(Figure 13.5c). In the early phase of post-PHx response, hepatocyte growth was a positive contributor to liver regeneration, while metabolic demand negatively affected the progression of regeneration. At this early time, metabolic demand acted in hepatocytes predominantly to induce apoptosis in damaged cells through high metabolic load, causing reduced liver mass. After approximately 12 h post-PHx, high metabolic load induced high response in nonparenchymal cells causing increased priming and regeneration. From approximately 12 to 86 h post-PHx, metabolic load and hepatocyte growth acted synergistically to promote liver regeneration. Near the termination stage of liver regeneration, however, hepatocyte growth inhibited liver regeneration by inducing hepatomegaly and decreasing the driving force for regeneration. Maximum absolute sensitivity values computed for each parameter revealed that both molecular and physiological parameters showed high sensitivity (Figure 13.5d). Model parameters related to production of factors from nonparenchymal cells as well as model parameters related to hepatocyte response to these factors showed high sensitivity, suggesting that parenchymal and nonparenchymal cell regulation are both important for governing liver regeneration. Hepatocyte responsiveness to nonparenchymal signaling regulated synchronous entry into the cell cycle and nonparenchymal signaling governed the timing and magnitude of regeneration response.

13.5 MODEL LIMITATIONS, FUTURE DIRECTIONS, AND EXPERIMENTAL INSIGHTS

The present physiologically based model, while accounting for mass recovery profiles and some of the hepatocyte molecular events, does not fully incorporate the nonparenchymal cell processes. It is widely recognized that nonparenchymal cells play important roles in modulating liver homeostatic function and dynamic regeneration. KCs, endothelial cells, HSCs, and extrahepatic immune cells have all been implicated in modulating the liver cytokine microenvironment during regeneration [19–22]. Likely, this modulation is the result of complex processes and networks active in these nonparenchymal cells. Our current model, however, represents these processes as simple linear responses to an as-yet incompletely characterized signal. Future models may need to incorporate a more detailed representations of the nonparenchymal cell regulation and potential recruitment of extrahepatic cells. This is likely to involve modifying the model to explicitly account for nonparenchymal cell states with the associated molecular process details. In this regard, we have made progress in developing a regulatory network model to describe activation of HSCs [23].

Another important liver process missing from the current model is hepatocyte metabolism. The current model assumes that hepatocyte metabolism and replication processes are largely independent. This is likely not the case. There may be interactions between hepatocyte metabolism and replication or growth ability. These interactions may cause altered regeneration profiles across the liver sinusoid because of the gradient of metabolic function occurring across sinusoids [24]. Model extensions integrating the metabolic processes with the physiological-scale information will be better placed to explore this possibility.

Additionally, it is becoming clear that populations of cells do not behave homogeneously but display high variability at both the transcriptional and protein levels [25–27]. It is highly likely that such variability of individual cellular responses is also present in the liver in both parenchymal and nonparenchymal cells. In such a scenario, it may be insufficient to model individual cell types as existing in a small set of discrete states. Population balance-based methods offer an avenue for explicitly considering the impact of variability and distributions of cell types on the overall mass recovery [28]. Alternatively, agent-based models could also account for the cellular scale variability of regulatory networks and molecular processes. In this scheme, individual cells would be modeled as driven by parameters drawn from the observed single-cell data distributions, or be responding to the particular local microenvironment such that the experimentally observed variability is recreated *in silico*. Physiologically based models could account for such variability in a more coarse-grained manner by using identified "subphenotypes" as representative states for each distinct cell type. Emerging results on single-cell transcriptional regulation point to such subphenotypes embedded in what would otherwise be considered as a highly variable population [27].

13.6 CONCLUSIONS

Computational modeling approaches are being successfully applied to study various aspects of liver regeneration. The computational modeling framework can organize disparate elements of our understanding of the processes and events involved in liver regeneration, as well as serving as an important tool for hypothesis generation and experimental prioritization. The computational approach further allows for comparison between experiments from different labs, protocols, and model systems to synthesize information into a unifying framework for studying liver regeneration across multiple disease conditions.

References

[1] Hoehme S, Brulport M, Bauer A, Bedawy E, Schormann W, Hermes M, et al. Prediction and validation of cell alignment along microvessels as order principle to restore tissue architecture in liver regeneration. Proc Natl Acad Sci USA 2010;107(23):10371–6. http://dx.doi.org/10.1073/pnas.0909374107.

[2] Gille C, Boelling C, Hoppe A, Bulik S, Hoffmann S, Huebner K, et al. HepatoNet1: a comprehensive metabolic reconstruction of the human hepatocyte for the analysis of liver physiology. Mol Syst Biol 2010;6:411. http://dx.doi.org/10.1038/msb.2010.62.

[3] Furchtgott LA, Chow CC, Periwal V. A model of liver regeneration. Biophys J 2009;96(10):3926–35. http://dx.doi.org/10.1016/j.bpj.2009.01.061.

[4] Sheikh-Bahaei S, Kim SH, Sheikhbahaei S, Hunt CA. Understanding the role of liver zonation in toxin elimination. Int J Intell Control Syst 2009;14:33–40.

[5] Schwen LO, Krauss M, Niederalt C, Gremse F, Kiessling F, Schenk A, et al. Spatio-temporal simulation of first pass drug perfusion in the liver. PLoS Comput Biol 2014;10(3):e1003499. http://dx.doi.org/10.1371/journal.pcbi.1003499.

[6] Dutta-Moscato J, Solovyev A, Mi Q, Nishikawa T, Soto-Gutierrez A, Fox IJ, et al. A multiscale agent-based in silico model of liver fibrosis progression. Syst Biol 2014;2:18.

[7] Naveau S, Chollet-Martin S, Dharancy S, Mathurin P, Jouet P, Piquet M, et al. A double-blind randomized controlled trial of infliximab associated with prednisolone in acute alcoholic hepatitis. Hepatology 2004;39(5):1390–7. http://dx.doi.org/10.1002/hep.20206.

[8] Diaz-Munoz M, Canedo-Merino R, Gutierrez-Salinas J, Hernandez-Munoz R. Modifications of intracellular calcium release channels and calcium mobilization following 70% hepatectomy. Arch Biochem Biophys 1998;349(1):105–12. http://dx.doi.org/10.1006/abbi.1997.0396.

[9] Grumm S, Cofan M, Juskeviciute E, Hoek JB. Adenine nucleotide changes in the remnant liver: an early signal for regeneration after partial hepatectomy. Hepatology 2008;48(3):898–908. http://dx.doi.org/10.1002/hep.22421.

[10] Michalopoulos G. Liver-regeneration—molecular mechanisms of growth-control. FASEB J 1990;4(2):176–87.

[11] Michalopoulos G, DeFrances M. Liver regeneration. Science 1997;276(5309):60–6. http://dx.doi.org/10.1126/science.276.5309.60.

[12] Monga S, Pediaditakis P, Mule K, Stolz D, Michalopoulos G. Changes in wnt/beta-catenin pathway during regulated growth in rat liver regeneration. Hepatology 2001;33(5):1098–109. http://dx.doi.org/10.1053/jhep.2001.23786.

[13] Taub R. Liver regeneration: from myth to mechanism. Nat Rev Mol Cell Biol 2004;5(10):836–47. http://dx.doi.org/10.1038/nrm1489.

[14] Michalopoulos GK. Liver regeneration after partial hepatectomy critical analysis of mechanistic dilemmas. Am J Pathol 2010;176(1):2–13. http://dx.doi.org/10.2353/ajpath.2010.090675.

[15] Lehmann K, Tschuor C, Rickenbacher A, Jang J, Oberkofler CE, Tschopp O. Liver failure after extended hepatectomy in mice is mediated by a p21-dependent barrier to liver regeneration. Gastroenterology 2012;143(6):1609. http://dx.doi.org/10.1053/j.gastro.2012.08.043.
[16] Xue F, Takahara T, Yata Y, Kuwabara Y, Shinno E, Nonome K, et al. Hepatocyte growth factor gene therapy accelerates regeneration in cirrhotic mouse livers after hepatectomy. Gut 2003;52(5):694–700. http://dx.doi.org/10.1136/gut.52.5.694.
[17] Tanoue S, Uto H, Kumamoto R, Arima S, Hashimoto S, Nasu Y, et al. Liver regeneration after partial hepatectomy in rat is more impaired in a steatotic liver induced by dietary fructose compared to dietary fat. Biochem Biophys Res Commun 2011;407(1):163–8. http://dx.doi.org/10.1016/j.bbrc.2011.02.131.
[18] Correnti JM, Cook D, Aksamitiene E, Swarup A, Ogunnaike B, Vadigepalli R, et al. Adiponectin fine-tuning of liver regeneration dynamics revealed through cellular network modeling. J Physiol 2014;593(2):365–83. http://dx.doi.org/10.1113/jphysiol.2014.284109 [pii].
[19] Friedman SL. Hepatic stellate cells: protean, multifunctional, and enigmatic cells of the liver. Physiol Rev 2008;88(1):125–72. http://dx.doi.org/10.1152/physrev.00013.2007.
[20] Kang L, Mars WM, Michalopoulos GK. Signals and cells involved in regulating liver regeneration. Cell 2012;1(4):1261–92.
[21] Malik R, Selden C, Hodgson H. The role of non-parenchymal cells in liver growth. Semin Cell Dev Biol 2002;13(6):425–31. http://dx.doi.org/10.1016/S1084-9521(02)00130-1.
[22] Meijer C, Wiezer M, Diehl A, Yang S, Schouten H, Meijer S, et al. Kupffer cell depletion by CI2MDP-liposomes alters hepatic cytokine expression and delays liver regeneration after partial hepatectomy. Liver 2000;20(1):66–77. http://dx.doi.org/10.1034/j.1600-0676.2000.020001066.x.
[23] Kuttippurathu L, Parrish A, Vadigepalli R. Integrated computational model of intracellular signaling and microRNA regulation predicts the network balances and timing constraints critical to the hepatic stellate cell activation process. Processes 2014;2(4):773–94.
[24] Jungermann K, Kietzmann T. Oxygen: modulator of metabolic zonation and disease of the liver. Hepatology 2000;31(2):255–60. http://dx.doi.org/10.1002/hep.510310201.
[25] Amir ED, Davis KL, Tadmor MD, Simonds EF, Levine JH, Bendall SC, et al. viSNE enables visualization of high dimensional single-cell data and reveals phenotypic heterogeneity of leukemia. Nat Biotechnol 2013;31(6):545. http://dx.doi.org/10.1038/nbt.2594.
[26] Antebi YE, Reich-Zeliger S, Hart Y, Mayo A, Eizenberg I, Rimer J, et al. Mapping differentiation under mixed culture conditions reveals a tunable continuum of T cell fates. PLoS Biol 2013;11(7):e1001616. http://dx.doi.org/10.1371/journal.pbio.1001616.
[27] Park J, Brureau A, Kernan K, Starks A, Gulati S, Ogunnaike B, et al. Inputs drive cell phenotype variability. Genome Res 2014;24(6):930–41. http://dx.doi.org/10.1101/gr.161802.113.
[28] Ramkrishna D, Singh MR. Population balance modeling: current status and future prospects. Annu Rev Chem Biomol Eng 2014;5(5):123–46. http://dx.doi.org/10.1146/annurev-chembioeng-060713-040241.

CHAPTER

14

Mitogen-Induced Cell Proliferation and Cancer Promotion in the Liver

Joseph Locker

Department of Pathology, University of Pittsburgh, Pittsburgh, Pennsylvania, USA

14.1 INTRODUCTION

Most of the liver mass consists of mature hepatocytes that rarely divide. Nevertheless, control of proliferation is facultative and cell loss or mitogens rapidly induce widespread proliferation. There is an important biological distinction between the two types of induction, because regeneration induces proliferation that compensates for cell loss, while mitogens induce proliferation without cell loss—hyperplasia—that increases the number of hepatocytes [1]. Both types of proliferation promote cancer, but the much stronger effect of hyperplasia has driven this field of research.

Liver regeneration, a response to injuries that cause hepatocyte loss, is a controlled replacement of hepatic number and cell mass while maintaining liver function [2,3]. There is considerable species and strain variation, but 2/3 partial hepatectomy (PH) in mouse will induce S phase in ~75% of hepatocytes, beginning at 36 h [4,5]. Regeneration is even faster in the rat, where post-hepatectomy S phase begins 12 h after PH [6]. Regeneration occurs after either PH or hepatocyte necrosis by toxic agents like CCl_4 and is therefore controlled by hepatocyte loss, not the mechanism of injury. A finely graded response, regeneration will replace losses ranging from a few cells to greater than 90% of the total liver. Recent studies have demonstrated that liver regeneration is entirely dependent on the signal transduction cascades downstream of the tyrosine kinase receptors EGFR or Met [7]. Nevertheless, rapid regeneration first requires early priming signals from TNFα or other ligands of TNFR1 with induction of NFκB [5,8]. Blocking these ligands, receptors, or NFκB signal responses greatly slows regeneration and markedly increases liver injury and mortality. In addition, rapid liver regeneration is lost in aged animals [9–11].

Numerous agents—hormones, cytokines, drugs, and xenobiotics—act as hepatocyte mitogens. Some of these agents also enlarge the liver mass, although such growth has often been considered a gradual adaptation without a direct link to proliferation. There are many biological distinctions between hyperplasia and regeneration, but the concept that mitogens act through different pathways was formerly controversial, because many drugs and chemicals induce

Liver Regeneration
http://dx.doi.org/10.1016/B978-0-12-420128-6.00014-2

hepatocyte injury and thus regeneration. In addition, some chemical mitogens induce TNFα and mimic the early signaling of liver regeneration. In contrast, numerous other chemicals act directly on hepatocytes without injury or cytokine induction. These *direct chemical mitogens* activate three nuclear receptor transcription factors, the constitutive androstane receptor (CAR), the peroxisome proliferator-activating receptor (PPAR), or the thyroid hormone receptor (THR) [12]. Each of these receptors activates genes with a common heterodimeric partner, the retinoid X receptor (RXR).

Although each nuclear receptor heterodimer binds a different DNA motif and activates a distinctive set of target genes, five common features distinguish nuclear receptor-activated proliferation responses from regeneration: (1) The mitogens induce *de novo* enlargement—a combination of hyperplasia and hypertrophy—of the intact normal liver, while regeneration only restores liver to its normal size. (2) They are independent on TNFα, cytokine, and NFκB signaling. (3) None of the nuclear receptor-activating agents is mitogenic in cell culture, in contrast to the EGFR and Met ligands that are the essential mitogens of liver regeneration. (4) Nuclear-receptor-activated mitogenesis does not decline with aging, in contrast to liver regeneration [10,13]. (5) The mitogenic effects of nuclear receptor activators were discovered via studies of their powerful promotion of liver cancer [14–16].

14.2 CANCER PROMOTION

Research on hyperplasia and compensatory proliferation has focused on their distinctive relationships to carcinogenesis, which is defined by two essential processes, *initiation* and *promotion*. The first process, initiation, occurs by direct mutation of cancer driver or suppressor genes following genotoxic injury [17,18]. Promotion, the second essential process, is independent of initiation, nonmutagenic, and induced by distinct agents that do not initiate. However, the definition of promotion is vague, because different mechanisms can increase carcinogenesis a few-fold or have vastly greater effects. Differentiating *hyperplastic* and *compensatory promotion* distinguishes these weak and strong mechanisms.

Since hepatocytes are normally quiescent, tumor progression requires induced mitogenesis for clonal expansion of initiated cells. Early studies applied phenobarbital (PB), a known hepatic mitogen [19], and demonstrated a powerful promotional effect [20]. Later studies demonstrated that PB activates a set of genes via a discrete DNA motif that binds CAR and that DNA motif even though it is not a direct ligand [21,22].

Despite these important findings, the relationship of CAR activators to carcinogenesis has been problematic for several reasons. First, the agents have multiple promotional effects, some independent of proliferation. For example, PB strongly induces p450 enzymes that can activate procarcinogens [23]. This induction, formerly considered the main cause of tumor promotion, is actually independent of promotion because CAR ligands do not require activation while most pathways that activate procarcinogens are independent of CAR.

Second, proliferation affects the early stages of carcinogenesis by acting to fix mutations, since cell division before repair of chemically modified bases increases mutation frequency [24,25]. A limited effect, fixation targets cells near the time of carcinogenesis. The mechanism is particularly important for *complete* carcinogens, i.e., carcinogens that do not require separate promoting agents because they combine DNA modification with injury that leads to cell death, inflammation, and compensatory proliferation. Such *fixational* promotion is modified by the magnitude of inflammation and NFκB-facilitated survival of damaged or initiated cells. For example, ablation of IKKβ in the mouse causes a fourfold decrease in tumor incidence

8 months after carcinogenesis with diethylnitrosamine (DEN) [26]. In contrast, concurrent administration of PB accelerates the appearance of tumors by several months, an independent and more powerful promotional effect.

Describing activators of CAR and especially PPAR as *nongenotoxic carcinogens* is another source of confusion. Although these agents can eventually induce cancer without an initiating carcinogen [27–29], the effect is actually powerful promotion of weak initiation by spontaneous, metabolically induced, or environmental mutations [25,30].

There are critical differences between *hyperplastic* and *compensatory promotion*. One difference is pragmatic, the relationship to research strategies, since hyperplastic promotion leads to more expedient experimental models. Nevertheless, this expediency does not reduce the biological importance of compensatory promotion since diseases may subject livers to years of damage and compensatory proliferation. A more critical difference is that hyperplasia represents distinct control pathways with many differences from compensatory proliferation. From the perspective of cancer promotion, the following sections will detail specific pathways of hyperplasia, starting with a group of cytokine-activating mitogens that actually stimulate the compensatory pathway. Subsequent sections will consider the highly promotional CAR and PPAR pathways, the antipromotional TR pathway, and minor hyperplasia effects that have not been linked to promotion.

14.3 CYTOKINE-ACTIVATING MITOGENS

Early studies did not distinguish different proliferation pathways, but discriminated promotional and antipromotional forms of mitogenesis, the latter by diverse agents. Among these, lead nitrate rapidly induces liver growth and proliferation with minimal injury [31]. A single dose in the rat induces cell proliferation with a mitotic peak at 48 h and a 70% increase in liver mass 72 h after treatment. Surprisingly, this treatment reduces DEN-induced carcinogenesis by 50% [32,33]. Such antipromotional effect is explained by the mechanism of regression following lead-induced hyperplasia. After a temporary increase in liver mass and cell number, a wave of apoptosis peaks at 5 days and restores normal liver mass and hepatocyte number by 15 days. Moreover, the culling of surplus cells selectively eliminates carcinogen-initiated cells. Such apoptotic regression is quite difference from the regression following hyperplasia induced by agents that activate PPAR or CAR (Ref. [34]; see below).

Lead nitrate mitogenesis resembles liver regeneration because it acts by stimulating nonparenchymal cells—especially Kupffer cells—to release TNFα, which activates NFκB [35–37]. However, the NFκB activation is stronger and more prolonged than after PH. Two other chemicals have similar effects, cadmium salts and ethylene dibromide [38–41].

These antipromotional mitogens share a common priming mechanism with liver regeneration. However, they are antipromotional while liver regeneration promotes carcinogenesis. A single dose of CCl_4, a single 2/3 PH, and multiple PH treatments all increase the size and number of tumors after chemical carcinogenesis [33,42,43]. Multiple 1/3 PH treatments are somewhat more effective [44]. Similarly, cell death and increased inflammatory signaling through NFκB also stimulate chemical carcinogenesis [26,45].

The cell proliferation following either cell loss or lead nitrate treatment shares TNFα-NFκB priming. The divergent effects on carcinogenesis by the same pathway reflect a balance among three processes that affect initiated cells. Mutation fixation produces them, proliferation increases them, and apoptosis eliminates them. All three processes have moderate effects, since they mediate a few-fold increase or decrease in tumor incidence but do not change the timing of carcinogenesis.

14.4 DRUG-INDUCED HYPERPLASIA

Systematic study of tumor promotion and nongenotoxic carcinogens led to discovery of numerous agents that promote cancer, primarily mitogens that act through specific nuclear receptor transcription factors [46]. In contrast to the moderate promotional and antipromotional effects of compensatory proliferation and mitogens that activate TNFα, promotion via nuclear receptor pathways can accelerate the appearance of tumors by several months. Moreover, knockout mutations that abolish the nuclear receptors block ligand-induced mitogenesis but do not affect liver regeneration [47–52].

There are two main classes of non-TNFα-activating liver mitogens. Both induce hepatomegaly and cell proliferation. One class of drugs, the *peroxisome proliferators*, stimulates biogenesis of peroxisomes and fatty acid oxidation. Although not mutagens or procarcinogens, the drugs also induce cancer without an additional carcinogenic agent, a process described as *nongenotoxic carcinogenesis.* The full carcinogenesis is a composite of promotion and oxidative DNA damage that results from increased peroxisomal β-oxidation [53]. The promotional component can be distinguished because peroxisome proliferators also strongly promote carcinogenesis by mutating chemical carcinogens [15,54–56].

Studies have shown that the peroxisome proliferators are ligands of PPARα, a transcription factor that activates genes of peroxisome biogenesis and oxidative lipid metabolism. A second class of non-TNFα-activating mitogens, including PB, activates a second nuclear receptor, CAR, which stimulates biogenesis of smooth endoplasmic reticulum and enzymes that detoxify xenobiotic compounds.

Both CAR and PPAR bind DNA as heterodimeric partners with RXR. Other nuclear receptors—the thyroid hormone (THR), retinoic-acid (RAR), and pregnane X receptors (PXR)—are also RXR partners associated with proliferative responses. Each of these nuclear receptors has a distinctive binding site and mediates a different set of gene responses. Within this group of receptors, hepatocytes express significant levels of CAR, PXR, PPARα, THRβ, and RXRα, with minimal expression of other isoforms and all RARs (Figure 14.1).

The regression of hyperplasia induced by PPAR and CAR differs significantly from that induced by lead nitrate. Withdrawal of lead nitrate and similar agents causes rapid apoptosis and restores the liver cell number and mass prior to treatment. In contrast, withdrawal of PPAR and CAR inducers causes the hyperplastic liver to revert to its normal size via cell shrinkage without significant cell death, so that the restored liver retains a higher DNA content than normal [34]. Indeed, PPAR and CAR inducers have an antiapoptotic effect that enhances carcinogenesis, presumably by protecting initiated cells [34,57,58].

14.5 PPAR-INDUCED PROLIFERATION

Clofibrate, a hypolipidemic agent, was the first compound demonstrated to cause hepatomegaly via an increase in peroxisomes [53,59]. Drugs that induce this response are ligands of PPARα and fall into several groups. The fibrate class of hypolipidemic drugs includes clofibrate and its close analog, nafenopin, as well as more potent analogs that are preferred for experimental models [4-chloro-6-(2,3-xylidino)-2-pyrimidinylthio]-acetic acid (Wy-14,643) and [4-chloro-6-(2,3-xylidino)-2-pyrimidinylthio-(*N*-beta-hydroxyethyl)-acetamide] (Br-931). The phthalate ester plasticizer di(2-ethylhexyl) phthalate (DEHP) and related compounds are also active ligands. In addition to these compounds, fatty acids and eicosanoids serve as natural endogenous ligands [60].

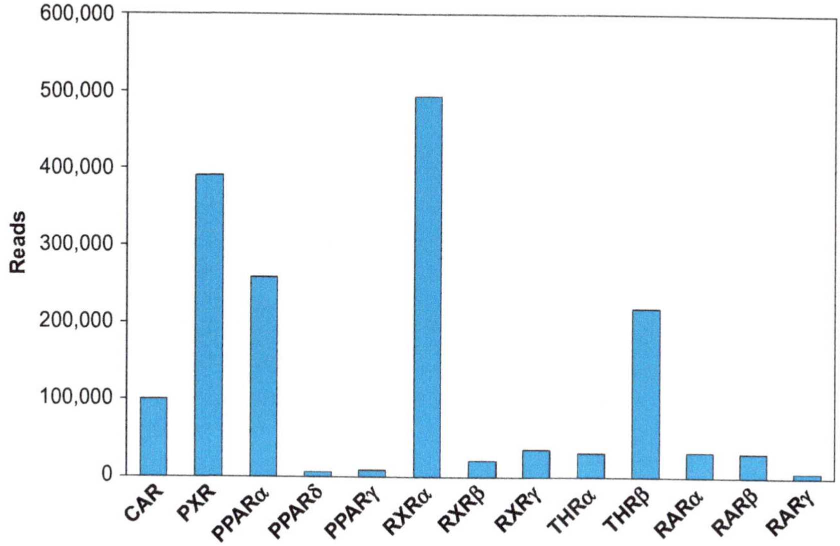

FIGURE 14.1 Hepatocyte nuclear receptors related to cell proliferation. Relative expression of mRNAs encoding nuclear receptors, displayed as reads from RNA-Seq analysis of isolated rat hepatocytes (unpublished results).

There are three PPARs, α, δ (or β), and γ, each of which can heterodimerize with RXRα, β, or γ. PPARα, δ, and γ are activated by distinct ligand sets. PPARα is the main form expressed in hepatocytes, and the peroxisome proliferators are all ligands of PPARα. Although PPARδ and γ expression is very low in hepatocytes, the PPARδ-knockout mouse has delayed liver regeneration [61] and increased resistance to toxicity of CCl_4 via an attenuated inflammatory response [62]. Whether these effects are cell-intrinsic to the hepatocyte is unresolved, and there are no reports of hepatocytic mitogenesis or promotion for PPARδ (or PPAR γ) when stimulated by specific ligands.

All PPARs bind DNA as heterodimers with RXR on binding sites generalized as direct repeats of 6 bp motifs separated by 1 bp (DR1) [63]. The targets are mainly genes that encode systems of mitochondrial, peroxisomal, and microsomal fatty acid oxidation [47].

Peroxisome proliferators are hepatocyte mitogens, but most studies of proliferation have analyzed delayed time intervals rather than short-term proliferation, so the levels of proliferation measured by these studies may reflect compensation for oxidative injury as well as direct proliferation stimulation. Thus, induced levels of mouse proliferation at 48 h after treatment with Wy-14643, one of the strongest inducers, were only moderate [64]. Earlier time intervals were studied in rats treated with BR-931, which showed the strongest documented proliferative response. This treatment gave BrdU labeling at approximately half the level observed with 2/3 PH, with S-phase induction at 12 h [37] that coincides with the beginning of S phase after PH. Like regeneration after PH, Cyclin D1 stimulation does not occur until the beginning of S phase [65]. The stimulation is independent of the priming required for liver regeneration, since neither BR-931 nor nafenopin induces TNFα or NFκB in association with hepatocyte proliferation [37,66]. Moreover, $TNFR1^{-/-}$ mutation blocks liver regeneration but does not affect PPARα-induced proliferation [64,67]. Thus, peroxisome proliferators are mitogens of intermediate strength that induce a rapid cell cycle without the priming required liver regeneration.

Study of the PPARα$^{-/-}$ mouse confirmed the central role of this receptor because the mutation abolished all changes including cell proliferation [29,47]. In addition, the knockout does not affect liver regeneration after PH [48], confirming that this process is independent of PPARα.

The direct mechanism by which PPARα signaling activates proliferation is not fully worked out but immediate-early expression of c-Myc is prominent after treatment of mice with Wy-14,643. Surprisingly, the effect does not result from transcriptional activation by PPARα. Instead, the activated PPARα inhibits micro-RNA let-7C, a Myc suppressor. Although inhibition of let-7c does not occur in the PPARα$^{-/-}$ mouse [68], the mechanism by which PPARα mediates this inhibition is not resolved. A second possible mechanism for cell-cycle stimulation is upregulation of CDK3—a regulator of the G1-S cell-cycle transition—since PPAR binds to the regulatory region of this gene [69].

14.6 CAR-INDUCED PROLIFERATION

Early studies showed that PB induces hepatocyte mitosis and hepatomegaly associated with a dramatic increase in smooth endoplasmic reticulum [19,70]. The mitogenic effect led to experiments demonstrating that PB is also a powerful promoter of chemical hepatocarcinogenesis [20,55]. PB induces a group of p450 oxidases—including characteristic induction of rat Cyp2b1 or its mouse equivalent,-Cyp2b10—a response that is also induced by numerous drugs and xenobiotic agents [71–75]. The smooth endoplasmic reticulum serves as a scaffold for these p450 oxidases. In mice, an even more potent inducer of Cyp2b10 is the hydrocarbon 1,4-bis[2-(3,5-dichloropyridyloxy)] benzene (TCPOBOP), subsequently shown to be a powerful tumor promoter that reproduces all effects of PB [76].

Further research showed that PB stimulates expression of Cyp2b10 by activating gene transcription through a specific enhancer [77], and that a nuclear receptor transcription factor, CAR, mediates this activation by binding DNA as a heterodimer with RXR [78]. Moreover, TCPOBOP is a specific ligand of CAR, verified by the demonstration that its gene activations and biological responses are blocked by CAR deletion [49–52,75,79,80].

Unlike most nuclear receptors, CAR is constitutively active if present in the nucleus because it can activate genes without a ligand. Thus, PB or CAR ligands act by causing translocation from cytoplasm to nucleus, where CAR dimerizes with RXR binds to bind DR4 response elements (6-base motif repeats separated by 4 bp) [49]. Another common tumor promoter, ethinyl estradiol, is also an agonist ligand of mouse CAR [55,81]. The range of CAR ligands is very broad, with considerable variation among species. TCPOBOP does not activate human CAR and there is no known human activator that mediates such a powerful response. The target genes induced by CAR overlap with those induced by PXR [72], a closely related nuclear receptor that also mediates responses to xenobiotic agents [82]. Despite this overlap, CAR and PXR have different binding sites—PXR-RXR dimers bind DR3 response elements—and selective as well as common targets [72].

Treatment of mice with TCPOBOP rapidly induces cell proliferation. The proliferation occurs more rapidly than after 2/3 PH, and the level of DNA labeling is comparable (Figure 14.2). PB induces proliferation with an identical temporal pattern but the magnitude is about half the response to TCPOBOP [83]; however, the stimulation of growth is about 80% of the TCPOBOP effect [83].

Early expression of Cyclin D1 also contrasts the TCPOBOP and PH responses. After TCPOBOP treatment, Cyclin D1 expression increases

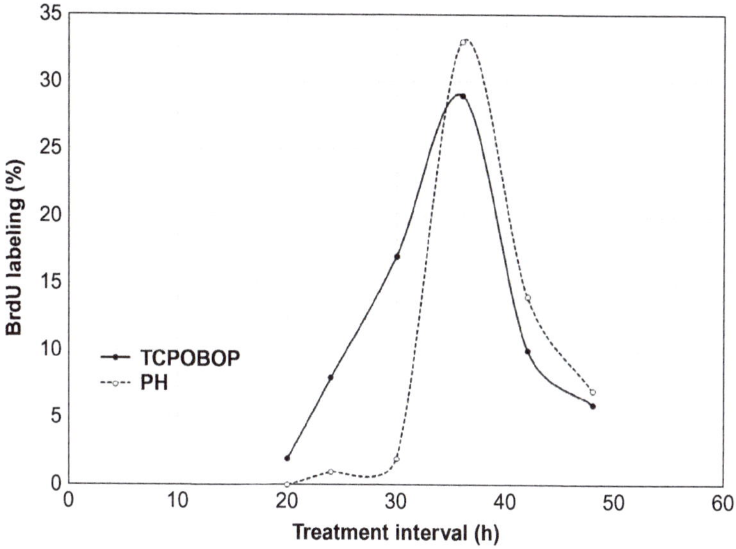

FIGURE 14.2 Comparison of proliferative responses. Mice received 2/3 partial hepatectomy or TCPOBOP by gavage, and a single intraperitoneal injection of BrdU 2 h before sacrifice.
The data are redrawn from Ref. [4].

in mid-G1, while the PH-mediated expression of Cyclin D1 occurs at the beginning of S phase [4]. As an additional difference from regeneration, TCPOBOP-mediated proliferation is independent of TNFα signaling, since it is not blocked by either TNFR1$^{-/-}$ or combined TNFR1$^{-/-}$ and TNFR2$^{-/-}$ mutation [67]. The response to PH is blunted in aged mice but TCPOBOP-induced hyperplasia is fully active, another distinction between the two processes [10].

CAR- and PPARα-induced hyperplasia are also distinct from each other. Thus, TCPOBOP-induced proliferation is enhanced in the PPARα$^{-/-}$ mouse [84], demonstrating that CAR-induced stimulation of proliferation does not act through metabolites that stimulate PPARα and there is no cross talk between the receptors. Indeed, the CAR- and PPARα-mediated proliferative responses appear antagonistic.

In addition to proliferation, CAR activation causes hepatomegaly. The growth is surprisingly fast—one dose of TCPOBOP can double the mass of a mouse liver in 18 h. Cyp2b10 is the gene most strongly activated by CAR, but the second strongest is Gadd45β, a small protein with many putative functions. TCPOBOP treatment causes an 80-fold increase of Gadd45β by 3 h after treatment [75,79,80,85]. To test the importance of this protein, TCPOBOP was applied to a Gadd45β$^{-/-}$ mouse. The mutation retarded growth and reduced early transcription of most CAR targets. Molecular analysis then explained this effect by showing that Gadd45β functions as a transcriptional coactivator of CAR. Gadd45β synthesis is therefore a feed-forward mechanism that accommodates rapid growth by enhancing the transcriptional capacity of the liver. This effect indicates that the rapid growth is an important adaptive process, presumably to accelerate the detoxification of xenobiotics. Surprisingly, loss of Gadd45β actually enhances proliferation, showing the independence of proliferation hyperplasia and growth within the CAR response [75,85].

The mechanism of PB stimulation has been intriguing, since the drug is not a ligand but nevertheless specifically activates nuclear translocation of CAR and the resultant transcriptional program [86]. A recent study has revealed a complex explanation. PB indirectly induces dephosphorylation of CAR residue Th^{38}—induced by EGFR signaling—which causes nuclear translocation. PB apparently acts by binding EGFR to prevent activation and signaling induced by EGF [87]. The mechanism is also consistent with an earlier study in which PB inhibited EGFR signaling of cultured hepatocytes [88]. Since EGFR stimulation is critical for liver regeneration [7], the explanation for PB activation of CAR suggests the interesting possibility that EGFR- and CAR-directed proliferation are mutually exclusive.

14.7 THR-INDUCED PROLIFERATION

Thyroid hormone (T3) also stimulates proliferation of rat hepatocytes with rapid induction comparable to PH, primarily through THRβ, and the effect is equivalent in young and old animals [13,89]. THR dimerizes with RXR to bind DR4 or DR0 sites [90], but these differ from the DR4 sites bound by CAR and there is little overlap between CAR- and THR-activated genes. T3 treatment does induce a cell cycle with early expression of Cyclin D1 [12,91], similar to the CAR- and different from the PH-induced cell cycle. However, the level of proliferation is significantly lower than observed with activators of CAR or PPARα [12,13]. The mitogenic stimulation coincides with protein kinase A-mediated phosphorylation of β-catenin and is absent from a liver-specific β-catenin-knock-out mouse, which fails to activate Cyclin D1 [92].

Despite mitogenic stimulation, T3 treatment has a paradoxical effect on carcinogenesis, because it inhibits promotion and causes regression of early precancerous lesions [93]. This intriguing effect, however, seems to act by causing preneoplastic foci to differentiate and thus has an unclear linkage to mitogenesis.

14.8 MINOR PROLIFERATIVE RESPONSES INDUCED BY NUCLEAR RECEPTORS

Two other ligands of nuclear receptors, 9-*cis* retinoic acid (cRA, an RXR ligand) and *trans*-retinoic acid (tRA, an RAR ligand) are weak hepatocyte mitogens in the rat [94,95]. Since RXR is the dimeric partner of CAR, RXR, and TR, the cRA stimulation probably represents the "dynamic partner effect" in which addition of an RXR ligand increases transcriptional stimulation by the heterodimer [96]. The weaker stimulation by tRA could represent the same effect via metabolic conversion to cRA, or a distinct mechanism. However, significance of these weak stimulations may be limited.

PXR is closely related to CAR and may also stimulate hepatocyte proliferation. Pure PXR ligands like pregnenolone 16a-carbonitrile (PCN) had a degree of mitogenic activity in early studies [97–99] and can potentiate the mitogenic activity of CAR or PPAR agonists [100]. However, most subsequent research focused on cyproterone acetate (CPA), a hepatocyte mitogen and tumor-promoting agent [55,101] that induces a "PCN pattern" of Cyp3A oxidases. This pattern represents PXR regulation, and CPA was shown to be an agonist ligand of PXR [102]. However, CPA also activates mouse CAR [81]. Moreover, metabolic activation of CPA *in vivo* induces DNA damage, injures the liver, and initiates carcinogenesis. Thus, CPA acts through multiple mechanisms, not solely from PXR activation. Since the demonstration that CPA is a carcinogen, there has been little research on the direct mitogenic response to PCN or other pure PXR agonists. Beyond direct mitogenesis, PXR contributes to normal liver regeneration [103], because the $PXR^{-/-}$ has normal growth kinetics but reduced late growth after PH.

14.9 HUMAN PROLIFERATIVE RESPONSES

The noncompensatory proliferation responses have been established in rat and mouse experimental models. Even within these two species, there is considerable difference in hyperplasia and liver regeneration and considerable variation among strains. Thus, extrapolation to human responses is difficult. Moreover, promotional effects in human liver cannot be tested with direct experiments although the potential impact is significant, because activating drugs like PB are used widely. Newer experimental models have allowed direct testing of many aspects of human responses, although the experiments do not recapitulate hepatocyte regulation within their natural environment, human liver. PPARα responses have been studied in a humanized mouse has ablated mPPARα genes and a transgene encoding hPPARα. In this mouse, treatment with Wy-14,643—a strong ligand of both mPPARα and hPPARα—induces peroxisome proliferation and expression of PPARα-target genes with kinetics and expression levels nearly identical to wild type. However, there is no proliferative response and minimal tumor promotion. The conclusions from this research are (1) the proliferation effect of PPARα is different from its direct transcriptional activation of target genes and (2) hPPARα does not mediate a critical interaction necessary for proliferation in mouse liver [47,104–106]. The nature of this interaction and its possible occurrence in human liver remain important problems for future research.

Whether CAR stimulates hepatocyte proliferation in humans is another significant experimental problem, considering the common use of PB for seizure disorders and other diseases. Surveys of these patients show little evidence that PB promotes cancer in humans [107,108], but this type of analysis is not equivalent to mechanistic research.

One complication in human research is that CAR ligand specificity is highly divergent among species and TCPOBOP does not activate human CAR. A human agonist has been characterized, 6-(4-chlorophenyl)imidazo[2,1-b][1,3] thiazole-5-carbaldehydeO-(3,4-dichlorobenzyl) oxime (CITCO) [109], although it does not stimulate human CAR as strongly as TCPOBOP stimulates mouse CAR. Both PB and CITCO stimulate CAR responses in a mouse with humanized CAR expression. In this mouse, PB induces proliferation and tumor promotion, but the effects are weaker than in wild-type mice [110–113]. Another study showed that PB had a limited effect on proliferation in a chimeric liver populated with human hepatocytes [114]. However, the stimulation of gene expression by CITCO or PB is considerably weaker than stimulation by TCPOBOP in wild-type mice, so the full effect of strong induction of hCAR has not been explored.

14.10 CONCLUSIONS

The mitogenic pathways are distinct from regeneration. They are also distinct from one other even though each acts through nuclear receptors. It is hard to rank the strength of these responses, because few have been compared under matched conditions. Moreover, even for a single pathway, the activating agents have different strengths. It appears, however, that the CAR pathway—via activators like PB and TCPOBOP—has the strongest mitogenic and promotional effects. In mouse, the CAR-mediated response is faster and comparable in magnitude to regeneration induced by 2/3 PH [4]. The strength of response and distinctive expression pattern of cell-cycle regulators make the CAR-activated cell cycle an important experimental model of hepatocyte regulation with a special relationship to cancer promotion.

Direct mitogenesis with a strong PPAR ligand is also fast but stimulates about half the level of cell division observed after PH [37]. Moreover, most research has focused on the more complex delayed effects of this pathway, which may reflect both mitogenesis and oxidative damage.

Supraphysiological doses of thyroid hormone or specific ligands of THRβ also induce rapid mitogenesis, but typically at only 10-20% of the level of PH. The most important aspect of this THR-mediated response is the antipromotional on carcinogenesis, which may be independent of its moderate mitogenesis.

At a more fundamental level, the models of noncompensatory and compensatory show that the hepatocyte has a variety of biologically distinct ways to enter and advance the cell cycle. The relationships to cancer promotion suggest that some proliferative responses are maladaptive, but the experiments models do not view the processes in settings of specific injuries or disease where each pathway could provide a different adaptive mechanism. The diverse proliferative pathways therefore have many potential applications for the treatment of liver disease.

References

[1] Columbano A, Shinozuka H. Liver regeneration versus direct hyperplasia. FASEB J 1996;10:1118–28.

[2] Michalopoulos GK. Liver regeneration after partial hepatectomy: critical analysis of mechanistic dilemmas. Am J Pathol 2010;176:2–13.

[3] Michalopoulos GK. Principles of liver regeneration and growth homeostasis. Compr Physiol 2013;3:485–513.

[4] Ledda-Columbano GM, Pibiri M, Loi R, Perra A, Shinozuka H, Columbano A. Early increase in cyclin-D1 expression and accelerated entry of mouse hepatocytes into S phase after administration of the mitogen 1,4-bis[2-(3,5-dichloropyridyloxy)] benzene. Am J Pathol 2000;156:91–7.

[5] Yamada Y, Kirillova I, Peschon JJ, Fausto N. Initiation of liver growth by tumor necrosis factor: deficient liver regeneration in mice lacking type I tumor necrosis factor receptor. Proc Natl Acad Sci USA 1997;94:1441–6.

[6] Bucher NL. Experimental aspects of hepatic regeneration. N Engl J Med 1967;277:738–46 [concluded].

[7] Bowen WC, Michalopoulos AW, Orr A, Ding MQ, Stolz DB, Michalopoulos GK. Development of a chemically defined medium and discovery of new mitogenic growth factors for mouse hepatocytes: mitogenic effects of FGF1/2 and PDGF. PLoS One 2014;9:e95487.

[8] Akerman P, Cote P, Yang SQ, et al. Antibodies to tumor necrosis factor-a inhibit liver regeneration after partial hepatectomy. Am J Physiol 1992;263:G579–85.

[9] Bucher NL, Glinos AD. The effect of age on regeneration of rat liver. Cancer Res 1950;10:324–32.

[10] Ledda-Columbano GM, Pibiri M, Cossu C, Molotzu F, Locker J, Columbano A. Aging does not reduce the hepatocyte proliferative response of mice to the primary mitogen TCPOBOP. Hepatology 2004;40:981–8.

[11] Krupczak-Hollis K, Wang X, Dennewitz MB, Costa RH. Growth hormone stimulates proliferation of old-aged regenerating liver through forkhead box m1b. Hepatology 2003;38:1552–62.

[12] Pibiri M, Ledda-Columbano GM, Cossu C, et al. Cyclin D1 is an early target in hepatocyte proliferation induced by thyroid hormone (T3). FASEB J 2001;15:1006–13.

[13] Columbano A, Simbula M, Pibiri M, et al. Triiodothyronine stimulates hepatocyte proliferation in two models of impaired liver regeneration. Cell Prolif 2008;41:521–31.

[14] Yamamoto Y, Moore R, Goldsworthy TL, Negishi M, Maronpot RR. The orphan nuclear receptor constitutive active/androstane receptor is essential for liver tumor promotion by phenobarbital in mice. Cancer Res 2004;64:7197–200.

[15] Lake BG. Mechanisms of hepatocarcinogenicity of peroxisome-proliferating drugs and chemicals. Annu Rev Pharmacol Toxicol 1995;35:483–507.

[16] Corton JC, Lapinskas PJ, Gonzalez FJ. Central role of PPARalpha in the mechanism of action of hepatocarcinogenic peroxisome proliferators. Mutat Res 2000;448:139–51.

[17] Pitot HC, Dragan YP. Facts and theories concerning the mechanisms of carcinogenesis. FASEB J 1991;5:2280–6.

[18] Pitot HC, Dragan YP, Neveu MJ, Rizvi TA, Hully JR, Campbell HA. Chemicals, cell proliferation, risk estimation, and multistage carcinogenesis. Prog Clin Biol Res 1991;369:517–32.

[19] Japundzic M, Knezevic B, Djordjevic-Camba V, Japundzic I. The influence of phenobarbital-NA on the mitotic activity of parenchymal liver cells during rat liver regeneration. Exp Cell Res 1967;48:163–7.

[20] Peraino C, Fry RJ, Staffeldt E. Reduction and enhancement by phenobarbital of hepatocarcinogenesis induced in the rat by 2-acetylaminofluorene. Cancer Res 1971;31:1506–12.

[21] Honkakoski P, Zelko I, Sueyoshi T, Negishi M. The nuclear orphan receptor CAR-retinoid X receptor

heterodimer activates the phenobarbital-responsive enhancer module of the CYP2B gene. Mol Cell Biol 1998;18:5652–8.
[22] Sueyoshi T, Kawamoto T, Zelko I, Honkakoski P, Negishi M. The repressed nuclear receptor CAR responds to phenobarbital in activating the human CYP2B6 gene. J Biol Chem 1999;274:6043–6.
[23] Hlavica P. Regulatory mechanisms in the activation of nitrogenous compounds by mammalian cytochrome P-450 isozymes. Drug Metab Rev 1994;26:325–48.
[24] Cohen SM, Ellwein LB. Genetic errors, cell proliferation, and carcinogenesis. Cancer Res 1991;51:6493–505.
[25] Cohen SM, Ellwein LB. Cell proliferation in carcinogenesis. Science 1990;249:1007–11.
[26] Maeda S, Kamata H, Luo JL, Leffert H, Karin M. IKK-beta couples hepatocyte death to cytokine-driven compensatory proliferation that promotes chemical hepatocarcinogenesis. Cell 2005;121:977–90.
[27] Uno Y, Matsuura K, Miyagawa M, et al. Rat liver in vivo replicative DNA synthesis test for short-term prediction of nongenotoxic (Ames-negative) hepatocarcinogenicity: a collaborative study of the Nongenotoxic Carcinogen Study Group of Japan. Toxicol Lett 1999;109:105–14.
[28] Seidel SD, Stott WT, Kan HL, Sparrow BR, Gollapudi BB. Gene expression dose–response of liver with a genotoxic and nongenotoxic carcinogen. Int J Toxicol 2006;25:57–64.
[29] Peters JM, Cattley RC, Gonzalez FJ. Role of PPAR alpha in the mechanism of action of the nongenotoxic carcinogen and peroxisome proliferator Wy-14,643. Carcinogenesis 1997;18:2029–33.
[30] Cohen SM, Purtilo DT, Ellwein LB. Ideas in pathology. Pivotal role of increased cell proliferation in human carcinogenesis. Mod Pathol 1991;4:371–82.
[31] Columbano A, Ledda GM, Sirigu P, Perra T, Pani P. Liver cell proliferation induced by a single dose of lead nitrate. Am J Pathol 1983;110:83–8.
[32] Columbano A, Ledda-Columbano GM, Coni PP, et al. Occurrence of cell death (apoptosis) during the involution of liver hyperplasia. Lab Invest 1985;52: 670–5.
[33] Columbano A, Endoh T, Denda A, et al. Effects of cell proliferation and cell death (apoptosis and necrosis) on the early stages of rat hepatocarcinogenesis. Carcinogenesis 1996;17:395–400.
[34] Bursch W, Grasl-Kraupp B, Wastl U, et al. Role of apoptosis for mouse liver growth regulation and tumor promotion: comparative analysis of mice with high (C3H/He) and low (C57Bl/6J) cancer susceptibility. Toxicol Lett 2004;149:25–35.
[35] Shinozuka H, Ohmura T, Katyal SL, Zedda AI, Ledda-Columbano GM, Columbano A. Possible roles of nonparenchymal cells in hepatocyte proliferation induced by lead nitrate and by tumor necrosis factor alpha. Hepatology 1996;23:1572–7.
[36] Kubo Y, Yasunaga M, Masuhara M, Terai S, Nakamura T, Okita K. Hepatocyte proliferation induced in rats by lead nitrate is suppressed by several tumor necrosis factor alpha inhibitors. Hepatology 1996;23:104–14.
[37] Ohmura T, Ledda-Columbano GM, Piga R, et al. Hepatocyte proliferation induced by a single dose of a peroxisome proliferator. Am J Pathol 1996;148:815–24.
[38] Ledda-Columbano GM, Columbano A, Coni PP, Vargiu M, Liguori C, Pani P. Stimulation of DNA synthesis after a single administration of cadmium nitrate. Toxicol Lett 1984;23:267–72.
[39] Waalkes MP, Diwan BA, Weghorst CM, et al. Further evidence of the tumor-suppressive effects of cadmium in the B6C3F1 mouse liver and lung: late stage vulnerability of tumors to cadmium and the role of metallothionein. J Pharmacol Exp Ther 1993;266:1656–63.
[40] Columbano A, Dessi S, Ledda-Columbano GM, et al. HMP-shunt and cholesterol metabolism in experimental models involving normal and preneoplastic liver growth. Toxicol Pathol 1987;15:43–50.
[41] Ledda-Columbano GM, Columbano A, Cannas A, et al. Dexamethasone inhibits induction of liver tumor necrosis factor-alpha mRNA and liver growth induced by lead nitrate and ethylene dibromide. Am J Pathol 1994;145:951–8.
[42] Shirai T, Lee MS, Wang CY, King CM. Effects of partial hepatectomy and dietary phenobarbital on liver and mammary tumorigenesis by two *N*-hydroxy-*N*-acylaminobiphenyls in female CD rats. Cancer Res 1981;41:2450–6.
[43] Barbason H, Rassenfosse C, Betz EH. Promotion mechanism of phenobarbital and partial hepatectomy in DENA hepatocarcinogenesis cell kinetics effect. Br J Cancer 1983;47:517–25.
[44] Barbason H. Promoting effect of liver regeneration in hepatocarcinogenesis initiated by diethylnitrosamine in the rat. Pathol Biol (Paris) 2007;55:168–70.
[45] Taniguchi K, Karin M. IL-6 and related cytokines as the critical lynchpins between inflammation and cancer. Semin Immunol 2014;26:54–74.
[46] Columbano A, Ledda-Columbano GM. Mitogenesis by ligands of nuclear receptors: an attractive model for the study of the molecular mechanisms implicated in liver growth. Cell Death Differ 2003;10(Suppl. 1): S19–21.
[47] Cheung C, Akiyama TE, Ward JM, et al. Diminished hepatocellular proliferation in mice humanized for the nuclear receptor peroxisome proliferator-activated receptor alpha. Cancer Res 2004;64:3849–54.

[48] Rao MS, Peters JM, Gonzalez FJ, Reddy JK. Hepatic regeneration in peroxisome proliferator-activated receptor alpha-null mice after partial hepatectomy. Hepatol Res 2002;22:52–7.

[49] Tzameli I, Pissios P, Schuetz EG, Moore DD. The xenobiotic compound 1,4-bis[2-(3,5-dichloropyridyloxy)] benzene is an agonist ligand for the nuclear receptor CAR. Mol Cell Biol 2000;20:2951–8.

[50] Wei P, Zhang J, Egan-Hafley M, Liang S, Moore DD. The nuclear receptor CAR mediates specific xenobiotic induction of drug metabolism. Nature 2000;407:920–3.

[51] Wei P, Zhang J, Dowhan DH, Han Y, Moore DD. Specific and overlapping functions of the nuclear hormone receptors CAR and PXR in xenobiotic response. Pharmacogenomics J 2002;2:117–26.

[52] Huang W, Zhang J, Washington M, et al. Xenobiotic stress induces hepatomegaly and liver tumors via the nuclear receptor constitutive androstane receptor. Mol Endocrinol 2005;19:1646–53.

[53] Warren JR, Lalwani ND, Reddy JK. Phthalate esters as peroxisome proliferator carcinogens. Environ Health Perspect 1982;45:35–40.

[54] Kraupp-Grasl B, Huber W, Putz B, Gerbracht U, Schulte-Hermann R. Tumor promotion by the peroxisome proliferator nafenopin involving a specific subtype of altered foci in rat liver. Cancer Res 1990;50:3701–8.

[55] Schulte-Hermann R, Schuppler J, Timmermann-Trosiener I, Ohde G, Bursch W, Berger H. The role of growth of normal and preneoplastic cell populations for tumor promotion in rat liver. Environ Health Perspect 1983;50:185–94.

[56] Peters JM, Shah YM, Gonzalez FJ. The role of peroxisome proliferator-activated receptors in carcinogenesis and chemoprevention. Nat Rev Cancer 2012;12:181–95.

[57] Bursch W, Chabicovsky M, Wastl U, et al. Apoptosis in stages of mouse hepatocarcinogenesis: failure to counterbalance cell proliferation and to account for strain differences in tumor susceptibility. Toxicol Sci 2005;85:515–29.

[58] Baskin-Bey ES, Huang W, Ishimura N, et al. Constitutive androstane receptor (CAR) ligand, TCPOBOP, attenuates Fas-induced murine liver injury by altering Bcl-2 proteins. Hepatology 2006;44:252–62.

[59] Reddy JK, Rao MS, Azarnoff DL, Sell S. Mitogenic and carcinogenic effects of a hypolipidemic peroxisome proliferator, [4-chloro-6-(2,3-xylidino)-2-pyrimidinylthio]acetic acid (Wy-14, 643), in rat and mouse liver. Cancer Res 1979;39:152–61.

[60] Peraza MA, Burdick AD, Marin HE, Gonzalez FJ, Peters JM. The toxicology of ligands for peroxisome proliferator-activated receptors (PPAR). Toxicol Sci 2006;90:269–95.

[61] Liu HX, Fang Y, Hu Y, Gonzalez FJ, Fang J, Wan YJ. PPARbeta regulates liver regeneration by modulating Akt and E2f signaling. PLoS One 2013;8:e65644.

[62] Shan W, Nicol CJ, Ito S, et al. Peroxisome proliferator-activated receptor-beta/delta protects against chemically induced liver toxicity in mice. Hepatology 2008;47:225–35.

[63] Chandra V, Huang P, Hamuro Y, et al. Structure of the intact PPAR-gamma-RXR-alpha nuclear receptor complex on DNA. Nature 2008;456:350–6.

[64] Anderson SP, Dunn CS, Cattley RC, Corton JC. Hepatocellular proliferation in response to a peroxisome proliferator does not require TNFalpha signaling. Carcinogenesis 2001;22:1843–51.

[65] Rininger JA, Goldsworthy TL, Babish JG. Time course comparison of cell-cycle protein expression following partial hepatectomy and WY14,643-induced hepatic cell proliferation in F344 rats. Carcinogenesis 1997;18:935–41.

[66] Menegazzi M, Carcereri-De Prati A, Suzuki H, et al. Liver cell proliferation induced by nafenopin and cyproterone acetate is not associated with increases in activation of transcription factors NF-kappaB and AP-1 or with expression of tumor necrosis factor alpha. Hepatology 1997;25:585–92.

[67] Ledda-Columbano GM, Curto M, Piga R, et al. In vivo hepatocyte proliferation is inducible through a TNF and IL-6-independent pathway. Oncogene 1998;17: 1039–44.

[68] Shah YM, Morimura K, Yang Q, Tanabe T, Takagi M, Gonzalez FJ. Peroxisome proliferator-activated receptor alpha regulates a microRNA-mediated signaling cascade responsible for hepatocellular proliferation. Mol Cell Biol 2007;27:4238–47.

[69] McMullen PD, Bhattacharya S, Woods CG, et al. A map of the PPARalpha transcription regulatory network for primary human hepatocytes. Chem Biol Interact 2014;209:14–24.

[70] Staubli W, Hess R, Weibel ER. Correlated morphometric and biochemical studies on the liver cell. II. Effects of phenobarbital on rat hepatocytes. J Cell Biol 1969;42:92–112.

[71] Moore LB, Maglich JM, McKee DD, et al. Pregnane X receptor (PXR), constitutive androstane receptor (CAR), and benzoate X receptor (BXR) define three pharmacologically distinct classes of nuclear receptors. Mol Endocrinol 2002;16:977–86.

[72] Timsit YE, Negishi M. CAR and PXR: the xenobiotic-sensing receptors. Steroids 2007;72:231–46.

[73] Honkakoski P, Lang MA. Mouse liver phenobarbital-inducible P450 system: purification, characterization, and differential inducibility of four cytochrome P450 isozymes from D2 mouse. Arch Biochem Biophys 1989;273:42–57.

[74] Negishi M, Fujii-Kuriyama Y, Tashiro Y, Imai Y. Site of biosynthesis of cytochrome P450 in hepatocytes of phenobarbital treated rats. Biochem Biophys Res Commun 1976;71:1153–60.
[75] Tian J, Huang H, Hoffman B, et al. Gadd45beta is an inducible coactivator of transcription that facilitates rapid liver growth in mice. J Clin Invest 2011;121: 4491–502.
[76] Diwan BA, Lubet RA, Ward JM, Hrabie JA, Rice JM. Tumor-promoting and hepatocarcinogenic effects of 1,4-bis[2-(3,5-dichloropyridyloxy)]benzene (TCPOBOP) in DBA/2NCr and C57BL/6NCr mice and an apparent promoting effect on nasal cavity tumors but not on hepatocellular tumors in F344/NCr rats initiated with *N*-nitrosodiethylamine. Carcinogenesis 1992;13:1893–901.
[77] Honkakoski P, Negishi M. Characterization of a phenobarbital-responsive enhancer module in mouse P450 Cyp2b10 gene. J Biol Chem 1997;272:14943–9.
[78] Baes M, Gulick T, Choi HS, Martinoli MG, Simha D, Moore DD. A new orphan member of the nuclear hormone receptor superfamily that interacts with a subset of retinoic acid response elements. Mol Cell Biol 1994;14:1544–52.
[79] Locker J, Tian J, Carver R, et al. A common set of immediate-early response genes in liver regeneration and hyperplasia. Hepatology 2003;38:314–25.
[80] Columbano A, Ledda-Columbano GM, Pibiri M, et al. Gadd45beta is induced through a CAR-dependent, TNF-independent pathway in murine liver hyperplasia. Hepatology 2005;42:1118–26.
[81] Repo S, Jyrkkarinne J, Pulkkinen JT, Laatikainen R, Honkakoski P, Johnson MS. Ligand specificity of constitutive androstane receptor as probed by induced-fit docking and mutagenesis. J Med Chem 2008;51:7119–31.
[82] Moore LB, Parks DJ, Jones SA, et al. Orphan nuclear receptors constitutive androstane receptor and pregnane X receptor share xenobiotic and steroid ligands. J Biol Chem 2000;275:15122–7.
[83] Lin CW, Mars WM, Paranjpe S, et al. Hepatocyte proliferation and hepatomegaly induced by phenobarbital and 1,4-bis [2-(3,5-dichloropyridyloxy)] benzene is suppressed in hepatocyte-targeted glypican 3 transgenic mice. Hepatology 2011;54:620–30.
[84] Columbano A, Ledda-Columbano GM, Pibiri M, Concas D, Reddy JK, Rao MS. Peroxisome proliferator-activated receptor-alpha mice show enhanced hepatocyte proliferation in response to the hepatomitogen 1,4-bis [2-(3,5-dichloropyridyloxy)] benzene, a ligand of constitutive androstane receptor. Hepatology 2001;34:262–6.
[85] Tian J, Locker J. Gadd45 in the liver: signal transduction and transcriptional mechanisms. Adv Exp Med Biol 2013;793:69–80.
[86] Zelko I, Sueyoshi T, Kawamoto T, Moore R, Negishi M. The peptide near the C terminus regulates receptor CAR nuclear translocation induced by xenochemicals in mouse liver. Mol Cell Biol 2001;21:2838–46.
[87] Mutoh S, Sobhany M, Moore R, et al. Phenobarbital indirectly activates the constitutive active androstane receptor (CAR) by inhibition of epidermal growth factor receptor signaling. Sci Signal 2013;6:ra31.
[88] Jirtle RL, Meyer SA. Liver tumor promotion: effect of phenobarbital on EGF and protein kinase C signal transduction and transforming growth factor-beta 1 expression. Dig Dis Sci 1991;36:659–68.
[89] Kowalik MA, Perra A, Pibiri M, et al. TRbeta is the critical thyroid hormone receptor isoform in T3-induced proliferation of hepatocytes and pancreatic acinar cells. J Hepatol 2010;53:686–92.
[90] Ramadoss P, Abraham BJ, Tsai L, et al. Novel mechanism of positive versus negative regulation by thyroid hormone receptor beta1 (TRbeta1) identified by genome-wide profiling of binding sites in mouse liver. J Biol Chem 2014;289:1313–28.
[91] Severino V, Locker J, Ledda-Columbano GM, Columbano A, Parente A, Chambery A. Proteomic characterization of early changes induced by triiodothyronine in rat liver. J Proteome Res 2011;10:3212–24.
[92] Fanti M, Singh S, Ledda-Columbano GM, Columbano A, Monga SP. Tri-iodothyronine induces hepatocyte proliferation by protein kinase A-dependent beta-catenin activation in rodents. Hepatology 2014;59:2309–20.
[93] Perra A, Kowalik MA, Pibiri M, Ledda-Columbano GM, Columbano A. Thyroid hormone receptor ligands induce regression of rat preneoplastic liver lesions causing their reversion to a differentiated phenotype. Hepatology 2009;49:1287–96.
[94] Ohmura T, Columbano GL, Columbano A, Katyal SL, Locker J, Shinozuka H. 9-cis retinoic acid is a direct hepatocyte mitogen in rats. Life Sci 1996;58:PL211–6.
[95] Ledda-Columbano GM, Pibiri M, Molotzu F, et al. Induction of hepatocyte proliferation by retinoic acid. Carcinogenesis 2004;25:2061–6.
[96] de Lera AR, Bourguet W, Altucci L, Gronemeyer H. Design of selective nuclear receptor modulators: RAR and RXR as a case study. Nat Rev Drug Discov 2007;6:811–20.
[97] Japundzic M, Garg BD, Kovac K, Japundcic I. Effect of pregnenolone-16alpha-carbonitrile on mitotic activity in the intact and regenerating rat liver. Experientia 1974;30:562–3.
[98] Schulte-Hermann R, Hoffman V, Parzefall W, Kallenbach M, Gerhardt A, Schuppler J. Adaptive responses of rat liver to the gestagen and anti-androgen cyproterone acetate and other inducers. II. Induction of growth. Chem Biol Interact 1980;31:287–300.

[99] Chen ZY, White CC, He CY, Liu YF, Eaton DL. Zonal differences in DNA synthesis activity and cytochrome P450 gene expression in livers of male F344 rats treated with five nongenotoxic carcinogens. J Environ Pathol Toxicol Oncol 1995;14:83–99.
[100] Shizu R, Benoki S, Numakura Y, et al. Xenobiotic-induced hepatocyte proliferation associated with constitutive active/androstane receptor (CAR) or peroxisome proliferator-activated receptor alpha (PPARalpha) is enhanced by pregnane X receptor (PXR) activation in mice. PLoS One 2013;8:e61802.
[101] Coni P, Pichiri-Coni G, Ledda-Columbano GM, et al. Stimulation of DNA synthesis by rat plasma following in vivo treatment with three liver mitogens. Cancer Lett 1992;61:233–8.
[102] Schuetz EG, Brimer C, Schuetz JD. Environmental xenobiotics and the antihormones cyproterone acetate and spironolactone use the nuclear hormone pregnenolone X receptor to activate the CYP3A23 hormone response element. Mol Pharmacol 1998;54:1113–7.
[103] Dai G, He L, Bu P, Wan YJ. Pregnane X receptor is essential for normal progression of liver regeneration. Hepatology 2008;47:1277–87.
[104] Morimura K, Cheung C, Ward JM, Reddy JK, Gonzalez FJ. Differential susceptibility of mice humanized for peroxisome proliferator-activated receptor alpha to Wy-14,643-induced liver tumorigenesis. Carcinogenesis 2006;27:1074–80.
[105] Zhou J, Zhai Y, Mu Y, et al. A novel pregnane X receptor-mediated and sterol regulatory element-binding protein-independent lipogenic pathway. J Biol Chem 2006;281:15013–20.
[106] Yang Q, Nagano T, Shah Y, Cheung C, Ito S, Gonzalez FJ. The PPAR alpha-humanized mouse: a model to investigate species differences in liver toxicity mediated by PPAR alpha. Toxicol Sci 2008;101:132–9.
[107] Elcombe CR, Peffer RC, Wolf DC, et al. Mode of action and human relevance analysis for nuclear receptor-mediated liver toxicity: a case study with phenobarbital as a model constitutive androstane receptor (CAR) activator. Crit Rev Toxicol 2014;44:64–82.
[108] La Vecchia C, Negri E. A review of epidemiological data on epilepsy, phenobarbital, and risk of liver cancer. Eur J Cancer Prev 2014;23:1–7.
[109] Maglich JM, Parks DJ, Moore LB, et al. Identification of a novel human constitutive androstane receptor (CAR) agonist and its use in the identification of CAR target genes. J Biol Chem 2003;278:17277–83.
[110] Scheer N, Ross J, Rode A, et al. A novel panel of mouse models to evaluate the role of human pregnane X receptor and constitutive androstane receptor in drug response. J Clin Invest 2008;118:3228–39.
[111] Ross J, Plummer SM, Rode A, et al. Human constitutive androstane receptor (CAR) and pregnane X receptor (PXR) support the hypertrophic but not the hyperplastic response to the murine nongenotoxic hepatocarcinogens phenobarbital and chlordane in vivo. Toxicol Sci 2010;116:452–66.
[112] Luisier R, Lempiainen H, Scherbichler N, et al. Phenobarbital induces cell cycle transcriptional responses in mouse liver humanized for constitutive androstane and pregnane X receptors. Toxicol Sci 2014;139:501–11.
[113] Braeuning A, Gavrilov A, Brown S, Wolf CR, Henderson CJ, Schwarz M. Phenobarbital-mediated tumor promotion in transgenic mice with humanized CAR and PXR. Toxicol Sci 2014;140:259–70.
[114] Yamada T, Okuda Y, Kushida M, et al. Human hepatocytes support the hypertrophic but not the hyperplastic response to the murine nongenotoxic hepatocarcinogen sodium phenobarbital in an in vivo study using a chimeric mouse with humanized liver. Toxicol Sci 2014;142:137–57.

CHAPTER

15

Metabolic Regulation of Liver Regeneration

David A. Rudnick,†, Jiansheng Huang**

*Department of Pediatrics, Washington University School of Medicine, St. Louis, Missouri, USA

†Department of Developmental Biology, Washington University School of Medicine, St. Louis, Missouri, USA

15.1 INTRODUCTION

Recovery from any serious hepatic injury depends on the liver's intrinsic ability to regenerate [1–6]. Such regeneration is also required for survival after liver tumor resection and live-donor liver transplantation. Moreover, impaired regeneration is associated with poor outcomes after acute liver injury and increased risk of hepatocellular carcinoma in chronic liver disease [7–9]. Thus, liver regeneration has been extensively investigated, with such studies identifying many physiological, cellular, and molecular mechanisms that control this process. Nevertheless, gaps remain both in knowledge about these mechanisms and in the translation of such knowledge into improved clinical care. Thus, additional research examining the mechanisms that regulate liver regeneration is necessary.

Experimental analyses of liver regeneration, using rodent partial hepatectomy (PH) and other models, show that hepatic regenerative capability is conserved in all vertebrates where it has been studied. Such capability mediates recovery of the unique, essential metabolic and other functions of this organ from liver injury [1]. PH has been the paradigm most commonly used to investigate liver regeneration. Typically, two-thirds of the mouse liver is surgically resected. A characteristic hepatocellular proliferative response ensues, restoring normal liver mass and architecture [10]. Hepatic function, which is disrupted by liver resection, normalizes coincident with regenerative recovery of liver mass, and the liver subsequently reverts to its preresection state of relative proliferative quiescence. Liver regeneration has also been investigated in models of hepatocellular injury induced by toxins (e.g., carbon tetrachloride (CCl_4), acetaminophen, and others [11–13]) or genetic manipulation (e.g., the PiZ transgenic mouse model of α1-antitrypsin deficiency liver disease [14]). Together with PH-based analyses, these studies reveal the virtually unlimited proliferative potential of hepatocytes [1–6] and establish those cells as the typical source of regenerated liver mass [15]. Thus, hepatic regeneration does not necessarily require a stem cell; rather, all of

Liver Regeneration
http://dx.doi.org/10.1016/B978-0-12-420128-6.00015-4

the normally quiescent cellular populations of the mature, healthy liver have potential to proliferate. However, bipotential stem cells can be induced to expand in liver under specific experimental circumstances [16]. These so-called "oval" cells have also been identified in human liver diseases [17]. Experimental analyses also show that liver regeneration is precisely regulated in its initiation and duration, terminating only when the original liver-to-body-mass ratio is restored. That observation has inferred the existence of a *hepatostat*, i.e., a master thermostat-like regulator that maintains the normal liver-to-body-mass ratio in health and directs regenerative recovery of the liver back to this set point following injury [3]. A recent study reported that mice with genetically based hypertrophy of skeletal muscle exhibit a lower liver-to-body-mass ratio than do their strain-matched controls. That finding indicates a previously unrecognized and still poorly characterized extra hepatic tissue specificity to hepatostatic control [18].

PH and other regenerative models have also been used to discover molecular mechanisms that control regeneration. The importance of circulating factors in such regulation was first established by elegant parabiotic analyses of PH-induced regeneration [19,20]. Those seminal observations have stimulated still-ongoing efforts to comprehensively identify the circulating factors and cellular targets of such factors that control liver regeneration. Many cytokines, growth and matrix-derived factors, secondary messengers, and transcriptional regulators have been identified. Moreover, analyses of pharmacologically and genetically manipulated animal models have established the essential functions of several of these signals during PH-induced liver regeneration [1–6]. Some but not all of these mechanisms are also conserved in toxin and genetically based experimental regenerative paradigms [21,22]. Nevertheless, despite the broad knowledge gained from these studies, the most proximal events that initiate regeneration and those distal signals that terminate this process remain incompletely defined and the essence of the hepatostat remains unknown. Moreover, translation of these basic research efforts into novel proregenerative clinical therapeutic interventions for human liver diseases has not yet been achieved.

In addition to stimulating tissue-specific hepatic regeneration, all serious liver injuries alter metabolism. Indeed, rodents subjected to PH or exposed to hepatotoxins develop a stereotypical sequence of metabolic changes that begins with hypoglycemia, followed by systemic catabolism, and then accumulation of specific metabolites in the circulation and the liver. Recent studies implicate these metabolic changes, themselves, as vital regulators of regeneration [23–30]. Those analyses, some of which were previously reviewed in Refs. [31–34], show that (i) disruption of normal metabolism precedes onset of regenerative hepatocellular proliferation and resolves with restoration of appropriate liver mass and (ii) many interventions that suppress elements of the metabolic response to liver injury also impair subsequent hepatic regenerative recovery. This chapter provides an update to our recent reviews [31–34] of this subject, including an overview of the evidence in support of the regenerative regulation and function of liver injury-induced alterations in metabolism, consideration of important areas for further study, and discussion of provocative clinical implications raised by these considerations.

15.2 THE METABOLIC RESPONSE TO HEPATIC INSUFFICIENCY

Following experimentally induced hepatic injury, the liver continues to perform essential metabolic and other functions necessary for survival. Nevertheless, studies show that hepatic regulation of systemic metabolism [35] is rapidly and stereotypically altered in response to such injury [31,32]. For example, mice subjected to PH develop hypoglycemia within 2 h of that surgery [28]. This finding is consistent with the liver's essential role in glucose homeostasis and likely results, at least in part, from the acute removal of hepatic glycogen stores and gluconeogenic capacity [35]. Indeed, glycogen content in the remnant liver is depleted within hours of PH [36]. CCl_4-treated- and α1-antitrypsin-deficient PiZ mice are also hypoglycemic relative to controls [11,12,14]. Thus, early adaptive responses to regeneration-inducing liver injuries do not entirely counteract acutely compromised hepatic metabolic function in these models. Nevertheless, some compensation occurs, including induction of gluconeogenic machinery and suppression of glycolytic activity [37]. Those changes are determined in part via transcriptional regulation [38] and limit the post-hepatectomy decline in blood glucose at the expense of glucose-derived hepatic ATP production. By 12 h after surgery, animals demonstrate evidence of a systemic catabolic response, characterized by specific patterns of decline in lean tissue mass and adipose tissue mass [29]. Systemic fat depletion also occurs in models of hepatotoxin-induced liver regeneration [39]. From 12 to 24 h after surgery, marked steatosis develops in the regenerating mouse liver [24]. As with alterations in glycemia, liver triglyceride accumulation coincides with hepatic induction of "adipogenic" gene expression [24]. This transient steatosis probably results from hepatic uptake of systemic adipose-derived fatty acids based on several observations: (i) serum-free fatty acids rise in animals subjected to PH compared to sham-surgery controls prior to accumulation of fat in regenerating liver [31]; (ii) fatty liver dystrophy (*fld*) mice, which have a genetically based paucity of systemic adipose, accumulate less liver fat after PH than controls [29]; and (iii) suppression of *de novo* hepatic lipogenesis (by liver-specific disruption of fatty acid synthase expression) does not prevent resection-induced hepatic steatosis [40]. Over this same span of time, specific amino acids, likely derived from systemic proteolysis, appear in serum and accumulate in regenerating liver [27]. Hepatic ATP content coincidentally declines and AMP rises in the postresection liver remnant [41] with fatty acid β-oxidation as the predominant source of ATP production in regenerating liver [31,42]. These metabolic alterations precede the onset of resection- or toxin-induced hepatocellular proliferation, which is subsequently promoted by induction of specific cell-cycle regulatory machinery [43]. In mice, the first wave of hepatocellular DNA replication, which reflects such proliferation, peaks 36 h after PH and is followed by mitotic progression 48 h after surgery. Regeneration proceeds to completion with subsequent waves of hepatocellular proliferation restoring the liver-to-body-mass ratio to its pre-injury hepatostatic set point [44]. Injury-induced perturbations in metabolism coincidentally resolve with such normalization of liver mass [24,28,29].

15.3 EVIDENCE FOR THE METABOLIC REGULATION OF LIVER REGENERATION

Several experimental observations implicate the metabolic response to hepatic insufficiency as a physiological determinant of normal liver regeneration.

15.3.1 PH-Induced Liver Regeneration Occurs in Proportion to Associated Alterations in Metabolism

Two-thirds PH results in significantly greater hypoglycemia, systemic catabolism, and accumulation of hepatic triglyceride as well as more robust hepatocellular proliferation than does one-third hepatectomy [29]. Thus, the metabolic and hepatocellular proliferative responses to liver injury occur in proportion to each other. Consistent with this consideration, experimental animals subjected to subtotal hepatectomy, in which 85-90% of the native liver is resected, show delayed recovery of liver mass [45] but more severe PH-induced hypoglycemia and augmented hepatocellular proliferation compared to animals subjected to 2/3 PH.

15.3.2 Glucose Supplementation Impairs Experimental Liver Regeneration

PH- and hepatotoxin-induced hepatocellular proliferation are inhibited by enteral glucose supplementation [28,42]. Such supplementation also suppresses the transient hepatic steatosis that occurs during regeneration. Conversely, dietary caloric restriction accelerates regenerative hepatocellular proliferation [11,46]. The mechanisms responsible for these effects have also been investigated. For example, enteral glucose supplementation augments hepatic expression of CCAAT-enhancer binding protein (C/EBP) α, whose level and activity initially decline during [28] and subsequently promote termination of [47] regeneration. Hepatic expression of the cyclin-dependent kinase inhibitors (CDKIs) $p21^{Cip1}$ and $p27^{Kip1}$ are also increased in glucose-supplemented animals; however, the mechanisms by which supplemental glucose promotes such CDKI expression and whether other signals contribute to delayed regeneration in glucose-supplemented mice is not yet known.

Circulating insulin levels decline in response to PH-induced hypoglycemia and are augmented by exogenous glucose supplementation [28]. However, systemic diversion of portal circulation (i.e., portacaval shunting) causes atrophy of the liver lobe from which such flow is diverted, insulin-supplementation reverses such atrophy [48], and the insulin receptor translocates to hepatocellular nuclei in regenerating liver [49]. Glycemic regulation of these changes in insulin (and other hormonal) signaling during liver regeneration require further characterization. Of note, the glucose supplementation regimens employed in the studies referenced above do not entirely prevent hypoglycemia or disrupt early hepatic induction of pro-proliferative cell-cycle machinery, and reduce but do not completely abrogate hepatocellular proliferation after PH [28,42]. Thus, the development of an experimental paradigm with which PH-induced hypoglycemia can be prevented (or at least delayed) will enlighten the mechanisms linking glycemia to liver regeneration. Nevertheless, these data implicate glycemic alterations in the regulation of liver regeneration.

15.3.3 Disrupting Regenerative Hepatic Steatosis Compromises Liver Regeneration

Suppression of PH-induced transient hepatic steatosis using various pharmacological [24,50,51] or genetic [24,29] strategies has been reported to inhibit liver regeneration. For example, *fld* mice, which as mentioned above have diminished systemic adipose stores and accumulate less hepatic fat after PH, exhibit impaired regeneration. However, studies using other models describe seemingly contradictory results. For example, regeneration in caveolin-1-null mice, which exhibit reduced PH-induced hepatic steatosis, has variably been reported as impaired or normal [52,53], and hepatic triglyceride accumulation is diminished but regeneration proceeds normally following PH in liver fatty acid binding protein- and intestine-specific microsomal triglyceride transfer protein-null mice [40]. Importantly, PH-induced fat accumulation is not entirely abrogated in either of those latter models, leading to speculation that a "threshold of adaptive lipogenesis" essential for regeneration is not crossed in those mice [31]. Interestingly, fat accumulates concomitantly with cell proliferation in primary hepatocytes in culture [54], and both dietary and parenteral administration of various lipid-based formulations accelerates resection- [42] and toxin- [11] induced hepatocellular proliferation in experimental animals. Together, these observations suggest that hepatocellular lipid accumulation promotes hepatocyte proliferation, at least in some circumstances, and they support the possibility that the antiregenerative effect of glucose supplementation is secondary to the suppressive action of such intervention on release of free fatty acids from adipose [42]. With these observations in mind, it is intriguing that impaired liver regeneration in *fld* mice, like that in glucose-supplemented animals, is associated with augmented hepatic $p21^{Cip1}$ expression [28,29].

15.3.4 Amino Acid Metabolism is Specifically Altered During Experimental Liver Regeneration

Regenerative changes in amino acid metabolism also occur. For example, α-NH_2-butyric acid (Aab) accumulates in mouse serum and liver after PH [27]. Aab is generated by transamination of 2-oxobutyrate, which is produced from methionine catabolism [55]. Methionine is primarily metabolized in the liver, and such metabolism is commonly deranged in chronic liver disease [56]. Furthermore, hepatic metabolism of the methyl donor *S*-adenosyl-methionine (SAM), which is synthesized by methionine-adenosyl transferase 1A (MAT1A), is highly regulated during liver regeneration, and such regulation is disrupted in experimental and clinical liver disease and liver cancer [56,57]. Moreover, genetic disruption of MAT1A expression inhibits mouse liver regeneration [58]. These findings link methionine metabolism with regulation of liver regeneration. Other studies show that the activities of (periportal) urea cycle and (centrilobular) glutamine synthase enzymes are downregulated during regeneration [59,60] and that disruption of hepatic neutral amino acid uptake, such as that of the gluconeogenic precursor alanine, inhibits regeneration [61]. Thus, hepatic insufficiency-induced alterations in uptake and metabolism of amino acids might be functionally important during regenerative hepatocellular proliferation.

15.3.5 Regenerative Regulation of Metabolism

The data summarized above implicate the specific alterations in metabolism that occur in response to liver injury as proregenerative. Conversely, hepatic regenerative signals also appear to have marked effects on metabolism. Although beyond the scope of this chapter, the interested reader is referred to a recent publication addressing this noteworthy aspect of mechanisms linking liver regeneration and metabolism [62].

15.4 CANDIDATE MECHANISMS LINKING METABOLISM TO REGENERATION

The specific molecular mechanisms that link alterations in metabolism to induction of liver regeneration have not yet been established; nevertheless, several prospects are worthy of consideration. One possibility is that the metabolic response to hepatic insufficiency provides the liver with substrate to meet increased regenerative energy requirements while concurrently accommodating the glucose demands of extrahepatic tissues. The liver might also meet systemic energy demand via ketone body synthesis. However, even though ketogenic amino acids accumulate in regenerating liver, serum and hepatic ketone body levels decline during regeneration [63,64]. These changes, like those of glycemia and lipid metabolism, are accompanied by concordant effects on ketogenic gene transcription [65]. Not exclusive of these considerations, altered metabolism during liver regeneration must also generate the anabolic macromolecular precursors necessary for regenerative cell proliferation. Consistent with this idea, the level of the less-active M2 isoform of the glycolytic enzyme pyruvate kinase (PK-M2) increases while that of the more active PK-L isoenzyme decreases in regenerating liver [66]. This switch should divert glucose-derived carbon from glycolytic ATP production to macromolecule precursor synthesis (or back to glucose). Increased PK-M2 also occurs as a part of the "Warburg effect," i.e., the aerobic glycolysis characteristic of cancer cell proliferation [67]. Thus, alterations in metabolism likely contribute to both the energy and macromolecular precursor demands of the hepatic regenerative response. However, this conclusion cannot account for the inhibitory effect of supplemental glucose on liver regeneration, or does it explain the precision with which liver-to-body-mass is restored by regeneration following hepatic injury. Those observations in particular implicate the metabolic response to hepatic insufficiency itself as the source of specific proregenerative signals.

Although the exact mechanisms connecting metabolism and regeneration have not been identified, a number of signaling molecules and pathways regulated by metabolism have also been implicated in the control of liver regeneration [1–6,31–34]. Examples include cytokines, epidermal growth factor receptor, glycogen synthase kinase 3, mammalian target of rapamycin, AMP kinase, xenobiotic nuclear steroid hormone receptors, and other specific transcription factors. Recent studies also suggest that altered metabolism might influence regeneration through epigenetic effects on histone acetylation. Space constraints do not permit comprehensive consideration of the evidence supporting each of these candidate links between metabolism and regeneration; therefore, an updated review of two examples (xenobiotic and other nuclear hormone receptors and epigenetic regulation) is provided here (Figure 15.1), and the interested reader is referred to our recent reviews for additional information on this subject [31–34]. It should also be noted that, in general, the essential regenerative function of the metabolic regulation of these or any other signals remains to be definitively established.

15.4.1 Nuclear Hormone Receptor-Dependent Hepatocellular Proliferation

Administration of certain xenobiotics to rodents induces hepatocellular hypertrophy and hyperplasia, thereby increasing liver mass in the absence of liver injury [68]. Thyroid hormone exerts similar effects [69], and withdrawal of these agents is followed by restoration of "normal" liver mass, providing additional evidence for the "hepatostat." Several nuclear hormone receptor transcription factors have been identified as specific and direct mediators of such xenobiotic- and hormonally induced hepatocellular proliferation. For example, peroxisome proliferator-activated receptor (PPAR) α expression is required for development of the hepatomegaly that occurs in rodents exposed to clofibrate and Wy-14,636 [70]. Similarly, the increase in liver mass that occurs with administration of 1,4-bis(2-(3,5-dichloropyridoxyloxy)) benzene (TCPOBOP) or phenobarbitol depends upon the constitutive androstane receptor (CAR) [71]. Interestingly, liver regeneration after PH and after toxic injury is altered in several models in which the expression or activity of specific xenobiotic- and other nuclear hormone activated transcription factors has been altered. Thus, the mechanisms that regulate mitogen-induced hepatomegaly overlap, at least to some degree, with those that direct hepatic insufficiency-induced liver regeneration. Conversely, disruption of specific upstream signals implicated in PH-induced regeneration does not necessarily prevent xenobiotic-dependent changes in liver mass [72–74]. Those observations suggest that these mitogen-activated pathways act downstream of or in parallel to at least some of the signals that promote such regeneration. Even so, characterization of hepatocellular proliferation and liver regeneration in models in which nuclear hormone receptor transcription factor activity has been genetically or pharmacologically manipulated

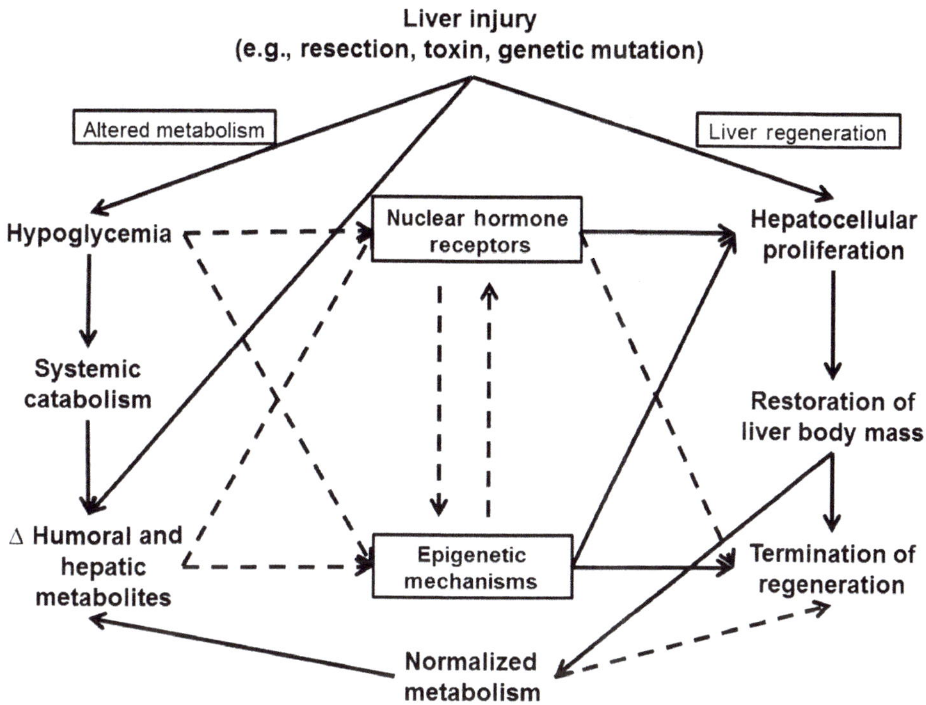

FIGURE 15.1 Metabolic regulation of liver regeneration. Liver injuries alter metabolism and induce regeneration. Solid arrows show established relationships, and dashed arrows show proposed mechanisms linking metabolism to liver regeneration, each of which are discussed in the text.

provides insight into the mechanisms that might link metabolism to liver regeneration. For example:

PPARα—The role of PPARα during regeneration remains somewhat enigmatic based on conflicting reports about the magnitude of impairment of PH-induced regeneration in mice genetically disrupted for PPARα or its co-activator PPAR-binding protein [40,75–78]. Nevertheless, data implicating endogenous lipid metabolites as ligand activators of PPARα [79] suggest that naturally occurring PPARα ligands might link hepatic lipid accumulation following PH to subsequent initiation of regenerative hepatocellular proliferation. Furthermore, other metabolites that accumulate in regenerating liver might serve as endogenous ligands for other xenobiotic-binding transcription factors, including those discussed below, to regulate regeneration.

PPARγ—Several studies now suggest that PPARγ activity negatively regulates and is downregulated during normal PH-induced liver regeneration. For example, thiazolidinediones, which activate PPARγ-suppress hepatic regeneration [26,80]. Consistent with this finding, liver-specific PPARγ-KO mice display modestly accelerated regeneration [30]. Furthermore, hepatic expression of PPARγ2 (but not the alternatively splice γ1) isoform is also downregulated during normal liver regeneration [30] and suppresses cyclin D1 expression in cultured hepatocytes [81]. Moreover, unsaturated fatty acids, prostaglandins, and other eicosanoids, which themselves affect liver regeneration, are all naturally occurring PPARγ ligands [23,82]. Nevertheless, the regenerative regulation of hepatocellular PPARγ activity (e.g., by changes in nuclear receptor expression or cellular levels of ligand regulators) has not been defined and the specific transcriptional targets of such regulation are unknown.

Farnesoid X receptor (FXR)—FXR is a bile acid-activated transcriptional regulator and global disruption of its expression impairs PH- and toxin-induced liver regeneration [83,84]. Furthermore, feeding bile acids to unoperated mice increases hepatocellular mitoses and induces hepatomegaly [83]. These observations suggest that the proportionately increased entero-hepatic delivery of bile acids to the postresection liver remnant might link altered metabolism to onset of regeneration. Consistent with this conclusion, surgical or genetic disruption of bile acid return to liver inhibits PH- and toxin-induced regeneration [85,86]. However, hepatic bile acid content declines after PH in wild-type mice [83]. FXR is expressed in both liver and intestine, which raises the question as to whether disruption of hepatic or intestinal expression (or both) contributes to impaired regeneration in global FXR-null animals. Recent studies employing tissue-specific FXR-deleted mice show that both liver- and intestine-specific FXR-nulls exhibit impaired regeneration in response to resection- and toxin-induced regenerative stimuli [87,88]. However, published analyses have not yet addressed whether global or tissue-specific FXR-knockout mice are resistant to the hepatomegaly-inducing effects of enteral bile acids. Intestinal FXR affects bile acid homeostasis by inducing intestinal epithelial expression of fibroblast growth factor 15 (FGF-15), which is transported via portal circulation to the liver where it suppresses bile acid synthesis [89], and FGF-15-null mice also exhibit impaired liver regeneration and reduced enteral bile acid-stimulated hepatomegaly [90,91]. Thus, FXR, FGF15, and entero-hepatic circulation of bile acids are candidate mechanisms involved in the metabolic regulation of liver regeneration.

Studies complementary to those described above have also explored the regenerative functions of TGR5 (the G-protein-coupled bile acid receptor [92]) and MRP3 (the membrane bile acid transporter [93]) using global KO mice. In each case, the results show impaired liver regeneration in the null mice. These new data provide additional support for the functional importance of FXR, FGF15, and entero-hepatic circulation of bile acids in the metabolic regulation of liver regeneration. However, questions remain about the mechanistic basis for these effects. For example, two recent studies reported that FXR-regulated transcription suppresses fasting-induced autophagy [94,95]. Another report showed that Atg5-null mice, with genetically disrupted autophagy, exhibit impaired liver regeneration [96]. These findings are difficult to reconcile with the impaired hepatic regenerative phenotype of FXR-KO mice described above. Further confounding this issue, yet another recent study reported that cultured FXR-KO and bile acid-supplemented hepatocytes display suppression of autophagy [97]. Thus, the links between bile acid-dependent FXR activation, autophagy, and metabolic regulation of liver regeneration require further elucidation.

Other xenobiotic-activated nuclear receptors—Mice also develop hepatomegaly when treated with CAR- and pregnane X receptor (PXR)-activating ligands, such as phenobarbitol and TCPOBOP [68,98]. Although these chemically induced responses are dependent upon expression of CAR or PXR [71,99], the corresponding null mice show only modestly impaired liver regeneration after PH [83,100]. Moreover, studies have not yet defined whether naturally occurring metabolite ligands of CAR or PXR accumulate in regenerating liver. Other xenobiotic-activated nuclear receptor transcription factor family members (in addition to PPARγ which was discussed above) might also negatively regulate liver regeneration.

For example, pharmacological activation of the liver X receptor (LXR) [101] suppresses regeneration. Oxysterols serve as endogenous LXR ligands [101], and one study suggested that glucose itself directly regulates LXR activity [102]. These data re-enforce the intriguing hypothesis that liver injury-induced alterations in hepatocellular metabolite levels affect regeneration by regulating xenobiotic-activated transcription factor activities (Figure 15.1).

15.4.2 Metabolic Influences on Epigenetic Regulation of Liver Regeneration

Regenerative regulation of liver histone acetylation—Newer data suggest that alterations in metabolism influence liver regeneration by altering hepatocellular patterns of histone acetylation. Indirect evidence for this hypothesis comes from studies showing that glucose supplementation promotes histone acetylation in mammalian cell culture, with fatty acids unable to substitute for glucose in those models [103,104]. Thus, PH-induced alterations in glycemia might alter histone acetylation, perhaps by reducing histone acetyltransferase acetyl-CoA substrate availability. A recently published study used acetyl histone H3K9 chromatin immunoprecipitation combined with next-generation DNA sequencing (i.e., ChIP-Seq) to characterize the histone acetylome of early regenerating liver [105]. The results showed that proregenerative changes in liver histone acetylation include both hyperacetylation of positive regulators of cell proliferation as well as deacetylation of antiproliferative factors. Another recent study investigated the regulation and regenerative function of hepatic zinc-dependent histone deacetylases (Zn-HDACs), showing that total Zn-HDAC activity increases and global liver histone acetylation declines in parallel with onset of hypoglycemia after PH [106]. That study also reported that liver regeneration is suppressed by suberoylanilide hydroxamic acid (SAHA), a broad inhibitor of Zn-HDAC activity [106]. Interestingly, SAHA also reverses both hyper- and deacetylating regenerative changes induced by PH in the liver histone acetylome [105]. The former finding (gene-specific increases in histone acetylation) corresponds to the predicted direct effect of Zn-HDAC inhibition, while the latter observation (histone deacetylation) must result from indirect effects. Regeneration is also inhibited in isoform-specific HDAC1- and HDAC2-null mice [107]. Together, these data implicate Zn-HDACs as important regulators of liver regeneration. Metabolic regulation of these enzymes during regeneration has been suggested by data showing that HDAC5, whose subcellular localization is affected by glycemia [108], undergoes PH-induced nuclear localization [106]. Furthermore, HDAC5 target-specific deacetylating activity depends on its interaction with HDAC3 [109], and HDAC3 suppresses hepatic PPARγ expression [110,111]. Thus, PH-induced changes in metabolism could influence liver regeneration by epigenetic regulation of hepatocellular PPARγ expression. It is also intriguing to note that the ketone body β-hydroxybutyrate was reported to inhibit HDAC3 (and other specific Zn-HDACs [112]), further supporting the provocative possibility that endogenous metabolites whose levels change as a part of the metabolic response to hepatic insufficiency might affect liver regeneration by influencing HDAC activity (Figure 15.1).

The regenerative function of NAD-dependent (i.e., sirtuin) HDACs has also been examined, with recent studies variably reporting that SIRT1 promotes [113] or suppresses [114] liver regeneration. Nevertheless, these data further support the possibility that altered metabolism influences epigenetic regulation of liver regeneration by affecting HDACs. With this in mind, it is also intriguing to recall older work reporting that levels of NAD, which is an essential sirtuin cofactor, decline in regenerating rat liver, and that supplementation with nicotinamide, which is enzymatically converted to NAD in the liver, increases hepatic NAD content and impairs liver regeneration [115]. Thus, the regenerative functions of SIRT1 and other NAD-dependent HDACs should be further characterized.

Liver regeneration and DNA methylation—Metabolic influences on epigenetic regulation of DNA methylation have been suggested by various older as well as more recent studies. Although not yet established, those mechanisms might also be relevant to the regulation of liver regeneration (and influenced by the regenerative effects on SAM metabolism discussed above). The interested reader is referred to our recent consideration of this subject [34].

MicroRNAs and liver regeneration—miRNAs regulate gene expression by affecting mRNA stability and translation, and miRNA regulation during liver regeneration is beginning to be investigated. Although beyond the space constraints of this chapter, the interested reader is referred to our recent review of this subject [34].

15.4.3 Other Considerations and Areas for Future Basic Research

Although less well studied than the mechanisms that initiate hepatic regeneration, the signals that terminate this response after recovery of normal liver-to-body mass have recently received increased experimental attention [6]. While the specific influences of metabolism on termination of liver regeneration also require further elucidation, published observations suggest that regenerative recovery of metabolic function likely contributes to such regulation. For example, hepatic C/EBPα gene acetylation [105] and mRNA expression [116] are downregulated during early normal regeneration. Such expression is also augmented by dextrose supplementation [28] and required for termination of regeneration [47]. Nevertheless, additional research investigating mechanisms connecting regenerative recovery of hepatic metabolic function to termination of liver regeneration should be pursued. Similarly, the influences of metabolism on oval cell-dependent recovery of the injured liver [16] are entirely unknown and should also be explored.

15.5 CLINICAL IMPLICATIONS

The data reviewed here implicate alterations in metabolism in response to hepatic insufficiency as the source of molecular signals that initiate experimental regenerative hepatocellular proliferation. These considerations suggest that metabolic influences might also affect hepatic regeneration in human liver diseases. For example, impaired regeneration in fatty liver disease, aging, and acute liver failure (ALF) might be related to perturbations in metabolism that accompany such conditions. These topics are briefly considered here.

15.5.1 Nonalcoholic Fatty Liver Disease

Unlike the transient steatosis that develops during early normal liver regeneration [24], chronic hepatic steatosis is associated with impaired liver regeneration after PH- and chemical-induced liver injury in many experimental models [31]. Those studies also suggest that the magnitude of such steatosis might determine its effect on regeneration. Nonalcoholic fatty liver disease is also linked to impaired regeneration [117] and adverse outcomes [118] after liver resection in humans. Thus, acute versus chronic hepatic steatosis, and accompanying differences in hepatic and systemic metabolism, might exert distinct effects on cytokine, growth factor, secondary messenger, genetic, or epigenetic signals and, thereby, have divergent influences on liver regeneration. Nevertheless, the specific mechanisms responsible remain uncertain and require further investigation.

15.5.2 Aging

The aged liver has reduced regenerative capacity in response to resection-induced hepatic insufficiency [119,120]. Although long recognized, the mechanisms responsible remain incompletely defined. Experiments show that specific antiproliferative factors whose hepatic expression is induced in association with impaired regeneration in glucose-supplemented mice [28], including C/EBPα and $p21^{Cip1}$ [28], are also upregulated in postresection livers from aged-animals [120]. Thus, age-related effects on glycemia might contribute to impaired liver regeneration in old animals. Further support for this idea comes from studies implicating Zn-HDACs as mediators of the antiregenerative influence of aging [120] together with previously mentioned data indicating that metabolism influences Zn-HDAC activity and subcellular localization during regeneration [106]. Surprisingly, old age does not suppress the hepatocellular proliferative response to certain toxins [11,121] or xenobiotics (e.g., TCPOBOP [122]). The relevance of these data to human liver regeneration warrants further study. In particular, whether diminished resection-induced regeneration of the aged liver contributes to reduced survival of transplanted liver grafts from older donors reported in some studies [120] should be examined.

15.5.3 Subtotal Hepatectomy

Experimental analyses in which more than two-thirds of the native liver is resected show delayed regenerative recovery of liver mass [45]. Unlike the effect on regeneration after two-thirds PH, glucose supplementation improves outcomes from experimental subtotal hepatectomy [123]. Elucidating the mechanisms responsible for these seemingly discordant effects of glycemia on the response to partial versus subtotal hepatectomy might inform strategies to improve management of patients with ALF, for which subtotal hepatectomy in rodents has been used as an experimental model.

15.5.4 Hepatic Regenerative Metabolomic Biomarkers

Based on considerations raised here, metabolic serum biomarkers of experimental liver regeneration could enable more reliable and non-invasive assessment of human liver regeneration. Such tools would be especially useful in the clinical management of ALF, a potentially devastating condition from which some patients die or undergo liver transplantation [124]. Others recover spontaneously, based in part on regeneration of the liver. However, early reliable distinction between ALF patients likely to survive spontaneously and those at increased risk of death without liver transplantation remains challenging. The data reviewed here suggest that unique patterns of circulating metabolites might distinguish progression of normal liver regeneration from an impaired response. Consistent with this idea, serum levels of α-NH_2 butyric acid (Aab) were recently identified as a sensitive and specific biomarker of mouse liver regeneration and subsequently shown to correlate with spontaneous survival in pediatric ALF patients [27]. Nonetheless, additional studies are needed to comprehensively define the humoral metabolomic signature of experimental regeneration and determine its value for reliable ALF outcomes prognostication.

15.6 SUMMARY AND CONCLUSIONS

Many studies implicate altered metabolism in response to hepatic insufficiency in the regulation of liver regeneration. Such analyses also suggest nuclear hormone receptor, epigenetic, and other mechanisms by which such linkage might occur. Together, these data imply that the hepatostat is regulated by metabolic functions of the liver. Ongoing and future studies to define specific functional relationships between altered metabolism and signaling pathways during experimental regeneration should enable translational investigations to determine whether derangement of these interactions contribute to impaired regeneration in human liver diseases. The ultimate goal of such work is to develop proregenerative metabolic strategies that improve outcomes of such diseases.

References

[1] Rudnick DA. Liver regeneration: the developmental biologists approach. In: Orlando G, Lerut JP, Soker S, Stratta RJ, editors. Regenerative medicine applications in organ transplantation. 1st ed. Waltham, MA: Elsevier/Academic Press; 2014. p. 353–74.

[2] Fausto N, Campbell JS, Riehle KJ. Liver regeneration. J Hepatol 2012;57(3):692–4.

[3] Michalopoulos GK. Liver regeneration after partial hepatectomy: critical analysis of mechanistic dilemmas. Am J Pathol 2010;176(1):2–13 [PMCID:PMC2797862].

[4] Michalopoulos GK. Liver regeneration. J Cell Physiol 2007;213(2):286–300.

[5] Michalopoulos GK, DeFrances MC. Liver regeneration. Science 1997;276(5309):60–6.

[6] Michalopoulos GK. Advances in liver regeneration. Expert Rev Gastroenterol Hepatol 2014;8(8):897–907.

[7] Vetelainen R, van Vliet A, Gouma DJ, van Gulik TM. Steatosis as a risk factor in liver surgery. Ann Surg 2007;245(1):20–30 [PMCID:PMC1867945].

[8] Miyazaki S, Takasaki K, Yamamoto M, Tsugita M, Otsubo T. Liver regeneration and restoration of liver function after partial hepatectomy: the relation of fibrosis of the liver parenchyma. Hepatogastroenterology 1999;46(29):2919–24.

[9] Donato MF, Arosio E, Del NE, Ronchi G, Lampertico P, Morabito A, et al. High rates of hepatocellular carcinoma in cirrhotic patients with high liver cell proliferative activity. Hepatology 2001;34(3):523–8.

[10] Higgins GM, Anderson RM. Experimental pathology of the liver. 1. Restoration of the liver of the white rat following partial surgical removal. Arch Pathol 1931;12:186–202.

[11] Mehendale HM. Tissue repair: an important determinant of final outcome of toxicant-induced injury. Toxicol Pathol 2005;33(1):41–51.

[12] Anand SS, Mehendale HM. Liver regeneration: a critical toxicodynamic response in predictive toxicology. Environ Toxicol Pharmacol 2004;18(2):149–60.

[13] Bhushan B, Walesky C, Manley M, Gallagher T, Borude P, Edwards G, et al. Pro-regenerative signaling after acetaminophen-induced acute liver injury in mice identified using a novel incremental dose model. Am J Pathol 2014;184(11):3013–25 [PMCID:PMC4215032].

[14] Rudnick DA, Liao Y, An JK, Muglia LJ, Perlmutter DH, Teckman JH. Analyses of hepatocellular proliferation in a mouse model of alpha-1-antitrypsin deficiency. Hepatology 2004;39(4):1048–55.

[15] Malato Y, Naqvi S, Schurmann N, Ng R, Wang B, Zape J, et al. Fate tracing of mature hepatocytes in mouse liver homeostasis and regeneration. J Clin Invest 2011;121(12):4850–60 [PMCID:PMC3226005].

[16] Duncan AW, Dorrell C, Grompe M. Stem cells and liver regeneration. Gastroenterology 2009;137(2):466–81 [PMCID:PMC3136245].

[17] Bird TG, Lorenzini S, Forbes SJ. Activation of stem cells in hepatic diseases. Cell Tissue Res 2008;331 (1):283–300, [PMCID:PMC3034134].

[18] Huang J, Glauber M, Qiu Z, Gazit V, Dietzen DJ, Rudnick DA. The influence of skeletal muscle on the regulation of liver:body mass and liver regeneration. Am J Pathol 2012;180(2):575–82 [PMCID:PMC3349872].

[19] Moolten FL, Bucher NLR. Regeneration of rat liver: transfer of humoral agent by cross circulation. Science 1967;158:272–4.

[20] Fisher B, Szuch P, Levine M, Fisher ER. A portal blood factor as the humoral agent in liver regeneration. Science 1971;171:575–7.

[21] Yamada Y, Fausto N. Deficient liver regeneration after carbon tetrachloride injury in mice lacking type 1 but not type 2 tumor necrosis factor receptor. Am J Pathol 1998;152(6):1577–89.

[22] Huh CG, Factor VM, Sanchez A, Uchida K, Conner EA, Thorgeirsson SS. Hepatocyte growth factor/c-met signaling pathway is required for efficient liver regeneration and repair. Proc Natl Acad Sci USA 2004;101 (13):4477–82.

[23] Rudnick DA, Perlmutter DH, Muglia LJ. Prostaglandins are required for CREB activation and cellular proliferation during liver regeneration. Proc Natl Acad Sci USA 2001;98(15):8885–90 [PMCID:PMC37530].

[24] Shteyer E, Liao Y, Muglia LJ, Hruz PW, Rudnick DA. Disruption of hepatic adipogenesis is associated with impaired liver regeneration in mice. Hepatology 2004;40(6):1322–32.

[25] Rudnick DA. Trimming the fat from liver regeneration. Hepatology 2005;42(5):1001–3.

[26] Turmelle YP, Shikapwashya O, Tu S, Hruz PW, Yan Q, Rudnick DA. Rosiglitazone inhibits mouse liver regeneration. FASEB J 2006;20:2609–11.

[27] Rudnick DA, Dietzen DJ, Turmelle YP, Shepherd R, Zhang S, Belle SH, et al. Serum alpha-NH-butyric acid

may predict spontaneous survival in pediatric acute liver failure. Pediatr Transplant 2009;13(2):223–30 [PMCID:PMC2659723].

[28] Weymann A, Hartman E, Gazit V, Wang C, Glauber M, Turmelle Y, et al. p21 is required for dextrose-mediated inhibition of mouse liver regeneration. Hepatology 2009;50:207–15 [PMCID: PMC2705473].

[29] Gazit V, Weymann A, Hartman E, Finck BN, Hruz PW, Tzekov A, et al. Liver regeneration is impaired in lipodystrophic fatty liver dystrophy mice. Hepatology 2010;52(6):2109–17 [PMCID:PMC2991544].

[30] Gazit V, Huang J, Weymann A, Rudnick DA. Analysis of the role of hepatic PPARgamma expression during mouse liver regeneration. Hepatology 2012;56 (4):1489–98 [PMCID:PMC3465497].

[31] Rudnick DA, Davidson NO. Functional relationships between lipid metabolism and liver regeneration. Int J Hepatol 2012;2012:549241 [PMCID:PMC3272806].

[32] Huang J, Rudnick DA. Elucidating the metabolic regulation of liver regeneration. Am J Pathol 2014;184 (2):309–21 [PMCID:PMC3906487].

[33] Huang J, Rudnick DA. Metabolic regulation of liver regeneration. In: Dufour J-F, Clavien P-A, editors. Signaling pathways in liver diseases. 3rd ed. Oxford, UK: Wiley-Blackwell; 2015.

[34] Huang J, Rudnick DA. Elucidating metabolic and epigenetic mechanisms that regulate liver regeneration. Curr Pathobiol Rep 2015;3:89–98.

[35] Felber JP, Golay A. Regulation of nutrient metabolism and energy expenditure. Metabolism 1995;44 (2 Suppl. 2):4–9.

[36] Murray AB, Strecker W, Silz S. Ultrastructural changes in rat hepatocytes after partial hepatectomy, and comparison with biochemical results. J Cell Sci 1981;50: 433–48.

[37] Brinkmann A, Katz N, Sasse D, Jungermann K. Increase of the gluconeogenic and decrease of the glycolytic capacity of rat liver with a change of the metabolic zonation after partial hepatectomy. Hoppe Seylers Z Physiol Chem 1978;359(11):1561–71.

[38] Wang H, Peiris TH, Mowery A, Le Lay J, Gao Y, Greenbaum LE. CCAAT/enhancer binding protein-beta is a transcriptional regulator of peroxisome-proliferator-activated receptor-gamma coactivator-1alpha in the regenerating liver. Mol Endocrinol 2008;22 (7):1596–605 [PMCID:PMC2453599].

[39] Klingensmith JS, Mehendale HM. Chlordecone-induced fat depletion in the male rat. J Toxicol Environ Health 1982;10(1):121–9.

[40] Newberry EP, Kennedy SM, Xie Y, Luo J, Stanley SE, Semenkovich CF, et al. Altered hepatic triglyceride content after partial hepatectomy without impaired liver regeneration in multiple murine genetic models. Hepatology 2008;48(4):1097–105.

[41] Crumm S, Cofan M, Juskeviciute E, Hoek JB. Adenine nucleotide changes in the remnant liver: an early signal for regeneration after partial hepatectomy. Hepatology 2008;48(3):898–908.

[42] Holecek M. Nutritional modulation of liver regeneration by carbohydrates, lipids, and amino acids: a review. Nutrition 1999;15(10):784–8.

[43] Crawford DF, Piwnica-Worms H. Regulation of the eukaryotic cell cycle. In: Arias IM, Boyer JL, Chisari FV, Schachter D, Schafritz DA, editors. The liver biology and pathobiology. 4th ed. Philadelphia: Lipincott Williams & Wilkins; 2001. p. 977–86.

[44] Zou Y, Bao Q, Kumar S, Hu M, Wang GY, Dai G. Four waves of hepatocyte proliferation linked with three waves of hepatic fat accumulation during partial hepatectomy-induced liver regeneration. PLoS One 2012;7(2):e30675 [PMCID:PMC3272022].

[45] Lehmann K, Tschuor C, Rickenbacher A, Jang JH, Oberkofler CE, Tschopp O, et al. Liver failure after extended hepatectomy in mice is mediated by a p21-dependent barrier to liver regeneration. Gastroenterology 2012;143(6):1609–19.

[46] Cuenca AG, Cress WD, Good RA, Marikar Y, Engelman RW. Calorie restriction influences cell cycle protein expression and DNA synthesis during liver regeneration. Exp Biol Med (Maywood) 2001;226 (11):1061–7.

[47] Jin J, Hong IH, Lewis K, Iakova P, Breaux M, Jiang Y, et al. Cooperation of C/EBP family proteins and chromatin remodeling proteins is essential for termination of liver regeneration. Hepatology 2015;61: 315–25.

[48] Starzl TE, Porter KA, Putnam CW. Intraportal insulin protects from the liver injury of portacaval shunt in dogs. Lancet 1975;2(7947):1241–2.

[49] Amaya MJ, Oliveira AG, Guimaraes ES, Casteluber MC, Carvalho SM, Andrade LM, et al. The insulin receptor translocates to the nucleus to regulate cell proliferation in liver. Hepatology 2014;59 (1):274–83 [PMCID:PMC3823683].

[50] Walldorf J, Hillebrand C, Aurich H, Stock P, Hempel M, Ebensing S, et al. Propranolol impairs liver regeneration after partial hepatectomy in C57Bl/6-mice by transient attenuation of hepatic lipid accumulation and increased apoptosis. Scand J Gastroenterol 2010;45:468–76.

[51] Srinivasan SR, Chow CK, Glauert HP. Effect of the peroxisome proliferator ciprofibrate on hepatic DNA synthesis and hepatic composition following partial hepatectomy in rats. Toxicology 1990;62(3): 321–32.

[52] Fernandez MA, Albor C, Ingelmo-Torres M, Nixon SJ, Ferguson C, Kurzchalia T, et al. Caveolin-1 is essential for liver regeneration. Science 2006;313(5793):1628–32.
[53] Mayoral R, Fernandez-Martinez A, Roy R, Bosca L, Martin-Sanz P. Dispensability and dynamics of caveolin-1 during liver regeneration and in isolated hepatic cells. Hepatology 2007;46(3):813–22.
[54] Michalopoulos G, Cianciulli HD, Novotny AR, Kligerman AD, Strom SC, Jirtle RL. Liver regeneration studies with rat hepatocytes in primary culture. Cancer Res 1982;42(11):4673–82.
[55] Newsholme EA, Leech AR. Amino acid metabolism. In: Newsholme EA, Leech AR, editors. Biochemistry for the medical sciences. New York: John Wiley & Sons; 1983. p. 382–441.
[56] Mato JM, Lu SC. Role of S-adenosyl-L-methionine in liver health and injury. Hepatology 2007;45(5):1306–12.
[57] Jung YS, Kim SJ, Kwon do Y, Jun DS, Kim YC. Significance of alterations in the metabolomics of sulfur-containing amino acids during liver regeneration. Biochimie 2013;95(8):1605–10.
[58] Chen L, Zeng Y, Yang H, Lee TD, French SW, Corrales FJ, et al. Impaired liver regeneration in mice lacking methionine adenosyltransferase 1A. FASEB J 2004;18(7):914–6.
[59] Brebnor LD, Balinsky JB. Changes in activities of urea cycle enzymes in early stages of liver regeneration after partial hepatectomy in rats. Life Sci 1983;32(12):1391–400.
[60] Gebhardt R. Altered acinar distribution of glutamine synthetase and different growth response of cultured enzyme-positive and -negative hepatocytes after partial hepatectomy. Cancer Res 1990;50(14):4407–10.
[61] Freeman TL, Ngo HQ, Mailliard ME. Inhibition of system A amino acid transport and hepatocyte proliferation following partial hepatectomy in the rat. Hepatology 1999;30(2):437–44.
[62] Hanse EA, Mashek DG, Becker JR, Solmonson AD, Mullany LK, Mashek MT, et al. Cyclin D1 inhibits hepatic lipogenesis via repression of carbohydrate response element binding protein and hepatocyte nuclear factor 4alpha. Cell Cycle 2012;11(14):2681–90 [PMCID:PMC3409010].
[63] Schofield PS, French TJ, Sugden MC. Ketone-body metabolism after surgical stress or partial hepatectomy. Evidence for decreased ketogenesis and a site of control distal to carnitine palmitoyltransferase I. Biochem J 1987;241(2):475–81 [PMCID: PMC1147585].
[64] Schofield PS, Sugden MC, Corstorphine CG, Zammit VA. Altered interactions between lipogenesis and fatty acid oxidation in regenerating rat liver. Biochem J 1987;241(2):469–74 [PMCID:PMC1147584].
[65] Asins G, Rosa JL, Serra D, Gil-Gomez G, Ayte J, Bartrons R, et al. Gene expression of enzymes regulating ketogenesis and fatty acid metabolism in regenerating rat liver. Biochem J 1994;299(Part 1):65–9.
[66] Chatzipanagiotou S, Nath A, Vogt B, Jungermann K. Alteration in the capacities as well as in the zonal and cellular distributions of pyruvate kinase L and M2 in regenerating rat liver. Biol Chem Hoppe Seyler 1985;366(3):271–80.
[67] Christofk HR, Vander Heiden MG, Harris MH, Ramanathan A, Gerszten RE, Wei R, et al. The M2 splice isoform of pyruvate kinase is important for cancer metabolism and tumour growth. Nature 2008;452(7184):230–3.
[68] Hall AP, Elcombe CR, Foster JR, Harada T, Kaufmann W, Knippel A, et al. Liver hypertrophy: a review of adaptive (adverse and non-adverse) changes—conclusions from the 3rd international ESTP expert workshop. Toxicol Pathol 2012;40(7):971–94.
[69] Fanti M, Singh S, Ledda-Columbano GM, Columbano A, Monga SP. Tri-iodothyronine induces hepatocyte proliferation by protein kinase A-dependent beta-catenin activation in rodents. Hepatology 2014;59(6):2309–20 [PMCID:PMC3979513].
[70] Lee SS, Pineau T, Drago J, Lee EJ, Owens JW, Kroetz DL, et al. Targeted disruption of the alpha isoform of the peroxisome proliferator-activated receptor gene in mice results in abolishment of the pleiotropic effects of peroxisome proliferators. Mol Cell Biol 1995;15(6):3012–22 [PMCID:PMC230532].
[71] Wei P, Zhang J, Egan-Hafley M, Liang S, Moore DD. The nuclear receptor CAR mediates specific xenobiotic induction of drug metabolism. Nature 2000;407(6806):920–3.
[72] Columbano A, Ledda-Columbano GM, Pibiri M, Piga R, Shinozuka H, De LV, et al. Increased expression of c-fos, c-jun and LRF-1 is not required for in vivo priming of hepatocytes by the mitogen TCPOBOP. Oncogene 1997;14(7):857–63.
[73] Menegazzi M, Carcereri-De Prati A, Suzuki H, Shinozuka H, Pibiri M, Piga R, et al. Liver cell proliferation induced by nafenopin and cyproterone acetate is not associated with increases in activation of transcription factors NF-kappaB and AP-1 or with expression of tumor necrosis factor alpha. Hepatology 1997;25(3):585–92.
[74] Ledda-Columbano GM, Curto M, Piga R, Zedda AI, Menegazzi M, Sartori C, et al. In vivo hepatocyte proliferation is inducible through a TNF and IL-6-independent pathway. Oncogene 1998;17(8):1039–44.
[75] Rao MS, Peters JM, Gonzalez FJ, Reddy JK. Hepatic regeneration in peroxisome proliferator-activated

receptor alpha-null mice after partial hepatectomy. Hepatol Res 2002;22(1):52–7.
[76] Wheeler MD, Smutney OM, Check JF, Rusyn I, Schulte-Hermann R, Thurman RG. Impaired Ras membrane association and activation in PPARalpha knockout mice after partial hepatectomy. Am J Physiol Gastrointest Liver Physiol 2003;284(2):G302–12.
[77] Anderson SP, Yoon L, Richard EB, Dunn CS, Cattley RC, Corton JC. Delayed liver regeneration in peroxisome proliferator-activated receptor-alpha-null mice. Hepatology 2002;36(3):544–54.
[78] Matsumoto K, Yu S, Jia Y, Ahmed MR, Viswakarma N, Sarkar J, et al. Critical role for transcription coactivator peroxisome proliferator-activated receptor (PPAR)-binding protein/TRAP220 in liver regeneration and PPARalpha ligand-induced liver tumor development. J Biol Chem 2007;282(23):17053–60.
[79] Chakravarthy MV, Lodhi IJ, Yin L, Malapaka RR, Xu HE, Turk J, et al. Identification of a physiologically relevant endogenous ligand for PPARalpha in liver. Cell 2009;138(3):476–88 [PMCID:PMC2725194].
[80] Yamamoto Y, Ono T, Dhar DK, Yamanoi A, Tachibana M, Tanaka T, et al. Role of peroxisome proliferator-activated receptor-gamma (PPARgamma) during liver regeneration in rats. J Gastroenterol Hepatol 2008;23(6):930–7.
[81] Sharma C, Pradeep A, Pestell RG, Rana B. Peroxisome proliferator-activated receptor gamma activation modulates cyclin D1 transcription via beta-catenin-independent and cAMP-response element-binding protein-dependent pathways in mouse hepatocytes. J Biol Chem 2004;279(17):16927–38.
[82] Rudnick DA, Muglia LJ. Eicosanoids and liver regeneration. In: Curtis-Prior P, editor. The eicosanoids. West Sussex, England: John Wiley and Sons; 2004. p. 415–22.
[83] Huang W, Ma K, Zhang J, Qatanani M, Cuvillier J, Liu J, et al. Nuclear receptor-dependent bile acid signaling is required for normal liver regeneration. Science 2006;312(5771):233–6.
[84] Meng Z, Wang Y, Wang L, Jin W, Liu N, Pan H, et al. FXR regulates liver repair after CCl4-induced toxic injury. Mol Endocrinol 2010;24(5):886–97 [PMCID: PMC3284774].
[85] Naugler WE. Bile acid flux is necessary for normal liver regeneration. PLoS One 2014;9(5):e97426 [PMCID:PMC4026228].
[86] Meng Z, Liu N, Fu X, Wang X, Wang YD, Chen WD, et al. Insufficient bile acid signaling impairs liver repair in CYP27(−/−) mice. J Hepatol 2011;55 (4):885–95.
[87] Zhang L, Wang YD, Chen WD, Wang X, Lou G, Liu N, et al. Promotion of liver regeneration/repair by farnesoid X receptor in both liver and intestine in mice. Hepatology 2012;56(6):2336–43 [PMCID: PMC3477501].
[88] Borude P, Edwards G, Walesky C, Li F, Ma X, Kong B, et al. Hepatocyte-specific deletion of farnesoid X receptor delays but does not inhibit liver regeneration after partial hepatectomy in mice. Hepatology 2012;56 (6):2344–52 [PMCID:PMC3469721].
[89] Inagaki T, Choi M, Moschetta A, Peng L, Cummins CL, McDonald JG, et al. Fibroblast growth factor 15 functions as an enterohepatic signal to regulate bile acid homeostasis. Cell Metab 2005;2(4):217–25.
[90] Uriarte I, Fernandez-Barrena MG, Monte MJ, Latasa MU, Chang HC, Carotti S, et al. Identification of fibroblast growth factor 15 as a novel mediator of liver regeneration and its application in the prevention of post-resection liver failure in mice. Gut 2013.
[91] Kong B, Huang J, Zhu Y, Li G, Williams J, Shen S, et al. Fibroblast growth factor 15 deficiency impairs liver regeneration in mice. Am J Physiol Gastrointest Liver Physiol 2014;306(10):G893–902 [PMCID: PMC4024724].
[92] Pean N, Doignon I, Garcin I, Besnard A, Julien B, Liu B, et al. The receptor TGR5 protects the liver from bile acid overload during liver regeneration in mice. Hepatology 2013;58(4):1451–60.
[93] Fernandez-Barrena MG, Monte MJ, Latasa MU, Uriarte I, Vicente E, Chang HC, et al. Lack of Abcc3 expression impairs bile-acid induced liver growth and delays hepatic regeneration after partial hepatectomy in mice. J Hepatol 2012;56(2):367–73.
[94] Seok S, Fu T, Choi SE, Li Y, Zhu R, Kumar S, et al. Transcriptional regulation of autophagy by an FXR-CREB axis. Nature 2014;516(7529):108–11 [PMCID: PMC4257899].
[95] Lee JM, Wagner M, Xiao R, Kim KH, Feng D, Lazar MA, et al. Nutrient-sensing nuclear receptors coordinate autophagy. Nature 2014;516(7529):112–5 [PMCID:PMC4267857].
[96] Toshima T, Shirabe K, Fukuhara T, Ikegami T, Yoshizumi T, Soejima Y, et al. Suppression of autophagy during liver regeneration impairs energy charge and hepatocyte senescence in mice. Hepatology 2014;60(1):290–300.
[97] Manley S, Ni HM, Kong B, Apte U, Guo G, Ding WX. Suppression of autophagic flux by bile acids in hepatocytes. Toxicol Sci 2014;137(2):478–90 [PMCID: PMC3908720].
[98] Kretschmer XC, Baldwin WS. CAR and PXR: xenosensors of endocrine disrupters? Chem Biol Interact 2005;155(3):111–28.
[99] Staudinger J, Liu Y, Madan A, Habeebu S, Klaassen CD. Coordinate regulation of xenobiotic

and bile acid homeostasis by pregnane X receptor. Drug Metab Dispos 2001;29(11):1467–72.
[100] Dai G, He L, Bu P, Wan YJ. Pregnane X receptor is essential for normal progression of liver regeneration. Hepatology 2008;47(4):1277–87.
[101] Lo SG, Celli N, Caboni M, Murzilli S, Salvatore L, Morgano A, et al. Down-regulation of the LXR transcriptome provides the requisite cholesterol levels to proliferating hepatocytes. Hepatology 2010;51: 1334–44.
[102] Mitro N, Mak PA, Vargas L, Godio C, Hampton E, Molteni V, et al. The nuclear receptor LXR is a glucose sensor. Nature 2007;445(7124):219–23.
[103] Wellen KE, Hatzivassiliou G, Sachdeva UM, Bui TV, Cross JR, Thompson CB. ATP-citrate lyase links cellular metabolism to histone acetylation. Science 2009;324 (5930):1076–80 [PMCID:PMC2746744].
[104] Rathmell JC, Newgard CB. Biochemistry. A glucose-to-gene link. Science 2009;324(5930):1021–2 [PMCID: PMC2788238].
[105] Huang J, Schriefer AE, Yang W, Cliften PF, Rudnick DA. Identification of an epigenetic signature of early mouse liver regeneration that is disrupted by Zn-HDAC inhibition. Epigenetics 2014;9(11):1521–31.
[106] Huang J, Barr E, Rudnick DA. Characterization of the regulation and function of zinc-dependent histone deacetylases during mouse liver regeneration. Hepatology 2013;57:1742–51.
[107] Xia J, Zhou Y, Ji H, Wang Y, Wu Q, Bao J, et al. Loss of histone deacetylases 1 and 2 in hepatocytes impairs murine liver regeneration through Ki67 depletion. Hepatology 2013;58(6):2089–98.
[108] Mihaylova MM, Vasquez DS, Ravnskjaer K, Denechaud PD, Yu RT, Alvarez JG, et al. Class IIa histone deacetylases are hormone-activated regulators of FOXO and mammalian glucose homeostasis. Cell 2011;145(4):607–21 [PMCID:PMC3117637].
[109] Fischle W, Dequiedt F, Hendzel MJ, Guenther MG, Lazar MA, Voelter W, et al. Enzymatic activity associated with class II HDACs is dependent on a multiprotein complex containing HDAC3 and SMRT/N-CoR. Mol Cell 2002;9(1):45–57.
[110] Knutson SK, Chyla BJ, Amann JM, Bhaskara S, Huppert SS, Hiebert SW. Liver-specific deletion of histone deacetylase 3 disrupts metabolic transcriptional networks. EMBO J 2008;27(7):1017–28 [PMCID: PMC2323257].
[111] Li D, Wang X, Ren W, Ren J, Lan X, Wang F, et al. High expression of liver histone deacetylase 3 contributes to high-fat-diet-induced metabolic syndrome by suppressing the PPAR-gamma and LXR-alpha-pathways in E3 rats. Mol Cell Endocrinol 2011;344 (1-2):69–80.
[112] Shimazu T, Hirschey MD, Newman J, He W, Shirakawa K, Le Moan N, et al. Suppression of oxidative stress by beta-hydroxybutyrate, an endogenous histone deacetylase inhibitor. Science 2013;339(6116): 211–4.
[113] Jin J, Iakova P, Jiang Y, Medrano EE, Timchenko NA. The reduction of SIRT1 in livers of old mice leads to impaired body homeostasis and to inhibition of liver proliferation. Hepatology 2011;54(3):989–98 [PMCID: PMC3242815].
[114] Garcia-Rodriguez JL, Barbier-Torres L, Fernandez-Alvarez S, Juan VG, Monte MJ, Halilbasic E, et al. SIRT1 controls liver regeneration by regulating BA metabolism through FXR and mTOR signaling. Hepatology 2013;59 (5):1972–83 [PMCID:PMC3999184].
[115] Ferris GM, Clark JB. Nicotinamide nucleotide synthesis in regenerating rat liver. Biochem J 1971;121 (4):655–62 [PMCID:PMC1176642].
[116] Liao Y, Shikapwashya ON, Shteyer E, Dieckgraefe BK, Hruz PW, Rudnick DA. Delayed hepatocellular mitotic progression and impaired liver regeneration in early growth response-1-deficient mice. J Biol Chem 2004;279(41):43107–16.
[117] Kele PG, van der Jagt EJ, Gouw AS, Lisman T, Porte RJ, de Boer MT. The impact of hepatic steatosis on liver regeneration after partial hepatectomy. Liver Int 2013;33(3):469–75.
[118] de Meijer VE, Kalish BT, Puder M, Ijzermans JN. Systematic review and meta-analysis of steatosis as a risk factor in major hepatic resection. Br J Surg 2010;97 (9):1331–9.
[119] Bucher NL, GLINOS AD. The effect of age on regeneration of rat liver. Cancer Res 1950;10(5):324–32.
[120] Timchenko NA. Aging and liver regeneration. Trends Endocrinol Metab 2009;20(4):171–6.
[121] Murali B, Korrapati MC, Warbritton A, Latendresse JR, Mehendale HM. Tolerance of aged Fischer 344 rats against chlordecone-amplified carbon tetrachloride toxicity. Mech Ageing Dev 2004;125(6):421–35.
[122] Ledda-Columbano GM, Pibiri M, Cossu C, Molotzu F, Locker J, Columbano A. Aging does not reduce the hepatocyte proliferative response of mice to the primary mitogen TCPOBOP. Hepatology 2004;40(4):981–8.
[123] Gaub J, Iversen J. Rat liver regeneration after 90% partial hepatectomy. Hepatology 1984;4(5):902–4.
[124] Lee WM. Acute liver failure in the United States. Semin Liver Dis 2003;23(3):217–26.

CHAPTER

16

Liver Regeneration: The Biliary Perspective

Debolina Ray[*,#], *Yuyan Han*[*,#], *Luca Maroni*[†], *Marco Marzioni*[†], *Fanyin Meng*[‡,§], *Gianfranco Alpini*[*,‡,§], *Shannon Glaser*[*,‡,§]

[*]Department of Internal Medicine, College of Medicine, Texas A&M University Health Science Center, Temple, Texas USA

[†]Clinic of Gastroenterology, Università Politecnica delle Marche, "Ospedali Riuniti" University Hospital, Ancona, Italy

[‡]Division of Research, Central Texas Veterans Health Care System, Temple, Texas, USA

[§]BaylorScott & White Digestive Disease Research Center, BaylorScott & White, Temple, Texas, USA

ABBREVIATIONS

CFTR cystic fibrosis transmembrane conductance regulator
PBG peribiliary gland
PBP peribiliary plexus
PBC primary biliary cirrhosis
PSC primary sclerosing cholangitis

16.1 INTRODUCTION

16.1.1 Biliary Function

The primary function of cholangiocytes is to modify bile of canalicular origin before it reaches the duodenum through a series of secretory and absorptive events [1–4]. One of the most important and well-studied functions of cholangiocytes is secretin-induced release of bicarbonate into the bile [5]. Under physiological conditions, bile secretion occurs in response to meal intake. The acidic pH caused by food promotes secretion of secretin from the S cells into the venous circulation. Secretin receptor on large cholangiocytes (the only cell type expressing this receptor) [6,7] leads to extrusion of Cl^- ions out of the epithelial cells, and subsequently, Cl^-/HCO_3^- exchanger on the apical membrane of cholangiocytes exchanges HCO_3^- into the bile driven by the Cl^- gradient [8]. The HCO_3^- released into the bile forms a biliary HCO_3^- umbrella, which is protective

[#] These authors contributed equally to this work.

http://dx.doi.org/10.1016/B978-0-12-420128-6.00016-6

for the biliary tract. This process also favors the digestive process by building up the intestinal luminal pool of HCO_3^- [8–10]. The HCO_3^- exchange occurs via cAMP or IP_3-mediated signaling mechanisms. The Cl^- channel cystic fibrosis transmembrane conductance regulator (CFTR) is regulated by increased intracellular cAMP levels [11]. On the other hand, stimulation of IP_3 levels results in the release of Ca^{2+}, ultimately leading to Cl^-/HCO_3^- exchange into the bile [12,13]. Under basal physiological conditions, Cl^- transport across the apical membrane of cholangiocytes is low which is enhanced by stimulation with cAMP to a significantly greater extent [11,14].

16.1.2 Anatomy and Heterogeneity of the Biliary Tree

The biliary epithelium is composed of three main components: intrahepatic bile ducts, extrahepatic bile ducts, and peribiliary glands (PBGs) [15]. The intrahepatic bile ducts, which originate in the liver at the canals of Hering as small ductules, can be classified as either small or large bile ducts [16–18]. Small bile ducts are generally defined as bile ducts with a diameter of $<15\ \mu m$ and large bile ducts are defined as bile ducts with a diameter $>15\ \mu m$ [1]. The human intrahepatic bile ducts can be classified into large ducts measuring approximately 800 μm in diameter, which subsequently taper off to zonal, septal, and interlobular ducts with a diameter ranging between 15 and 100 μm. The intrahepatic bile ducts end in ductules, which measure less than 15 μm in diameter [19]. The extrahepatic bile ducts include the right and left hepatic ducts and ends as the common bile duct, which drains into the duodenum [15,17]. The extrahepatic bile ducts are lined by high columnar epithelial cells and together with the large intrahepatic bile ducts contain the PBGs [15].

Electron microscopic studies have shown that small and large cholangiocytes have a multilobulated nucleus, numerous vesicles at the apical domain, tight junctions, high density of microvilli and lysosomes [20]. While small cholangiocytes exhibit a high nucleus to cytoplasm ratio, large cholangiocytes display a low nucleus to cytoplasm ratio. The small cholangiocytes, which are activated by Ca^{2+} signaling, are more resistant to insults such as bile duct ligation (BDL) and carbon tetrachloride (CCl_4)-induced liver injury and can repair the biliary epithelium when the large cholangiocytes are damaged [2,21,22]. On the other hand, large cholangiocytes express secretin receptor (SR), CFTR, Cl^-/HCO_3^- AE2 and respond to secretin via cAMP-dependent pathway indicating a state of more differentiated and senescent features. This could help explain why large cholangiocytes are more susceptible to insults and actively respond to different agonists, such as secretin and melatonin [23–25].

A group of peptides/hormones [26] shows stimulatory or inhibitory effects in modulating the function of small and large cholangiocytes. Studies have shown that somatostatin receptors ($SSTR_2$) are expressed only in large cholangiocytes and function to inhibit secretin-induced choleresis [25,27]. NFAT2 and NFAT4 are found to be expressed dominantly in small ducts and stimulate small duct proliferation [28]. Inhibition of NFAT2 or NFAT4 decreased the phenylephrine-induced proliferation in small cholangiocytes. The cytochrome P4502E1, an enzyme that could inhibit CCl_4 hepatotoxicity, is only expressed in large but not small cholangiocytes [21,29].

16.2 CHARACTERISTICS OF BILIARY REGENERATION

Cholangiocytes are quiescent in the normal physiological state but respond to specific types of injuries (e.g., BDL and partial hepatectomy) under experimental conditions as well as in human cholangiopathies by proliferation

[30]. The type of injury activates a specific cholangiocyte population to start to proliferate. In most conditions, proliferation of cholangiocytes forms the main component of ductular reaction [31], although newer ductules could originate from activated and differentiated progenitor cells or from cells that entered from the circulation and differentiate into liver cells [30,32]. Considering the variability in the regenerative potential of cholangiocytes and different categories of biliary epithelial cells undergoing proliferation, the proliferative processes have been categorized in literature [27].

16.2.1 Type I or "Typical" Cholangiocyte Proliferation

This results in increased number of intrahepatic bile ducts arising from older ducts confined to portal spaces. Evidence supporting this concept comes from studies on rat BDL model (one of the established *in vivo* models for studying Type I proliferation). In these studies, it was observed that the proliferating ducts retained immunohistochemical and functional characteristics of normal cholangiocytes [30]. Further, thymidine labeling indicated proliferation from pre-existing ducts instead of evolution of brand new ducts [33]. Typical cholangiocyte proliferation is also observed under several other experimental conditions such as acute CCl_4 treatment, feeding of α-napthylisothiocyanate as well as bile salts and 70% partial hepatectomy. Studies in rat models have demonstrated preferential proliferation of either large or small bile ducts depending on the type of liver injury applied. For example, BDL induces the proliferation of large bile ducts, whereas partial hepatectomy results in the proliferation of both large (cAMP-dependent) and small (Ca^{2+}-dependent) cholangiocytes [2,34]. On the other hand, CCl_4 administration is followed by the *de novo* proliferation of small cholangiocytes to compensate for the damage of large bile ducts. Contrary to the classic definition of typical proliferation, a study has shown that cholangiocyte hyperplasia can also be associated with sprouting of new side branches rather than enlargement of pre-existing ducts [35]. Additionally, some studies have shown the expression of progenitor cell-type markers in proliferating cholangiocytes, which opened up the possibility of the involvement of the stem cell compartment in typical cholangiocyte proliferation [36].

16.2.2 Type II or "Atypical" Cholangiocyte Proliferation

This type of proliferation occurs in chronic liver injury such as chronic administration of D-galactosamine [37] or CCl_4 in rodents [21]; alcoholic liver diseases, chronic extrahepatic biliary obstruction, and cholestatic liver diseases in humans (PBC, PSC) [38]. Studies have shown that Type II proliferation results from transformation of hepatic liver cell cords implicating a "retrodifferentiation" of hepatocytes into cholangiocytes. This atypical proliferation is characterized by irregular proliferation of intrahepatic bile ducts that are not confined to portal areas and spread into periportal and parenchymal regions creating an irregular portal-parenchymal interface. These random structures had poorly developed lumens leading to edema and neutrophil infiltration. In addition, ultrastructural studies in human liver sections from primary biliary cirrhosis (PBC) and primary sclerosing cholangitis (PSC) patients identified this population of cells, which represented characteristics of both hepatocytes and cholangiocytes [39]. Similarly, in immunohistochemical studies of liver sections from patients with hepatitis, alcohol injury, PBC and PSC, proliferating hepatic parenchymal cells with intermediate (hepatocytic/biliary) morphological features and combined immuno-phenotype were observed [32]. These studies suggest that atypical proliferation can occur from a combination of both pre-existing ductule proliferation as well as a result of the expansion of the "hepatic progenitor cell" compartment [30,32,40]. In a particular study, a 3D culture system of a liver progenitor cell line developed cholangiocyte features *in vitro* [41]. Similarly, a recent study has shown that induction of progenitor stem cell properties via forced expression of *Hnf1β* and *Foxa3* in mouse embryonic fibroblasts gives rise to a bipotent population of stem cells similar to hepatic stem cells. These cells retain their capacity of self-renewal and differentiation into both hepatocytes and cholangiocytes [42].

There are also studies indicating that post-injury of large bile ducts, the small cholangiocytes compensate for the damage of large bile ducts by initiating proliferation (by activation of Ca^{2+}-dependent signaling) [12,13] in addition to acquiring large cholangiocyte features *de novo*. For example, after damage of large cholangiocytes by administration of GABA or CCl_4, small cholangiocytes in BDL rodents *de novo* express SR, CFTR, and AE2 and secrete water and electrolytes to compensate for large biliary damage [43,44].

16.2.3 Type III or Oval Cell Proliferation

The oval cell proliferation is the first step toward carcinogenesis in rat liver [45,46]. Oval cells constitute a subpopulation of nonparenchymal cells that express markers such as α-fetoprotein, albumin, and glucose-6-phospahatase [47]. These cells have the capability to differentiate into hepatocytes, cholangiocytes as well as exocrine pancreatic cells [46,48]. This type of proliferation differs from typical cholangiocyte proliferation with respect to duct structure, which is disorganized and randomly spread into hepatic lobules. This leads to a distorted hepatic architecture [47].

16.2.4 Factors Involved in Cholangiocyte Proliferation

Several hormones and peptides regulate biliary mass during liver injury. While some of these factors could promote cholangiocyte proliferation, others display inhibitory properties. The signaling mechanisms associated with regulation of cholangiocyte proliferation by individual factors are mediated by either cAMP or IP_3 signaling and may affect either large or small cholangiocyte proliferation. Some of the major factors that regulate biliary mass are discussed below.

16.2.4.1 Gastrointestinal Hormones

The autocrine role of secretin in the regulation of large cholangiocyte proliferation has been recently elucidated in BDL rats where we observed enhanced expression of secretin in cholangiocytes, which promotes biliary hyperplasia [49]. Apart from this, the paracrine role of secretin has been described by other studies [3,50]. The inhibitory role of gastrin and somatostatin on BDL-induced cholangiocyte hyperplasia occurs by downregulation of secretin-stimulated cAMP signaling [25,51].

16.2.4.2 Angiogenic Factors

Angiogenic factors such as VEGF-A and VEGF-C have been shown to mediate pro-proliferative actions on cholangiocytes both by autocrine and paracrine mechanisms by binding to VEGFR2 and VEGFR3. For example, immunoneutralization of VEGF-A and VEGF-C in BDL rodents decreased biliary proliferation [52]. The concept of the peribiliary plexus's (PBP) (source of VEGFs) role in modulating cholangiocyte proliferation came from a study where blockage of the hepatic artery by ligation leads to the disappearance of the PBP, diminished biliary growth associated with enhanced apoptosis, and reduced secretion of VEGF by cholangiocytes [53].

16.2.4.3 Bile Acids

There are several factors that maintain biliary homeostasis by both autocrine and paracrine pathways. Bile acids have been found to activate nuclear receptors (mainly farnesoid X-receptor or FXR), the G-protein-coupled receptor TGR5 and cell signaling pathways including c-jun JNK 1/2, AKT, and ERK 1/2 in cells of the liver and gastrointestinal tract [54]. Studies in BDL rat models have shown that increased biliary bile acids promote cholangiocyte proliferation and secretion, which is blocked by wortmannin [55]. More recent studies have shown ursodeoxycholic acid to be of therapeutic potential against biliary diseases [56]. It has been shown that mice with hepatic knockout of the primary bile acid-sensing receptor FXR have delayed liver regeneration after partial hepatectomy [57]. Similarly, there was delayed liver repair following CCl_4-induced liver injury in $FXR^{-/-}$ mice compared to the control mice due to decreased regenerative DNA synthesis and expression of liver repair genes [58].

16.2.4.4 Neurotransmitters

Studies looking at the regulation of cholangiocyte growth/loss during biliary diseases by neurotransmitters have been performed in rodent models of liver damage such as BDL and partial hepatectomy. A study related to the role of opioids in cholangiocyte proliferation has shown that their increased expression during cholestasis prevented growth of cholangiocytes [59]. Serotonin expressed by cholangiocytes reduces biliary proliferation during the course of cholestasis in an autocrine fashion [60]. Other neurotransmitters (such as acetylcholine) act via M3-ACh receptors to potentiate the effect of secretin, which in turn is responsible for cholangiocyte proliferation during cholestasis [61]. Apart from this, GABA-induced liver damage is accompanied by the repair of the biliary tree by small cholangiocytes by *de novo* acquisition of large cholangiocyte features [43]. In addition, Caveolin-1 plays an important role in aiding liver regeneration after partial hepatectomy and mice with disrupted *Caveolin* gene display impaired repair process [62].

16.2.4.5 Inflammatory Cytokines

Emerging evidence has shown that proinflammatory and inflammatory cytokines play a critical role in the initiation and progression of liver regeneration. During the cholangiocyte regeneration, they also play a critical role in regulating cholangiocyte repopulation. During the repopulation of biliary tree, we have demonstrated that the expression of stem cell factor (SCF) and granulocyte-macrophage colony stimulating factor (GM-CSF) increased in total liver and isolated rat cholangiocytes until 7 days after partial hepatectomy. With the stimulation of SCF and GM-CSF, the expression of the remodeling molecule, S100A4, and miR-181b as well as the fibrosis-related molecules matrix metalloproteinases (MMP-2 and MMP-9) was elevated in small but not in large murine cholangiocytes. In addition, the study showed that TGF-β treatment stimulates the secretion of SCF and GM-CSF in small murine cholangiocytes. The upregulation of miR-181b, MMP-2, and MMP-9 was also shown in isolated rat cholangiocytes [63]. All together, the repopulation of cholangiocytes during liver regeneration was found to be related to the TGFβ signaling pathway that could stimulate the secretion of SCF and GM-CSF leading to the enhanced expression of miR-181b and MMP-2, MMP-9, which could potentially increase biliary fibrosis.

IL-6 is a key factor for liver regeneration after partial hepatectomy [64,65]. Although the DNA-synthesis response seems normal in non-parenchymal cells from IL-$6^{-/-}$ mice, another *in vitro* study has demonstrated that mouse cholangiocytes are able to produce, secrete, and respond to IL-6 via upregulation and activation of the IL-6 receptors in autocrine and paracrine mechanisms [66].

16.2.4.6 microRNAs

microRNAs (miRNAs) are a family of 21-25 nucleotide small noncoding RNAs that can bind to the mRNAs to regulate gene expression in a sequence-specific manner. They exert their function by base pairing to partially or completely complementary sites of the target genes and inhibiting their translation. miRNAs are well conserved between different species. Just like miRNA plays important roles in normal condition, their dysregulation is found in various conditions including liver regeneration [67]. Studies on the involvement of microRNA began 2010 [68]. miR-21 was found to be upregulated during the early stage of liver regeneration by targeting Pellino 1. Let-7 family, miR-127, miR-26a, miR-34a, and miR-23b were all found dysregulated during liver regeneration [67]. Specifically, the combination of cytokine SCF and GM-CSF with miR-181 decreased TIMP-3 expression.

16.2.4.7 Other Factors

Although the exact roles of Notch and Jagg-1 in biliary regeneration remain to be elucidated, Notch and Jagg-1 were found to be expressed in cholangiocytes, where they may play a role in the differentiation and/or proliferation of cholangiocytes [69].

In adult liver, Ep-CAM was found to be absent in hepatocytes but present in cholangiocytes. During regeneration, Ep-CAM was activated concomitant with ductular proliferation [70].

16.2.5 Contribution of the PBGs to Biliary Regeneration

The presence of glandular elements, called PBGs, around large intrahepatic and extrahepatic bile ducts, has been known for years [71,72]. PBGs of the intrahepatic bile ducts are generally divided into intramural or extramural with respect to the bile duct wall. The intramural glands, composed by mucinous elements, are present in the fibrous duct wall and are directly connected to the bile duct lumen [73]. The extramural glands, on the contrary, expand in the periductal tissue and are thought to connect to the bile duct lumen through their own conduits [73]. These glands are composed by variable proportions of serous and mucinous acini and, in addition, by sparse pancreatic exocrine cells [74]. PBGs are also present along the extrahepatic bile duct, where they form groups of tubulo-alveolar glands within the wall of the duct [72]. Interestingly, PBGs undergo a number of morphological alterations during the course of human diseases of the liver [75]. Necroinflammation and formation of cystic structures have been described in association with biliary obstruction with bacterial infection and in portal hypertension, respectively [74]. Moreover, hyperplasia of PBGs is present in a variety of liver diseases, such as cirrhosis, submassive hepatic necrosis, cholangitis, systemic infection, and extrahepatic biliary obstruction. The possible subsequent alterations of bile flow and viscosity have been proposed as predisposing factors for hepatolithiasis [76]. Pathophysiological functions of PBGs have been poorly studied. Classically, intrahepatic PBGs have been suggested to participate in mucin and enzymatic secretion, while extrahepatic PBGs were thought to have mainly absorptive functions [73,77]. In recent years, however, PBGs have again attracted the interest of researchers as a source of biliary tree stem cells. Carpino et al. recognized a classic phenotypic trait of stem/progenitor cells in human PBGs [78]. The authors evaluated the expression in PBGs within the several markers of definitive endoderm, such as FOXa2, OCT4, and CXCR4, [79] and of pancreatobiliary precursors, such as SOX17 and PDX1 [80]. They found that about 10% of cells within the PBGs coexpress PDX1, SOX17, and FOXa2, therefore showing phenotypic traits consistent with a stem cell population. Interestingly, the organization of the stem cells in PBG closely resembled the structure of the intestinal stem cell niche. The markers for the most undifferentiated cells (Ep-CAM$^{+/-}$, PDX1^{+}, SOX17^{+}, Lgr5^{+}) were indeed found at the base, while transit-amplifying progenitors (Ep-CAM^{+}, PDX1$^{+/-}$, SOX17$^{+/-}$, Lgr5^{-}) concentrated in the body of the glands. As a confirmation, the highest number of PCNA-positive cells was located at the bottom of the acini, with a progressive decrease toward the lumen of the duct [78]. The specific mature subpopulation derived from the stem cells of PBGs depends on the localization of the gland along the biliary tree. A relative predominance of cholangiocyte-specific markers, such as the SR, was found approximately in the middle of the extrahepatic biliary tree. Expression of albumin, indicative of a hepatocyte-like differentiation, was present proximally, while the glands close to the hepatopancreatic ampulla showed a pancreatic differentiation and were occasionally positive for insulin. In a parallel work of the same group, Cardinale et al. showed that cell suspensions obtained from different regions of the biliary tree can effectively replicate in a serum-free medium specifically formulated for endodermal progenitors [81]. The colonies formed on plastic cultures expressed early endodermal transcription factors, such as PDX1 and SOX17, and were consistently negative for a variety of markers for mesenchymal, vascular, and hematopoietic cells. Moreover, by varying the culturing conditions of the progenitor cell colonies, the authors could induce them to differentiate *in vitro* into hepatocytes, cholangiocytes, and pancreatic islet-like structures [81]. As a proof of principle, transplantation of human stem cells of PBG in livers of SCID mice or diabetic mice resulted in the presence of human hepatocytes and cholangiocytes or in the correction of hyperglycemia, respectively [81].

The identification of stem cells in the human biliary tree has multiple and important implications. The human fetal biliary tree contains a population of mutipotent stem cells that closely resembles that of adults [82]. The proliferation and differentiation of human biliary tree stem cells may therefore contribute to ongoing organogenesis of the liver, biliary epithelium, and pancreas in physiological conditions [83]. Moreover, the activation of the PBG stem cell compartment could play essential roles in a number of pathological states that are characterized by cholangiocyte proliferation. PBGs of the extrahepatic bile duct of patients with severe ischemic-type biliary lesions showed a marked proliferation, assessed by Ki-67 staining, as compared to controls [84]. In cases of mild epithelial damage caused by cholangitis, Ki-67^{+} cells were present at the basal layer of the duct, and only sporadically in PBGs, suggesting that the latter only activate in case of severe epithelial loss [84]. DiPaola et al. developed a new technique to perform immunostaining in mice while preserving the anatomy of the extrahepatic bile duct. Using double immunofluorescence for CK-19 and BrdU, the authors demonstrated that PBGs proliferate in response to viral-induced injury or obstructive cholestasis [85]. At last, stem cells of PBGs have been regarded as a possible site of origin of mucin-producing cholangiocarcinoma [83]. In particular, PBGs may be involved in the origin of the intraductal papillary neoplasm of the bile duct, an early neoplastic lesion of cholangiocarcinoma [86,87].

To conclude, the recent identification of a stem cell compartment in PBGs opens novel opportunities to expand the current knowledge on various pathophysiological processes involving cholangiocyte biology. Biliary tree stem cells may not only prove essential in ongoing cellular turnover or cancer development but also take part in the regeneration of the liver, biliary tree, and pancreas, with possible future applications in regenerative medicine [88] (Figure 16.1).

16.3 CONCLUSIONS AND HUMAN PERSPECTIVES

The human biliary epithelium is the target of heterogeneous group of genetic and acquired diseases called cholangiopathies [89]. Among them, PSC and PBC are the prototype of immune-mediated biliary disorders. Despite being originally selective for the biliary epithelium, most of the cholangiopathies slowly but constantly progress, and liver cirrhosis often develops as a consequence of chronic inflammation [89,90]. To date, treatment options for cholangiopathies and cirrhosis are exceptionally limited. Liver transplantation is the sole curative approach for end-stage liver diseases but its applicability is complicated by shortage of liver donors and by important side effects of the life-long immunosuppression therapy [91,92]. In this context, the studies of the mechanisms that drive biliary proliferation and regeneration in response to damage appear particularly attractive under different points of view. The neocholangiogenesis observed in cases of biliary injury is thought to represent an attempt to regenerate the damaged bile ducts [30]. However, the reactive proliferation of cholangiocytes often fails to generate functional bile ducts and is one of the major factors responsible for fibrogenesis [30,93]. Molecular therapies directed at sustaining a more physiological proliferation and differentiation of reactive cholangiocytes are therefore likely to be tested. To this extent, we have recently proposed that the fine-tuning of the Notch\Hes-1 pathway, which favors biliary differentiation, and of PDX-1\Ngn-3, which is responsible for proliferation, may prove beneficial in the treatment of cholangiopathies [94–96].

The potential of stem/progenitors cells located in the canal of Hering in replenishing both cholangiocytes and hepatocytes in a variety of liver diseases opens new possibilities to develop therapeutic strategy in order to modulate *in situ* the regeneration of the liver [97]. Of note, recent evidence has showed that in Zebrafish mature biliary cells too can transdifferentiate into hepatocytes during extreme liver damage [98,99].

Additionally, the regenerative capacities of biliary stem cells may prove useful as a way to generate hepatocyte to be used for transplantation [100]. So far, the accessibility to a source of human hepatic stem cells has been limited because available livers are usually used for transplantation. The recent identification of stem cells in PBGs opens novel opportunities not only to expand the current knowledge on various pathophysiological processes involving cholangiocyte biology but also as a more accessible source of stem cells [88].

Further studies are needed to clarify the complex molecular pathways and the cellular interactions that govern the enormous regenerative capacity of the hepatobiliary system. These studies will undoubtedly provide essential insight in both the physiology and the pathology of the biliary tree, and possibly open doors to new effective therapeutic options to treat its diseases.

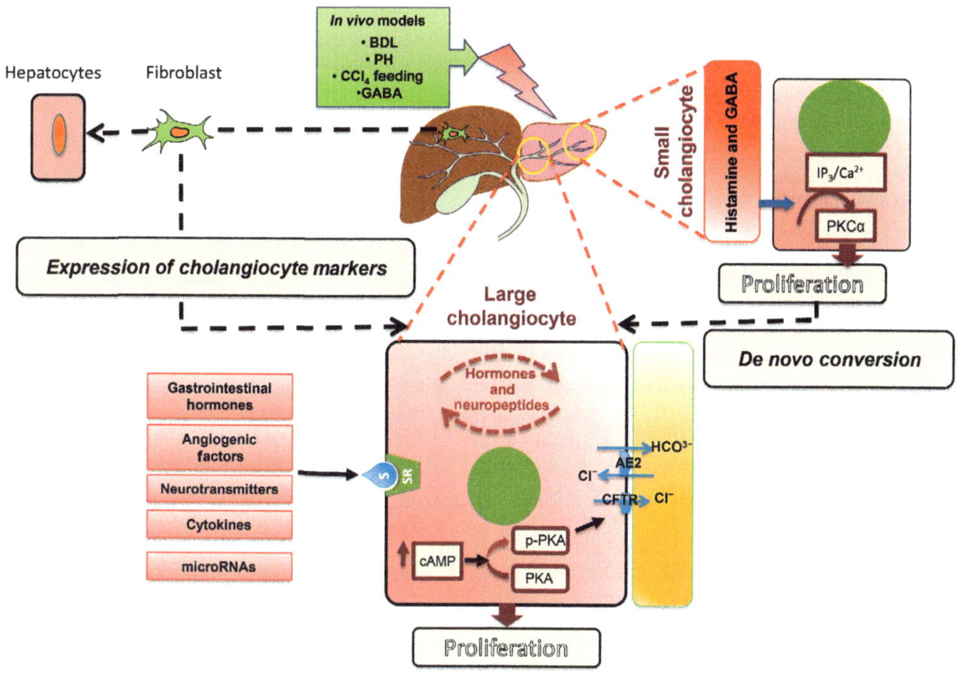

FIGURE 16.1 Overview of biliary regeneration. Intrahepatic bile ducts are lined by both large and small cholangiocytes. Liver injury to *in vivo* animal models initiated by bile duct ligation (BDL), partial hepatectomy (PH), feeding of CCl_4, and administration of GABA promotes biliary proliferation. Under physiological conditions, cholangiocytes (large cholangiocytes preferentially) modify ductal bile by a sequence of secretory and absorptive processes mediated by membrane exchangers, AE2 and CFTR. This modification mainly leads to the formation of bicarbonate rich bile. This process is enhanced by stimulation with secretin and cAMP, which increase in response to liver insult. Liver behaves as a neuroendocrine compartment in response to injury and starts to respond to hormones and peptides in an autocrine as well as paracrine manner. Liver injury is subsequently followed by large cholangiocyte proliferation under the influence of these factors (neurotransmitters, gastrointestinal peptides, steroids). Occasionally, in response to specific events such as administration of histamine, small cholangiocytes proliferate by an IP_3-mediated signaling pathway. Agents such as GABA promotes *de novo* conversion of small cholangiocytes into large. Liver also harbors hepatofibroblasts or hepatic progenitor cells which can give rise to both hepatocytes and cholangiocytes by expression of specific markers in response to injury.

Acknowledgments

This work was supported by the Dr. Nicholas C. Hightower Centennial Chair of Gastroenterology from Scott & White Hospital, a VA Research Career Scientist Award, a VA Merit award to Dr. Alpini, a VA Merit Award to Dr. Glaser, the NIH Grants DK58411 and DK07698 to Drs. Alpini and Glaser, by a MIUR Grant PRIN 2009—Prot. 2009X84L84_003 and Ministero della Salute Grant GR-2010-2306996 to Dr. Marzioni. This material is the result of work supported by resources at the Central Texas Veterans Health Care System. The views expressed in this article are those of the authors and do not necessarily represent the views of the Department of Veterans Affairs.

References

[1] Alpini G, Glaser S, Robertson W, Rodgers RE, Phinizy JL, Lasater J, et al. Large but not small intrahepatic bile ducts are involved in secretin-regulated ductal bile secretion. Am J Physiol Gastrointest Liver Physiol 1997;272:G1064–74.

[2] Alpini G, Glaser SS, Ueno Y, Pham L, Podila PV, Caligiuri A, et al. Heterogeneity of the proliferative capacity of rat cholangiocytes after bile duct ligation. Am J Physiol Gastrointest Liver Physiol 1998;274:G767–75.

[3] Alpini G, Lenzi R, Sarkozi L, Tavoloni N. Biliary physiology in rats with bile ductular cell hyperplasia. Evidence for a secretory function of proliferated bile ductules. J Clin Invest 1988;81:569–78.

[4] Alpini G, Lenzi R, Zhai WR, Slott PA, Liu MH, Sarkozi L, et al. Bile secretory function of intrahepatic biliary epithelium in the rat. Am J Physiol 1989;257: G124–33.
[5] Ulrich Ii CD, Holtmann M, Miller LJ. Secretin and vasoactive intestinal peptide receptors: members of a unique family of G protein-coupled receptors. Gastroenterology 1998;114:382–97.
[6] Alpini G, Roberts S, Kuntz SM, Ueno Y, Gubba S, Podila PV, et al. Morphological, molecular, and functional heterogeneity of cholangiocytes from normal rat liver. Gastroenterology 1996;110:1636–43.
[7] Glaser SS, Gaudio E, Rao A, Pierce LM, Onori P, Franchitto A, et al. Morphological and functional heterogeneity of the mouse intrahepatic biliary epithelium. Lab Invest 2009;89:456–69.
[8] Kanno N, LeSage G, Glaser S, Alpini G. Regulation of cholangiocyte bicarbonate secretion. Am J Physiol Gastrointest Liver Physiol 2001;281:G612–25.
[9] Beuers U, Hohenester S, de Buy Wenniger LJ, Kremer AE, Jansen PL, Elferink RP. The biliary HCO (3)(−) umbrella: a unifying hypothesis on pathogenetic and therapeutic aspects of fibrosing cholangiopathies. Hepatology 2010;52:1489–96.
[10] Kanno N, LeSage G, Glaser S, Alvaro D, Alpini G. Functional heterogeneity of the intrahepatic biliary epithelium. Hepatology 2000;31:555–61.
[11] Fitz JG, Basavappa S, McGill J, Melhus O, Cohn JA. Regulation of membrane chloride currents in rat bile duct epithelial cells. J Clin Invest 1993;91:319–28.
[12] Francis H, Glaser S, Demorrow S, Gaudio E, Ueno Y, Venter J, et al. Small mouse cholangiocytes proliferate in response to H1 histamine receptor stimulation by activation of the IP_3/CaMK I/CREB pathway. Am J Physiol Cell Physiol 2008;295:C499–513.
[13] Minagawa N, Ehrlich BE, Nathanson MH. Calcium signaling in cholangiocytes. World J Gastroenterol 2006;12:3466–70.
[14] McGill JM, Basavappa S, Gettys TW, Fitz JG. Secretin activates Cl^- channels in bile duct epithelial cells through a cAMP-dependent mechanism. Am J Physiol Gastrointest Liver Physiol 1994;266:G731–6.
[15] Nakanuma Y. Tutorial review for understanding of cholangiopathy. Int J Hepatol 2012;2012:547840. http://dx.doi.org/10.1155/2012/547840 [Epub 2011 August 17].
[16] Alpini G, Phillips JO, LaRusso NF. The biology of the biliary epithelia. In: Arias I, Boyer JL, Fausto N, Jakoby W, Schachter D, Shafritz DA, editors. The liver: biology & pathobiology. 3rd ed. New York: Raven; 1994. p. 623–53.
[17] Healey Jr. JE, Schroy PC. Anatomy of the biliary ducts within the human liver; analysis of the prevailing pattern of branchings and the major variations of the biliary ducts. AMA Arch Surg 1953;66:599–616.
[18] Schaffner F, Popper H. Electron microscopic studies of normal and proliferated bile ductules. Am J Pathol 1961;38:393–410.
[19] Carruthers JS, Steiner JW. Studies on the fine structure of proliferated bile ductules. I. Changes of cytoarchitecture of biliary epithelial cells. Can Med Assoc J 1961;85:1223–36.
[20] Benedetti A, Bassotti C, Rapino K, Marucci L, Jezequel AM. A morphometric study of the epithelium lining the rat intrahepatic biliary tree. J Hepatol 1996;24:335–42.
[21] LeSage G, Benedetti A, Glaser S, Marucci L, Tretjak Z, Caligiuri A, et al. Acute carbon tetrachloride feeding selectively damages large, but not small, cholangiocytes from normal rat liver. Hepatology 1999;29:307–19.
[22] LeSage G, Glaser S, Marucci L, Benedetti A, Phinizy JL, Rodgers R, et al. Acute carbon tetrachloride feeding induces damage of large but not small cholangiocytes from BDL rat liver. Am J Physiol Gastrointest Liver Physiol 1999;276:G1289–301.
[23] Alpini G, RT P, LaRusso NF. The pathobiology of biliary epithelia. In: Arias I, Boyer JL, Chrisari F, Fausto N, Schachter D, Shafritz D, editors. The liver: biology & pathobiology. 4th ed. Philadephia, PA: Lippincott Williams & Wilkins; 2001. p. 421–35.
[24] Renzi A, Glaser S, Demorrow S, Mancinelli R, Meng F, Franchitto A, et al. Melatonin inhibits cholangiocyte hyperplasia in cholestatic rats by interaction with MT1 but not MT2 melatonin receptors. Am J Physiol Gastrointest Liver Physiol 2011;301: G634–43.
[25] Tietz PS, Alpini G, Pham LD, Larusso NF. Somatostatin inhibits secretin-induced ductal hypercholeresis and exocytosis by cholangiocytes. Am J Physiol Gastrointest Liver Physiol 1995;269:G110–8.
[26] Han Y, Glaser S, Meng F, Francis H, Marzioni M, McDaniel K, et al. Recent advances in the morphological and functional heterogeneity of the biliary epithelium. Exp Biol Med 2013;238:549–65.
[27] Alvaro D, Gigliozzi A, Attili AF. Regulation and deregulation of cholangiocyte proliferation. J Hepatol 2000;33:333–40.
[28] Alpini G, Franchitto A, Demorrow S, Onori P, Gaudio E, Wise C, et al. Activation of alpha(1)-adrenergic receptors stimulate the growth of small mouse cholangiocytes via calcium-dependent activation of nuclear factor of activated T cells 2 and specificity protein 1. Hepatology 2011;53:628–39.
[29] Lakehal F, Wendum D, Barbu V, Becquemont L, Poupon R, Balladur P, et al. Phase I and phase II drug-metabolizing enzymes are expressed and

heterogeneously distributed in the biliary epithelium. Hepatology 1999;30:1498–506.

[30] Alvaro D, Mancino MG, Glaser S, Gaudio E, Marzioni M, Francis H, et al. Proliferating cholangiocytes: a neuroendocrine compartment in the diseased liver. Gastroenterology 2007;132:415–31.

[31] Popper H, Kent G, Stein R. Ductular cell reaction in the liver in hepatic injury. J Mt Sinai Hosp NY 1957;24:551–6.

[32] Tan J, Hytiroglou P, Wieczorek R, Park YN, Thung SN, Arias B, et al. Immunohistochemical evidence for hepatic progenitor cells in liver diseases. Liver 2002;22:365–73.

[33] Grisham JW. A morphologic study of deoxyribonucleic acid synthesis and cell proliferation in regenerating rat liver; autoradiography with thymidine-H3. Cancer Res 1962;22:842–9.

[34] Glaser S, Lam IP, Franchitto A, Gaudio E, Onori P, Chow BK, et al. Knockout of secretin receptor reduces large cholangiocyte hyperplasia in mice with extrahepatic cholestasis induced by bile duct ligation. Hepatology 2010;52:204–14.

[35] Masyuk TV, Ritman EL, LaRusso NF. Quantitative assessment of the rat intrahepatic biliary system by three-dimensional reconstruction. Am J Pathol 2001;158:2079–88.

[36] Omori M, Evarts RP, Omori N, Hu Z, Marsden ER, Thorgeirsson SS. Expression of alpha-fetoprotein and stem cell factor/c-kit system in bile duct ligated young rats. Hepatology 1997;25:1115–22.

[37] Kuhlmann WD, Wurster K. Correlation of histology and alpha 1-fetoprotein resurgence in rat liver regeneration after experimental injury by galactosamine. Virchows Arch A Pathol Anat Histol 1980;387:47–57.

[38] Yamada S, Howe S, Scheuer PJ. Three-dimensional reconstruction of biliary pathways in primary biliary cirrhosis: a computer-assisted study. J Pathol 1987;152:317–23.

[39] De Vos R, Desmet V. Ultrastructural characteristics of novel epithelial cell types identified in human pathologic liver specimens with chronic ductular reaction. Am J Pathol 1992;140:1441–50.

[40] Francis HL, Demorrow S, Franchitto A, Venter JK, Mancinelli RA, White MA, et al. Histamine stimulates the proliferation of small and large cholangiocytes by activation of both IP_3/Ca^{2+} and cAMP-dependent signaling mechanisms. Lab Invest 2012;92:282–94.

[41] Tanimizu N, Miyajima A, Mostov KE. Liver progenitor cells develop cholangiocyte-type epithelial polarity in three-dimensional culture. Mol Biol Cell 2007;18:1472–9.

[42] Yu B, He ZY, You P, Han QW, Xiang D, Chen F, et al. Reprogramming fibroblasts into bipotential hepatic stem cells by defined factors. Cell Stem Cell 2013;13:328–40.

[43] Mancinelli R, Franchitto A, Gaudio E, Onori P, Glaser S, Francis H, et al. After damage of large bile ducts by gamma-aminobutyric acid, small ducts replenish the biliary tree by amplification of calcium-dependent signaling and *de novo* acquisition of large cholangiocyte phenotypes. Am J Pathol 2010;176:1790–800.

[44] Mancinelli R, Franchitto A, Glaser S, Meng F, Onori P, Demorrow S, et al. GABA induces the differentiation of small into large cholangiocytes by activation of $Ca(^{2+})$/CaMK I-dependent adenylyl cyclase 8. Hepatology 2013;58:251–63.

[45] Petropoulos CJ, Yaswen P, Panzica M, Fausto N. Cell lineages in liver carcinogenesis: possible clues from studies of the distribution of alpha-fetoprotein RNA sequences in cell populations isolated from normal, regenerating, and preneoplastic rat livers. Cancer Res 1985;45:5762–8.

[46] Farber E. Similarities in the sequence of early histological changes induced in the liver of the rat by ethionine, 2-acetylamino-fluorene, and 3′-methyl-4-dimethylaminoazobenzene. Cancer Res 1956;16:142–8.

[47] Desmet VJ, van Eyken P, Roskams T. Histopathology of vanishing bile duct diseases. Adv Clin Pathol 1998;2:87–99.

[48] Sirica AE, Cihla HP. Isolation and partial characterizations of oval and hyperplastic bile ductular cell-enriched populations from the livers of carcinogen and noncarcinogen-treated rats. Cancer Res 1984;44:3454–66.

[49] Glaser S, Meng F, Han Y, Onori P, Chow BK, Francis H, et al. Secretin stimulates biliary cell proliferation by regulating expression of microRNA 125b and microRNA let7a in mice. Gastroenterology 2014;146:1795–808, e12.

[50] Alvaro D, Cho WK, Mennone A, Boyer JL. Effect of secretion on intracellular pH regulation in isolated rat bile duct epithelial cells. J Clin Invest 1993;92:1314–25.

[51] Glaser SS, Rodgers RE, Phinizy JL, Robertson WE, Lasater J, Caligiuri A, et al. Gastrin inhibits secretin-induced ductal secretion by interaction with specific receptors on rat cholangiocytes. Am J Physiol Gastrointest Liver Physiol 1997;273:G1061–70.

[52] Gaudio E, Barbaro B, Alvaro D, Glaser S, Francis H, Ueno Y, et al. Vascular endothelial growth factor stimulates rat cholangiocyte proliferation via an autocrine mechanism. Gastroenterology 2006;130:1270–82.

[53] Gaudio E, Barbaro B, Alvaro D, Glaser S, Francis H, Franchitto A, et al. Administration of r-VEGF-A

prevents hepatic artery ligation-induced bile duct damage in bile duct ligated rats. Am J Physiol Gastrointest Liver Physiol 2006;291:G307–17.

[54] Hylemon PB, Zhou H, Pandak WM, Ren S, Gil G, Dent P. Bile acids as regulatory molecules. J Lipid Res 2009;50:1509–20.

[55] Alpini G, Glaser S, Alvaro D, Ueno Y, Marzioni M, Francis H, et al. Bile acid depletion and repletion regulate cholangiocyte growth and secretion by a phosphatidylinositol 3-kinase-dependent pathway in rats. Gastroenterology 2002;123:1226–37.

[56] Poupon R. Ursodeoxycholic acid and bile-acid mimetics as therapeutic agents for cholestatic liver diseases: an overview of their mechanisms of action. Clin Res Hepatol Gastroenterol 2012;36(Suppl. 1): S3–S12.

[57] Borude P, Edwards G, Walesky C, Li F, Ma X, Kong B, et al. Hepatocyte-specific deletion of farnesoid X receptor delays but does not inhibit liver regeneration after partial hepatectomy in mice. Hepatology 2012;56:2344–52.

[58] Meng Z, Wang Y, Wang L, Jin W, Liu N, Pan H, et al. FXR regulates liver repair after CCl_4-induced toxic injury. Mol Endocrinol 2010;24:886–97.

[59] Marzioni M, Alpini G, Saccomanno S, de Minicis S, Glaser S, Francis H, et al. Endogenous opioids modulate the growth of the biliary tree in the course of cholestasis. Gastroenterology 2006;130:1831–47.

[60] Marzioni M, Glaser S, Francis H, Marucci L, Benedetti A, Alvaro D, et al. Autocrine/paracrine regulation of the growth of the biliary tree by the neuroendocrine hormone serotonin. Gastroenterology 2005;128:121–37.

[61] LeSag EG, Alvaro D, Benedetti A, Glaser S, Marucci L, Baiocchi L, et al. Cholinergic system modulates growth, apoptosis, and secretion of cholangiocytes from bile duct-ligated rats. Gastroenterology 1999;117:191–9.

[62] Fernandez MA, Albor C, Ingelmo-Torres M, Nixon SJ, Ferguson C, Kurzchalia T, et al. Caveolin-1 is essential for liver regeneration. Science 2006;313:1628–32.

[63] Meng F, Francis H, Glaser S, Han Y, DeMorrow S, Stokes A, et al. Role of stem cell factor and granulocyte colony-stimulating factor in remodeling during liver regeneration. Hepatology 2012;55:209–21.

[64] Cressman DE, Greenbaum LE, DeAngelis RA, Ciliberto G, Furth EE, Poli V, et al. Liver failure and defective hepatocyte regeneration in interleukin-6-deficient mice. Science 1996;274:1379–83.

[65] Taub R. Liver regeneration: from myth to mechanism. Nat Rev Mol Cell Biol 2004;5:836–47.

[66] Yokomuro S, Lunz 3rd JG, Sakamoto T, Ezure T, Murase N, Demetris AJ. The effect of interleukin-6 (IL-6)/gp130 signalling on biliary epithelial cell growth, *in vitro*. Cytokine 2000;12:727–30.

[67] Finch ML, Marquardt JU, Yeoh GC, Callus BA. Regulation of microRNAs and their role in liver development, regeneration and disease. Int J Biochem Cell Biol 2014;54:288–303.

[68] Marquez RT, Wendlandt E, Galle CS, Keck K, McCaffrey AP. MicroRNA-21 is upregulated during the proliferative phase of liver regeneration, targets Pellino-1, and inhibits NF-kappaB signaling. Am J Physiol Gastrointest Liver Physiol 2010;298:G535–41.

[69] Kohler C, Bell AW, Bowen WC, Monga SP, Fleig W, Michalopoulos GK. Expression of Notch-1 and its ligand Jagged-1 in rat liver during liver regeneration. Hepatology 2004;39:1056–65.

[70] de Boer CJ, van Krieken JH, Janssen-van Rhijn CM, Litvinov SV. Expression of Ep-CAM in normal, regenerating, metaplastic, and neoplastic liver. J Pathol 1999;188:201–6.

[71] Masuko K, Rubin E, Popper H. Proliferation of bile ducts in cirrhosis. Arch Pathol 1964;78:421–31.

[72] Spitz L, Petropoulos A. The development of the glands of the common bile duct. J Pathol 1979;128: 213–20.

[73] Nakanuma Y, Hoso M, Sanzen T, Sasaki M. Microstructure and development of the normal and pathologic biliary tract in humans, including blood supply. Microsc Res Tech 1997;38:552–70.

[74] Terada T, Nakanuma Y. Pathological observations of intrahepatic peribiliary glands in 1,000 consecutive autopsy livers. II. A possible source of cholangiocarcinoma. Hepatology 1990;12:92–7.

[75] Nakanuma Y, Katayanagi K, Terada T, Saito K. Intrahepatic peribiliary glands of humans. I. Anatomy, development and presumed functions. J Gastroenterol Hepatol 1994;9:75–9.

[76] Terada T, Nakanuma Y. Pathologic observations of intrahepatic peribiliary glands in 1,000 consecutive autopsy livers: IV. Hyperplasia of intramural and extramural glands. Hum Pathol 1992;23:483–90.

[77] Nakanuma Y, Sasaki M, Terada T, Harada K. Intrahepatic peribiliary glands of humans. II. Pathological spectrum. J Gastroenterol Hepatol 1994;9:80–6.

[78] Carpino G, Cardinale V, Onori P, Franchitto A, Berloco PB, Rossi M, et al. Biliary tree stem/progenitor cells in glands of extrahepatic and intraheptic bile ducts: an anatomical in situ study yielding evidence of maturational lineages. J Anat 2012;220:186–99.

[79] Zaret KS, Grompe M. Generation and regeneration of cells of the liver and pancreas. Science 2008;322:1490–4.

[80] Spence JR, Lange AW, Lin SC, Kaestner KH, Lowy AM, Kim I, et al. Sox17 regulates organ lineage segregation of ventral foregut progenitor cells. Dev Cell 2009;17:62–74.

[81] Cardinale V, Wang Y, Carpino G, Cui CB, Gatto M, Rossi M, et al. Multipotent stem/progenitor cells in

human biliary tree give rise to hepatocytes, cholangiocytes, and pancreatic islets. Hepatology 2011;54: 2159–72.

[82] Semeraro R, Carpino G, Cardinale V, Onori P, Gentile R, Cantafora A, et al. Multipotent stem/progenitor cells in the human foetal biliary tree. J Hepatol 2012;57:987–94.

[83] Cardinale V, Wang Y, Carpino G, Mendel G, Alpini G, Gaudio E, et al. The biliary tree—a reservoir of multipotent stem cells. Nat Rev Gastroenterol Hepatol 2012; 9:231–40.

[84] Sutton ME, op den Dries S, Koster MH, Lisman T, Gouw AS, Porte RJ. Regeneration of human extrahepatic biliary epithelium: the peribiliary glands as progenitor cell compartment. Liver Int 2012;32: 554–9.

[85] Dipaola F, Shivakumar P, Pfister J, Walters S, Sabla G, Bezerra JA. Identification of intramural epithelial networks linked to peribiliary glands that express progenitor cell markers and proliferate after injury in mice. Hepatology 2013;58:1486–96.

[86] Nakanuma Y, Sato Y. Cystic and papillary neoplasm involving peribiliary glands: a biliary counterpart of branch-type intraductal papillary mucinous [corrected] neoplasm? Hepatology 2012;55:2040–1.

[87] Sato Y, Harada K, Sasaki M, Nakanuma Y. Cystic and micropapillary epithelial changes of peribiliary glands might represent a precursor lesion of biliary epithelial neoplasms. Virchows Arch 2014;464: 157–63.

[88] Lanzoni G, Oikawa T, Wang Y, Cui CB, Carpino G, Cardinale V, et al. Concise review: clinical programs of stem cell therapies for liver and pancreas. Stem Cells 2013;31:2047–60.

[89] Lazaridis KN, Strazzabosco M, Larusso NF. The cholangiopathies: disorders of biliary epithelia. Gastroenterology 2004;127:1565–77.

[90] Imam MH, Talwalkar JA, Lindor KD. Clinical management of autoimmune biliary diseases. J Autoimmun 2013;46:88–96.

[91] Adam R, Hoti E. Liver transplantation: the current situation. Semin Liver Dis 2009;29:3–18.

[92] Porrett PM, Hashmi SK, Shaked A. Immunosuppression: trends and tolerance? Clin Liver Dis 2014;18:687–716.

[93] Xia X, Demorrow S, Francis H, Glaser S, Alpini G, Marzioni M, et al. Cholangiocyte injury and ductopenic syndromes. Semin Liver Dis 2007;27:401–12.

[94] Lozier J, McCright B, Gridley T. Notch signaling regulates bile duct morphogenesis in mice. PLoS One 2008;3:e1851.

[95] Marzioni M, Saccomanno S, Agostinelli L, Rychlicki C, De Minicis S, Pierantonelli I, et al. PDX-1/Hes-1 interactions determine cholangiocyte proliferative response to injury in rodents: possible implications for sclerosing cholangitis. J Hepatol 2013;58:750–6.

[96] Marzioni M, Agostinelli L, Candelaresi C, Saccomanno S, De Minicis S, Maroni L, et al. Activation of the developmental pathway neurogenin-3/microRNA-7a regulates cholangiocyte proliferation in response to injury. Hepatology 2014;60:1324–35.

[97] Katoonizadeh A, Poustchi H, Malekzadeh R. Hepatic progenitor cells in liver regeneration: current advances and clinical perspectives. Liver Int 2014;34: 1464–72.

[98] He J, Lu H, Zou Q, Luo L. Regeneration of liver after extreme hepatocyte loss occurs mainly via biliary transdifferentiation in zebrafish. Gastroenterology 2014;146:789–800.

[99] Choi TY, Ninov N, Stainier DY, Shin D. Extensive conversion of hepatic biliary epithelial cells to hepatocytes after near total loss of hepatocytes in zebrafish. Gastroenterology 2014;146:776–88.

[100] Strom S, Fisher R. Hepatocyte transplantation: new possibilities for therapy. Gastroenterology 2003;124: 568–71.

ROLE OF HEPATIC PROGENITOR CELLS IN LIVER REGENERATION

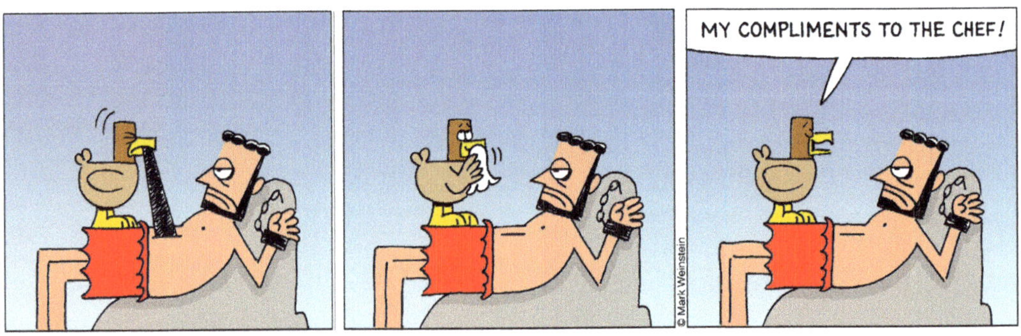
MY COMPLIMENTS TO THE CHEF!
© Mark Weinstein

CHAPTER

17

Introduction to Hepatic Progenitor Cells

Paulette M. Robinson, Liya Pi, Alicia R. Brown, David C. Sullivan, Bryon E. Petersen

Pediatric Stem Cell Research and Hepatic Disorders, Child Health Research Institute, Department of Pediatrics, University of Florida, Gainesville, Florida, USA

17.1 INTRODUCTION

Stem cells are defined as cells that are undifferentiated, capable of self-renewal, and have the potential to differentiation into multiple lineages. The liver contains a small subset of hepatic progenitor cells (HPCs, also termed oval cells in rodents) that have the ability to proliferate clonogenically and differentiate into several lineages, including bile ductal epithelia, hepatocytes, intestinal epithelia, and exocrine pancreas [1]. Morphologically, oval cells are small in size (approximately 10 μm in diameter) with a large nuclear to cytoplasm ratio and an oval-shaped nucleus (hence their name) [2]. Although debate continues as to the site of origin of these cells, they are universally considered to be the resident hepatic stem cell [3].

17.2 ANIMAL MODELS

In normal liver tissue, oval cells are undetectable. Also, oval cells do not participate in the regenerative response to partial hepatectomy (PH), in which the regeneration is mainly driven through hepatocyte proliferation. The most common and established protocol to induce oval cells in rats is *N*-2-acteylaminofluorene (2-AAF) or retrorsine in combination PH. Robust ductular reaction (DR) also can be induced in response to D-galactosamine and allyl alcohol in rats. In these models, hepatocytes are prevented from initiating the regenerative response by inhibiting DNA synthesis and expression of cyclins. When hepatocyte proliferation is blocked, the oval cell compartment is activated and induces a compensatory oval cell-mediated regenerative process [4–6]. The 2AAF/PH model shows that these cells expand from the Canal of Hering [3], from a stem cell niche and/or from the bone marrow [7], and enter hepatic parenchyma in the periportal regions where oval cells proliferate and differentiate into hepatocytes and bile duct cells [8]. When the bile ductular epithelia are damaged in periportal zones, oval cell proliferation is reduced [9]. While HPCs play a compensatory role in liver regeneration when hepatocyte proliferation is inhibited, HPCs can also cause fibrosis. The panlobular fibrosis model consists of repetitive injection of carbon

Liver Regeneration
http://dx.doi.org/10.1016/B978-0-12-420128-6.00017-8

tetrachloride (CCl_4) for 4 weeks causing massive oxidative stress leading to severe hepatocyte necrosis and panlobular fibrosis with nodule formation similar to what is found in most human chronic liver diseases. Repeated exposures to CCl_4 causes metabolism of highly reactive CCl_3 radicals by cytochrome P450. This metabolism leads to central vein damage and necrosis. There is subsequent periportal damage, micro nodules, excessive lipid build up (steatosis), abundant connective tissue development, disruption in calcium homeostasis, streaming cords of fibrous tissue, cirrhosis, oval cell and biliary cell infiltration [10–13]. Treating rats with 2AAF and CCl_4 at the same time for at total of 6 weeks causes a marked increase in oval cell activation and fibrotic extracellular matrix (ECM) protein compared to 2AAF or CCl_4 alone [14].

Unfortunately, the 2-AAF/PH model does not stimulate oval cell activation in the mice for reasons not clearly understood. Alternatively, the administration of 3,5-diethoxycarbonyl-1,4-dihydro-collidine (DDC) provokes DRs in mice and has been utilized to investigate HPCs activation and associated biliary fibrosis. The chronic damaging effect of the DDC diet to the liver results in suppression of hepatocyte proliferation and expansion of the HPCs compartment [15]. Administration of CCl_4 in combination of PHx has also been used to stimulate HPCs and panlobular fibrosis in mice [16]. Both the DDC and CCl_4/PHx models are used to understand the role of HPCs in fibrosis. It is also unclear whether oval cells that arise in different species or as a result of different insults are truly comparable.

17.3 ORIGINS

There is a substantial debate as to the origin of oval cells. Some researchers believe that the oval cells exist in very small numbers within the periportal region, specifically the canals of Hering, and emerge from that niche in response to liver damage [17,18]. Petersen et al. exposed rats to methylene dianiline (DAPM) 24 h prior to hepatic damage (2-AAF/hepatic injury, PH or CCl_4). Under these conditions, the bile ductular epithelium was destroyed, and the oval cell response was severely inhibited. This study was the first to elucidate a direct association between the requirement of an intact bile ductular epithelium and the ability to mount an oval cell response [9]. Oval cells have also repeatedly been shown to originate from bone-marrow-derived stem cells. Transplanted donor bone-marrow-derived stem cells transiently populated the hepatic oval cell pool and differentiated into mature hepatocytes. When there was no injury to the liver, donor-derived hepatocytes were scattered throughout the hepatic parenchyma [19]. In contrast, in the rodent oval cell activation model, the donor-derived bone marrow stem cells appear to enter the parenchyma as oval-like cells, and subsequently expand and differentiate producing clusters of donor-derived hepatocytes [20]. Cell fusion is also believed to be a source of bone-marrow-derived hepatocytes, but the fusion phenomenon has only been demonstrated in the fumarylacetoacetate hydrolase (Fah)-deficient mouse model [21]. Identifying exactly where oval cells originate from is extremely difficult, and the field is continually working to better understand the origin of oval cells.

Oval cell isolation has been a difficult process, which was initially accomplished by density gradient centrifugation or by counter-flow centrifugation elutriation [22–24]. Unfortunately, these methods had mediocre purity and viability. Florescence-activated cell sorting and magnetic activation cell sorting produce higher purity and viability, but these methods use cell-surface markers to identify and isolate the oval cell. The cell-surface markers of oval cells are highly debated, because oval cells comprise a heterogeneous cell population that can rapidly change morphology and phenotypic expression markers. Petersen et al. showed that by using the Thy1 (CD90) marker, they were able to elucidate a purity of 90% from the 2AAF/PH rat [7]. Other investigators have used EpCam, CD133, claudin-7, and other cell-surface markers to identify and isolate oval cells [25–28]. More recently, Dorrell et al. performed extensive studies to identify additional markers to label rodent progenitor cells including MIC1-1C3 and OC2-1D11[29]. Since oval cells are extremely dynamic, continued exploration is necessary to identify an agreed upon subset of surface markers definitively tied to the oval cell population.

17.4 CHARACTERISTICS

Oval cells possess characteristics similar to ductular cells in their distinct isoenzyme profiles intermediate filaments (BSD7, OC2, OC3, OV-1, and OV-6), ECM proteins (CK8, 19), enzymes, and secreted proteins (alpha-fetoprotein, γ-glutamyl transferase) [30–32]. They also possess primary hepatic functions, such as albumin production. They express some hematopoietic stem cell markers such as CD-34, c-kit, Thy1, and Flt-3 [7,33–35]. When oval cells are present in the liver, they differentiate toward both hepatic and bile ductular epithelial lineages. When going toward hepatocyte formation, oval cells first differentiate into basophilic small hepatocytes and then into mature adult hepatocytes [36,37]. Oval cells may constitute more than 50% of the liver during regeneration caused by administration of 2AAF/PH. The cells may form a transient amplifying compartment that includes undifferentiated progenitors, medially differentiated transient cells, and newly differentiated hepatocytes [4]. Oval cells are similar to hepatocytes in that both require growth factors for cell cycle progression, and both cell types also require a "priming" process in order to respond to these stimuli [38]. Oval cells express c-met, the receptor for HGF, and are highly responsive to high levels of HGF during regeneration [39].

17.5 MOLECULAR SIGNALING FACTORS INVOLVED IN OVAL CELLS-MEDIATED LIVER REGENERATION

Our laboratory has identified several molecular factors that play key roles in oval cell-mediated liver regeneration via the activation and differentiation of oval cells during 2AAF/PH. Insulin-like growth factor-binding protein 3 (IGFBP3) was identified in the rat 2AAF-PH model as a potent activator of oval cells. *In vitro*, oval cells migrate in response to an IGFBP3 gradient [40].

Connective tissue growth factor (CTGF) is another protein that plays a major role in the activation of oval cells and was found to be significantly upregulated during response to a 2-AAF/PH injury. Inhibition of CTGF upregulation by Iloprost significantly decreased the number of oval cells that responded to the 2-AAF/PH injury, indicating the importance of CTGF in oval cell-mediated liver regeneration [41,42].

We have also shown that somatostatin acts as a strong chemotactic factor during oval cell migration within the regenerating liver [43]. The CXC chemokine, stromal cell-derived factor-1 (SDF-1), is upregulated following 2AAF-PH, but there is no upregulation of SDF-1 following non-oval cell-inducing CCl4 injury or PH alone. Oval cells will migrate in response to a SDF-1 concentration gradient in culture [44,45].

Finally, Notch signaling was found to be upregulated at the peak of oval cell induction during 2AAF-PH alone. Treatment with gamma secretase inhibitor (GSIXX) led to interruption of the Notch signal, as shown by a decrease in expression of Hes1. While there was a robust oval cell response seen at day 11 post-PH, there was a measurable delay in differentiation when, Notch was inhibited [46].

17.6 HEPATIC PROGENITOR CELL ACTIVATION AND ASSOCIATED WITH LIVER FIBROSIS DURING CHRONIC LIVER DISEASE

The expansion of the hepatic progenitor compartment (HPC), which is also known as DRs, involves a heterogeneous population of transient amplifying cells capable of differentiating into either biliary or hepatocytic lineages. Besides the epithelial elements, myofibroblast cells, macrophages/Kupffer cells, endothelial cells, and associated ECM are also constituents of HPC niches. HPCs are often accompanied with myofibroblast activation as well as excessive deposition of ECM around the portal areas, suggesting a direct correlation between HPC activation and periportal fibrosis (Figure 17.1) [47].

This reactive process occurs not only in biliary disorders such as cholangiocarcinoma but also in many acute and chronic liver diseases resulting from insults such as alcohol toxicity and submassive parenchymal necrosis [48], ECM remodeling has been shown to be a requirement for the hepatic progenitor cell response as failure of the ECM to remodel after chronic fibrotic liver injury hinders the ability of the liver to activate HPCs [49,50].

Several signaling pathways have been shown to regulate HPC activation and associated liver fibrogenic response. Tumor necrosis factor-like weak inducer of apoptosis (TWEAK) proteins stimulate HPC mitosis through its Fn14 receptor via the NFκB-dependent pathway [49,51,52]. Inhibition of HPC proliferation using anti-TWEAK antibody prevents the fibrogenic response and improves liver regeneration [16]. Notch and Wnt signaling regulate HPCs via their interactions with activated myofibroblasts or macrophages [53]. Expression of Jagged 1 by myofibroblasts promotes Notch signaling in HPCs and thus their biliary specification to cholangiocytes during biliary regeneration. Conversely, macrophage engulfment of hepatocyte debris induces Wnt3a expression and promotes hepatocyte regeneration [53]. In addition, Wnt/β-catenin signaling mediates the oval cell response since inhibition of β-catenin decreases oval cell activation in rodents [54]. Hedgehog ligands also appear to be critical signals that are activated in various types of liver injury. Constitutive activation of hedgehog signals in the liver is associated with progenitor and/or cholangiocyte proliferation and fibrosis [55]. A member of the fibroblast growth factor (FGF) family, FGF7, has been reported to be secreted from Thy1$^+$ mesenchymal cells and stimulates HPC expansion in animal models [56]. Galectin-3 (Gal-3) is a β-galactoside-binding lectin that binds to laminin and is expressed in injured liver in mice and humans. Gal-3 knockout mice fail to form a HPC niche and have reduced laminin formation [57]. TGF-β does not affect HPCs proliferation due to glycosylation of the TGFβRII but does inhibit hepatocytes proliferation [58]. However, TGF-β on its own is not sufficient to induce a fibrogenic DR since significant fibrogenesis occurs only with the use of CCl_4 and 2-AAF in a rodent study (Figure 17.2) [14].

CTGF of the CCN (Cyr61/CTGF/Nov) protein family, also termed CCN2, is transcriptionally activated by TGF-β in various types of fibrotic disorders [59]. CTGF can bind to TGF-β and potentiate the TGF-β signaling [60]. In our previous studies, we have demonstrated that CTGF is induced during oval cell-mediated liver regeneration (Figure 17.2). Administration of iloprost, a prostacyclin analogue that reduces production of CTGF, impairs rat oval cell response following 2-AAF/PHx [41]. Moreover, CTGF interacts with integrin α5β1 and fibronectin and promotes the migration and adhesion of primary rat oval cells [61].

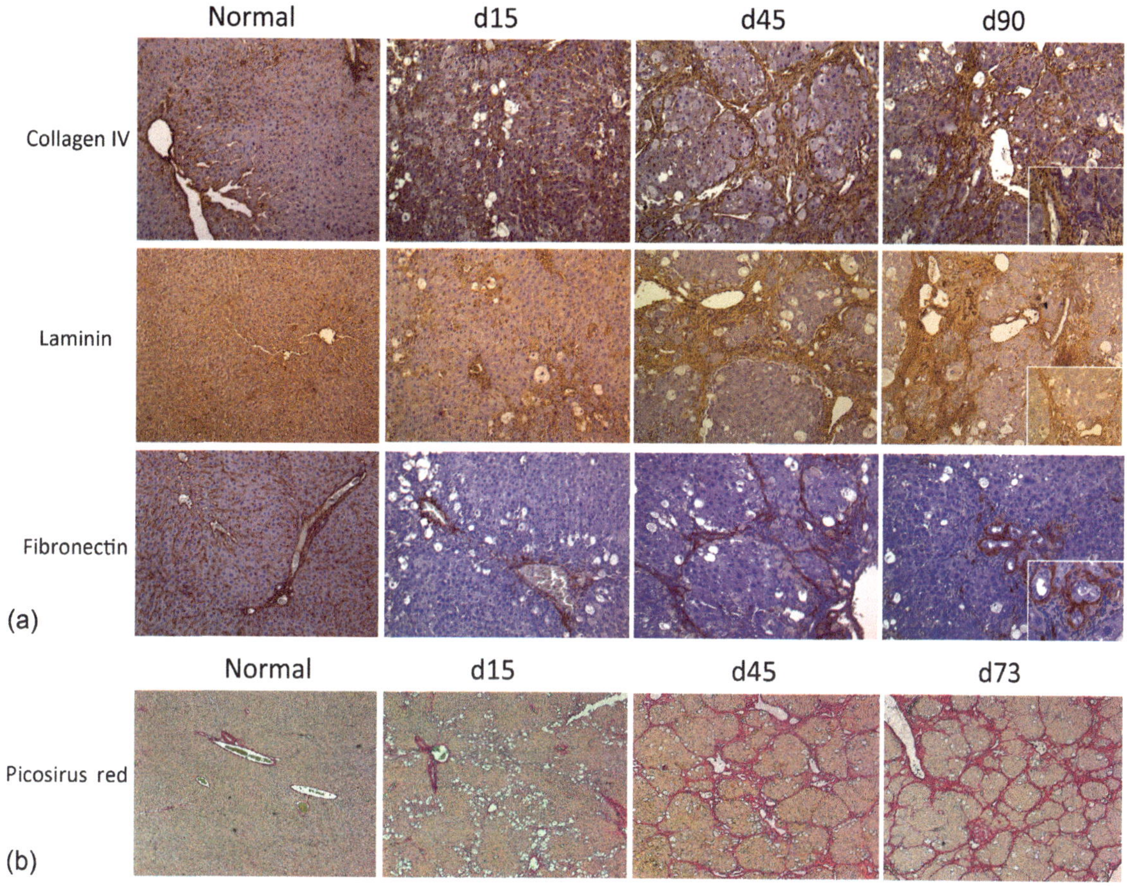

FIGURE 17.1 Analysis of extracellular matrix components by immunohistochemistry in rat livers treated with CCl_4. (a) Extracellular matrix proteins, collagen IV (200×, higher magnification inset=600×), laminin (200×, higher magnification inset=600×), and fibronectin (200×, higher magnification inset=400×) were evaluated in normal and CCl_4-treated rat livers at d15, d45, and d90, respectively. In normal rat livers, collagen IV was concentrated in bordering vessels and within the basement membrane of the biliary tree. In CCl_4 rat livers, collagen IV had progressive pericentral deposition and became more concentrated at the later time points. In the CCl_4 rat livers, there was progressive laminin deposition in the pericentral fibrotic regions compared to ordered cords of hepatocytes in normal rat livers. In normal rat livers, fibronectin is found surrounding all vessels and ducts, and in CCl_4-treated rat livers, fibronectin was observed within the basement membrane surrounding areas of proliferating bile ducts. (b) Picosirus red was used to identify collagens I and III (50×) in normal and CCl_4-treated rat livers at d15, d45, and d72. In normal rat livers, the Picosirus red staining was concentrated in bordering vessels and within the basement membrane of the biliary tree. Picosirus red staining of CCl_4-treated rat livers had a gradual increase into the pericental region of the liver.

Our recent study has shown that CTGF is highly expressed together with integrin αvβ6 in HPCs of cirrhotic human livers and cholangiocarcinoma as well as mouse oval cells activated by DDC, whereas deletion of *Ctgf* exon 4 inhibits DDC-induced oval cell activation and biliary fibrosis [62].

In summary, HPC-mediated liver regeneration is an important component of regenerative capacity of the liver. Whereas HPCs provide

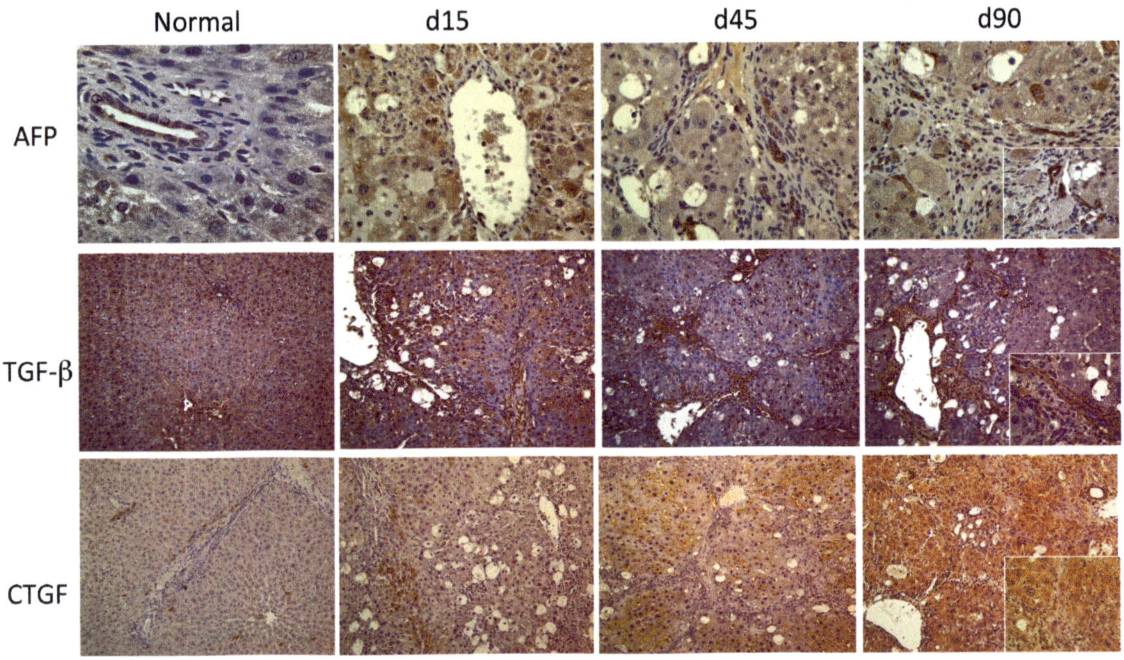

FIGURE 17.2 Identification of oval cells and profibrotic proteins by immunohistochemistry in rat livers treated with CCl_4. AFP was used to locate and identify oval cells in rat livers at d15, 45, and 90, respectively (600×, higher magnification inset = 1000×). The normal adult liver contains very few AFP^+ cells. Oval cells were found scattered throughout the fibrotic cord structures, and there were several AFP^+ transitional hepatocytes apparent at later time points. Profibrotic growth factors, TGFβ, and CTGF were evaluated in normal and CCl_4-treated rat livers at d15, d45, and d90, respectively (200×, higher magnification inset = 600×). Normal liver demonstrated very low levels of TGFβ and CTGF. TGFβ is strongly expressed by nonparenchymal cells that are in intimate contact with fibrotic cords. CTGF continually increased over the time course of injury and expressed both, hepatocytes and nonparenchymal cells.

a reliable backup plan for recovery from acute liver injury, they may play a pathogenic role in chronic liver injury model. HPC-mediated liver regeneration involves a diverse variety of autocrine/paracrine signals that mediate niche-progenitor cell and matrix-progenitor cell interactions. Additional studies focused at identifying, purifying, and further utilizing HPCs for the treatment of liver diseases is the major focus of today's liver research. Understanding the molecular mechanism underlying HPC activation and associated liver fibrosis would help the identification of therapeutic targets for the treatment of related liver diseases.

References

[1] Oh SH, Muzzonigro TM, Bae SH, LaPlante JM, Hatch HM, Petersen BE. Adult bone marrow-derived cells trans-differentiating into insulin-producing cells for the treatment of type I diabetes. Lab Invest 2004;84:607–17.

[2] Fausto N. Liver stem cells. In: Arias IM, Boyer JL, Fausto N, Jakoby WB, Schachter DA, Shafritz DA, editors. The liver biology and pathobiology. New York: Raven Press; 1994. p. 1501–18.

[3] Farber E. Similarities in the sequence of early histological changes induced in the liver of the rat by ethionine, 2-acetylamino-fluorene, and 3′-methyl-4-dimethylaminoazobenzene. Cancer Res 1956;16:142–8.

[4] Fausto N, Campbell JS. The role of hepatocytes and oval cells in liver regeneration and repopulation. Mech Dev 2003;120:117–30.

[5] Evarts RP, Nakatsukasa H, Marsden ER, Hsia CC, Dunsford HA, Thorgeirsson SS. Cellular and molecular changes in the early stages of chemical hepatocarcinogenesis in the rat. Cancer Res 1990;50:3439–44.
[6] Fausto N, Lemire JM, Shiojiri N. Oval cells in liver carcinogenesis; cell lineages in hepatic development and identification of stem cells in normal live. Boca Raton: CRC Press; 1992.
[7] Petersen BE, Goff JP, Greenberger JS, Michalopoulos GK. Hepatic oval cells express the hematopoietic stem cell marker Thy-1 in the rat. Hepatology 1998;27:433–45.
[8] Oh SH, Hatch HM, Petersen BE. Hepatic oval 'stem' cell in liver regeneration. Semin Cell Dev Biol 2002; 13:405–9.
[9] Petersen BE, Zajac VF, Michalopoulos GK. Bile ductular damage induced by methylene dianiline inhibits oval cell activation. Am J Pathol 1997;151:905–9.
[10] Friedman LS, Martin P, Munoz SJ. Laboratory evaluation of the patient with liver disease. In: Zakim D, Boyer TD, editors. Hepatology: a textbook of liver disease, vol. 1. New York: Saunders; 2003. p. 661–709.
[11] Bucher NLR, Farmer SR. Liver regeneration following partial hepatectomy: genes and metabolism. In: Strain AJ, Diehl AM, editors. Liver growth and repair. New York: Chapman and Hall; 1998. p. 3–27.
[12] Smuckler EA. Structural and functional changes in acute liver injury. Environ Health Perspect 1976;15:13–25.
[13] Weber LW, Boll M, Stampfl A. Hepatotoxicity and mechanism of action of haloalkanes: carbon tetrachloride as a toxicological model. Crit Rev Toxicol 2003;33:105–36.
[14] Chobert MN, Couchie D, Fourcot A, Zafrani ES, Laperche Y, Mavier P, et al. Liver precursor cells increase hepatic fibrosis induced by chronic carbon tetrachloride intoxication in rats. Lab Invest 2012;92:135–50.
[15] Wang X, Foster M, Al-Dhalimy M, Lagasse E, Finegold M, Grompe M. The origin and liver repopulating capacity of murine oval cells. Proc Natl Acad Sci USA 2003;100(Suppl. 1):11881–8.
[16] Kuramitsu K, Sverdlov DY, Liu SB, Csizmadia E, Burkly L, Schuppan D, et al. Failure of fibrotic liver regeneration in mice is linked to a severe fibrogenic response driven by hepatic progenitor cell activation. Am J Pathol 2013;183:182–94.
[17] Golding M, Sarraf C, Lalani EN, Alison MR. Reactive biliary epithelium: the product of a pluripotential stem cell compartment? Hum Pathol 1996;27:872–84.
[18] Menthena A, Deb N, Oertel M, Grozdanov PN, Sandhu J, Shah S, et al. Bone marrow progenitors are not the source of expanding oval cells in injured liver. Stem Cells 2004;22:1049–61.
[19] Gupta S, Rajvanshi P, Irani AN, Palestro CJ, Bhargava KK. Integration and proliferation of transplanted cells in hepatic parenchyma following D-galactosamine-induced acute injury in F344 rats. J Pathol 2000;190:203–10.
[20] Theise ND, Nimmakayalu M, Gardner R, Illei PB, Morgan G, Teperman L, et al. Liver from bone marrow in humans. Hepatology 2000;32:11–6.
[21] Wang X, Willenbring H, Akkari Y, Torimaru Y, Foster M, Al-Dhalimy M, et al. Cell fusion is the principal source of bone-marrow-derived hepatocytes. Nature 2003;422:897–901.
[22] Sirica AE, Mathis GA, Sano N, Elmore LW. Isolation, culture, and transplantation of intrahepatic biliary epithelial cells and oval cells. Pathobiology 1990;58:44–64.
[23] Yaswen P, Hayner NT, Fausto N. Isolation of oval cells by centrifugal elutriation and comparison with other cell types purified from normal and preneoplastic livers. Cancer Res 1984;44:324–31.
[24] Hayner NT, Braun L, Yaswen P, Brooks M, Fausto N. Isozyme profiles of oval cells, parenchymal cells, and biliary cells isolated by centrifugal elutriation from normal and preneoplastic livers. Cancer Res 1984;44:332–8.
[25] Yovchev MI, Grozdanov PN, Zhou H, Racherla H, Guha C, Dabeva MD. Identification of adult hepatic progenitor cells capable of repopulating injured rat liver. Hepatology 2008;47:636–47.
[26] Okabe M, Tsukahara Y, Tanaka M, Suzuki K, Saito S, Kamiya Y, et al. Potential hepatic stem cells reside in EpCAM+ cells of normal and injured mouse liver. Development 2009;136:1951–60.
[27] Yovchev MI, Grozdanov PN, Joseph B, Gupta S, Dabeva MD. Novel hepatic progenitor cell surface markers in the adult rat liver. Hepatology 2007;45:139–49.
[28] Herrera MB, Bruno S, Buttiglieri S, Tetta C, Gatti S, Deregibus MC, et al. Isolation and characterization of a stem cell population from adult human liver. Stem Cells 2006;24:2840–50.
[29] Dorrell C, Erker L, Lanxon-Cookson KM, Abraham SL, Victoroff T, Ro S, et al. Surface markers for the murine oval cell response. Hepatology 2008;48:1282–91.
[30] Shinozuka H, Lombardi B, Sell S, Iammarino RM. Early histological and functional alterations of ethionine liver carcinogenesis in rats fed a choline-deficient diet. Cancer Res 1978;38:1092–8.
[31] Hixson DC, Chapman L, McBride A, Faris R, Yang L. Antigenic phenotypes common to rat oval cells, primary hepatocellular carcinomas and developing bile ducts. Carcinogenesis 1997;18:1169–75.
[32] Nagy P, Kiss A, Schnur J, Thorgeirsson SS. Dexamethasone inhibits the proliferation of hepatocytes and oval cells but not bile duct cells in rat liver. Hepatology 1998;28:423–9.

[33] Omori M, Evarts RP, Omori N, Hu Z, Marsden ER, Thorgeirsson SS. Expression of alpha-fetoprotein and stem cell factor/c-kit system in bile duct ligated young rats. Hepatology 1997;25:1115–22.
[34] Omori N, Omori M, Evarts RP, Teramoto T, Miller MJ, Hoang TN, et al. Partial cloning of rat CD34 cDNA and expression during stem cell-dependent liver regeneration in the adult rat. Hepatology 1997;26:720–7.
[35] Fujio K, Evarts RP, Hu Z, Marsden ER, Thorgeirsson SS. Expression of stem cell factor and its receptor, c-kit, during liver regeneration from putative stem cells in adult rat. Lab Invest 1994;70:511–6.
[36] Golding M, Sarraf CE, Lalani EN, Anilkumar TV, Edwards RJ, Nagy P, et al. Oval cell differentiation into hepatocytes in the acetylaminofluorene-treated regenerating rat liver. Hepatology 1995;22:1243–53.
[37] Evarts RP, Nagy P, Nakatsukasa H, Marsden E, Thorgeirsson SS. *In vivo* differentiation of rat liver oval cells into hepatocytes. Cancer Res 1989;49:1541–7.
[38] Nagy P, Bisgaard HC, Santoni-Rugiu E, Thorgeirsson SS. *In vivo* infusion of growth factors enhances the mitogenic response of rat hepatic ductal (oval) cells after administration of 2-acetylaminofluorene. Hepatology 1996;23:71–9.
[39] Hasuike S, Ido A, Uto H, Moriuchi A, Tahara Y, Numata M, et al. Hepatocyte growth factor accelerates the proliferation of hepatic oval cells and possibly promotes the differentiation in a 2-acetylaminofluorene/partial hepatectomy model in rats. J Gastroenterol Hepatol 2005;20:1753–61.
[40] Steiger-Luther NC, Darwiche H, Oh SH, Williams JM, Petersen BE. Insulin-like growth factor binding protein-3 is required for the regulation of rat oval cell proliferation and differentiation in the 2AAF/PHX model. Hepat Med 2010;2010:13–32.
[41] Pi L, Oh SH, Shupe T, Petersen BE. Role of connective tissue growth factor in oval cell response during liver regeneration after 2-AAF/PHx in rats. Gastroenterology 2005;128:2077–88.
[42] Pintilie DG, Shupe TD, Oh SH, Salganik SV, Darwiche H, Petersen BE. Hepatic stellate cells' involvement in progenitor-mediated liver regeneration. Lab Invest 2010;90:1199–208.
[43] Jung Y, Oh SH, Witek RP, Petersen BE. Somatostatin stimulates the migration of hepatic oval cells in the injured rat liver. Liver Int 2012;32:312–20.
[44] Hatch HM, Zheng D, Jorgensen ML, Petersen BE. SDF-1alpha/CXCR4: a mechanism for hepatic oval cell activation and bone marrow stem cell recruitment to the injured liver of rats. Cloning Stem Cells 2002;4:339–51.
[45] Zheng D, Oh SH, Jung Y, Petersen BE. Oval cell response in 2-acetylaminofluorene/partial hepatectomy rat is attenuated by short interfering RNA targeted to stromal cell-derived factor-1. Am J Pathol 2006;169:2066–74.
[46] Darwiche H, Oh SH, Steiger-Luther NC, Williams JM, Pintilie DG, Shupe TD, et al. Inhibition of Notch signaling affects hepatic oval cell response in rat model of 2AAF-PH. Hepat Med 2011;3:89–98.
[47] Greenbaum LE, Wells RG. The role of stem cells in liver repair and fibrosis. Int J Biochem Cell Biol 2011;43:222–9.
[48] Gouw AS, Clouston AD, Theise ND. Ductular reactions in human liver: diversity at the interface. Hepatology 2011;54:1853–63.
[49] Van Hul NK, Abarca-Quinones J, Sempoux C, Horsmans Y, Leclercq IA. Relation between liver progenitor cell expansion and extracellular matrix deposition in a CDE-induced murine model of chronic liver injury. Hepatology 2009;49:1625–35.
[50] Kallis YN, Robson AJ, Fallowfield JA, Thomas HC, Alison MR, Wright NA, et al. Remodelling of extracellular matrix is a requirement for the hepatic progenitor cell response. Gut 2011;60:525–33.
[51] Jakubowski A, Ambrose C, Parr M, Lincecum JM, Wang MZ, Zheng TS, et al. TWEAK induces liver progenitor cell proliferation. J Clin Invest 2005;115:2330–40.
[52] Tirnitz-Parker JE, Viebahn CS, Jakubowski A, Klopcic BR, Olynyk JK, Yeoh GC, et al. Tumor necrosis factor-like weak inducer of apoptosis is a mitogen for liver progenitor cells. Hepatology 2010;52:291–302.
[53] Boulter L, Govaere O, Bird TG, Radulescu S, Ramachandran P, Pellicoro A, et al. Macrophage-derived Wnt opposes Notch signaling to specify hepatic progenitor cell fate in chronic liver disease. Nat Med 2012;18:572–9.
[54] Apte U, Thompson MD, Cui S, Liu B, Cieply B, Monga SP. Wnt/beta-catenin signaling mediates oval cell response in rodents. Hepatology 2008;47:288–95.
[55] Omenetti A, Porrello A, Jung Y, Yang L, Popov Y, Choi SS, et al. Hedgehog signaling regulates epithelial-mesenchymal transition during biliary fibrosis in rodents and humans. J Clin Invest 2008;118:3331–42.
[56] Takase HM, Itoh T, Ino S, Wang T, Koji T, Akira S, et al. FGF7 is a functional niche signal required for stimulation of adult liver progenitor cells that support liver regeneration. Genes Dev 2013;27:169–81.
[57] Hsieh WC, Mackinnon AC, Lu WY, Jung J, Boulter L, Henderson NC, et al. Galectin-3 regulates hepatic progenitor cell expansion during liver injury. Gut 2014;64:312–21.
[58] Nguyen LN, Furuya MH, Wolfraim LA, Nguyen AP, Holdren MS, Campbell JS, et al. Transforming growth

factor-beta differentially regulates oval cell and hepatocyte proliferation. Hepatology 2007;45:31–41.
[59] Gressner OA, Gressner AM. Connective tissue growth factor: a fibrogenic master switch in fibrotic liver diseases. Liver Int 2008;28:1065–79.
[60] Abreu JG, Ketpura NI, Reversade B, De Robertis EM. Connective-tissue growth factor (CTGF) modulates cell signalling by BMP and TGF-beta. Nat Cell Biol 2002;4:599–604.
[61] Pi L, Ding 21X, Jorgensen M, Pan JJ, Oh SH, Pintilie D, et al. Connective tissue growth factor with a novel fibronectin binding site promotes cell adhesion and migration during rat oval cell activation. Hepatology 2008;47:996–1004.
[62] Pi L, Robinson PM, Jorgensen M, Oh SH, Brown AR, Weinreb PH, et al. Connective tissue growth factor and integrin αvβ6: a new pair of regulators critical for ductular reaction and biliary fibrosis. Hepatology 2014;2:678–91.

CHAPTER

18

Hepatocyte to Biliary Transdifferentiation: To Be(*come*) or Not to *Be*(*come*)?

Vishakha Bhave

Philadelphia College of Osteopathic Medicine, School of Pharmacy, Suwanee, Georgia, USA

18.1 WHAT IS TRANSDIFFERENTIATION?

Transdifferentiation, also known as lineage reprogramming [1], is a process where one mature somatic cell transforms into another mature somatic cell without undergoing an intermediate pluripotent state [2]. It is a type of metaplasia, which includes all cell fate switches, including the interconversion of stem cells. Current uses of transdifferentiation include disease modeling, drug discovery and in the future may include gene therapy and regenerative medicine [3]. The term "transdifferentiation" was originally coined by Selman and Kafatos in 1974 to describe a change in cell properties as cuticle-producing cells became salt-secreting cells in silk moths undergoing metamorphosis [4]. In the liver, biliary epithelium cells transform into hepatocytes via formation of oval cells [5–7]. However, the first use of the term "transdifferentiation" in the liver was by Michalopoulos et al. who demonstrated that hepatocyte to biliary transdifferentiation (HBT) is conversion of hepatocytes to biliary cells in a bile duct injury model [8].

18.2 TRANSDIFFERENTIATION IN THE LIVER

Experimental evidence from animal models and human liver tissue documents the following types of transdifferentiations:

1. *Biliary to hepatocyte*: This pathway is activated when there is a major hepatocyte loss and the resident hepatocytes are unable to divide. Biliary cells in this pathway form an intermediate form known as "oval cells" before eventually differentiating into hepatocytes. A classic rodent model to study biliary to hepatocyte transdifferentiation is the 2/3 partial hepatectomy (PHx) + 2-acetylaminofluorine (AAF) rat model. This model was first developed to study oval cell response. However, Thorgeirsson and colleagues

Liver Regeneration
http://dx.doi.org/10.1016/B978-0-12-420128-6.00018-X

identified biliary epithelial cells as the origin of oval cells using thymidine pulse labeling [5,6,9]. This discovery led to this model now being considered as a transdifferentiation rather than a mere oval cell generation model. Soon after their appearance, oval cells continue to express biliary markers, while at the same time initiating albumin and alpha-fetoprotein expression. This is associated with expression of hepatocyte-specific transcription factors in the biliary epithelial cells. Toxic damage to biliary cells before initiation of oval cell process eliminates the appearance of oval cells [7], further confirming the origin of oval cells as biliary.

2. *Hepatocyte to biliary*: This pathway is activated when there is extensive loss of functional biliary cells, and the resident biliary cells are unable/insufficient to proliferate to compensate for the loss. Under such situations, hepatocytes can transdifferentiate to biliary cells and help regenerate the bile ducts. There is now growing evidence for existence of this pathway, and current and future studies in this area are geared toward identifying the mechanisms and drivers that bring about this change. This pathway is the prime focus of this chapter.
3. *Epithelial to mesenchymal transition (EMT)*: EMT is a kind of transdifferentiation involving conversion of epithelial cells (hepatocytes and/or biliary) to mesenchymal phenotype that enables enhanced migratory capacity, invasiveness, elevated resistance to apoptosis, and greatly increased production of extracellular matrix component [10]. Evidence for this kind of transdifferentiation in the liver is provided by Diehl and collaborators [11,12]. The involvement of Hedgehog signaling pathway in this process has also been described [13].
4. *Mesenchymal to epithelial transition (MET)*: MET in the liver is a kind of transdifferentiation that involves conversion of mesenchymal cells particularly stellate and endothelial cells to epithelial cells. Stellate cells have been reported to express CD133, which also happens to be a progenitor cell marker [14]. Moreover, application of cytokines that guide stem cells to develop hepatocytes lead to the appearance of rotund cells and expression of hepatocyte markers albumin and α-fetoprotein [14]. This theory is further supported by the finding that hematopoetic progenitor cells characterized by CD133, CD14, CD45, and CXCR4 appear in the peripheral blood of healthy humans after PHx, and these cells differentiate into hepatocytes *in vitro* [15]. However, whether this pathway is functional during human liver disease and the extent of its contribution to liver regeneration and repair is not completely understood.

18.3 WHEN DOES TRANSDIFFERENTIATION OCCUR?

The two major epithelial compartments of the liver include hepatocytes and the biliary cells (also known as cholangiocytes). These two cell types combined are responsible for the majority of functions expected from the liver as an organ. These functions include regulation of lipid and glucose metabolism, and biotransformation of drugs and toxicants, among others. Following loss of liver mass such as after PHx or recovery from CCl_4-induced chemical injury, regeneration of the liver proceeds through a limited set of proliferative events in which hepatocytes generate more hepatocytes, biliary cells generate more biliary cells [16]. In this process, all liver cells participate in regeneration without the involvement of stem cells or progenitor cells [16–18]. However, recent evidence suggests a possible contribution of endothelial progenitor cells from the bone marrow in liver regeneration following PHx [19].

When the capacity of one cell compartment to proliferate is impaired when there is a need/stimulus for that compartment to proliferate, transdifferentiation comes into the picture. In the absence of a resident stem cell population in the liver, a variety of studies have established that biliary cells and hepatocytes, the two mature epithelial cells of the liver, although fully differentiated, retain the potential to interchange their phenotype, on demand [8,20–24]. Under normal circumstances, there is sufficient evidence to support that each compartment is able to make new cells of the same type when those cells are damaged [17,25]. For example, bile duct ligation induces biliary proliferation without damage to the resident biliary cells. Very low or no transdifferentiation of hepatocytes to biliary cells is observed under such a situation [26–28]. Similarly, following PHx, existing hepatocytes proliferate to produce new hepatocytes to compensate for lost cells. However, if the existing hepatocytes are prohibited from proliferating (by administration of a DNA adduct forming chemical like AAF or due to very severe liver injury), biliary cells transdifferentiate to form hepatocytes via formation of transient "oval cell" population. In situations, where the damage is such that the cells cannot proliferate to make up for the loss, cells from the other compartment transdifferentiate to form these cells. Transdifferentiation between hepatocytes and biliary cells is a very important phenomenon because it occurs in liver failure in humans and may be a factor in determining survival. Clinical conditions such as massive hepatic necrosis and hepatitis C viral-infected human livers have documented transdifferentiation of biliary cell to hepatocyte. Similarly, primary biliary cirrhosis and biliary obstruction in human livers have documented hepatocyte to biliary cell transdifferentiation. Whereas in normal liver, the expression of hepatocyte- or biliary cell-associated single-genes is localized to specific cell types; cells with expression of both hepatocyte and biliary cells single-gene markers (biphenotypic) are often seen in chronic liver diseases [29–33]. For example, expression of biliary-associated hepatocyte nuclear factor 3β (HNF-3β, biliary-specific marker) in hepatocytes during chronic biliary disease (primary biliary cirrhosis and biliary obstruction) indicates that phenotypic reprogramming of hepatocytes to biliary cells is operative in the human liver [34]. Thus, hepatocytes act as facultative stem cells to rescue biliary cells in times of need. The appearance of marker genes from biliary cells in hepatocytes in acute and chronic biliary damage in humans and rodents indicates that a reprogramming of gene expression is taking place, as an adaptation to a specific pathobiologic or physiological challenge.

Majority of scientific evidence suggests that transdifferentiation, more specifically HBT, requires fulfillment of two conditions: (1) stimulus for biliary proliferation and (2) extensive injury to resident biliary cells rendering them incapable/insufficient of proliferating (Figure 18.1).

Various studies have demonstrated that bipotential progenitors may be present in the liver, which are capable of giving rise to hepatocytes and biliary epithelial cells [35–38]. In a previous lineage-tracing study Willenbring et al. showed that newly formed biliary cells in a bile duct ligation model were not derived from hepatocytes [39]. However, these studies were performed in situations in which there was no injury to the resident biliary epithelium, and thus, it was able to repair itself via biliary proliferation without requiring contribution from hepatocytes [40]. In a later study, the same group reported that hepatocytes are a major source of neoplastic biliary cells and further suggested that the conversion of hepatocytes to biliary cells is much more frequent than envisioned in their previous study [41]. Recently, different scientific groups [24,28,42], through exhaustive lineage-tracing experiments, have established that induction of chronic liver injury in mice involves HBT. These studies strongly establish the phenomenon of HBT in rodent models and human chronic liver diseases.

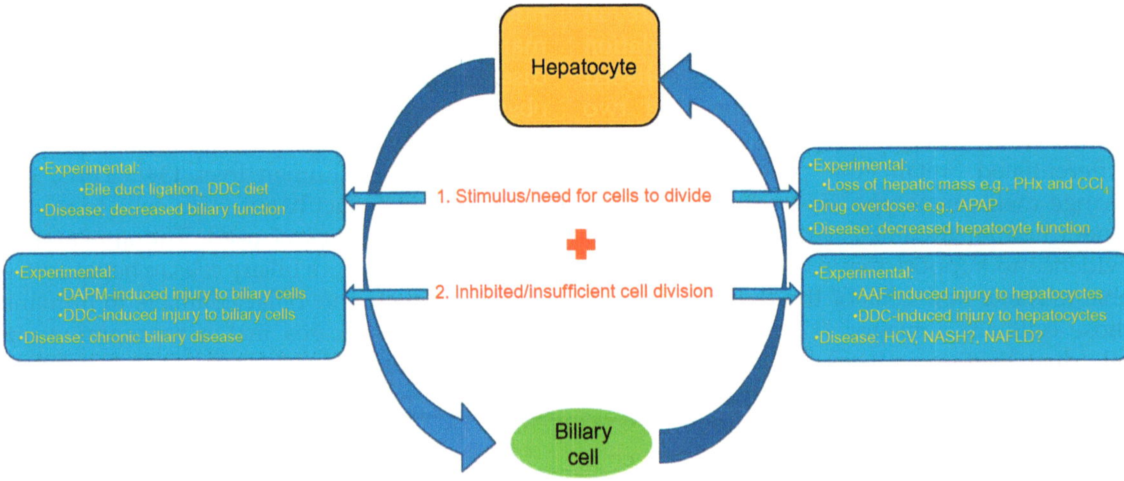

FIGURE 18.1 Conditions that govern transdifferentiation. A schema describing the two conditions that need fulfillment for transdifferentiation to occur. These conditions may be fulfilled via experimental procedures or disease conditions.

18.4 EVIDENCE FOR HBT IN HUMAN DISEASE

HBT has been documented in various human liver diseases. Limaye et al. reported that HNF3β, which is normally expressed only in biliary cells in normal adult liver, was also expressed in hepatocytes in primary biliary cirrhosis and chronic biliary obstruction livers [34]. Sekiya and Suzuki have reported that Hes family transcription factor 1 (Hes1) is essential for conversion of hepatocytes into biliary lineage cells at the onset of intrahepatic colangiocarcinoma [43]. Yanger et al. have reported existence of abundant cells coexpressing hepatocyte (HNF4α) and biliary (Sox9) transcription factors in human liver disease sections. Periportal hepatocytes in patients with cholestatic or biliary autoimmune disorders were shown to express biliary-specific markers [44]. Other findings have shown that dual Notch and AKT signaling in hepatocytes can lead to their conversion into biliary cells that eventually progress into cholangiocarcinomas, a malignancy typically associated with ductal origin [45].

18.5 ANIMAL MODELS FOR STUDYING HBT

Currently, three rodent models of HBT exist. The first one is an *in vitro* model of organoid culture established by Michalopoulos et al. [46,47]. This unique model is an important tool for mechanistic studies involving HBT. The second is an *in vivo* rat model of chronic biliary injury (induced by biliary selective toxin coupled with bile duct ligation) in which the hepatocytes transdifferentiate to biliary cells to restore the lost biliary structure and function [8]. The third model is an *in vivo* mouse model of chronic liver injury induced by feeding mice with 3,5-diethoxycarbonyl-1,4-dihydrocollidine (DDC) containing diet [24,48].

18.5.1 *In Vitro* Model of Transdifferentiation (Organoid Culture)

Freshly isolated primary hepatocytes (>98% purity) when plated on collagen-coated pleated surface roller bottles and cultured in chemically defined hepatocyte growth medium containing hepatocyte growth factor (HGF), epidermal growth factor (EGF), insulin, nicotinamide, and dexamethasone form characteristic and reproducible tissue architecture composed of a superficial layer of biliary epithelial cells, an intermediate layer of connective tissue and hepatocytes, and a basal layer of endothelial cells [47]. Histologic analysis of cultures at early stages does not demonstrate the presence of biliary epithelium (negative for HNF-3β and CK19). In this model, the hepatocytes start transdifferentiating into bile duct epithelial cells between days 6 and 8. The full histologic maturation of the organoid cultures occurs between 15 and 21 days in culture. The cellular origin of the biliary epithelium in this culture could be either (1) derived from contaminating biliary epithelial cells at the time of perfusion and/or (2) hepatocytes undergoing transdifferentiation. However, the cellular origin of the biliary epithelium in this culture has been determined to be hepatocytic using the retrorsine/dipeptidyl peptidase-4 (DPP4) chimeric liver system [46] described below and not resulting from proliferation of contaminating biliary epithelial cells in the culture.

18.5.2 *In Vivo* Rat Model of HBT

Selective biliary toxin 4,4′-diaminodiphenyl methane (DAPM) dissolved in DMSO at a concentration of 50 mg/ml is given intraperitoneally to rats. Two days later, rats are subjected to a mid-abdominal incision 3 cm long, under general anesthesia. The common bile duct is ligated in two adjacent positions approximately 1 cm from the porta hepatis. The duct is then severed between the two sites of ligation. Bile duct ligation in this model provides a stimulus for biliary proliferation, and DAPM selectively inhibits resident biliary from proliferation, thus leading to a high rate of transdifferentiation of hepatocytes to biliary cells. This protocol when used in the DPP4 chimeric rats (described below) is an excellent tool to track biliary epithelial cells derived from hepatocytes (stain positive for DPP4). This model mimics chronic biliary liver diseases wherein hepatocytes transdifferentiate to biliary cells to restore lost biliary function and postpone/prevent liver failure [8].

Retrorsine/DPP4 chimeric liver system: This technique results in a liver of hybrid hepatocyte composition in which the donor hepatocytes (stainable red for DPP4) colonize most of the liver. The recipient hepatocytes remain "gray" (negative for DPP4). The biliary epithelial cells remain DPP4 negative as well. DPP4-positive Fisher and DPP4-negative German Fisher rats are utilized to generate the chimeric livers. Protocol first established by Laconi et al. [49] and later modified by Michalopoulos and colleagues [8] is used for generating the chimeric rats. German Fisher rats (DPP4-negative), weighing 200 g, are given two injections with retrorsine, an inhibitor of hepatocyte proliferation (30 mg/kg, i.p. in water) 15 days apart. A month after the last injection, the rats are subjected to PHx. During the PHx surgery, the rats are also injected with 3.5 million hepatocytes isolated immediately before the operation from a DPP4-positive Fisher rat. The animals are allowed to recover for 3 months. About 40-50% of hepatocytes (but not biliary cells) in these rats become DPP4 positive. These chimeric rats having tagged hepatocytes can then be used to track biliary cells derived from hepatocytes (stain red for DPP4). When these chimeric rats are subjected to significant biliary injury using DAMP, they exhibit appearance of DDPIV-positive biliary cells, which come of HBT of DDPIV-positive hepatocytes.

18.5.3 *In Vivo* Mouse Model of HBT

Lineage tagged mice: Use of lineage tracing has proved to be a very important tool in studying transdifferentiation *in vivo*. Hepatocyte-specific expression of Cre recombinase ensures specificity to hepatocytes [50]. Several Cre reporter strains of mice have been described, in which a *lacZ* gene is turned on in cells expressing Cre recombinase, as well as their daughter cells, following Cre-mediated excision of a loxP-flanked transcriptional "stop" sequence [51]. The green fluorescent protein and variants such as EYFP (enhanced yellow fluorescence protein) and ECFP (enhanced cyan fluorescence protein) offer an advantage over lacZ as a reporter, in that they can be easily visualized without additional assays required to visualize beta-gal in living tissue. Costantini et al. have constructed a generic *ROSA26* targeting vector [25]. Mouse reporter lines have been generated by inserting EYFP or ECFP cDNAs into the *ROSA26* locus, preceded by a loxP-flanked stop sequence. These strains are crossed with transgenic strains expressing tamoxifen-driven Cre in a ubiquitous (β-actin-Cre) or a cell-specific pattern (albumin-Cre and CK19-Cre). Tamoxifen is administered to induce the activity of Cre in these mice.

When these lineage-tagged mice are fed a diet containing 1% DDC for 21 days, they show symptoms of chronic liver injury marked by cholestasis and increased serum bile levels [48]. In this model, the mouse livers present with appearance of atypical ductular proliferation (a.k.a. ductular reaction). These ductular cells originate from the portal tracts, spread into the hepatic lobule, and are associated closely with appearance of hepatocytes harboring an antigen (A6), which is normally expressed by biliary epithelium. The origin of these atypical ductules was always thought to be either progenitor derived or biliary derived. The possibility that these ductular cells could have a hepatocyte origin was tested by Yanger et al. Using this DDC model in lineage-tagged mice, they found evidence suggesting that these atypical ductular cells expressed a biphenotypic molecular signature and were hepatocyte derived [28]. More recently Sekiya and Suzuki, using Cre-based genetic lineage-tracing analysis, established that hepatocytes are a major source of these primitive ductular cells (that arise to compensate for the compromised biliary function) confirming HBT in this model [24].

18.6 WHAT ARE THE DIFFERENT SOURCES OF THE NEW BILIARY CELLS?

In situations where the biliary compartment is damaged, the first and major source of new cells is the remaining resident biliary population [17,25]. In situations where the resident biliary population is unable to proliferate (due to injury), at least two different sources have been documented to replenish biliary cells: (1) hepatic progenitors that reside in Canals of Herring [38,52–55] and (2) hepatocytes that transdifferentiate into biliary cells. The first-ever report suggesting that hepatocytes could transdifferentiate to form biliary cells came from Michalopoulos and colleagues in 2002 [46]. This finding was documented in an *in vitro* model of organoid culture and eventually in an *in vivo* rat model [8]. *In vitro* HBT was duplicated independently by Nishikawa et al. [20].

18.7 WHICH HEPATOCYTES IN THE LIVER TRANSDIFFERENTIATE INTO BILIARY CELLS?

Histological evidence from animal models of HBT identifies periportal areas of the liver to be active in HBT [38,56]. More recently, lineage-tracing experiments have confirmed that periportal hepatocytes preferentially undergo HBT. Whereas "oval cells" formed during biliary to hepatocyte transdifferentiation appear as a morphologically distinct population of transient cells, the transdifferentiation of hepatocytes to biliary cells appears to be limited to periportal hepatocytes and there is no apparent distinct population of this intermediate form of cells [8]. This is further supported by a recent study by Yanger et al. in which ectopic induction of Notch signaling in hepatocytes led to transdifferentiation of hepatocytes to biliary cells from all over the liver except zone 3 hepatocytes located near the central vein (a.k.a. centrilobular or pericentral) [28]. This and another study reported that when the lineage-tagged mice were subjected to chronic liver injury via DDC-fed diet, the biphenotypic intermediate cells (expressing both hepatocyte and biliary markers) were most abundant in the periportal region [24,28]. In normal human livers, such biphenotypic cells occur in Canals of Herring, where biliary epithelial cells are in close direct proximity to hepatocytes [57]. Evidence from human chronic liver disease sections also demonstrates that the intermediate cells that stain positive for both hepatocyte and biliary markers (biphenotypic) are situated in the periportal regions [34,41].

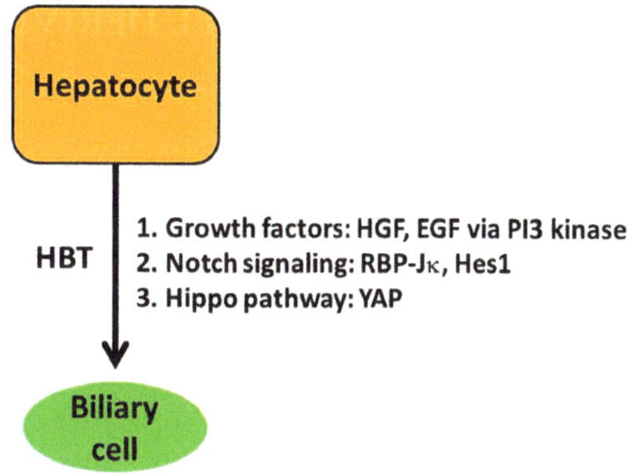

FIGURE 18.2 Signaling for HBT. Three signaling pathways and the individual factors known to play a role in HBT.

18.8 WHAT IS THE EXTENT OF HBT?

It is important to note that a mere decrease in biliary cell number is not sufficient to elicit a HBT response. This is exemplified by the fact that no significant HBT is observed after PHx. Among animal models of liver injury, bile duct ligation by itself has shown minimum (1.75%) HBT [8,39]. Rationale behind this is that the resident biliary cells are unharmed and readily proliferate to make up the lost biliary cell number and function [40]. However, bile duct ligation when coupled with selective injury to resident biliary cells induced by DAPM has shown to significantly increase the rate of HBT (45%), due to the fact that the DAPM-injured biliary cells are unable/insufficient to proliferate and make up for the loss [8,40]. DDC diet-induced chronic liver injury in mice involves significant HBT as well [24,28].

18.9 MECHANISMS, DRIVERS, AND SIGNALING FOR HBT

At least three pathways have been implicated to play a role in HBT (Figure 18.2). Michalopoulos and colleagues reported that growth factors HGF and EGF promoted HBT via AKT-independent PI3 kinase signaling [58]. Later, Yanger et al. identified the role of Notch signaling in HBT via RBP-Jκ, an essential component of the canonical Notch pathway [28]. Sekiya and Suzuki further identified Hes1 as the target for RBP-Jκ in this pathway [24]. Although these studies support the role of Notch signaling in HBT observed in DDC-fed mouse model, regeneration of intrahepatic bile ducts has been shown to occur even without Notch signaling involving RBP-Jκ and HNF6 [59]. These newly formed intrahepatic bile duct cells post ductular reaction are negative for RBP-Jκ and HNF6 but positive for Sox9 and hepatocyte markers [59]. Most recently, Camargo et al. reported that Hippo pathway influences the liver cell fate via Notch signaling. Using a combination of lineage tracing, clonal analysis, and organoid culture approaches, they documented that acute inactivation of Hippo pathway signaling (via YAP) dedifferentiates hepatocytes to biphenotypic or "progenitor cell" characteristics [42]. These studies shed some much awaited light on the mechanisms involved in HBT.

18.10 PROGENITOR CELL DERIVED OR HBT?

Current literature is inconclusive on the existence of a well-established pool of liver stem cells. A primary reason for this is that unlike skin and intestine, which have a resident stem cell population that can be identified using standard microscopy techniques, no such population of cells is visible in quiescent liver [40,60]. However, scientific evidence does support the emergence of bipotential or progenitor cells after infliction of certain disease conditions (in human livers) and experimental models (in animals). These cells typically express markers of both lineages. Various studies have identified "liver progenitor cells" that have differentiated into hepatocytes or biliary cells depending on the cell injury model. These progenitor cells are identified based on expression of one or more markers including CD133 [14], Foxl1 [36,37], AFP [61], Sox9 [38,53], and OV6 [62], among others. Grompe et al. have utilized cell surface-marking antibodies, including MIC1-1C3, to isolate putative adult liver progenitor cells and tested their capacity to form bilineage colonies *in vitro* [63]. *In vivo* lineage tracing using Sox9creER(T2)-R26R(YFP) mice revealed that the cells that proliferate during progenitor-driven liver regeneration are progeny of Sox9-expressing precursors [63]. More recent evidence suggests that expression of new genes like *Lgr5* [64] and *TWEAK* [65] (of the TNF family) may regulate the proliferation of these so-called progenitor cells. Additionally, osteopontin has also been shown to regulate progenitor cell response [66] in the liver. Whether these progenitor cells arise from a specific pool of cells existing at all times in the liver or are a transient phenotype in the process of transdifferentiation, or both, is a much debated topic and an area of extensive scrutiny in liver biology at present.

On the other hand, lineage-tracing experiments in liver injury models suggest that these bipotential cells were at some point in time, differentiated hepatocytes or biliary cells (based on expression of lineage-specific transcription factors) providing evidence for transdifferentiation of lineage-committed differentiated cells [5,7,24,28,42,63]. Considering that both hepatocytes and biliary cells are derived from the same cell type (hepatoblast) during embryonic development, and that their differentiation into specific lineages takes place considerably late in rodents and mid-trimester in human gestation [34,38,56,67], the phenomenon of transdifferentiation is not at all technically improbable or surprising.

18.11 IMPACT OF HBT IN LIVER DISEASE AND THERAPY

Current scientific evidence suggests involvement of multiple pathways for regeneration and repair of the biliary compartment. These include (1) clonal expansion of differentiated biliary cells, (2) progenitor cell derived, and (3) HBT. The extent of contribution of each of these pathways is speculated to be dependent on the nature and extent of the inflicted injury.

Inadequate biliary function is observed in many disease conditions. These include paucity of intrahepatic bile ducts and lost biliary structure and function during chronic biliary liver disease, among others. Current therapy for chronic biliary disease includes symptomatic relief via Ursadiol [68] and liver transplant. Liver transplant is a complex and expensive procedure including challenges like long wait times, host rejection, and lifelong immunosuppressant therapy. Although stem cell therapy is a potentially promising alternative for the future, current challenges with stem cell research battle with availability, extensive resetting of epigenetic marks, and possibility of teratoma formation, among others [69].

Induction of lineage reprogramming in the form of HBT can be a very novel and unique strategy, considering that (1) both hepatocytes and biliary cells originate from a common developmental cell-type hepatoblast (hence the amount of reprogramming required is less extensive as compared to generation from stem or induced pluripotent stem cells; this decreases the possibility of lineage reversion or teratoma formation), (2) this process occurs naturally at a low scale in chronic biliary disease (the required signaling mechanisms are already in place), and (3) host rejection is a nonissue.

Some potential applications for this phenomenon could be the treatment of chronic biliary diseases like cholestasis and Alagille syndrome. Induction of HBT in the hepatocyte compartment of the patient's liver is a potential therapeutic intervention in conjugation with current therapies to postpone/prevent liver fibrosis in (1) paucity of intrahepatic bile ducts in pediatric patients and (2) chronic biliary liver diseases. Such a therapy that focuses on manipulation of the host's native healthy cell compartment (hepatocytes) to make cells of damaged compartment (biliary cells in this case) is an innovative approach that could preclude/decrease the complications involved with liver transplantation or stem cell therapy.

18.12 FUTURE DIRECTION

Are hepatocytes that transdifferentiate to biliary cells a special population of cells expressing a different molecular signature? Do they exist in quiescent liver at all times? Are these hepatocytes metabolically similar to those that do not transdifferentiate? What percent of hepatocytes constitute these cells at any given time in a liver? These questions point at the future direction of this area of research. More characterization of correlation between the type of liver disease and extent of transdifferentiation is required. Extracellular factors that act as drivers for specific hepatocytes to initiate transdifferentiation to biliary cells need to be identified as these will pave the way for future therapeutic interventions.

Mechanistic studies to investigate HBT started with experimental models utilizing biliary-specific toxin (DAPM) paired with bile duct ligation. Identification of transdifferentiation in this model relied heavily on dual staining of both the lineage markers in the same cells. Generation of chimeric rat livers in which only the hepatocytes but not the biliary cells expressed DPP4 marker was the first elegant method to identify biliary cells that originated from hepatocytes *in vivo*. More recently, lineage-tagged mice have provided with a very important tool to establish and study this phenomenon *in vivo*. Developments in animal models have allowed the creation of mice with genetic alterations that cause hepatocyte damage that results, over time, in the loss of native hepatocytes. If donor human hepatocytes are transplanted into these animals, they repopulate the host liver, frequently replacing over 70% of the native liver with human cells. Immunodeficient mice that overexpress urokinase-type plasminogen activator (uPA) and, alternatively, with a knockout of the fumarylacetoacetate hydrolase (Fah) genes are the two most common mouse models for these studies. These mice are called chimeric or "humanized" because the liver is now partially repopulated with human cells [70]. These mice have opened a new and promising possibility to investigate how human hepatocytes would behave in different injury models. Documentation of HBT in these humanized mice would validate the usefulness of this phenomenon in human chronic biliary disease.

In recent years, many efforts have been aimed at generating pluripotent stem cells from somatic cells by inducing high expression of a combination of transcription (reprogramming) factors including Sox2, OCT4, Nanog, Klf4, and cMyc [71–75]. While these reprogramming factors are expressed in stem cells, their expression in adult somatic cells with high potential for clonal expansion such as hepatocytes has been less explored. Bhave et al. have documented that adult rat liver expresses some of these reprogramming factors in hepatocytes and biliary cells during hepatocyte proliferation [76]. They have further demonstrated that inhibition of these reprogramming factor(s) affects the survival and proliferation of hepatocytes in culture [76]. However, whether one or more reprogramming factors are involved in HBT (a process involving somatic cell reprogramming) when the need arises is not known and is an interesting avenue for further investigation.

References

[1] Orkin SH, Zon LI. Hematopoiesis: an evolving paradigm for stem cell biology. Cell 2008;132:631–44.

[2] Graf T, Enver T. Forcing cells to change lineages. Nature 2009;462:587–94.

[3] Pournasr B, Khaloughi K, Salekdeh GH, Totonchi M, Shahbazi E, Baharvand H. Concise review: alchemy of biology: generating desired cell types from abundant and accessible cells. Stem Cells 2011;29:1933–41.

[4] Selman K, Kafatos FC. Transdifferentiation in the labial gland of silk moths: is DNA required for cellular metamorphosis? Cell Differ 1974;3:81–94.

[5] Evarts RP, Hu Z, Omori N, Omori M, Marsden ER, Thorgeirsson SS. Precursor-product relationship between oval cells and hepatocytes: comparison between tritiated thymidine and bromodeoxyuridine as tracers. Carcinogenesis 1996;17:2143–51.

[6] Nagy P, Bisgaard HC, Thorgeirsson SS. Expression of hepatic transcription factors during liver development and oval cell differentiation. J Cell Biol 1994;126:223–33.

[7] Petersen BE, Zajac VF, Michalopoulos GK. Bile ductular damage induced by methylene dianiline inhibits oval cell activation. Am J Pathol 1997;151:905–9.

[8] Michalopoulos GK, Barua L, Bowen WC. Transdifferentiation of rat hepatocytes into biliary cells after bile duct ligation and toxic biliary injury. Hepatology 2005;41:535–44.

[9] Hu Z, Evarts RP, Fujio K, Omori N, Omori M, Marsden ER, et al. Expression of transforming growth factor alpha/epidermal growth factor receptor, hepatocyte growth factor/c-met and acidic fibroblast growth factor/fibroblast growth factor receptors during hepatocarcinogenesis. Carcinogenesis 1996;17:931–8.

[10] Kalluri R, Weinberg RA. The basics of epithelial-mesenchymal transition. J Clin Invest 2009;119:1420–8.

[11] Choi SS, Diehl AM. Epithelial-to-mesenchymal transitions in the liver. Hepatology 2009;50:2007–13.

[12] Xie G, Diehl AM. Evidence for and against epithelial-to-mesenchymal transition in the liver. Am J Physiol Gastrointest Liver Physiol 2013;305:G881–90.

[13] Choi SS, Omenetti A, Witek RP, Moylan CA, Syn WK, Jung Y, et al. Hedgehog pathway activation and epithelial-to-mesenchymal transitions during myofibroblastic transformation of rat hepatic cells in culture and cirrhosis. Am J Physiol Gastrointest Liver Physiol 2009;297:G1093–106.

[14] Kordes C, Sawitza I, Muller-Marbach A, Ale-Agha N, Keitel V, Klonowski-Stumpe H, et al. $CD133^+$ hepatic stellate cells are progenitor cells. Biochem Biophys Res Commun 2007;352:410–7.

[15] Gehling UM, Willems M, Dandri M, Petersen J, Berna M, Thill M, et al. Partial hepatectomy induces mobilization of a unique population of haematopoietic progenitor cells in human healthy liver donors. J Hepatol 2005;43:845–53.

[16] Michalopoulos GK, DeFrances MC. Liver regeneration. Science 1997;276:60–6.

[17] Michalopoulos GK. Principles of liver regeneration and growth homeostasis. Compr Physiol 2013;3:485–513.

[18] Michalopoulos GK. Liver regeneration. J Cell Physiol 2007;213:286–300.

[19] Wang L, Wang X, Xie G, Wang L, Hill CK, DeLeve LD. Liver sinusoidal endothelial cell progenitor cells promote liver regeneration in rats. J Clin Invest 2012;122:1567–73.

[20] Nishikawa Y, Doi Y, Watanabe H, Tokairin T, Omori Y, Su M, et al. Transdifferentiation of mature rat hepatocytes into bile duct-like cells in vitro. Am J Pathol 2005;166:1077–88.

[21] Cable EE, Isom HC. Exposure of primary rat hepatocytes in long-term DMSO culture to selected transition metals induces hepatocyte proliferation and formation of duct-like structures. Hepatology 1997;26:1444–57.

[22] Desmet V, Roskams T, Van Eyken P. Ductular reaction in the liver. Pathol Res Pract 1995;191:513–24.

[23] Sirica AE. Ductular hepatocytes. Histol Histopathol 1995;10:433–56.

[24] Sekiya S, Suzuki A. Hepatocytes, rather than cholangiocytes, can be the major source of primitive ductules in the chronically injured mouse liver. Am J Pathol 2014;184:1468–78.

[25] Srinivas S, Watanabe T, Lin CS, William CM, Tanabe Y, Jessell TM, et al. Cre reporter strains produced by targeted insertion of EYFP and ECFP into the ROSA26 locus. BMC Dev Biol 2001;1:4.

[26] Stocker E, Heine WD. Regeneration of liver parenchyma under normal and pathological conditions. Beitr Pathol 1971;144:400–8.

[27] Fausto N. Liver regeneration and repair: hepatocytes, progenitor cells, and stem cells. Hepatology 2004;39:1477–87.

[28] Yanger K, Zong Y, Maggs LR, Shapira SN, Maddipati R, Aiello NM, et al. Robust cellular reprogramming occurs spontaneously during liver regeneration. Genes Dev 2013;27:719–24.

[29] Nijjar SS, Wallace L, Crosby HA, Hubscher SG, Strain AJ. Altered Notch ligand expression in human liver disease: further evidence for a role of the Notch signaling pathway in hepatic neovascularization and biliary ductular defects. Am J Pathol 2002;160: 1695–703.

[30] Nijjar SS, Crosby HA, Wallace L, Hubscher SG, Strain AJ. Notch receptor expression in adult human liver: a possible role in bile duct formation and hepatic neovascularization. Hepatology 2001;34:1184–92.

[31] Crosby HA, Hubscher SG, Joplin RE, Kelly DA, Strain AJ. Immunolocalization of OV-6, a putative progenitor cell marker in human fetal and diseased pediatric liver. Hepatology 1998;28:980–5.

[32] Crosby HA, Hubscher S, Fabris L, Joplin R, Sell S, Kelly D, et al. Immunolocalization of putative human liver progenitor cells in livers from patients with end-stage primary biliary cirrhosis and sclerosing cholangitis using the monoclonal antibody OV-6. Am J Pathol 1998;152:771–9.

[33] Strain AJ, Crosby HA, Nijjar S, Kelly DA, Hubscher SG. Human liver-derived stem cells. Semin Liver Dis 2003;23:373–84.

[34] Limaye PB, Alarcon G, Walls AL, Nalesnik MA, Michalopoulos GK, Demetris AJ, et al. Expression of specific hepatocyte and cholangiocyte transcription factors in human liver disease and embryonic development. Lab Invest 2008;88:865–72.

[35] Dorrell C, Erker L, Schug J, Kopp JL, Canaday PS, Fox AJ, et al. Prospective isolation of a bipotential clonogenic liver progenitor cell in adult mice. Genes Dev 2011;25:1193–203.

[36] Shin S, Walton G, Aoki R, Brondell K, Schug J, Fox A, et al. Foxl1-Cre-marked adult hepatic progenitors have clonogenic and bilineage differentiation potential. Genes Dev 2011;25:1185–92.

[37] Sackett SD, Li Z, Hurtt R, Gao Y, Wells RG, Brondell K, et al. Foxl1 is a marker of bipotential hepatic progenitor cells in mice. Hepatology 2009;49:920–9.

[38] Carpentier R, Suner RE, van Hul N, Kopp JL, Beaudry JB, Cordi S, et al. Embryonic ductal plate cells give rise to cholangiocytes, periportal hepatocytes, and adult liver progenitor cells. Gastroenterology 2011;141:1432–8, e1431-4.

[39] Malato Y, Naqvi S, Schurmann N, Ng R, Wang B, Zape J, et al. Fate tracing of mature hepatocytes in mouse liver homeostasis and regeneration. J Clin Invest 2011;121:4850–60.

[40] Michalopoulos GK. The liver is a peculiar organ when it comes to stem cells. Am J Pathol 2014;184:1263–7.

[41] Fan B, Malato Y, Calvisi DF, Naqvi S, Razumilava N, Ribback S, et al. Cholangiocarcinomas can originate from hepatocytes in mice. J Clin Invest 2012;122:2911–5.

[42] Yimlamai D, Christodoulou C, Galli GG, Yanger K, Pepe-Mooney B, Gurung B, et al. Hippo pathway activity influences liver cell fate. Cell 2014;157:1324–38.

[43] Sekiya S, Suzuki A. Intrahepatic cholangiocarcinoma can arise from Notch-mediated conversion of hepatocytes. J Clin Invest 2012;122:3914–8.

[44] Gouw AS, Clouston AD, Theise ND. Ductular reactions in human liver: diversity at the interface. Hepatology 2011;54:1853–63.

[45] Komuta M, Govaere O, Vandecaveye V, Akiba J, Van Steenbergen W, Verslype C, et al. Histological diversity in cholangiocellular carcinoma reflects the different cholangiocyte phenotypes. Hepatology 2012;55:1876–88.

[46] Michalopoulos GK, Bowen WC, Mule K, Lopez-Talavera JC, Mars W. Hepatocytes undergo phenotypic transformation to biliary epithelium in organoid cultures. Hepatology 2002;36:278–83.

[47] Michalopoulos GK, Bowen WC, Mule K, Stolz DB. Histological organization in hepatocyte organoid cultures. Am J Pathol 2001;159:1877–87.

[48] Preisegger KH, Factor VM, Fuchsbichler A, Stumptner C, Denk H, Thorgeirsson SS. Atypical ductular proliferation and its inhibition by transforming growth factor beta1 in the 3,5-diethoxycarbonyl-1,4-dihydrocollidine mouse model for chronic alcoholic liver disease. Lab Invest 1999;79:103–9.

[49] Laconi E, Oren R, Mukhopadhyay DK, Hurston E, Laconi S, Pani P, et al. Long-term, near-total liver replacement by transplantation of isolated hepatocytes in rats treated with retrorsine. Am J Pathol 1998;153:319–29.

[50] Kellendonk C, Opherk C, Anlag K, Schutz G, Tronche F. Hepatocyte-specific expression of Cre recombinase. Genesis 2000;26:151–3.

[51] Akagi K, Sandig V, Vooijs M, Van der Valk M, Giovannini M, Strauss M, et al. Cre-mediated somatic site-specific recombination in mice. Nucleic Acids Res 1997;25:1766–73.

[52] Greenbaum LE. The ductal plate: a source of progenitors and hepatocytes in the adult liver. Gastroenterology 2011;141:1152–5.

[53] Furuyama K, Kawaguchi Y, Akiyama H, Horiguchi M, Kodama S, Kuhara T, et al. Continuous cell supply from a Sox9-expressing progenitor zone in adult liver, exocrine pancreas and intestine. Nat Genet 2011; 43:34–41.

[54] Cardinale V, Wang Y, Carpino G, Mendel G, Alpini G, Gaudio E, et al. The biliary tree—a reservoir of multipotent stem cells. Nat Rev Gastroenterol Hepatol 2012;9:231–40.

[55] Carpino G, Cardinale V, Onori P, Franchitto A, Berloco PB, Rossi M, et al. Biliary tree stem/progenitor cells in glands of extrahepatic and intraheptic bile ducts: an anatomical in situ study yielding evidence of maturational lineages. J Anat 2012;220:186–99.

[56] Clotman F, Jacquemin P, Plumb-Rudewiez N, Pierreux CE, Van der Smissen P, Dietz HC, et al. Control of liver cell fate decision by a gradient of TGF beta signaling modulated by Onecut transcription factors. Genes Dev 2005;19:1849–54.

[57] Isse K, Lesniak A, Grama K, Maier J, Specht S, Castillo-Rama M, et al. Preexisting epithelial diversity in normal human livers: a tissue-tethered cytometric analysis in portal/periportal epithelial cells. Hepatology 2013;57:1632–43.

[58] Limaye PB, Bowen WC, Orr AV, Luo J, Tseng GC, Michalopoulos GK. Mechanisms of hepatocyte growth factor-mediated and epidermal growth factor-mediated signaling in transdifferentiation of rat hepatocytes to biliary epithelium. Hepatology 2008;47:1702–13.

[59] Walter TJ, Vanderpool C, Cast AE, Huppert SS. Intrahepatic bile duct regeneration in mice does not require Hnf6 or Notch signaling through Rbpj. Am J Pathol 2014;184:1479–88.

[60] Michalopoulos GK. Advances in liver regeneration. Expert Rev Gastroenterol Hepatol 2014;8:897–907.

[61] Li J, Xin J, Zhang L, Wu J, Jiang L, Zhou Q, et al. Human hepatic progenitor cells express hematopoietic cell markers CD45 and CD109. Int J Med Sci 2014;11:65–79.

[62] Oh SH, Hatch HM, Petersen BE. Hepatic oval 'stem' cell in liver regeneration. Semin Cell Dev Biol 2002;13:405–9.

[63] Tarlow BD, Pelz C, Naugler WE, Wakefield L, Wilson EM, Finegold MJ, et al. Bipotential adult liver progenitors are derived from chronically injured mature hepatocytes. Cell Stem Cell 2014;15:605–18.

[64] Huch M, Dorrell C, Boj SF, van Es JH, Li VS, van de Wetering M, et al. In vitro expansion of single $Lgr5^+$ liver stem cells induced by Wnt-driven regeneration. Nature 2013;494:247–50.

[65] Jakubowski A, Ambrose C, Parr M, Lincecum JM, Wang MZ, Zheng TS, et al. TWEAK induces liver progenitor cell proliferation. J Clin Invest 2005;115:2330–40.

[66] Coombes JD, Swiderska-Syn M, Dolle L, Reid D, Eksteen B, Claridge L, et al. Osteopontin neutralisation abrogates the liver progenitor cell response and fibrogenesis in mice. Gut 2014, pii: utjnl-2013-306484, doi:10.1136/gutjnl-2013-306484.

[67] Desmet VJ. Congenital diseases of intrahepatic bile ducts: variations on the theme "ductal plate malformation" Hepatology 1992;16:1069–83.

[68] Mousa HS, Lleo A, Invernizzi P, Bowlus CL, Gershwin ME. Advances in pharmacotherapy for primary biliary cirrhosis. Expert Opin Pharmacother 2014;16:1–11.

[69] Harding J, Mirochnitchenko O. Preclinical studies for induced pluripotent stem cell-based therapeutics. J Biol Chem 2014;289:4585–93.

[70] Strom SC, Davila J, Grompe M. Chimeric mice with humanized liver: tools for the study of drug metabolism, excretion, and toxicity. Methods Mol Biol 2010;640:491–509.

[71] Lowry WE, Richter L, Yachechko R, Pyle AD, Tchieu J, Sridharan R, et al. Generation of human induced pluripotent stem cells from dermal fibroblasts. Proc Natl Acad Sci USA 2008;105:2883–8.

[72] Park IH, Zhao R, West JA, Yabuuchi A, Huo H, Ince TA, et al. Reprogramming of human somatic cells to pluripotency with defined factors. Nature 2008;451:141–6.

[73] Sullivan GJ, Hay DC, Park IH, Fletcher J, Hannoun Z, Payne CM, et al. Generation of functional human hepatic endoderm from human induced pluripotent stem cells. Hepatology 2010;51:329–35.

[74] Takahashi K, Tanabe K, Ohnuki M, Narita M, Ichisaka T, Tomoda K, et al. Induction of pluripotent stem cells from adult human fibroblasts by defined factors. Cell 2007;131:861–72.

[75] Ye Z, Zhan H, Mali P, Dowey S, Williams DM, Jang YY, et al. Human-induced pluripotent stem cells from blood cells of healthy donors and patients with acquired blood disorders. Blood 2009;114:5473–80.

[76] Bhave VS, Paranjpe S, Bowen WC, Donthamsetty S, Bell AW, Khillan JS, et al. Genes inducing iPS phenotype play a role in hepatocyte survival and proliferation in vitro and liver regeneration in vivo. Hepatology 2011;54:1360–70.

CHAPTER

19

Hepatocyte-Like Cells Derived from Pluripotent Stem Cells

Yanfeng Li[*,†], *Zsuzsanna Polgar*[*,†], *Namita Roy-Chowdhury*[*,†,‡], *Jayanta Roy-Chowdhury*[*,†,‡]

***Department of Medicine, Division of Gastroenterology and Hepatology, Albert Einstein College of Medicine, Bronx, New York, USA**

[†]Marion Bessin Liver Research Center, Albert Einstein College of Medicine, Bronx, New York, USA

[‡]Department of Genetics, Albert Einstein College of Medicine, Bronx, New York, USA

19.1 INTRODUCTION

The liver plays a central role in protein synthesis, metabolism, detoxification, and excretion. The major parenchymal liver cells, the hepatocytes, conduct many of these diverse functions. Therefore, availability of viable, functional hepatocytes could greatly benefit pharmacological evaluation, creating cellular models of liver diseases for pathophysiological analysis, generating bioartificial liver support, and regenerative therapy of the liver. In the intact liver, the viability, proliferation, and function of hepatocytes is supported by the intricate microanatomy of the organ and interaction with nonparenchymal cells. Maintaining long-term survival and reproducing most of the *in vivo* functions of hepatocytes *in vitro* culture continues to be a major challenge. Liver transplantation has radically improved the prognosis of acute and chronic liver failure, as well as monogenic liver diseases, such as urea cycle disorders, Crigler-Najjar Syndrome type 1, α-1 antitrypsin deficiency, and primary hyperoxaluria. As liver transplantation is complex and expensive, has significant mortality and morbidity, and is dependent on immediate availability of a transplantable organ, investigators have been exploring the transplantation of isolated hepatocytes as a minimally invasive alternative for some of these disorders. Unfortunately, additional research and universal application of this promising technology is severely constrained by the universal shortage of donor livers, which are, quite reasonably, prioritized for organ transplantation. Rapid deterioration of the function of cultured primary hepatocytes and the inconsistency of their viability after cryopreservation present additional hurdles. Thus, a renewable source of human hepatocytes is urgently needed. A variety of stem cells, including mesenchymal and hematopoietic stem cells, liver progenitor cells,

Liver Regeneration
http://dx.doi.org/10.1016/B978-0-12-420128-6.00019-1

and induced pluripotent stem cells (iPSCs), are being evaluated as sources of human hepatocytes. Here, we will discuss the approaches for reprogramming normal or patient-specific somatic cells to iPSCs and their differentiation into hepatocyte-like cells (Figure 19.1).

19.2 PLURIPOTENT STEM CELLS FOR DISEASE MODELING AND REGENERATIVE MEDICINE

Unlimited proliferation capacity and pluripotency of human embryonic stem cells [1] have enabled much progress in disease modeling, drug testing, and regenerative medicine. In their landmark work, Yamanaka and associates were able to reprogram somatic cells to iPSCs [1,2] via relatively short-term expression of only four genes. The iPSCs offer several potential advantages over using ESCs. As no embryo needs to be destroyed in this process, ethical concerns are mitigated. Second, iPSCs generated from individual patients, allow the development of cell-based models of genetic diseases, so that insights can be gained regarding the variation of clinical presentation of diseases resulting from lesions of the same gene, or even identical gene mutations. Finally, autologous cells generated by differentiation of iPSCs from patients with monogenic disorders could be corrected genetically and then transplanted back into the donor, thereby circumventing the need for immune suppression.

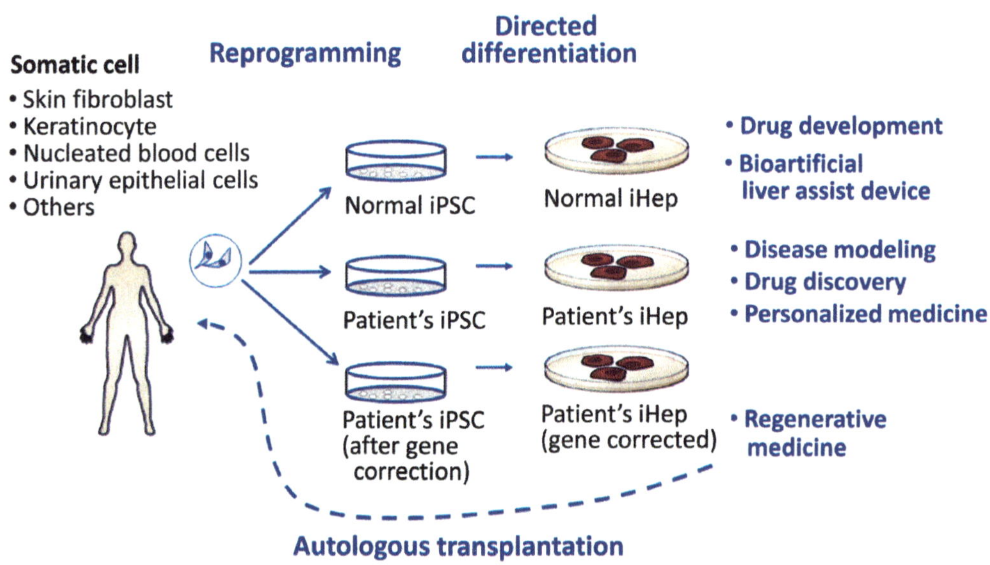

FIGURE 19.1 The different types of somatic cells commonly used for reprogramming to iPSCs are listed. The cells may be obtained from normal subjects or patients with specific inherited diseases. It appears that any mitotically active cell can be reprogrammed to iPSCs. Reprogramming of somatic cells to iPSCs is achieved by expressing the four Yamanaka pluripotency factors, OCT3/4, SOX2, KLF4, and MYC or, alternatively, by expressing the miR 316/302 cluster (see text). Once iPSC clones are isolated, grown, and characterized, they are subjected to a multistep differentiation strategy for generating iHep cells. Normal iHep cells could be used potentially for drug testing or be utilized in bioartificial liver assist devices. iHeps derived from patients are being utilized for disease modeling and drug discovery. Finally, iPSCs derived from individual patients could be corrected genotypically and transplanted back into the donor as autograft after directed differentiation to iHeps.

19.3 GENERATING iPSCs

Extending on the method originally reported by Yamanaka and associates in 2006, a variety of methods have been reported for reprogramming somatic cells to iPSCs. In general, these methods include overexpression of the four pluripotency factors, Oct3/4, Sox2, Klf4, and c-Myc, which are often termed Yamanaka factors. An alternative approach based on the expression of specific microRNAs that may regulate the expression of multiple genes has also been discussed later in this chapter.

19.3.1 Selection of Somatic Cells for Reprogramming

It is not clear whether the starting somatic cells provide any special benefit for generation and subsequent differentiation of iPSCs. Skin fibroblasts were used by Yamanaka et al. in their initial work on reprogramming [2]. In subsequent studies, other investigators have also reprogrammed other cells including peripheral blood cells [3,4], chord blood endothelia [5], adipose-derived stem cells [6], neural stem cells [7], hepatocytes [8], keratinocytes [9], pancreatic β cells [10], amniotic cells [11], or renal tubular epithelial cells shed in the urine [12] for reprogramming into iPSCs. Technically, any proliferative cell type can be reprogrammed. However, several issues need to be considered in selecting the somatic cell source. Although the epigenetic memory of a differentiated cell is mostly erased during reprogramming to iPSCs, some methylation marks may persist, leading to the retention of some characteristics of the original somatic cell. A possible implication of this is that it might be easier to start with a somatic cell, which is similar in gene expression to the differentiated cells that are intended to be generated. The ease of collecting the cells should also be considered. For example, umbilical chord blood epithelial cells would be a highly desirable cell type for reprogramming; however, such cells are usually not available in many patients by the time the diagnosis of an inherited disease is made. Tissue stem cells derived from adipose tissue and bone marrow are also excellent cells to reprogram, but invasive procedures are required to obtain them. Skin fibroblasts are the most common cells used for iPSC generation. These cells are obtained by skin biopsy, which is a microinvasive procedure. A less invasive and convenient source is lymphocytes present in venous blood. However, the random rejoining of the variable, diverse, and joining gene segments [V(D)J recombination] in T and B lymphocytes remains a concern. Therefore, the use of peripheral blood hematopoietic stem cells, which are normally present in very small number, is being explored. Another consideration is the probability of acquired genetic mutations or rearrangements in the somatic cells from which the iPSCs are developed. For example, the lifelong exposure of the skin cells to environmental pollutants, chemicals, and UV rays may potentially cause somatic cell mutations in skin fibroblasts. The incidence of this has not been determined systematically. Recently, renal tubular epithelial cells shed in the urine have been utilized for iPSC generation. Collection of these cells is completely noninvasive. The renal tubular epithelial cells exhibit significant similarities of gene expression to hepatocytes, which may be a potential advantage. In a limited number of observations, no chromosome break or translocation was observed in iPSCs generated from these cells. It should be noted, however, that all somatic cells can acquire mutations, which then persist in the derived iPSCs, even after removing the epigenetic marks. To what extent such mutations may affect the ability of the iPSCs to be differentiated to cell of interest remains to be determined.

19.3.2 Approaches to Reprogramming

The most common approach to reprogramming is based on the original work by Yamanaka and associates [2], in which transient overexpression of only four transcription factors,

OCT3/4, KLF4, SOX2, and MYC (now called Yamanaka factors, Y4) can reprogram somatic cells to iPSCs. A radically different method is based on overexpressing a combination of specific miRNAs, which affect the expression of multiple genes simultaneously.

19.3.2.1.1 Reprogramming Based on Integration of Yamanaka Factors

In their original study, Yamanaka et al. used Moloney's murine leukemia retroviral vectors (MuLVs) that integrate into the cellular genome at random sites. As recombinant MuLVs infect only dividing cells [13], the reprogramming efficiency was low. Expression of the exogenous reprogramming factors is silenced subsequently, but the reprogrammed cells continue to express endogenous pluripotency factors. The transgenes do remain integrated into the host cellular genome at random sites, so that there is continued risk of reactivation of expression of the factors, two of which (cMyc and Klf4) are oncogenes. In contrast, vectors based on a different class of retroviruses, termed lentiviruses, can infect nondividing cells and exhibit a higher reprogramming efficiency [14]. To reduce the possibility of activation of oncogenes, lentiviral vectors have been designed with flanking *Plox* sites to permit the removal of the exogenous genes using *cre* recombinase [15]. However, removal of the exogenous genes may not be complete. Another highly efficient integration-based method utilizes the piggyback transposons, which require only an active transposase and the 13bp inverted terminal repeats for insertion and excision of the reprogramming cassette. The transgenes can be eliminated seamlessly after reprogramming, without leaving any residual nucleic acid sequences in the genome [16].

19.3.2.1.2 Nonintegrating Vectors for Delivering Yamanaka Factors

Despite these design improvements, random integration of transposons or proviruses may be undesirable for future clinical applications, because it increases the potential risk of tumor development [17]. Therefore, other methods have been designed to provide transgene-free iPSCs. Such methods include transfection of modified mRNAs, transfection of "minicircles" from which the bacterial component of plasmids is removed [18] or episomal plasmids containing Epstein-Barr viral sequences, infection with recombinant episomal viruses (e.g., adenoviruses and Sendaï viruses), and delivery of recombinant transcription factor proteins.

Recombinant adenoviruses have a large packaging space, can transduce a wide range of cells from various species, and can be generated at high transduction titers [19]. As the adenoviral vectors are episomal, they are lost rapidly in dividing cells, and repeated infection is needed. Integration of the transgenes transferred by these vectors into the host genome is extremely infrequent, but does occur rarely [20]. Unfortunately, the efficiency for generating iPS cells from primary human cells is much lower compared to mouse fibroblasts [21], which may be related in part to the species difference in the cell-surface expression of the adenoviral receptor (Coxsackie adenovirus receptor) [22].

Another convenient method of transferring pluripotency factor genes consists of transfecting episomal plasmid vectors. Following transfection, conventional plasmids are lost from dividing cells, necessitating repeated transfections and yielding a low reprogramming efficiency [23–25]. To overcome this, episomal vectors containing oriP/Epstein-Barr nuclear-antigen 1 have been developed that can replicate during cell division for about six cycles [26]. Cotransfection of a set of three plasmids expressing the *OCT3/4, SOX2, KLF4, L-MYC and LIN28*, in addition to a shRNA that promotes cell division by suppressing p53 expression [24] results in a relatively high reprogramming efficiency.

Although integration of episomal viruses and plasmids is very infrequent, the chance of integration into the host genome is not eliminated entirely. Therefore, investigators have explored reprogramming using RNAs. Plews et al. [27] reprogrammed somatic cells by transfecting *in vitro*-transcribed modified mRNAs encoding five pluripotency factors. As mRNAs consisting of unmodified ribonucleotides are rapidly lost from the cells because of cellular interferon response, to increase the half life of the transfected mRNAs, cytidine and uridine residues were replaced with 5-methylcytidine and pseudouridine, respectively. In addition, an interferon receptor mimetic, B18R/B19R was coexpressed to reduce interferon binding to its receptor. To further improve the efficiency of this system, internal ribosomal entry site sequences and strong translational initiation signals were added in the 5′UTR, and a polyA signal was added at the 3′UTR [28]. However, despite the efforts at increasing the dwell time of the mRNAs, repeated transfections are still needed for reprogramming, which may be harmful for more sensitive primary cells.

A highly efficient nonintegrative method for iPSC generation has been developed using recombinant Sendaï virus (Hemagglutinating Virus of Japan) [29]. The entire replication cycle of this single-stranded RNA virus of the paramyxovirus family occurs within the cytosol, eliminating the possibility of integration into the genome. A single infection with recombinant Sendaï viruses expressing the Yamanaka pluripotency factors results in a high frequency of iPSC generation from human somatic cells [30–33].

Finally, reprogramming has been achieved also by delivery of the Yamanaka pluripotency factor proteins. Cell-penetrating peptides that are rich in cationic amino acids, such as arginine or lysine (CPP) derived from HIV transactivator of transcription [34,35], were tagged on to the pluripotency transcription factors for transmembrane delivery. The reprogramming efficiency of this method was low [7], probably because of a relatively short dwell time of the proteins in the dividing cells.

Of the various methods for iPSC generation, recombinant Sendai virus has been the most efficient, especially when a small number of the somatic cells are available for reprogramming. For convenience and economy, episomal plasmid vectors are used by a large number of investigators.

19.3.2.2 Reprogramming Methods Based on Overexpression of MicroRNAs

A radically different approach is based on the expression of selected microRNAs (miRNAs), which are able to regulate the expression of multiple genes simultaneously. Several miRNAs expressed preferentially in embryonic stem cells are thought to help maintain the pluripotent cell phenotype [36,37]. The *miR302/367* cluster has been reported to enhance the efficiency of Yamanaka transcription factor-based reprogramming of somatic cells [38]. This cluster, which is located in intron 8 of the *LARP 7* gene on chromosome 3, is highly expressed in pluripotent cells, consists of five miRNAs that are transcribed as a single polycistronic transcript [39]. Four of these miRNAs (*miR301a, b, c, and d*) have identical seed sequences. The *miR302/367* cluster is conserved across species and its expression is induced by the core pluripotency transcription factors, OCT3/4 and SOX2. Expression of the miRNAs from this cluster activates endogenous OCT3/4 expression during reprogramming. The role of these miRNAs in mesenchymal-to-epithelial transition (MET), cell cycle, and epigenetic regulation is suggested by pathway analysis. The *miR302-367* targets transforming growth factor β (TGF-β) receptor 2, promotes E-cadherin expression, accelerates MET, and promotes cell division [40]. Chromatin remodeling is essential in the reprogramming process. Consistent with this, inhibition of histone deacetylase 2 by valproic acid facilitates *miR302/367*-mediated reprogramming [38,40,41].

19.4 DIFFERENTIATION OF iPSs INTO HEPATOCYTE-LIKE CELLS (iHep)

Differentiating iPSCs to iHeps is not only relevant in drug development and regenerative medicine but also permits the understanding of inherited liver diseases by providing cell-based pathophysiological models from individual patients. iHeps generated from patients with various monogenic liver diseases reflected several aspects of the pathologic phenotype of patients. Therefore, this approach may provide deeper insights into the disease processes, identify new therapeutic targets, and even has a potential for the advancement of personalized medicine [42].

Elucidation of the molecular correlates of liver development has contributed to the development of *in vitro* differentiation methods [43,44]. In mammalian embryos, the liver originates from the ventral foregut endoderm. Sequential development of definitive endoderm, hepatic progenitors, and finally mature hepatocytes during mammalian liver development has been used as a model for directed differentiation of iPSCs to iHeps. For the various successful methods for producing iHeps [45], the initial step of generating definitive endoderm cells is crucial in differentiation toward hepatocytes. The interplay of activin A, Wnt3a, fibroblast growth factors (FGFs), and bone morphogenic protein 4 (BMP4) plays a major role in determining the early cell fate of pluripotent stem cells toward definitive endoderm. This is achieved *in vitro* by exposing the iPSCs to two members of the TGF-β superfamily, activin A and BMP4 [46–49]. Wnt 3a is expressed at early stages of mammalian liver development and interacts with activin A [50]. A short exposure to Wnt3a increases the efficiency of definitive endoderm generation. FGFs, together with BMP4, contribute to the definitive endoderm cell commitment at a later stage of embryonic development [51,52]. Interestingly, the effect of all factors that stimulate early mammalian development is inhibited by the presence of fetal bovine serum [53]. SRY (sex-determining region Y)-box 17 (*SOX17*) and forkhead box A2 (*FOXA2*) are specific marker genes that are expressed during embryonic development of definitive endoderm. These proteins specify foregut endoderm, which subsequently gives rise to pancreatic and hepatic cells [54].

In the next step, the definitive endoderm cells are differentiated further to hepatic progenitor cells or hepatoblasts. This is accomplished by adding specific growth factors, of which hepatocyte growth factor appears to be the critical [55,56]. The transcription factor, hepatocyte nuclear factor 4 alpha (HNF4α) is expressed first in the developing hepatic diverticulum. Increased expression of HNF4α during liver development marks the differentiation toward hepatocyte lineage during *in vitro* differentiation [57]. The serum glycoprotein, alpha-fetoprotein (AFP) is an important marker of hepatic progenitor cells and primitive hepatocytes [58].

Maturation of the hepatic progenitor cells to iHeps is induced in culture by adding oncostatin M, an interleukin-6 family cytokine, in combination with glucocorticoids [59] to the cell culture medium. Differentiation to hepatocyte-like cells is associated with the appearance of hepatocyte-like morphology and intracellular glycogen accumulation. This step requires simulation of the molecular/physiological changes that occur during perinatal life. Some hepatocyte-specific functions, such as albumin and apolipoprotein synthesis, and urea production, are near the adult levels during late fetal life. Some other functions, such as UGT1A1-mediated bilirubin glucuronidation and α-1 antitrypsin synthesis are at a very low level at birth and increase rapidly after birth. These changes in gene expression are thought to be the result of perinatal hormonal changes, increased portal blood flow and oxygen tension after the closure of the ductus venosus, elevation of plasma glucagon and catecholamine levels, and possibly other factors. Epigenetic modification of DNA during perinatal life affects the expression of adult hepatocyte-specific genes [60]. Levels of several detoxification enzymes, such as some forms of UDP-glucuronosyltransferase and sulfotransferases, attain adult levels only at adolescence, coincident with the surge of sex hormones. The lack of a comprehensive understanding of the mechanisms of hepatocyte maturation makes it challenging to effectively simulate perinatal changes *in vitro*. However, the experimentally gained information on differentiating hepatoblasts to mature hepatocyte-like cells may help in understanding the process of hepatocyte maturation.

Marker gene expression for evaluating iHep maturation generally includes albumin, CK18, cytochrome p450 enzymes (CyP), α1-antitrypsin (ATT), asialoglycoprotein receptor 1, C/EBPα, UGT1A1, and Prox1. On the other hand, expression of AFP is expected to be markedly downregulated at this stage. Further evaluation of hepatocyte functions includes function of the iHeps, such as urea production, the uptake of indocyanin green, inducible cytochrome P450 activity, and secretion of albumin and α-1 antitrypsin into the medium. Primary hepatocytes are usually used for comparison in these *in vitro* assays. However, a high quality of primary human hepatocytes needs to be assured for this purpose [61]. Han et al. have reviewed the cell culture components used by various investigators for directed differentiation of iPSCs to iHeps [45].

19.4.1 Improving Differentiation of iPSC to iHep

Despite the effort of many laboratories, generation of iHeps that are equivalent to mature primary hepatocytes has not been achieved. It is often stated in the literature that the iHeps are equivalent to hepatic epithelial cells or fetal hepatocytes. In reality, these cells have some characteristics of fetal hepatocytes along with expression of mature hepatocyte markers such as *UGT1A1* and ATT (*SERPINA1*). Expression of most hepatocyte-preferred genes is quantitatively lower in iHeps than in primary hepatocytes. Upon transplantation into the liver of immunodeficient or immunosuppressed hosts, these cells engraft with a lower efficiency than do primary hepatocytes. Another problem that iHeps share with cultured primary hepatocytes is that the expression of liver-specific genes decline rapidly upon prolonged culturing. Therefore, more effort is underway to improve the differentiation of iHep cells and maintaining them in culture.

19.4.2 Forced Expression of Transcription Factors

In addition to manipulation of the media, overexpression of stage-specific transcription factors has been utilized at various steps of differentiation. Thus, adenovector-mediated expression of SOX17 for inducing definitive endoderm [62], HEX for generating hepatoblasts [63], and HNF4α for generating mature iHeps [64] has been used to improve the differentiation. This suggests that sequential transient expression of transcription factors, or perhaps miRNAs, might yield a more desirable phenotype of the iHeps. Until an ideal iHep cells that are very similar to adult primary hepatocytes can be generated, it may be necessary to pay special attention to characteristics required for a specific application. For example, transduction with the nuclear receptor, constitutive androstane receptor, which regulates the expression of multiple detoxification genes at a late stage of iHep differentiation, has been reported to yield iHeps that exhibited high levels of xenobiotic detoxification functions [65].

19.4.3 Maintenance of the iHeps in Culture

Interaction with the liver matrix and other hepatocytes, as well as cross-talk with nonparenchymal cells within the three-dimensional liver structure, plays an important role in maintaining the viability and gene expression characteristics of hepatocytes. Not unexpectedly, neither primary hepatocytes nor iHeps retain hepatocyte-specific functions in "minimalistic" monolayer cultures. As a partial solution to this problem, investigators are attempting to recreate the spatial organization of the liver to support the function and viability of iHeps [66]. Primary hepatocytes and liver sinusoidal epithelial cells appear to support each other in 3D coculture [67]. Therefore, iHeps have been cocultured with endothelial and stromal cells to improve the maturation of the iHeps via cell-cell contact or through paracrine factors [68]. Hepatocyte-specific definitive endoderm has been cocultured with human umbilical vein endothelial cells and mesenchymal stem cells, generating three-dimensional clusters, in which the iHeps expressed α-fetoprotein, albumin, and other hepatocyte-preferred genes, indicating that cluster formation promoted maturation toward a hepatocyte phenotype. When implanted intracranialy, in the small bowel mesentery, or under the renal capsule of immune-deficient mice, the clusters became vascularized and proliferated for 2 months. Secretion of human serum albumin and ATT into the host plasma, expresssion of cytochrome P450 activity, and improved the survival of mice subjected to toxic hepatic injury indicated long-term survival of the transplanted cells. This partially recreated liver organoid did not possess a drainage system, namely, the bile duct. Other investigators have attempted to provide an anatomically complete liver-like structure by decellularizing whole livers by detergent perfusion, and then populating the scaffold with primary human hepatocytes and endothelial cells [69,70]. This decellularization/recellularization approach could be employed potentially to support the survival and function of iHeps.

19.5 USE OF IHEPS AS CELLULAR MODELS OF LIVER DISEASES AND IN LIVER REPOPULATION

Hepatocytes are the primary site of disease expression in many inherited metabolic disorders. However, the difficulty of obtaining primary hepatocytes from patients for *in vitro* studies and the fact that such cells cannot be expanded significantly in culture limits the use of cultured primary hepatocytes as cell-based disease models. As a large variety of somatic cells can now be reprogrammed into iPSCs and then differentiated to iHeps, the latter provides a viable model for genetic disorders. For example, iPSCs derived from patients with Wilson's disease, which is caused by mutations in the copper transport gene *ATP7B* was differentiated to iHeps [71]. The iHeps exhibited mislocalization of mutant ATP7B proteins and a copper transport defect. Transduction with a lentiviral vector expressing the wild-type ATP7B corrected the copper transport defect. In addition, curcumin partially restored the localization of mutant ATP7B proteins, suggesting that its potential as a therapeutic agent in some patients with Wilson's disease. In another study, iHeps generated from a patient with alpha-1 antitrypsin deficiency (ATD) due to a misfolding mutation exhibited accumulation of the mutant protein, which was cleared after treatment with a lentiviral knockdown vector targeting *SERPINA1* [72]. In another study, iHeps from ATD patients with lung disease but no liver disease were found to be different from iHeps derived from ATD patients with liver disease, based on their ability to clear the mutant protein [73]. The ability to generate cellular models for individual patients may permit the identification of factors other than the primary disease-causing mutations that may result in the diversity of clinical presentation and penetrance of a disease caused by various mutations of a single gene, or even identical mutations.

19.6 IHEPS FOR TRANSPLANTATION

Transplanting human iHeps into the liver of rodents has faced hurdles including low engraftment levels and inefficient subsequent proliferation of the transplanted cells. However, recently, preparative X-irradiation of a single liver lobe of recipient rats before transplanting human iHep cells permitted proliferation of the cells in the preconditioned lobe, resulting in reduction of serum bilirubin levels in the Gunn rat model of Crigler-Najjar syndrome type 1 [74].

19.7 SUMMARY

Recent breakthroughs in reprogramming somatic cells to iPSCs and directed differentiation of these cells to iHeps offer great promise for application in generating cellular models of diseases, drug development, and regenerative medicine. Differentiation of the iPSCs toward the hepatocyte phenotype needs further improvement, but research by many laboratories around the world is producing creative solutions for overcoming the existing hurdles to translation of exciting scientific developments to pharmacological, pathophysiological, and therapeutic applications.

Acknowledgments

This work was supported in part by NIDDK 1PO1 DK 096990-01 (to J. R.-C., P.D.: D. Perlmutter); NIDDK DK092469 (to N. R.-C.), and New York Stem Cell Foundation CO26440 (to J. R.-C.).

Conflict of Interest

Yanfeng Li, Zsuzsanna Polgar, Namita Roy-Chowdhury, and Jayanta Roy-Chowdhury declare that they have no conflict of interest.

Human and Animal Rights and Informed Consent

This chapter does not contain any studies with human or animal subjects performed by any of the authors.

References

[1] Takahashi K, Tanabe K, Ohnuki M, Narita M, Ichisaka T, Tomoda K, et al. Induction of pluripotent stem cells from adult human fibroblasts by defined factors. Cell 2007;131:861–72. http://dx.doi.org/10.1016/j.cell.2007.11.019, S0092-8674(07)01471-7 [pii].

[2] Takahashi K, Yamanaka S. Induction of pluripotent stem cells from mouse embryonic and adult fibroblast cultures by defined factors. Cell 2006;126:663–76. http://dx.doi.org/10.1016/j. cell .2006.07.024, S0092-8674(06)00976-7 [pii].

[3] Loh YH, Hartung O, Li H, Guo C, Sahalie JM, Manos PD, et al. Reprogramming of T cells from human peripheral blood. Cell Stem Cell 2009;7:15–9. http://dx.doi.org/10.1016/j.stem.2010.06.004, S1934-5909(10)00286-9 [pii].

[4] Staerk J, Dawlaty MM, Gao Q, Maetzel D, Hanna J, Sommer CA, et al. Reprogramming of human peripheral blood cells to induced pluripotent stem cells. Cell Stem Cell 2010;7:20–4. http://dx.doi.org/10.1016/j.stem.2010.06.002, S1934-5909(10)00284-5 [pii].

[5] Haase A, Olmer R, Schwanke K, Wunderlich S, Merkert S, Hess C, et al. Generation of induced pluripotent stem cells from human cord blood. Cell Stem Cell 2009;5:434–41. http://dx.doi.org/10.1016/j.stem.2009.08.021, S1934-5909(09)00445-7 [pii].

[6] Sun N, Panetta NJ, Gupta DM, Wilson KD, Lee A, Jia F, et al. Feeder-free derivation of induced pluripotent stem cells from adult human adipose stem cells. Proc Natl Acad Sci USA 2009;106:15720–5. http://dx.doi.org/10.1073/pnas.0908450106, 0908450106 [pii].

[7] Kim JB, Greber B, Arauzo-Bravo MJ, Meyer J, Park KI, Zaehres H, et al. Direct reprogramming of human neural stem cells by OCT4. Nature 2009;461:649–53. http://dx.doi.org/10.1038/nature08436, nature08436 [pii].

[8] Liu H, Ye Z, Kim Y, Sharkis S, Jang YY. Generation of endoderm-derived human induced pluripotent stem cells from primary hepatocytes. Hepatology 2010;51:1810–9. http://dx.doi.org/10.1002/hep.23626.

[9] Aasen T, Raya A, Barrero MJ, Garreta E, Consiglio A, Gonzalez F, et al. Efficient and rapid generation of induced pluripotent stem cells from human keratinocytes. Nat Biotechnol 2008;26:1276–84. http://dx.doi.org/10.1038/nbt.1503, nbt.1503 [pii].

[10] Bar-Nur O, Russ HA, Efrat S, Benvenisty N. Epigenetic memory and preferential lineage-specific differentiation in induced pluripotent stem cells derived from human pancreatic islet beta cells. Cell Stem Cell 2011;9:17–23. http://dx.doi.org/10.1016/j.stem.2011.06.007, S1934-5909(11)00293-1 [pii].

[11] Li H, Collado M, Villasante A, Strati K, Ortega S, Canamero M, et al. The Ink4/Arf locus is a barrier for iPS cell reprogramming. Nature 2009;460:1136–9. http://dx.doi.org/10.1038/nature08290, nature08290 [pii].

[12] Zhou T, Benda C, Duzinger S, Huang Y, Li X, Li Y, et al. Generation of induced pluripotent stem cells from urine. J Am Soc Nephrol 2011;22:1221–8. http://dx.doi.org/10.1681/ASN.2011010106, ASN.2011010106 [pii].

[13] Kitamura T, Koshino Y, Shibata F, Oki T, Nakajima H, Nosaka T, et al. Retrovirus-mediated gene transfer and expression cloning: powerful tools in functional genomics. Exp Hematol 2003;31:1007–14, S0301472X03002601 [pii].

[14] Picanco-Castro V, de Sousa Russo-Carbolante EM, Tadeu CD. Advances in lentiviral vectors: a patent review. Recent Pat DNA Gene Seq 2012;6:82–90, DNAG-EPUB-20120605-8 [pii].
[15] Soldner F, Hockemeyer D, Beard C, Gao Q, Bell GW, Cook EG, et al. Parkinson's disease patient-derived induced pluripotent stem cells free of viral reprogramming factors. Cell 2009;136:964–77. http://dx.doi.org/10.1016/j.cell.2009.02.013, S0092-8674(09)00151-2 [pii].
[16] Woltjen K, Michael IP, Mohseni P, Desai R, Mileikovsky M, Hamalainen R, et al. piggyBac transposition reprograms fibroblasts to induced pluripotent stem cells. Nature 2009;458:766–70. http://dx.doi.org/10.1038/nature07863, nature07863 [pii].
[17] Mikkers H, Berns A. Retroviral insertional mutagenesis: tagging cancer pathways. Adv Cancer Res 2003;88:53–99.
[18] Jia F, Wilson KD, Sun N, Gupta DM, Huang M, Li Z, et al. A nonviral minicircle vector for deriving human iPS cells. Nat Methods 2010;7:197–9. http://dx.doi.org/10.1038/nmeth.1426, nmeth.1426 [pii].
[19] Graham FL, Prevec L. Adenovirus-based expression vectors and recombinant vaccines. Biotechnology 1992;20:363–90.
[20] Stephen SL, Sivanandam VG, Kochanek S. Homologous and heterologous recombination between adenovirus vector DNA and chromosomal DNA. J Gene Med 2008;10:1176–89. http://dx.doi.org/10.1002/jgm.1246.
[21] Zhou W, Freed CR. Adenoviral gene delivery can reprogram human fibroblasts to induced pluripotent stem cells. Stem Cells 2009;27:2667–74. http://dx.doi.org/10.1002/stem.201.
[22] Roy-Chowdhury J, Horwitz MS. Evolution of adenoviruses as gene therapy vectors. Mol Ther 2002;5:340–4. http://dx.doi.org/10.1006/mthe.2001.0575, S1525001601905751 [pii].
[23] Gonzalez F, Barragan Monasterio M, Tiscornia G, Montserrat Pulido N, Vassena R, Batlle Morera L, et al. Generation of mouse-induced pluripotent stem cells by transient expression of a single nonviral polycistronic vector. Proc Natl Acad Sci USA 2009;106:8918–22. http://dx.doi.org/10.1073/pnas.0901471106, 0901471106 [pii].
[24] Okita K, Nakagawa M, Hyenjong H, Ichisaka T, Yamanaka S. Generation of mouse induced pluripotent stem cells without viral vectors. Science 2008;322:949–53. http://dx.doi.org/10.1126/science.1164270, 1164270 [pii].
[25] Si-Tayeb K, Noto FK, Sepac A, Sedlic F, Bosnjak ZJ, Lough JW, et al. Generation of human induced pluripotent stem cells by simple transient transfection of plasmid DNA encoding reprogramming factors. BMC Dev Biol 2010;10:81. http://dx.doi.org/10.1186/1471-213X-10-81, 1471-213X-10-81 [pii].
[26] Yu J, Hu K, Smuga-Otto K, Tian S, Stewart R, Slukvin II, et al. Human induced pluripotent stem cells free of vector and transgene sequences. Science 2009;324:797–801. http://dx.doi.org/10.1126/science.1172482, 1172482 [pii].
[27] Plews JR, Li J, Jones M, Moore HD, Mason C, Andrews PW, et al. Activation of pluripotency genes in human fibroblast cells by a novel mRNA based approach. PLoS One 2010;5:e14397. http://dx.doi.org/10.1371/journal.pone.0014397.
[28] Warren L, Manos PD, Ahfeldt T, Loh YH, Li H, Lau F, et al. Highly efficient reprogramming to pluripotency and directed differentiation of human cells with synthetic modified mRNA. Cell Stem Cell 2010;7:618–30. http://dx.doi.org/10.1016/j.stem.2010.08.012, S1934-5909(10)00434-0 [pii].
[29] Li HO, Zhu YF, Asakawa M, Kuma H, Hirata T, Ueda Y, et al. A cytoplasmic RNA vector derived from nontransmissible Sendai virus with efficient gene transfer and expression. J Virol 2000;74:6564–9.
[30] Tokusumi T, Iida A, Hirata T, Kato A, Nagai Y, Hasegawa M. Recombinant Sendai viruses expressing different levels of a foreign reporter gene. Virus Res 2002;86:33–8, S0168170202000473 [pii].
[31] Fusaki N, Ban H, Nishiyama A, Saeki K, Hasegawa M. Efficient induction of transgene-free human pluripotent stem cells using a vector based on Sendai virus, an RNA virus that does not integrate into the host genome. Proc Jpn Acad Ser B Phys Biol Sci 2009;85:348–62, JST.JSTAGE/pjab/85.348 [pii].
[32] Macarthur CC, Fontes A, Ravinder N, Kuninger D, Kaur J, Bailey M, et al. Generation of human-induced pluripotent stem cells by a nonintegrating RNA Sendai virus vector in feeder-free or xeno-free conditions. Stem Cells Int 2012;2012:564612. http://dx.doi.org/10.1155/2012/564612.
[33] Nishimura K, Sano M, Ohtaka M, Furuta B, Umemura Y, Nakajima Y, et al. Development of defective and persistent Sendai virus vector: a unique gene delivery/expression system ideal for cell reprogramming. J Biol Chem 2011;286:4760–71. http://dx.doi.org/10.1074/jbc.M110.183780, M110.183780 [pii].
[34] Frankel AD, Pabo CO. Cellular uptake of the tat protein from human immunodeficiency virus. Cell 1988;55:1189–93, 0092-8674(88)90263-2 [pii].
[35] Ziegler A, Nervi P, Durrenberger M, Seelig J. The cationic cell-penetrating peptide CPP(TAT) derived from the HIV-1 protein TAT is rapidly transported into living fibroblasts: optical, biophysical, and metabolic evidence. Biochemistry 2005;44:138–48. http://dx.doi.org/10.1021/bi0491604.
[36] Wang Y, Baskerville S, Shenoy A, Babiarz JE, Baehner L, Blelloch R. Embryonic stem cell-specific microRNAs regulate the G1-S transition and promote rapid proliferation. Nat Genet 2008;40:1478–83. http://dx.doi.org/10.1038/ng.250, ng.250 [pii].

[37] Wang Y, Blelloch R. Cell cycle regulation by microRNAs in embryonic stem cells. Cancer Res 2009;69:4093–6. http://dx.doi.org/10.1158/0008-5472.CAN-09-0309, 0008-5472.CAN-09-0309 [pii].

[38] Judson RL, Babiarz JE, Venere M, Blelloch R. Embryonic stem cell-specific microRNAs promote induced pluripotency. Nat Biotechnol 2009;27:459–61. http://dx.doi.org/10.1038/nbt.1535, nbt.1535 [pii].

[39] Card DA, Hebbar PB, Li L, Trotter KW, Komatsu Y, Mishina Y, et al. Oct4/Sox2-regulated miR-302 targets cyclin D1 in human embryonic stem cells. Mol Cell Biol 2008;28:6426–38. http://dx.doi.org/10.1128/MCB.00359-08, MCB.00359-08 [pii].

[40] Liao B, Bao X, Liu L, Feng S, Zovoilis A, Liu W, et al. MicroRNA cluster 302–367 enhances somatic cell reprogramming by accelerating a mesenchymal-to-epithelial transition. J Biol Chem 2011;286:17359–64. http://dx.doi.org/10.1074/jbc.C111.235960, C111.235960 [pii].

[41] Anokye-Danso F, Trivedi CM, Juhr D, Gupta M, Cui Z, Tian Y, et al. Highly efficient miRNA-mediated reprogramming of mouse and human somatic cells to pluripotency. Cell Stem Cell 2011;8:376–88. http://dx.doi.org/10.1016/j.stem.2011.03.001, S1934-5909(11)00111-1 [pii].

[42] Park IH, Arora N, Huo H, Maherali N, Ahfeldt T, Shimamura A, et al. Disease-specific induced pluripotent stem cells. Cell 2008;134:877–86. http://dx.doi.org/10.1016/j. cell .2008.07.041, S0092-8674(08)01001-5 [pii].

[43] Irion S, Nostro MC, Kattman SJ, Keller GM. Directed differentiation of pluripotent stem cells: from developmental biology to therapeutic applications. Cold Spring Harb Symp Quant Biol 2008;73:101–10. http://dx.doi.org/10.1101/sqb.2008.73.065, sqb.2008.73.065 [pii].

[44] Murry CE, Keller G. Differentiation of embryonic stem cells to clinically relevant populations: lessons from embryonic development. Cell 2008;132:661–80. http://dx.doi.org/10.1016/j. cell .2008.02.008, S0092-8674(08)00216-X [pii].

[45] Han S, Bourdon A, Hamou W, Dziedzic N, Goldman O, Gouon-Evans V. Generation of functional hepatic cells from pluripotent stem cells. J Stem Cell Res Ther 2012; S10:008:http://dx.doi.org/10.4172/2157-7633.S10-008.

[46] Touboul T, Hannan NR, Corbineau S, Martinez A, Martinet C, Branchereau S, et al. Generation of functional hepatocytes from human embryonic stem cells under chemically defined conditions that recapitulate liver development. Hepatology 2010;51:1754–65. http://dx.doi.org/10.1002/hep.23506.

[47] Yamanaka Y, Ralston A. Early embryonic cell fate decisions in the mouse. Adv Exp Med Biol 2010;695:1–13. http://dx.doi.org/10.1007/978-1-4419-7037-4_1.

[48] Vallier L, Touboul T, Chng Z, Brimpari M, Hannan N, Millan E, et al. Early cell fate decisions of human embryonic stem cells and mouse epiblast stem cells are controlled by the same signalling pathways. PLoS One 2009;4:e6082. http://dx.doi.org/10.1371/journal.pone.0006082.

[49] Teo AK, Ali Y, Wong KY, Chipperfield H, Sadasivam A, Poobalan Y, et al. Activin and BMP4 synergistically promote formation of definitive endoderm in human embryonic stem cells. Stem Cells 2012;30:631–42. http://dx.doi.org/10.1002/stem.1022.

[50] Toivonen S, Lundin K, Balboa D, Ustinov J, Tamminen K, Palgi J, et al. Activin A and Wnt-dependent specification of human definitive endoderm cells. Exp Cell Res 2013;319:2535–44. http://dx.doi.org/10.1016/j.yexcr.2013.07.007, S0014-4827(13)00296-6 [pii].

[51] Hansson M, Olesen DR, Peterslund JM, Engberg N, Kahn M, Winzi M, et al. A late requirement for Wnt and FGF signaling during activin-induced formation of foregut endoderm from mouse embryonic stem cells. Dev Biol 2009;330:286–304. http://dx.doi.org/10.1016/j.ydbio.2009.03.026, S0012-1606(09)00230-9 [pii].

[52] Xu X, Browning VL, Odorico JS. Activin, BMP and FGF pathways cooperate to promote endoderm and pancreatic lineage cell differentiation from human embryonic stem cells. Mech Dev 2011;128:412–27. http://dx.doi.org/10.1016/j.mod.2011.08.001, S0925-4773(11)00093-1 [pii].

[53] Johansson BM, Wiles MV. Evidence for involvement of activin A and bone morphogenetic protein 4 in mammalian mesoderm and hematopoietic development. Mol Cell Biol 1995;15:141–51.

[54] Norrman K, Strombeck A, Semb H, Stahlberg A. Distinct gene expression signatures in human embryonic stem cells differentiated towards definitive endoderm at single-cell level. Methods 2013;59:59–70. http://dx.doi.org/10.1016/j.ymeth.2012.03.030, S1046-2023(12)00086-2 [pii].

[55] Kamiya A, Kinoshita T, Miyajima A. Oncostatin M and hepatocyte growth factor induce hepatic maturation via distinct signaling pathways. FEBS Lett 2001;492:90–4, S0014-5793(01)02140-8 [pii].

[56] Schmidt C, Bladt F, Goedecke S, Brinkmann V, Zschiesche W, Sharpe M, et al. Scatter factor/hepatocyte growth factor is essential for liver development. Nature 1995;373:699–702. http://dx.doi.org/10.1038/373699a0.

[57] Duncan SA, Manova K, Chen WS, Hoodless P, Weinstein DC, Bachvarova RF, et al. Expression of transcription factor HNF-4 in the extraembryonic endoderm, gut, and nephrogenic tissue of the developing mouse embryo: HNF-4 is a marker for primary endoderm in the implanting blastocyst. Proc Natl Acad Sci USA 1994;91:7598–602.

[58] Gualdi R, Bossard P, Zheng M, Hamada Y, Coleman JR, Zaret KS. Hepatic specification of the gut endoderm in vitro: cell signaling and transcriptional control. Genes Dev 1996;10:1670–82.

[59] Kamiya A, Kinoshita T, Ito Y, Matsui T, Morikawa Y, Senba E, et al. Fetal liver development requires a paracrine action of oncostatin M through the gp130 signal transducer. EMBO J 1999;18:2127–36. http://dx.doi.org/10.1093/emboj/18.8.2127.

[60] Thomassin H, Flavin M, Espinas ML, Grange T. Glucocorticoid-induced DNA demethylation and gene memory during development. EMBO J 2001;20:1974–83. http://dx.doi.org/10.1093/emboj/20.8.1974.

[61] Hengstler JG, Brulport M, Schormann W, Bauer A, Hermes M, Nussler AK, et al. Generation of human hepatocytes by stem cell technology: definition of the hepatocyte. Expert Opin Drug Metab Toxicol 2005;1:61–74. http://dx.doi.org/10.1517/17425255.1.1.61.

[62] Takayama K, Inamura M, Kawabata K, Tashiro K, Katayama K, Sakurai F, et al. Efficient and directive generation of two distinct endoderm lineages from human ESCs and iPSCs by differentiation stage-specific SOX17 transduction. PLoS One 2011;6:e21780. http://dx.doi.org/10.1371/journal.pone.0021780, PONE-D-11-02464 [pii].

[63] Inamura M, Kawabata K, Takayama K, Tashiro K, Sakurai F, Katayama K, et al. Efficient generation of hepatoblasts from human ES cells and iPS cells by transient overexpression of homeobox gene HEX. Mol Ther 2011;19:400–7. http://dx.doi.org/10.1038/mt.2010.241, mt2010241 [pii].

[64] Takayama K, Inamura M, Kawabata K, Katayama K, Higuchi M, Tashiro K, et al. Efficient generation of functional hepatocytes from human embryonic stem cells and induced pluripotent stem cells by HNF4alpha transduction. Mol Ther 2012;20:127–37. http://dx.doi.org/10.1038/mt.2011.234, mt2011234 [pii].

[65] Funakoshi N, Duret C, Pascussi JM, Blanc P, Maurel P, Daujat-Chavanieu M, et al. Comparison of hepatic-like cell production from human embryonic stem cells and adult liver progenitor cells: CAR transduction activates a battery of detoxification genes. Stem Cell Rev 2011;7:518–31. http://dx.doi.org/10.1007/s12015-010-9225-3.

[66] Decaens C, Durand M, Grosse B, Cassio D. Which in vitro models could be best used to study hepatocyte polarity? Biol Cell 2008;100:387–98. http://dx.doi.org/10.1042/BC20070127, BC20070127 [pii].

[67] Kim Y, Rajagopalan P. 3D hepatic cultures simultaneously maintain primary hepatocyte and liver sinusoidal endothelial cell phenotypes. PLoS One 2010;5:e15456. http://dx.doi.org/10.1371/journal.pone.0015456.

[68] Stevens KR, Ungrin MD, Schwartz RE, Ng S, Carvalho B, Christine KS, et al. InVERT molding for scalable control of tissue microarchitecture. Nat Commun 2013;4:1847. http://dx.doi.org/10.1038/ncomms2853, ncomms2853 [pii].

[69] Baptista PM, Siddiqui MM, Lozier G, Rodriguez SR, Atala A, Soker S. The use of whole organ decellularization for the generation of a vascularized liver organoid. Hepatology 2011;53:604–17. http://dx.doi.org/10.1002/hep.24067.

[70] Uygun BE, Soto-Gutierrez A, Yagi H, Izamis ML, Guzzardi MA, Shulman C, et al. Organ reengineering through development of a transplantable recellularized liver graft using decellularized liver matrix. Nat Med 2010;16:814–20. http://dx.doi.org/10.1038/nm.2170, nm.2170 [pii].

[71] Zhang S, Chen S, Li W, Guo X, Zhao P, Xu J, et al. Rescue of *ATP7B* function in hepatocyte-like cells from Wilson's disease induced pluripotent stem cells using gene therapy or the chaperone drug curcumin. Hum Mol Genet 2011;20:3176–87.

[72] Eggenschwiler R, Loya K, Wu G, Sharma AD, Sgodda M, Herr C, et al. Sustained knockdown of a disease-causing gene in patient-specific induced pluripotent stem cells using lentiviral vector-based gene therapy. Stem Cells Transl Med 2013;2:641–54.

[73] Tafaleng EN, Chakraborty S, Han B, Hale P, Wu W, Soto-Gutierrez A, et al. Induced pluripotent stem cells model personalized variations in liver disease due to α1-antitrypsin deficiency. Hepatology 2015, in press.

[74] Chen Y, Li Y, Wang X, Zhang W, Sauer V, Chang C-J, et al. Amelioration of hyperbilirubinemia in Gunn rats after transplantation of human induced pluripotein stem cell-derived hepatocytes. Stem Cell Reports 2015, in press.

CHAPTER

20

Hepatic Progenitor Cell Transplantation: Relevant Models and Mechanisms

David A. Shafritz

Marion Bessin Liver Research Center, Division of Gastroenterology and Liver Diseases, Albert Einstein College of Medicine, Bronx, NY, USA

20.1 OVERVIEW

The expectation that liver cell transplantation will become a therapeutic reality in the near future is based on three special properties of this organ: (1) the ability of the liver to regenerate and restore function after massive chemical (toxic) injury or surgical reduction in hepatic mass (2/3 partial hepatectomy), (2) the ability of the liver to remodel into a normal lobular structure following hyperplastic nodular transformation induced by chemical agents, and (3) the unique portal to central venous circulation of this organ that allows transplanted cells introduced through the portal venous system to perfuse through the liver sinusoids and widely seed the hepatic parenchyma. It also needs to be emphasized that restoration of liver mass through proliferation of endogenous hepatocytes, e.g., following chemical injury or surgical reduction in hepatic mass, proceeds rapidly in the normal liver to restore the hepatic cell number within 2-3 days and total liver mass within 1-2 weeks [1–3]. During this process, the bulk of remaining hepatocytes (75-95%) undergoes DNA synthesis, proliferate, and participate in the regenerative/repopulation process [4,5]. It is also well established that hepatic stem or progenitor cells play no significant role in hepatocyte replenishment during liver regeneration in the normal liver [2,3,6].

In studies of liver repopulation by transplanted cells, however, the starting point for repopulation by the transplanted cells is quite different. Because of limitations in the number of cells that can be infused into the liver without causing massive ischemia, liver injury, and hepatic infarction, the number of cells that can be experimentally engrafted represents at most 1-2% of total hepatic mass. The transplanted cells then have to "compete" against a 50-100-fold greater number of host hepatocytes that are also induced to proliferate during the regeneration/repopulation process. Under these circumstances, for significant liver repopulation to occur, the transplanted cells need to have a

Liver Regeneration
http://dx.doi.org/10.1016/B978-0-12-420128-6.00020-8

selective advantage over the host cells they are replacing, and the repopulation process will be much slower than that which occurs following 2/3 PH in the normal liver. Although hepatic progenitor cells presumably have a higher proliferative potential than adult hepatocytes, they begin the repopulation process at a very low percentage of total liver mass, and there may be host mechanisms that restrict their proliferation or their ability to repopulate the liver. Therefore, it may require weeks to months to observe significant liver repopulation by transplanted hepatic progenitor cells. This principle is well established in studies of bone marrow replacement by hematopoietic stem cells in which little repopulation occurs unless the host bone marrow is first ablated, and reconstitution of the bone marrow by transplanted hematopoietic stem cells also occurs very slowly, requiring many months to be completed [7].

A number of experimental models have been developed to repopulate the liver with cells having different phenotypic characteristics and a description of these model systems, and their specific uses are presented in this chapter. What is most encouraging in studies of liver repopulation is that the transplanted cells become structurally and functionally incorporated into the host liver tissue.

20.2 LIVER REPOPULATION BY TRANSPLANTED HEPATOCYTES

The first study reporting the function of transplanted isolated hepatocytes in a recipient liver was in 1976 by Matas et al. [8]. These investigators transplanted a suspension of normal hepatocytes into the portal vein of UDP-glucuronosyl transferase-deficient (Gunn) rats and demonstrated a transient reduction in serum bilirubin. Subsequently, in 1980, Jirtle et al. [9] showed retention of morphologic, ultrastructural, and biological properties of hepatocytes transplanted into the dorsal fat pad of Fischer (F) 344 rats. The first study directly demonstrating that hepatocytes transplanted into the spleen or infused into the liver through the portal vein engraft into the hepatic parenchyma and function long-term was reported in 1990 by Gupta et al. [10]. These investigators transplanted hepatocytes isolated from transgenic mice containing the human Hepatitis B virus surface antigen (*HBsAg*) gene into wild-type (WT) syngeneic C57B16 mice and demonstrated incorporation of HBsAg expressing cells into the periportal region of the hepatic parenchyma and secretion of HBsAg into the serum for up to 36 weeks. These findings were confirmed the following year by Ponder et al. [11], who transplanted either β-gal or human α1-antitrypsin (hAAT) expressing mouse hepatocytes into WT syngeneic C57B16 mice and observed β-gal expressing hepatocytes in the liver tissue or secretion of hAAT into the serum for 6 months. Other studies showed a transient reduction in serum lipids in low-density lipoprotein (LDL) receptor-deficient Watanabe rabbits after transplanting WT hepatocytes under cyclosporine immunosuppression [12].

From the above studies, it was clear that transplanted hepatocytes can functionally engraft into the liver and that this approach could possibly lead to a new method to treat monogeneic liver diseases by transplanting hepatic cells. However, as indicated in the overview, the maximum percentage of liver mass that can be achieved through hepatocyte transplantation is 1-2% and the amount required to achieve effective therapy in inherited monogeneic metabolic disorders of the liver, such as Crigler-Najjar syndrome Type 1 (CN1), familial hypercholesterolemia (LDL-receptor deficiency), factor IX deficiency, or ornithione transcarbamylase deficiency is 3-5% [13,14]. In patients with CN1 [15,16] and familial hypercholesterolemia [17], there has been a partial therapeutic response when $\sim 1.0 \times 10^{10}$ hepatocytes were infused into the liver. However, the therapeutic response was only temporary, and the most likely explanation is that the transplanted cells did not proliferate after their transplantation and were lost over time either by normal cell turnover or by immunorejection [18].

Attempts have been made to increase the number of transplanted hepatocytes in the repopulated liver by stimulating liver regeneration (for example, through the use of partial hepatectomy (PH) or carbon tetrachloride (CCl_4)-induced hepatic necrosis), but these studies have generally been unsuccessful in the normal liver. This is not surprising, because the residual hepatocytes need to undergo only one or two rounds of cell division to replace all of the liver mass removed by two-thirds PH or CCl_4 injection [4,19], and both endogenous and transplanted hepatocytes should contribute equally to this limited proliferative response. However, a modest increase in the level of liver repopulation by transplanted hepatocytes has been achieved by repeated injections of CCl_4 or multiple cell transplantations [20,21], but these procedures are not sufficient for effective cell therapy, and it was necessary to develop alternative approaches to increase liver repopulation by transplanted normal hepatocytes.

20.3 SPECIAL ANIMAL MODELS TO AUGMENT REPOPULATE THE LIVER BY TRANSPLANTED HEPATOCYTES

20.3.1 Acute Massive Liver Injury Models

For many years, the prevailing view was that although mature hepatocytes could undergo 1-2 divisions to restore liver mass after acute toxic injury or 2/3 PH, these cells are "terminally" differentiated and incapable of further proliferation [19,22]. However, based on landmark studies beginning in the early to mid-1990s, this view has been radically changed, as it has been demonstrated repeatedly that the phenotypic behavior of hepatocytes is context dependent, and that the tissue microenvironment plays a major role in regulating the properties and function of hepatocytes, as well as other "differentiated" cell types, in the liver. By manipulating the host microenvironment, several rodent models have been developed in which transplanted normal hepatocytes exhibit high proliferative capability and can extensively repopulate the liver. In 1991, Sandgren et al. [23] reported that transgenic mice expressing the urokinase plasminogen activator (*uPA*) gene under control of the albumin promoter, show massive liver injury and die within 4-6 weeks after birth. However, some mice survived, and in these mice, nodules of normal liver tissue of varying sizes were noted throughout the hepatic parenchyma (Figure 20.1a). These nodules were comprised of normal hepatocytes in which the *uPA* transgene had been deleted, and the cells, which were now free from uPA injury, expanded clonally into large clusters to replace damaged tissue. Based on these findings, Rhim et al. [24] transplanted normal hepatocytes, marked with a β-galactosidase transgene, into uPA mice and observed massive liver repopulation by these cells (Figure 20.1b). In these studies, they estimated that each transplanted hepatocyte underwent ~12-14 cell divisions to fully repopulate the liver [24].

Through targeted disruption of fumarylacetoacetate hydrolase (Fah), the last gene in tyrosine catabolism, Grompe and colleagues developed a gene-knockout model that also permitted transplanted WT mature hepatocytes to totally repopulate the Fah null liver [25]. Deletion of Fah leads to accumulation of fumarylacetoacetate in liver tissue, which causes massive and continuous liver injury. A unique advantage of the Fah-null mouse is that the level of liver injury can be regulated by administration of 2 (2-nitro-4-trifluoromethylbenzoyl)-cyclohexane-1,3-dione (NTBC), a pharmacologic inhibitor of tyrosine catabolism upstream of Fah that prevents accumulation of fumarylacetoacetate, allowing animals to be bred and studied in a controlled fashion.

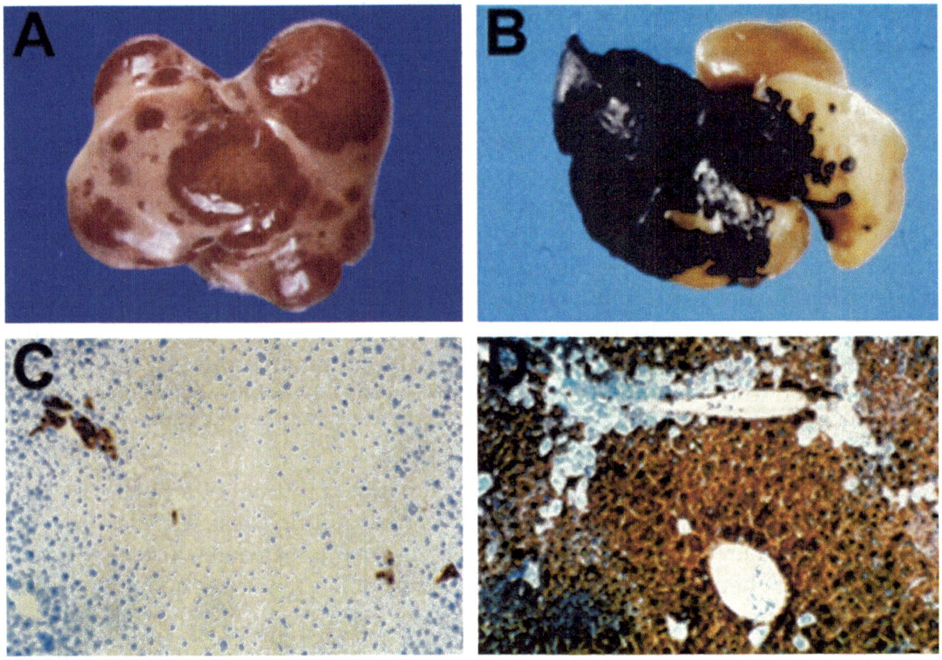

FIGURE 20.1 Alb-uPA transgenic and Fah$^{-/-}$ mouse models for liver repopulation by transplanted cells. (a) Spontaneous regeneration of the uPA transgenic mouse liver by endogenous revertant hepatocytes [23]. Hepatocytes that have deleted the uPA transgene replace uPA transgenic hepatocytes because they are not undergoing liver injury and have a selective advantage over uPA-expressing hepatocytes that are continually being destroyed. (b) Repopulation of the liver of a uPA transgenic mouse by transplanted WT hepatocytes expressing a marker gene, β-galactosidase [24]. Transplanted β-gal expressing, but otherwise normal, mouse hepatocytes have a selective advantage for survival compared to endogenous uPA-expressing hepatocytes that are continuously undergoing destruction by expression of the toxic uPA transgene. (c) Transplanted WT Fah^{+} hepatocyte in the Fah$^{-/-}$ mouse liver in animals maintained on NTBC or in the first several days after NTBC withdrawal [25]. Note the presence of isolated Fah^{+} cells or scattered small aggregates of FAH^{+} hepatocytes. (d) Fah$^{-/-}$ mouse transplanted with WT FAH^{+} hepatocytes 6 weeks after NTBC withdrawal [25]. Under these highly selective conditions, there is massive replacement of the liver by transplanted FAH^{+} hepatocytes.
Adapted from Ref. [26] with permission.

Grompe and coworkers showed that when NTBC is administered to Fah-null mice, host hepatocytes do not die, there is no selective advantage for WT hepatocytes to survive over host Fah-null hepatocytes, and liver repopulation by transplanted WT hepatocytes does not occur. If NTBC administration is continued after cell transplantation, initially only single or small aggregates of Fah-positive hepatocytes are seen within the parenchyma (Figure 20.1c). However, discontinuation of NTBC treatment leads to massive liver injury, and the transplanted WT hepatocytes that have normal tyrosine catabolism and are resistant to liver injury, proliferate extensively, form large clusters within 3 weeks and replace most of the liver mass within 6 weeks (Figure 20.1d). Fah-null mice with livers repopulated by WT hepatocytes remain healthy, have normal liver function tests, and show a relatively normal liver structure for many months after hepatocyte transplantation [25]. These studies were the first to show that liver repopulation can effectively cure a metabolic disease, namely, the mouse equivalent to the human disease, Hereditary Tyrosinemia Type 1[27].

In Fah-null mice, not only do transplanted WT hepatocytes replace Fah-null hepatocytes, but the transplanted cells can also be serially transplanted through seven consecutive Fah-null mice, while retaining full ability to proliferate and replace injured host hepatocytes [28]. In these studies, it was calculated that each serially transplanted hepatocyte underwent an average of at least 69 cell divisions. Under these most adverse experimental conditions in the host liver, transplanted WT murine hepatocytes exhibit two essential features of stem cells, self-renewal and serial transplantability. Thus, transplanted normal adult hepatocytes extensively repopulate the liver and restore function under circumstances in which there is (1) both massive and continuous liver injury and (2) the transplanted hepatocytes have a significant selective advantage for survival compared to host hepatocytes.

20.3.2 Impaired Proliferation of Host Hepatocytes

An alternative method to obtain a high level of liver repopulation by transplanted hepatocytes is to impair the proliferative capacity of endogenous hepatocytes and then transplant normal hepatocytes in conjunction with a liver proliferative stimulus. This was first achieved by treating rats with retrorsine, a plant alkaloid that cross-links cellular DNA and disrupts hepatocyte division [29,30]. When retrorsine or a closely related compound, monocrotaline, is administered to rats or mice [29–34], there is a long-lasting inhibition of hepatocyte proliferation. However, essential metabolic functions of the DNA-damaged hepatocytes are maintained, and the animals survive. After the acute effects of DNA damage had subsided (2-4 weeks), the animals are subjected to two-thirds PH or CCl_4 administration in conjunction with transplantation of hepatocytes from normal animals [31]. As in uPA or Fah-null mice, this leads to a brisk regenerative response by transplanted hepatocytes, leading to extensive repopulation of the liver (Figure 20.2a). Most surprisingly, in the retrorsine/PH model, the transplanted hepatocytes do not develop into hyperplastic nodules, i.e., the liver remodels and the transplanted hepatocytes become fully integrated into the hepatic parenchymal plates (Figure 20.2a) [31]. Transplanted hepatocytes form hybrid canaliculi with neighboring host hepatocytes (Figure 20.2b), a normal liver structure is maintained and at 6-9 months after cell transplantation, more than 90% of host hepatocytes can be replaced (Figure 20.2c and d) [31]. In the retrorsine/PH model, there is no regeneration or repopulation of the bile ducts by transplanted WT hepatocytes. Using the retrorsine/PH model, transplanted WT allogenic hepatocytes can fully repopulate the liver of albumin deficient, Nagase Analbuminemic Rats (NARs) under an immunosuppression protocol [35]. In repopulated NAR, there was 60-80% replacement of albumin-deficient host hepatocytes and a 7000-fold increase in serum albumin to levels within the normal range [35].

Another method to achieve effective liver repopulation by transplanted hepatocytes is to induce DNA damage by selective liver x-irradiation in conjunction with hepatocyte transplantation and either two-thirds PH, CCl_4 administration, or ischemic liver injury [36,37]. In the retrorsine/PH model, thyroid hormone (a hepatocyte mitogen) can partially replace PH as an inducer of liver repopulation [38] and most recently, administration of GC-1, a thyroid hormone receptor-b-selective agonist produces a similar response [39]. In x-irradiated mice, administration of HGF has been used to replace PH as a liver regenerative stimulus [40]. With retrorsine or monocrotaline pretreatment of the host, transplanted hepatocytes have a strong proliferative advantage over host hepatocytes, as DNA cross-linked host hepatocytes cannot undergo cell division. After retrorsine administration, host hepatocytes also exhibit

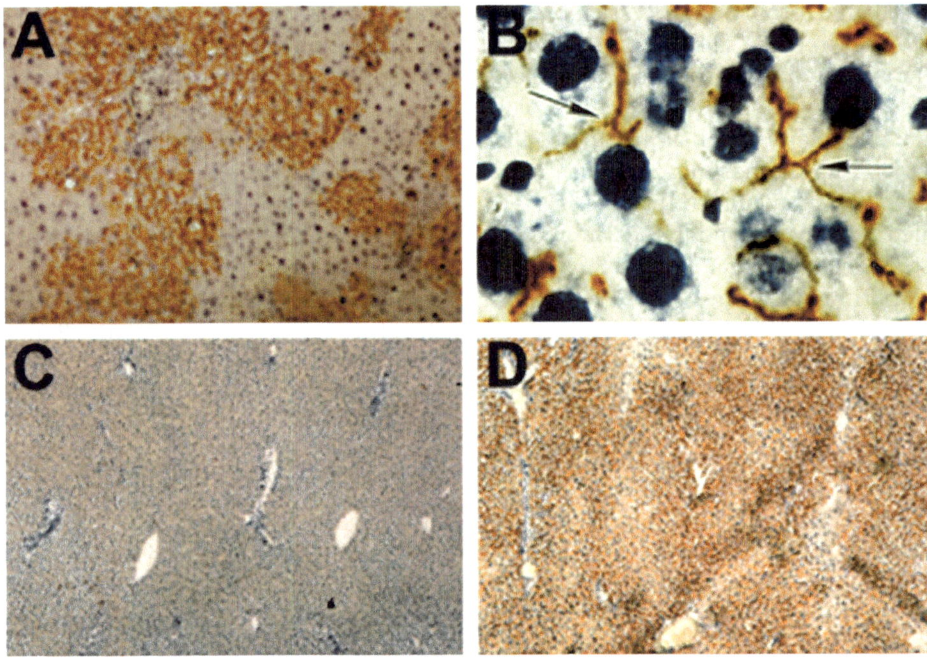

FIGURE 20.2 Retrorsine/PH model for liver repopulation. (a) Repopulation of the liver in a DPPIV$^-$ rat pretreated with two doses of retrorsine before cell transplantation and two-thirds PH at the time of cell transplantation. At 2 months after hepatocyte transplantation, 60-80% liver repopulation is achieved by DPPIV$^+$ hepatocytes without formation of hyperplastic nodules, compression of surrounding host parenchyma, or other abnormalities of host liver tissue [31]. (b) Transplanted hepatocytes from DPPIV$^+$ hybrid bile canaliculi with adjacent DPPIV$^-$ host hepatocytes and are completely integrated into the host parenchyma [31]. (c) Lack of DPPIV expression in hepatectomized host liver tissue removed by two-thirds pH at the time of cell transplantation [31]. (d) 99% replacement of host hepatocytes by DPPIV$^+$-transplanted hepatocytes in the same rat as in (c) at 9 months after transplantation of 1×10^6 DPPIV$^+$ hepatocytes [31].
Adapted from Ref. [26] with permission.

an increased level of apoptosis [38], and removal of these cells further augments liver repopulation by transplanted cells.

20.3.3 Augmented Repopulation Potential of Transplanted Hepatocytes

Other methods to enhance liver repopulation are to transplant donor hepatocytes that have augmented proliferative potential or a selective survival advantage into a host with a normal liver and then stimulate cycles of regeneration by repeated liver injury. Hepatocytes have been isolated from genetically modified mice in which the proliferative activity of hepatocytes has been increased either by deleting a gene that restricts cell proliferation ($p27^{Kip1}$) or introducing a gene that stimulates cell proliferation (Fox M1B). Deleting the Kip1 cyclin kinase inhibitor ($p27^{kip1}$) gene increases cell cycling in all tissues and such animals exhibit increased body weight and multiorgan hyperplasia, including the liver [41–43]. Transplanting $p27^{Kip1}$-null hepatocytes into a normal WT liver produces modest repopulation (~3.27%) [44]. However, this occurs only when the host is subjected to a low level of continuous liver injury/regeneration by administering small doses of

CCl_4 once weekly for 8 weeks [44]. Liver repopulation by $p27^{Kip1}$-null hepatocytes also occurs more rapidly compared to WT hepatocytes in Fah-null mice, which have a much higher level of continuous liver injury [45]. Hepatocytes from transgenic mice over expressing Fox M1B, a winged helix/forkhead transcription factor that enhances cell-cycle progression, also exhibit augmented liver repopulation, but this occurs only in uPA/Scid mice that exhibit continuous liver injury [46]. These studies provide the basis for liver repopulation by progenitor cells, which should exhibit higher repopulation potential than adult hepatocytes.

Another method to augment repopulation by transplanted cells would be to render them resistant to apoptosis. This has been achieved by transplanting hepatocytes from Bcl-2 transgenic mice into the liver of WT mice [47]. Augmented liver repopulation in this model requires repeated administration of anti-FAS antibody, which induces apoptosis in host hepatocytes to which transplanted Bcl-2-expressing hepatocytes are resistant. This approach produced 2-16% liver repopulation by Bcl-2 transgenic hepatocyte in normal host recipients [47]. However, this model has not been used by others because of systemic toxicity of anti-FAS antibody.

20.4 LIVER REPOPULATION BY ENDOGENOUS HEPATIC PROGENITOR CELLS

In 1956, Farber [48] introduced the term "oval cells" to describe small cells with scant basophilic cytoplasm and a small pale blue oval-shaped nucleus that proliferated in the periportal region of the hepatic lobule after treating rats with carcinogenic agents, such as 2-acetylaminofluorene (2-AAF), ethionine, or 3-methyl-4-diethylaminobenzene. Proliferation of "oval cells" also occurs in rats treated with D-galactosamine [49,50], or in mice treated with a choline-deficient/ethionine-augmented (CDE) diet [51,52], allyl alcohol administration [53], dipin, a carcinogenic agent [54], or 3,5-diethyoxy carbonyl-1,4-dihydrocollidine (DDC), which disrupts porphyrin metabolism and is hepatotoxic [55]. Although Farber did not think that "oval cells" are hepatocyte progenitors [56], Thorgeirsson and coworkers demonstrated by pulse-chase studies with ^{3}H-thymidine that "oval cells" proliferating in the periportal region in the rat 2-AAF/PH model differentiated into clusters of morphologically mature hepatocytes [57,58].

Through elegant ultrastructural and immunohistochemical studies, Thorgeirsson and colleagues [59] further demonstrated in the 2-AAF/PH model that "oval cells" are derived from undifferentiated cells in the Canals of Hering. These cells pass through discontinuities in the laminar basement membrane of the ductal limiting plate, enter the hepatic parenchyma, couple with stellate cells, proliferate, and subsequently differentiate into mature hepatocytes. However, most of the ^{3}H-thymidine-labeled "oval cells" are eliminated from the tissue and the contribution of these cells to liver regeneration in the 2-AAF/PH model was not determined [57,58]. Label retaining studies conducted by Kuwahara et al. [60] further support the origin of "oval cells," as well as hepatocytes, from undifferentiated cells in the Canals of Hering.

More recent studies using lineage-tracing tools have demonstrated that hepatic progenitor cells induced to proliferate following bile duct ligation, CDE diet, or DDC feeding of mice can differentiate into both hepatocytes and bile duct epithelial cells [61,62], supporting earlier studies in rats using the 2-AAF/PH model [57–59]. It has also been demonstrated that transplanted hepatocytes can differentiate into bile duct epithelial cells following bile duct ligation coupled with DAPM treatment which completely blocks proliferation of endogenous bile duct cells [63]. Through lineage-tracing studies, it has also been shown in mice treated with bile duct ligation, a CDE diet, or DDC feeding that hepatocytes retrodifferentiate into biphenotypic (Sox9$^+$/CK-19$^+$/OV6$^+$/HNF1b$^+$) hepatic epithelial cells [62,64,65]. Overexpression of Yap and Notch signaling appear to mediate this process [64]. Nonetheless, very few if any of the "oval cells" induced by the DDC or CDE diet differentiate into hepatocytes during a posttreatment recovery period [66] or after their transplantation into the liver of untreated mice [67]. Thus, the role of hepatic epithelial progenitor cells in liver regeneration and normal liver physiology remains controversial [68–70].

Many recent studies, including those conducted in the liver, have shown that the phenotypic expression, proliferative capability, and differentiation properties of a specific cell type, e.g., the hepatocyte, may vary depending on the tissue context in which that particular cell is located. Therefore, the designation of a particular cell as a stem or progenitor cell is a vast simplification. Grompe and colleagues have argued that hepatocytes exhibit properties of stem or progenitor cells under highly selective circumstances, e.g., in the Fah-null mouse [28,70]. Other studies also suggest that epithelial-to-mesenchymal transition occurs in the liver under specific pathologic or pathophysiology circumstances, as well as mesenchymal-to-epithelial transition, although this is highly controversial [71]. As new liver progenitor cell marker genes are established, these and other related issues can be addressed. In other tissues that are rapidly turning over, progenitor cells have also been termed "transit amplifying cells" [72]. "Oval cells" that have been activated to proliferate in various rodent model systems exhibit many features of "transit amplifying cells" and thus may represent the liver counterpart to progenitor cells identified in other organs, such as the skin and intestinal epithelium. However, it should be noted that tissue repopulation by "transit amplifying cells" is generally a rapid, but only a short-term process [73].

20.5 TRANSPLANTATION OF HEPATIC EPITHELIAL PROGENITOR CELLS

Since hepatocytes do not effectively repopulate the normal adult liver, an obvious alternative would be to transplant hepatic progenitor (oval) cells that should have a higher repopulation potential compared to adult hepatocytes. "Oval cells" isolated from rats treated with D-galactosamine proliferate and differentiate into hepatocytes after transplantation into normal rats subjects to two-thirds PH to initiate the repopulation process [74]. However, in this study, which was conducted in a nonselective tissue environment, liver repopulation by transplanted D-galactosamine-induced "oval cells" was very low. Individual clusters of transplanted cells with the morphologic appearance of mature hepatocytes were fully integrated into the hepatic parenchyma and contained up to ~20 cells/cluster in two-dimensional sections [74]. In a three-dimensional reconstruction, if these 20 cells spanned the diameter of a sphere, there would be ~100 total cells within the sphere. If these cells were derived from a single transplanted progenitor cell, this would represent ~7 cell divisions of the transplanted cell. With adult hepatocytes transplanted into the normal liver, the cluster size rarely exceeded 2-4 cells in two-dimensional sections [26,75], so that D-galactosamine-induced "oval cells" may have a very small proliferative advantage over adult hepatocytes, which was not sufficient to produce significant liver repopulation in a normal host liver microenvironment.

In later studies using D-galactosamine-induced "oval cells" in the highly selective retrorsine/PH model, Yovchev et al. [76] obtained 60-80% liver repopulation, similar to that observed with isolated hepatocytes. "Oval cells" isolated from the liver of DDC-fed mice also repopulate the liver of Fah null mice [77]. Similarly, "oval cells" from GFP transgenic mice maintained on a DDC diet repopulate the liver of WT mice treated with monocrotaline in conjunction with two-thirds PH [78]. "Oval cells" have been isolated from normal mouse and dog liver [79–81]. These cells exhibit properties characteristic of hepatic progenitor cells in culture, but *in vivo* repopulation data are very limited.

Numerous studies have reported the isolation and *in vitro* passage of "oval cells" and "oval cell" lines from mice and rats, as well as from humans. These cells are clonal, bipotent, and exhibit other stem and progenitor cell properties *in vitro* and *in vivo* [37,80,82–90]. However, liver repopulation by "oval cells" and "oval cell" lines has generally been very low, even under conditions in which the host liver is highly selective. "Oval cells" that have been isolated from DDC-treated mice and expanded in organoid culture give modest repopulation (~1%) after transplantation into the Fah-null mouse liver [91]. Despite these limited findings, *in vitro* expanded hepatic progenitor cells derived from adult human liver have been transplanted into a female child with ornithine transcarbamylase deficiency and at 100 days after cell transplantation, Y chromosome analysis showed 3-5% male donor cells in a liver biopsy specimen from this cell transplant recipient [92]. Reid and colleagues [93] have isolated, cultured, and expanded progenitor cells from the human biliary tree, and these cells produced human hepatocytes in the parenchyma and human biliary epithelial cells in the bile ducts of nonpreconditioned SCID mice 1 month after cell transplantation. The cells were injected directly into the hepatic parenchyma and showed ~6.5% of human hepatocytes and 12.7% of human biliary epithelial cells in the regions where the cells were injected. Cardinale et al. [94] have also transplanted human fetal biliary tree progenitor cells into the liver of two patients with advanced cirrhosis through the hepatic artery, and both patients showed biochemical and clinical improvement 6 months after cell transplantation.

20.6 LIVER REPOPULATION WITH EPITHELIAL PROGENITOR CELLS FROM THE FETAL LIVER

Another potential source of hepatic epithelial progenitor cells is the fetal liver (Figure 20.3). On embryonic day (ED) 8.0 in the mouse, when the endoderm becomes positioned next to the developing heart, the endodermal cells begin to proliferate [95–97]. Specification toward hepatic epithelial lineages occurs at ED8.5 and requires signaling by fibroblast growth factor (FGF), which is produced by the adjacent cardiac mesoderm [98], as well as bone morphogenic protein (BMP), which is produced and secreted by mesenchymal cells residing in the septum transversum [99]. These cells begin to express a series of transcription factors including HNF4α on ED 9.0-9.5, followed almost immediately by liver-specific genes, α-fetoprotein (AFP), and albumin (Alb) [96,97,100]. The cells are now hepatic-specified cells and referred to as hepatoblasts, which then proliferate extensively and invade the septum transversum mesenchyme containing stellate cells and sinusoidal endothelial cells that secrete a variety of cytokines and growth factors known to be involved in liver development, such as EGF, FGF, HGF, TGFβ, BMPs TNFα, and IL-6 [96–99,101]. The hepatoblasts expand rapidly between ED11-ED16, express genes that are specific for both the hepatocytic (AFP or albumin) and cholangiocytic (cytokeratin-19) lineages and are considered bipotential or biphenotypic. Just prior to ED16, hepatoblasts diverge along two epithelial lineages, hepatocytes and cholangiocytes [100,102]. Notch signaling promotes cholangiocytic differentiation and oncostatin M promotes hepatocytic differentiation. The liver gene expression program changes abruptly between ED16-17 [103], the percentage of bipotential cells falls dramatically, and the unipotent progeny are committed to either the hepatocytic or cholangiocytic lineage [75,103,104]. Beginning on ~ED17, intrahepatic bile ducts are formed in

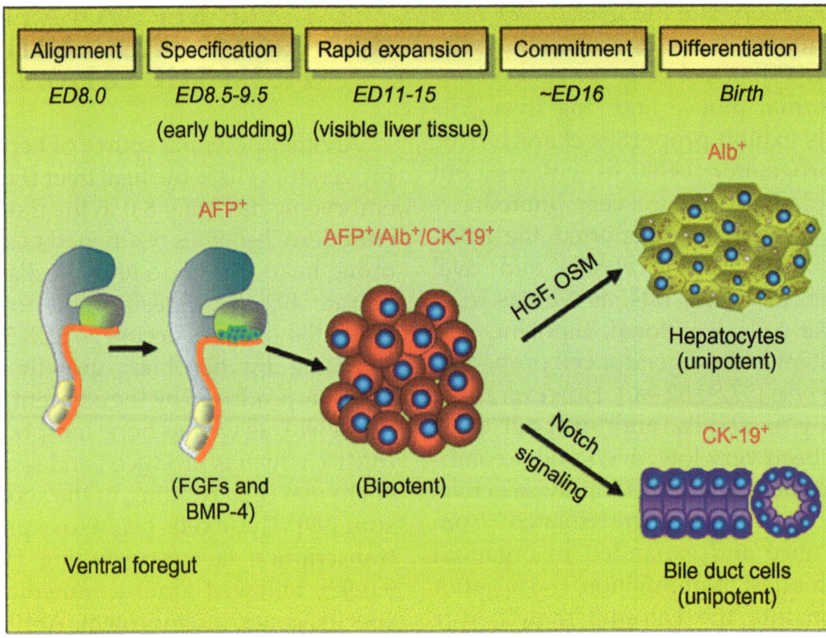

FIGURE 20.3 Schematic diagram of liver development in the mouse, indicating the cells used for repopulation of the normal adult liver.
Adapted from Ref. [26] with permission.

the vicinity of large portal vein branches and organogenesis proceeds toward an adult liver structure [105].

Rapidly proliferating hepatic epithelial progenitor cells from ED14 rats and ED12-13 mice have been isolated and transplanted into various animal models [33,75,104,106,107]. Fetal liver epithelial progenitor cells or stem/progenitor cells [106] exhibit superior repopulation properties compared to either mature hepatocytes or adult liver progenitor cells when they are transplanted into the normal adult liver under nonselective conditions [26,75,106]. However, under highly selective conditions with massive and continuous liver injury, all three cell types (fetal liver stem/progenitor cells, adult progenitor cells, and adult hepatocytes) efficiently repopulate the liver, and there is no evidence to date that progenitor cells outperform mature hepatocytes in any of these injury models [26,75,77,107]. In contrast, only epithelial progenitor cells from the fetal liver, which may be at an earlier stage of lineage progression than progenitor cells from the adult liver, effectively repopulate the liver in a normal host microenvironment [26,75,106].

Sandhu et al. [75] reported 5-10% repopulation of DPPIV$^-$ mutant F344 rat liver by transplanting WT ED14 fetal liver epithelial cells in conjunction with two-thirds PH. The transplanted cells were integrated into the host parenchyma, forming hybrid canaliculi with host hepatocytes, and the bulk of the repopulating clusters contained both hepatocytes and mature bile duct cells [75]. Liver repopulation by transplanted ED14 fetal liver cells increased slowly and progressively over 6 months. These results were obtained in a normal (nonselective) tissue environment, requiring only two-thirds PH to initiate the repopulation process.

The level of liver repopulation by ED14 fetal liver cells under nonselective conditions can also

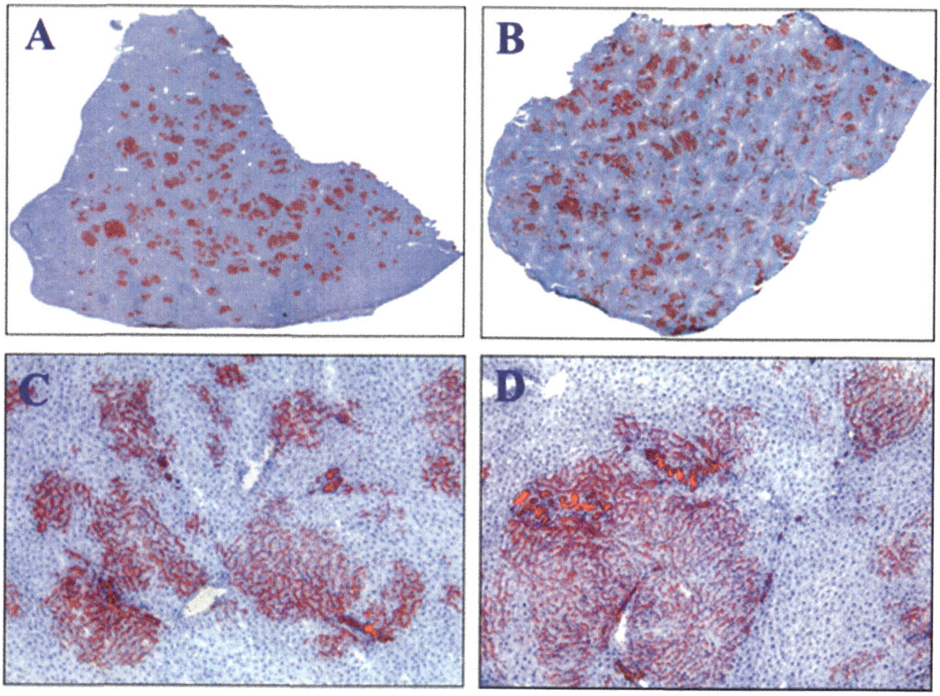

FIGURE 20.4 Repopulation of the DPPIV$^-$ mutant rat liver by DPPIV$^+$ fetal liver hepatic epithelial stem/progenitor cells. (a and b) 20-25% liver repopulation in whole liver lobes from two DPPIV$^-$ rats 6 months after transplantation of 50×10^6 unfractionated DPPIV$^+$ ED14 fetal liver cells [106]. Similar results were obtained with 1.8×10^6 95% purified fetal liver stem/progenitor cells obtained by DLK-1 antibody selection using Miltenyi immunomagnetic beads [108]. (c and d) Photomicrographs of selected fields from (a) and (b) at 100 × original magnification, showing large repopulation clusters containing both hepatocytes in normal parenchymal cord structures and mature bile ducts derived from transplanted ED14 DPPIV$^+$ fetal liver epithelial stem progenitor cells [106].
Adapted from Ref. [106] with permission.

be increased to 20-25% simply by increasing the number of ED14 fetal liver cells transplanted (Figure 20.4) [106]. Repopulation continues to increase for up to 1 year, reaching an average of ~30% for the total liver (Figure 20.5). Repopulation remains stable up to 18 months after cell transplantation (the duration of experiments conducted to date), which is consistent with the slow turnover of parenchymal cells in this organ. In this normal rat model, there is a several 1000-fold amplification of transplanted fetal liver epithelial progenitor cells in the repopulated normal host liver [106]. Both hepatic parenchymal cords and mature bile ducts are formed by transplanted fetal liver cells, and the progeny of the transplanted cells express normal levels of hepatocytic and cholangiocytic genes in the respective cell types [106]. The mechanism for liver repopulation by rat fetal liver stem/progenitor cells has been shown to be cell competition between the transplanted cells and host hepatocytes [106], a process originally described in *Drosophila* during wing development [109,110]. During cell competition, transplanted cells with higher proliferative activity cause apoptosis of more slowly growing host cells at their common boundary and in the immediately adjacent host area by a mechanism which has not yet been elucidated [106]. The more rapidly growing cells also exhibit much

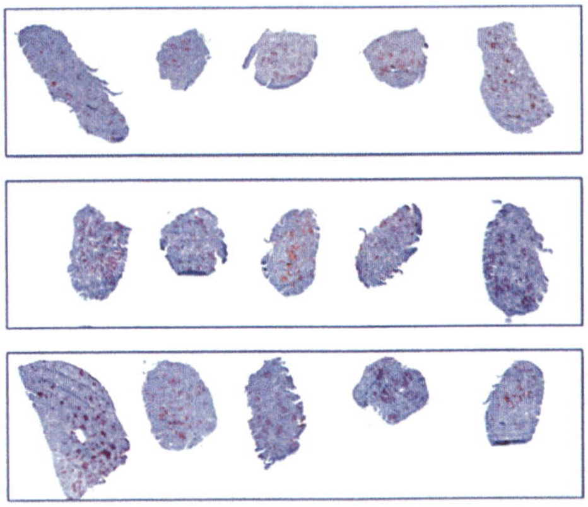

FIGURE 20.5 Long-term repopulation of DPPIV$^-$ rat liver with DPPIV$^+$ fetal liver epithelial stem/progenitor cells. Repopulation levels of 20-25% were obtained reproducibly at 6 months in different animals and 25-30% at 12 months, which remained stable at 18 months after cell transplantation (the duration of studies conducted to date).

lower levels of apoptosis than host hepatocytes in the immediately adjacent or more distant host parenchyma, as well as in age-matched, non-transplanted normal control liver.

Liver repopulation by cell competition has also been reported by WT hepatocytes transplanted into a transgenic mouse model of α_1-antitrypsin (AAT) deficiency expressing the AAT-Z mutant protein. In this model, the proliferation of AAT-Z-expressing hepatocytes is reduced, there is a low level of continuous liver injury, and the mice exhibit mild hepatic fibrosis [111]. Rat fetal liver stem/progenitor cells have been cryopreserved with full ability to repopulate the normal adult liver after thawing [112], and they have been enriched to 95% purity by selection with immunomagnetic beads [108].

20.7 LIVER REPOPULATION BY EPITHELIAL PROGENITOR CELLS FROM HUMAN FETAL LIVER

Nowak et al. [113] isolated expandable hepatic progenitor cells from human fetal liver at 6-10 weeks gestation and transplanted these cells into nude mice, pretreated 36 h earlier with D-galactosamine, which causes acute hepatic necrosis together with transient inhibition of host hepatocyte proliferation due to depletion of uridine ribonucleotides in the host liver. At 4 weeks after cell transplantation, both freshly isolated and passaged human hepatic epithelial progenitor cells gave 4-5% host liver replacement, although the photomicrographs included in the chapter would suggest that this may be an overestimate of their findings. Dan et al. [90] isolated and expanded multipotent progenitor cells from human fetal liver, and these cells showed 1-2% repopulation in retrorsine/CCl_4-pretreated Rag2$^{-/-}$/IL-2R$^{\gamma}$c$^{-/-}$ mice at 1 month after cell transplantation. Reid and colleagues [114,115] identified and extensively characterized fetal and adult human liver progenitor cells and found specific subpopulations at different stages of lineage progression. These cells have been expanded and transplanted into NOD/Scid mice and showed hepatic engraftment of transplanted cells at 2 days after transplantation with expansion after 1 week following CCl_4 administration. However, estimates of liver repopulation by these cells were not given.

Most recently, Huch et al. [116] have expanded bile duct-derived bipotent progenitor cells from human liver in organoid culture and after transplantation into the liver of retrorsine/CCl_4-treated mice, organoid cultured cells formed repopulation clusters of hepatocytes and secreted low levels of human albumin and α_1-antitrypsin into the circulation. Most importantly, however, these authors were able to establish long-term organoid cultures of bipotent progenitor cells from liver biopsies taken from patients with α_1-antrypsin deficiency and Alagille Syndrome. These cells can now be used to model these diseases *in vitro* and ultimately lead to new therapeutic approaches in regenerative medicine, cell and gene therapy. An extensive literature has been compiled on transplantation of mesenchymal progenitor or stem/progenitor cells from bone marrow and adipose tissue, differentiation of these cells into hepatocytes and repopulation of the liver. Once again, the level of liver repopulation has generally been very low, except in the $Fah^{-/-}$ mouse, where transplanted mesenchymal progenitor cells fuse with host hepatocytes [117].

20.8 LIVER REPOPULATION BY INDUCED PLURIPOTENT STEM CELLS

Because of their extensive proliferative capacity, pluripotent stem cells are an attractive potential source of transplantable hepatocytes or hepatic progenitor cells. Induced pluripotent stem (iPS) cells can not only divide extensively, but also retain the ability to differentiate into many different mature cell types [118], including cells with a hepatocyte-like phenotype. A detailed description of the cell sources, preparation, and properties of iPS cells that have been differentiated along the hepatic epithelial lineage pathway is provided in Chapter 19. The first study reporting that iPS-derived cells with a hepatocyte-like phenotype can be transplanted into the liver and exhibit hepatocytic function was conducted by Si-Tayeb et al. [119]. These investigators injected hepatocyte-differentiated human iPS directly into the hepatic parenchyma of newborn mice, and these mice expressed human albumin, determined 7 days after cell transplantation. Studies by other investigators failed to obtain positive results after splenic or portal vein infusion of hepatocyte-differentiated iPS cells and poor engraftment appears to be a factor in limiting repopulation. However, positive results were obtained by Liu et al. [120] who prepared human iPS cells differentiated along the hepatocytic lineage from differentiated cells representative of all the three gene layers and transplanted these cells into the liver of NOD/Scid/IL-$2R^{\gamma}c^{-/-}$ mice by infusion through the tail vein. Before cell transplantation, chronic liver injury was induced in recipients by treatment for 4 weeks with dimethylnitrosamine (DMN), which causes liver cirrhosis. All hepatocyte-directed iPS cells that were expanded and used for hepatic cell transplantation gave comparable levels of liver repopulation (~10-15% at 8 weeks after cell transplantation), regardless of their germ layer origin or stage in hepatic epithelial cell lineage progression. These studies used immunohistochemical detection of human albumin as the cellular marker for liver repopulation. For reasons that are unclear, the mouse serum levels of human albumin, transferrin, AAT, and fibrinogen were far lower than might be expected from the level of liver repopulation reported. However, this has also been found in the previous studies demonstrating much lower levels of human serum albumin than anticipated after liver repopulation of $Fah^{-/-}$/$Rag2^{-/-}$/IL-$2R^{\gamma}c^{-/-}$ mice by human hepatocytes [121].

Other studies have reported several different sets of *in vitro* culture conditions for long-term maintenance, clonal expansion, passage, and differentiation of human iPS or ES cell-derived hepatic epithelial progenitor cells or hepatoblast-like progenitor cells. [122–124]. In one of these studies [124], unpassaged or P10 iPSC-derived human hepatoblast-like cells were transplanted into CCl_4-pretreated $Rag2^{-/-}$/IL-$2R^{\gamma}c^{-/-}$ mice, and clusters of cells expressing human albumin were identified in mouse liver tissue, as well as low levels of human albumin in the serum with either unpassaged or P10 cells. However, details concerning the number of cells transplanted, the route of cell transplantation, and the level of liver repopulation were not provided.

In a very novel and innovative study, Zhu et al. [125] reprogrammed human fibroblasts into multipotent progenitor cells (iMPC) without first generating induced pluripotent stem cells (iPSC). iMPC can be transplanted either as endodermal progenitor cells (EPCs) or after their differentiation along the hepatocytic lineage into hepatocytes (Heps). Microarray analysis showed that iMPC-Heps exhibited an immature hepatocytic gene expression profile quite similar, but not identical, to fetal hepatoblasts. Human iMPC-Heps repopulated the $Fah^{-/-}/Rag2^{-/-}/IL\text{-}2R^{\gamma}c^{-/-}$ (FRG) mouse liver to the 2% range at 9 months after cell transplantation [125], which is substantially less repopulation than that obtained with adult primary human hepatocytes (a HEPs). Under the same conditions, iPSC-derived human hepatocyte-like cells did not repopulate the FRG mouse liver. Further studies demonstrated that in the highly selective FRG mouse liver, adult hepatocytes repopulated the liver much more rapidly than iMPC-EPC, but over time, the iMPC-EPCs differentiated into mature hepatocytes (demonstrated by laser-capture microdissection and gene expression microarray analysis). The reduced rate or level of repopulation by iMPC, however, does not diminish the importance of these findings, because iMPC-EPCs may be able to repopulate the host liver under less selective or nonselective conditions under which adult hepatocytes are not effective or only temporarily effective (as found in clinical cell therapy trials in CN1 [15,16] or familial hypercholesterolemia [17]). Since human iMPC-EPCs have been passaged more than 25 times, generating more than 10^{16} cells from a single colony [125], these cells can be very effectively expanded and may have the potential for therapeutic application. However, it will be necessary to find methods to improve engraftment and/or other steps that will lead to creation of a viable niche for the transplanted cells to proliferate and identify host factors that control liver repopulation by transplanted iMPC-EPC.

20.9 FUTURE PERSPECTIVE

The goal of current research on the properties of hepatic epithelial progenitor cells is to find effective means to utilize these cells to repopulate the liver and restore function. In cell culture, hepatic epithelial progenitor cells exhibit higher proliferative activity than hepatocytes isolated from the normal adult liver and such cells should be more effective than hepatocytes for therapeutic liver repopulation. However, repopulation of the liver with hepatic epithelial progenitor cells isolated from the adult liver is much lower in the normal host liver than that achieved with hepatic epithelial progenitor cells obtained from the fetal liver before the cells diverge along the hepatocytic or biliary lineage. Under conditions of substantial host liver injury, hepatic epithelial progenitor cells from the adult liver do exhibit modest levels of repopulation, but this is still much lower than that achieved with fetal liver hepatic epithelial cells under comparable conditions.

From many recent studies, it is clear that mammalian cells, including those in the liver, exhibit a high degree of phenotypic plasticity extending across germ layers, and that previous distinctions between stem cells, progenitor cells, and mature, differentiated cells (i.e., hepatocytes) are somewhat arbitrary. Based on the experimental conditions under which cells are tested to identify their phenotype, proliferative and specific differentiated functions, either *in vitro* or *in vivo*, hepatic epithelial cells, including progenitor cells, exhibit different patterns of gene expression, resulting in different cellular behavior. From studies conducted recently, it is clearly established that the functional properties of both hepatocytes and hepatic epithelial progenitor cells depend upon the context in which the cells are maintained and external signals they receive. Therefore, the ability of such cells to repopulate the liver will depend not only on their innate properties but also on the tissue microenvironment in which the cells are located. For effective tissue repopulation to be achieved, it will be important to identify the signals, pathways, and mechanisms that regulate the specific properties of the cells and to determine how the behavior of the cells can be modified by external signals, and much work is currently ongoing in this area.

Concerning the tissue microenvironment into which the cells are transplanted, this chapter focused on whether the host environment is either selective or nonselective for repopulation by a particular cell, either an hepatic progenitor or a differentiated hepatocyte, at what stage the cell is in lineage progression at the time of transplantation, how well the cell will engraft and begin to proliferate, what signals are needed to initiate or augment this process, and how these factors impact on the level of liver repopulation achieved. The role of underlying or ongoing liver injury, as well as whether there is accompanying hepatic fibrosis, may also impact heavily on the specific cell type (progenitor or differentiated hepatocyte) that will be most effective in a given clinical situation. For example, some monogenic inherited metabolic disorders of the liver with major health consequences, such as CN1, ornithine transcarbamylase deficiency, familial hypercholesterolemia, factor IX deficiency, phenylketonuria, primary hyperoxaluria, and others, have no underlying or ongoing liver injury and effective cell therapy for these disorders might be best achieved with hepatic progenitor cells that have higher proliferative activity than host hepatocytes. Alternatively, it might be possible to achieve effective liver repopulation by transplanting mature hepatocytes that have been genetically modified to increase their proliferative potential. In the latter case, it will be necessary to regulate the proliferative potential of the transplanted cells, so that their proliferative activity can be reduced to normal levels once effective liver repopulation is achieved. In the case of monogenic liver diseases with ongoing liver injury, such as in Wilson's disease, α1-antitrypsin deficiency, and hereditary hemochromatosis progressive familial intrahepatic cholestasis-types 1, 2, and 3, transplantation of mature normal hepatocytes which have a selective advantage over genetically deficient host hepatocytes may be sufficient for effective cell therapy. This has been demonstrated in an animal model for α1-antitrypsin deficiency, the AAT-Z transgenic mouse [111]. Studies in the thioacetamide model of hepatic fibrosis in rats have also demonstrated greater repopulation of the liver by fetal liver stem/progenitor cells compared to adult hepatocytes [126]. Therefore, in the future, it may be possible to combine hepatic epithelial progenitor cells therapy with the use of antifibrotic agents to treat hepatic fibrosis. Finding an effective marriage between the cell to be transplanted and the diseased host liver will represent a major challenge for clinicians. The concept of cell therapy to treat liver diseases may be wishful thinking or total fantasy, but this may become a reality sooner than we think.

Acknowledgments

Research cited in this chapter from the author's laboratory was supported by NIH R01 DK17609 (D.A.S, PI) and NIH P30 DK41296 (Allan W. Wolkoff, PI) and is currently supported by NIH R01 DK100490 (D.A.S, PI).

Conflict of Interest

David A. Shafritz declares that he has no conflict of interest.

References

[1] Bucher NL, Swaffield MN. The rate of incorporation of labeled thymidine into the deoxyribonucleic acid of regenerating rat liver in relation to the amount of liver excised. Cancer Res 1964;24:1611–25.
[2] Michalopoulos GK, DeFrances MC. Liver regeneration. Science 1997;276(5309):60–6.
[3] Fausto N, Campbell JS, Riehle KJ. Liver regeneration. Hepatology 2006;43(2 Suppl. 1):S45–53.
[4] Grisham JW. A morphologic study of deoxyribonucleic acid synthesis and cell proliferation in regenerating rat liver; autoradiography with thymidine-H3. Cancer Res 1962;22:842–9.
[5] Stocker E, Heine WD. Regeneration of liver parenchyma under normal and pathological conditions. Beitr Pathol 1971;144(4):400–8.
[6] Grisham JW, Thorgeirsson SS. Liver stem cells. In: Potten CS, editor. Stem cells. London: Academic Press; 1997. p. 233–82.
[7] Harrison DE, Astle CM, Lerner C. Number and continuous proliferative pattern of transplanted primitive

immunohematopoietic stem cells. Proc Natl Acad Sci USA 1988;85(3):822–6.
[8] Matas AJ, Sutherland DE, Steffes MW, Mauer SM, Sowe A, Simmons RL, et al. Hepatocellular transplantation for metabolic deficiencies: decrease of plasma bilirubin in Gunn rats. Science 1976;192(4242):892–4.
[9] Jirtle RL, Biles C, Michalopoulos G. Morphologic and histochemical analysis of hepatocytes transplanted into syngeneic hosts. Am J Pathol 1980;101(1):115–26.
[10] Gupta S, Chowdhury NR, Jagtiani R, Gustin K, Aragona E, Shafritz DA, et al. A novel system for transplantation of isolated hepatocytes utilizing HBsAg-producing transgenic donor cells. Transplantation 1990;50(3):472–5.
[11] Ponder KP, Gupta S, Leland F, Darlington G, Finegold M, DeMayo J, et al. Mouse hepatocytes migrate to liver parenchyma and function indefinitely after intrasplenic transplantation. Proc Natl Acad Sci USA 1991;88(4):1217–21.
[12] Wiederkehr JC, Kondos GT, Pollak R. Hepatocyte transplantation for the low-density lipoprotein receptor-deficient state. A study in the Watanabe rabbit. Transplantation 1990;50(3):466–71.
[13] Fisher RA, Strom SC. Human hepatocyte transplantation: worldwide results. Transplantation 2006;82(4):441–9.
[14] Fitzpatrick E, Mitry RR, Dhawan A. Human hepatocyte transplantation: state of the art. J Intern Med 2009;266(4):339–57.
[15] Fox IJ, Chowdhury JR, Kaufman SS, Goertzen TC, Chowdhury NR, Warkentin PI, et al. Treatment of the Crigler-Najjar syndrome type I with hepatocyte transplantation. N Engl J Med 1998;338(20):1422–6.
[16] Ambrosino G, Varotto S, Strom SC, Guariso G, Franchin E, Miotto D, et al. Isolated hepatocyte transplantation for Crigler-Najjar syndrome type 1. Cell Transplant 2005;14(2-3):151–7.
[17] Grossman M, Rader DJ, Muller DW, Kolansky DM, Kozarsky K, Clark 3rd BJ, et al. A pilot study of ex vivo gene therapy for homozygous familial hypercholesterolaemia. Nat Med 1995;1(11):1148–54.
[18] Fox IJ, Daley GQ, Goldman SA, Huard J, Kamp TJ, Trucco M. Stem cell therapy. Use of differentiated pluripotent stem cells as replacement therapy for treating disease. Science 2014;345(6199):1247391.
[19] Bucher NL, Swaffield MN, Ditroia JF. The influence of age upon the incorporation of thymidine-2-C^{14} into the DNA of regenerating rat liver. Cancer Res 1964;24:509–12.
[20] Rajvanshi P, Kerr A, Bhargava KK, Burk RD, Gupta S. Studies of liver repopulation using the dipeptidyl peptidase IV-deficient rat and other rodent recipients: cell size and structure relationships regulate capacity for increased transplanted hepatocyte mass in the liver lobule. Hepatology 1996;23(3):482–96.
[21] Rajvanshi P, Kerr A, Bhargava KK, Burk RD, Gupta S. Efficacy and safety of repeated hepatocyte transplantation for significant liver repopulation in rodents. Gastroenterology 1996;111(4):1092–102.
[22] Michalopoulos GK. Liver regeneration. J Cell Physiol 2007;213(2):286–300.
[23] Sandgren EP, Palmiter RD, Heckel JL, Daugherty CC, Brinster RL, Degen JL. Complete hepatic regeneration after somatic deletion of an albumin-plasminogen activator transgene. Cell 1991;66(2):245–56.
[24] Rhim JA, Sandgren EP, Degen JL, Palmiter RD, Brinster RL. Replacement of diseased mouse liver by hepatic cell transplantation. Science 1994;263(5150):1149–52.
[25] Overturf K, Al-Dhalimy M, Tanguay R, Brantly M, Ou CN, Finegold M, et al. Hepatocytes corrected by gene therapy are selected in vivo in a murine model of hereditary tyrosinaemia type I. Nat Genet 1996;12(3):266–73.
[26] Shafritz DA, Oertel M. Model systems and experimental conditions that lead to effective repopulation of the liver by transplanted cells. Int J Biochem Cell Biol 2011;43(2):198–213.
[27] Lindstedt S, Holme E, Lock EA, Hjalmarson O, Strandvik B. Treatment of hereditary tyrosinaemia type I by inhibition of 4-hydroxyphenylpyruvate dioxygenase. Lancet 1992;340(8823):813–7.
[28] Overturf K, al-Dhalimy M, Ou CN, Finegold M, Grompe M. Serial transplantation reveals the stem-cell-like regenerative potential of adult mouse hepatocytes. Am J Pathol 1997;151(5):1273–80.
[29] Jago MV. The development of the hepatic megalocytosis of chronic pyrrolizidine alkaloid poisoning. Am J Pathol 1969;56(3):405–21.
[30] Laconi E, Sarma DS, Pani P. Transplantation of normal hepatocytes modulates the development of chronic liver lesions induced by a pyrrolizidine alkaloid, lasiocarpine. Carcinogenesis 1995;16(1):139–42.
[31] Laconi E, Oren R, Mukhopadhyay DK, Hurston E, Laconi S, Pani P, et al. Long-term, near-total liver replacement by transplantation of isolated hepatocytes in rats treated with retrorsine. Am J Pathol 1998;153(1):319–29.
[32] Guo D, Fu T, Nelson JA, Superina RA, Soriano HE. Liver repopulation after cell transplantation in mice treated with retrorsine and carbon tetrachloride. Transplantation 2002;73(11):1818–24.
[33] Nierhoff D, Ogawa A, Oertel M, Chen YQ, Shafritz DA. Purification and characterization of mouse fetal liver epithelial cells with high in vivo repopulation capacity. Hepatology 2005;42(1):130–9.

[34] Witek RP, Fisher SH, Petersen BE. Monocrotaline, an alternative to retrorsine-based hepatocyte transplantation in rodents. Cell Transplant 2005;14(1):41–7.
[35] Oren R, Dabeva MD, Petkov PM, Hurston E, Laconi E, Shafritz DA. Restoration of serum albumin levels in nagase analbuminemic rats by hepatocyte transplantation. Hepatology 1999;29(1):75–81.
[36] Guha C, Sharma A, Gupta S, Alfieri A, Gorla GR, Gagandeep S, et al. Amelioration of radiation-induced liver damage in partially hepatectomized rats by hepatocyte transplantation. Cancer Res 1999;59(23):5871–4.
[37] Malhi H, Gorla GR, Irani AN, Annamaneni P, Gupta S. Cell transplantation after oxidative hepatic preconditioning with radiation and ischemia-reperfusion leads to extensive liver repopulation. Proc Natl Acad Sci USA 2002;99(20):13114–9.
[38] Oren R, Dabeva MD, Karnezis AN, Petkov PM, Rosencrantz R, Sandhu JP, et al. Role of thyroid hormone in stimulating liver repopulation in the rat by transplanted hepatocytes. Hepatology 1999;30 (4):903–13.
[39] Zhang W, Asp P, Vaitheesvaran B, Liu L, Kabarriti R, Yaffe H, et al. A thyroid hormone receptor beta-selective T3 analog promotes liver repopulation in apolipoprotein E-deficient mice, correcting hypercholesterolemia. Hepatology 2014;60:241A[#86].
[40] Landis CS, Yamanouchi K, Zhou H, Mohan S, Roy-Chowdhury N, Shafritz DA, et al. Noninvasive evaluation of liver repopulation by transplanted hepatocytes using 31P MRS imaging in mice. Hepatology 2006;44(5):1250–8.
[41] Fero ML, Rivkin M, Tasch M, Porter P, Carow CE, Firpo E, et al. A syndrome of multiorgan hyperplasia with features of gigantism, tumorigenesis, and female sterility in p27(Kip1)-deficient mice. Cell 1996;85(5):733–44.
[42] Nakayama K, Ishida N, Shirane M, Inomata A, Inoue T, Shishido N, et al. Mice lacking p27(Kip1) display increased body size, multiple organ hyperplasia, retinal dysplasia, and pituitary tumors. Cell 1996;85 (5):707–20.
[43] Hayashi E, Yasui A, Oda K, Nagino M, Nimura Y, Nakanishi M, et al. Loss of p27(Kip1) accelerates DNA replication after partial hepatectomy in mice. J Surg Res 2003;111(2):196–202.
[44] Yuan RH, Ogawa A, Ogawa E, Neufeld D, Zhu L, Shafritz DA. p27Kip1 inactivation provides a proliferative advantage to transplanted hepatocytes in DPPIV/Rag2 double knockout mice after repeated host liver injury. Cell Transplant 2003;12(8):907–19.
[45] Karnezis AN, Dorokhov M, Grompe M, Zhu L. Loss of p27(Kip1) enhances the transplantation efficiency of hepatocytes transferred into diseased livers. J Clin Invest 2001;108(3):383–90.
[46] Brezillon N, Lambert-Blot M, Morosan S, Couton D, Mitchell C, Kremsdorf D, et al. Transplanted hepatocytes over-expressing FoxM1B efficiently repopulate chronically injured mouse liver independent of donor age. Mol Ther 2007;15(9):1710–5.
[47] Mignon A, Guidotti JE, Mitchell C, Fabre M, Wernet A, De La Coste A, et al. Selective repopulation of normal mouse liver by Fas/CD95-resistant hepatocytes. Nat Med 1998;4(10):1185–8.
[48] Farber E. Similarities of the sequence of the early histological changes induced in the liver of the rat by ethionine, 2-acetylaminofluorene, and 3′-methyl-4-dimethylaminoazobenzene. Cancer Res 1956;16:142–8.
[49] Lemire JM, Shiojiri N, Fausto N. Oval cell proliferation and the origin of small hepatocytes in liver injury induced by D-galactosamine. Am J Pathol 1991;139(3):535–52.
[50] Dabeva MD, Shafritz DA. Activation, proliferation, and differentiation of progenitor cells into hepatocytes in the D-galactosamine model of liver regeneration. Am J Pathol 1993;143(6):1606–20.
[51] Sells MA, Katyal SL, Shinozuka H, Estes LW, Sell S, Lombardi B. Isolation of oval cells and transitional cells from the livers of rats fed the carcinogen DL-ethionine. J Natl Cancer Inst 1981;66(2):355–62.
[52] Akhurst B, Croager EJ, Farley-Roche CA, Ong JK, Dumble ML, Knight B, et al. A modified choline-deficient, ethionine-supplemented diet protocol effectively induces oval cells in mouse liver. Hepatology 2001;34(3):519–22.
[53] Yin L, Lynch D, Sell S. Participation of different cell types in the restitutive response of the rat liver to periportal injury induced by allyl alcohol. J Hepatol 1999;31(3):497–507.
[54] Factor VM, Radaeva SA, Thorgeirsson SS. Origin and fate of oval cells in dipin-induced hepatocarcinogenesis in the mouse. Am J Pathol 1994;145(2):409–22.
[55] Preisegger KH, Factor VM, Fuchsbichler A, Stumptner C, Denk H, Thorgeirsson SS. Atypical ductular proliferation and its inhibition by transforming growth factor beta1 in the 3,5-diethoxycarbonyl-1,4-dihydrocollidine mouse model for chronic alcoholic liver disease. Lab Investig 1999;79(2):103–9.
[56] Tatematsu M, Ho RH, Kaku T, Ekem JK, Farber E. Studies on the proliferation and fate of oval cells in the liver of rats treated with 2-acetylaminofluorene and partial hepatectomy. Am J Pathol 1984;114(3):418–30.
[57] Evarts RP, Nagy P, Marsden E, Thorgeirsson SS. A precursor-product relationship exists between oval cells and hepatocytes in rat liver. Carcinogenesis 1987;8(11):1737–40.
[58] Evarts RP, Nagy P, Nakatsukasa H, Marsden E, Thorgeirsson SS. In vivo differentiation of rat liver oval cells into hepatocytes. Cancer Res 1989;49(6):1541–7.

[59] Paku S, Schnur J, Nagy P, Thorgeirsson SS. Origin and structural evolution of the early proliferating oval cells in rat liver. Am J Pathol 2001;158(4):1313–23.
[60] Kuwahara R, Kofman AV, Landis CS, Swenson ES, Barendswaard E, Theise ND. The hepatic stem cell niche: identification by label-retaining cell assay. Hepatology 2008;47(6):1994–2002.
[61] Sackett SD, Li Z, Hurtt R, Gao Y, Wells RG, Brondell K, et al. Foxl1 is a marker of bipotential hepatic progenitor cells in mice. Hepatology 2009;49(3):920–9.
[62] Tarlow BD, Pelz C, Naugler WE, Wakefield L, Wilson EM, Finegold MJ, et al. Bipotential adult liver progenitors are derived from chronically injured mature hepatocytes. Cell Stem Cell 2014;15 (5):605–18.
[63] Michalopoulos GK, Barua L, Bowen WC. Transdifferentiation of rat hepatocytes into biliary cells after bile duct ligation and toxic biliary injury. Hepatology 2005;41(3):535–44.
[64] Yanger K, Zong Y, Maggs LR, Shapira SN, Maddipati R, Aiello NM, et al. Robust cellular reprogramming occurs spontaneously during liver regeneration. Genes Dev 2013;27(7):719–24.
[65] Sekiya S, Suzuki A. Hepatocytes, rather than cholangiocytes, can be the major source of primitive ductules in the chronically injured mouse liver. Am J Pathol 2014;184(5):1468–78.
[66] Schaub JR, Malato Y, Gormond C, Willenbring H. Evidence against a stem cell origin of new hepatocytes in a common mouse model of chronic liver injury. Cell Rep 2014;8(4):933–9.
[67] Tarlow BD, Finegold MJ, Grompe M. Clonal tracing of Sox9+ liver progenitors in mouse oval cell injury. Hepatology 2014;60(1):278–89.
[68] Greenbaum LE. The ductal plate: a source of progenitors and hepatocytes in the adult liver. Gastroenterology 2011;141(4):1152–5.
[69] Michalopoulos GK. Phenotypic fidelity (or not?) of epithelial cells in the liver. Hepatology 2012;55 (6):2024–7.
[70] Grompe M. Liver stem cells, where art thou? Cell Stem Cell 2014;15(3):257–8.
[71] Xie G, Diehl AM. Evidence for and against epithelial-to-mesenchymal transition in the liver. Am J Physiol Gastrointest Liver Physiol 2013;305(12):G881–90.
[72] Marshman E, Booth C, Potten CS. The intestinal epithelial stem cell. Bioessays 2002;24(1):91–8.
[73] Potten CS. Stem cells. London: Academic Press; 1997.
[74] Dabeva MD, Hwang SG, Vasa SR, Hurston E, Novikoff PM, Hixson DC, et al. Differentiation of pancreatic epithelial progenitor cells into hepatocytes following transplantation into rat liver. Proc Natl Acad Sci USA 1997;94(14):7356–61.
[75] Sandhu JS, Petkov PM, Dabeva MD, Shafritz DA. Stem cell properties and repopulation of the rat liver by fetal liver epithelial progenitor cells. Am J Pathol 2001;159 (4):1323–34.
[76] Yovchev MI, Grozdanov PN, Joseph B, Gupta S, Dabeva MD. Novel hepatic progenitor cell surface markers in the adult rat liver. Hepatology 2007;45 (1):139–49.
[77] Wang X, Foster M, Al-Dhalimy M, Lagasse E, Finegold M, Grompe M. The origin and liver repopulating capacity of murine oval cells. Proc Natl Acad Sci USA 2003;100(Suppl. 1):11881–8.
[78] Song S, Witek RP, Lu Y, Choi YK, Zheng D, Jorgensen M, et al. Ex vivo transduced liver progenitor cells as a platform for gene therapy in mice. Hepatology 2004;40(4):918–24.
[79] Wright N, Samuelson L, Walkup MH, Chandrasekaran P, Gerber DA. Enrichment of a bipotent hepatic progenitor cell from naive adult liver tissue. Biochem Biophys Res Commun 2008;366 (2):367–72.
[80] Suzuki A, Sekiya S, Onishi M, Oshima N, Kiyonari H, Nakauchi H, et al. Flow cytometric isolation and clonal identification of self-renewing bipotent hepatic progenitor cells in adult mouse liver. Hepatology 2008;48(6):1964–78.
[81] Arends B, Vankelecom H, Vander Borght S, Roskams T, Penning LC, Rothuizen J, et al. The dog liver contains a "side population" of cells with hepatic progenitor-like characteristics. Stem Cells Dev 2009;18 (2):343–50.
[82] Suzuki A, Zheng Y, Kondo R, Kusakabe M, Takada Y, Fukao K, et al. Flow-cytometric separation and enrichment of hepatic progenitor cells in the developing mouse liver. Hepatology 2000;32(6):1230–9.
[83] Kubota H, Reid LM. Clonogenic hepatoblasts, common precursors for hepatocytic and biliary lineages, are lacking classical major histocompatibility complex class I antigen. Proc Natl Acad Sci USA 2000;97 (22):12132–7.
[84] Coleman WB, McCullough KD, Esch GL, Faris RA, Hixson DC, Smith GJ, et al. Evaluation of the differentiation potential of WB-F344 rat liver epithelial stem-like cells in vivo. Differentiation to hepatocytes after transplantation into dipeptidylpeptidase-IV-deficient rat liver. Am J Pathol 1997;151(2):353–9.
[85] Yasui O, Miura N, Terada K, Kawarada Y, Koyama K, Sugiyama T. Isolation of oval cells from Long-Evans Cinnamon rats and their transformation into hepatocytes in vivo in the rat liver. Hepatology 1997;25 (2):329–34.
[86] Suzuki A, Zheng YW, Kaneko S, Onodera M, Fukao K, Nakauchi H, et al. Clonal identification and

characterization of self-renewing pluripotent stem cells in the developing liver. J Cell Biol 2002;156 (1):173–84.

[87] Strick-Marchand H, Morosan S, Charneau P, Kremsdorf D, Weiss MC. Bipotential mouse embryonic liver stem cell lines contribute to liver regeneration and differentiate as bile ducts and hepatocytes. Proc Natl Acad Sci USA 2004;101(22):8360–5.

[88] Fougere-Deschatrette C, Imaizumi-Scherrer T, Strick-Marchand H, Morosan S, Charneau P, Kremsdorf D, et al. Plasticity of hepatic cell differentiation: bipotential adult mouse liver clonal cell lines competent to differentiate in vitro and in vivo. Stem Cells 2006;24 (9):2098–109.

[89] Mahieu-Caputo D, Allain JE, Branger J, Coulomb A, Delgado JP, Andreoletti M, et al. Repopulation of athymic mouse liver by cryopreserved early human fetal hepatoblasts. Hum Gene Ther 2004;15(12):1219–28.

[90] Dan YY, Riehle KJ, Lazaro C, Teoh N, Haque J, Campbell JS, et al. Isolation of multipotent progenitor cells from human fetal liver capable of differentiating into liver and mesenchymal lineages. Proc Natl Acad Sci USA 2006;103(26):9912–7.

[91] Huch M, Dorrell C, Boj SF, van Es JH, Li VS, van de Wetering M, et al. In vitro expansion of single Lgr5+ liver stem cells induced by Wnt-driven regeneration. Nature 2013;494(7436):247–50.

[92] Sokal EM, Stephenne X, Ottolenghi C, Jazouli N, Clapuyt P, Lacaille F, et al. Liver engraftment and repopulation by in vitro expanded adult derived human liver stem cells in a child with ornithine carbamoyltransferase deficiency. JIMD Rep 2014;13:65–72.

[93] Cardinale V, Wang Y, Carpino G, Cui CB, Gatto M, Rossi M, et al. Multipotent stem/progenitor cells in human biliary tree give rise to hepatocytes, cholangiocytes, and pancreatic islets. Hepatology 2011;54 (6):2159–72.

[94] Cardinale V, Carpino G, Gentile R, Napoletano C, Rahimi H, Franchitto A, et al. Transplantation of human fetal biliary tree stem/progenitor cells into two patients with advanced liver cirrhosis. BMC Gastroenterol 2014;14(1):204.

[95] DuBois A. The embryonic liver. New York: Academic Press; 1963.

[96] Zhao R, Duncan SA. Embryonic development of the liver. Hepatology 2005;41(5):956–67.

[97] Zaret KS, Grompe M. Generation and regeneration of cells of the liver and pancreas. Science 2008;322 (5907):1490–4.

[98] Jung J, Zheng M, Goldfarb M, Zaret KS. Initiation of mammalian liver development from endoderm by fibroblast growth factors. Science 1999;284(5422):1998–2003.

[99] Rossi JM, Dunn NR, Hogan BL, Zaret KS. Distinct mesodermal signals, including BMPs from the septum transversum mesenchyme, are required in combination for hepatogenesis from the endoderm. Genes Dev 2001;15(15):1998–2009.

[100] Shiojiri N, Lemire JM, Fausto N. Cell lineages and oval cell progenitors in rat liver development. Cancer Res 1991;51(10):2611–20.

[101] Fausto N. Liver regeneration. J Hepatol 2000;32 (1 Suppl.):19–31.

[102] Marceau N, Blouin M-J, Noel M, Torok N, Loranger A. The role of cell types in hepatocarcinogenesis. Boca Raton, FL: CRC Press; 1992.

[103] Petkov PM, Zavadil J, Goetz D, Chu T, Carver R, Rogler CE, et al. Gene expression pattern in hepatic stem/progenitor cells during rat fetal development using complementary DNA microarrays. Hepatology 2004;39(3):617–27.

[104] Dabeva MD, Petkov PM, Sandhu J, Oren R, Laconi E, Hurston E, et al. Proliferation and differentiation of fetal liver epithelial progenitor cells after transplantation into adult rat liver. Am J Pathol 2000;156 (6):2017–31.

[105] Van Eyken P, Sciot R, Desmet V. Intrahepatic bile duct development in the rat: a cytokeratin-immunohistochemical study. Lab Investig 1988;59(1):52–9.

[106] Oertel M, Menthena A, Dabeva MD, Shafritz DA. Cell competition leads to a high level of normal liver reconstitution by transplanted fetal liver stem/progenitor cells. Gastroenterology 2006;130(2):507–20 [quiz 90].

[107] Haridass D, Yuan Q, Becker PD, Cantz T, Iken M, Rothe M, et al. Repopulation efficiencies of adult hepatocytes, fetal liver progenitor cells, and embryonic stem cell-derived hepatic cells in albumin-promoter-enhancer urokinase-type plasminogen activator mice. Am J Pathol 2009;175(4):1483–92.

[108] Oertel M, Menthena A, Chen YQ, Teisner B, Jensen CH, Shafritz DA. Purification of fetal liver stem/progenitor cells containing all the repopulation potential for normal adult rat liver. Gastroenterology 2008;134(3):823–32.

[109] Moreno E, Basler K. dMyc transforms cells into super-competitors. Cell 2004;117(1):117–29.

[110] de la Cova C, Abril M, Bellosta P, Gallant P, Johnston LA. Drosophila myc regulates organ size by inducing cell competition. Cell 2004;117(1):107–16.

[111] Ding J, Yannam GR, Roy-Chowdhury N, Hidvegi T, Basma H, Rennard SI, et al. Spontaneous hepatic repopulation in transgenic mice expressing mutant human alpha1-antitrypsin by wild-type donor hepatocytes. J Clin Invest 2011;121(5):1930–4.

[112] Oertel M, Menthena A, Chen YQ, Shafritz DA. Properties of cryopreserved fetal liver stem/progenitor cells

that exhibit long-term repopulation of the normal rat liver. Stem Cells 2006;24(10):2244–51.

[113] Nowak G, Ericzon BG, Nava S, Jaksch M, Westgren M, Sumitran-Holgersson S. Identification of expandable human hepatic progenitors which differentiate into mature hepatic cells in vivo. Gut 2005;54(7):972–9.

[114] Schmelzer E, Wauthier E, Reid LM. The phenotypes of pluripotent human hepatic progenitors. Stem Cells 2006;24(8):1852–8.

[115] Schmelzer E, Zhang L, Bruce A, Wauthier E, Ludlow J, Yao HL, et al. Human hepatic stem cells from fetal and postnatal donors. J Exp Med 2007;204(8):1973–87.

[116] Huch M, Gehart H, van Boxtel R, Hamer K, Blokzijl F, Verstegen MM, et al. Long-term culture of genome-stable bipotent stem cells from adult human liver. Cell 2015;160(1-2):299–312.

[117] Wang X, Willenbring H, Akkari Y, Torimaru Y, Foster M, Al-Dhalimy M, et al. Cell fusion is the principal source of bone-marrow-derived hepatocytes. Nature 2003;422(6934):897–901.

[118] Slack JM. Origin of stem cells in organogenesis. Science 2008;322(5907):1498–501.

[119] Si-Tayeb K, Noto FK, Nagaoka M, Li J, Battle MA, Duris C, et al. Highly efficient generation of human hepatocyte-like cells from induced pluripotent stem cells. Hepatology 2010;51(1):297–305.

[120] Liu H, Kim Y, Sharkis S, Marchionni L, Jang YY. In vivo liver regeneration potential of human induced pluripotent stem cells from diverse origins. Sci Transl Med 2011;3(82):82ra39.

[121] Azuma H, Paulk N, Ranade A, Dorrell C, Al-Dhalimy M, Ellis E, et al. Robust expansion of human hepatocytes in $Fah^{-/-}/Rag2^{-/-}/Il2rg^{-/-}$ mice. Nat Biotechnol 2007;25(8):903–10.

[122] Yanagida A, Ito K, Chikada H, Nakauchi H, Kamiya A. An in vitro expansion system for generation of human iPS cell-derived hepatic progenitor-like cells exhibiting a bipotent differentiation potential. PLoS One 2013;8(7):e67541.

[123] Zhao D, Chen S, Cai J, Guo Y, Song Z, Che J, et al. Derivation and characterization of hepatic progenitor cells from human embryonic stem cells. PLoS One 2009;4(7):e6468.

[124] Takayama K, Nagamoto Y, Mimura N, Tashiro K, Sakurai F, Tachibana M, et al. Long-term self-renewal of human ES/iPS-derived hepatoblast-like cells on human laminin 111-coated dishes. Stem Cell Rep 2013;1(4):322–35.

[125] Zhu S, Rezvani M, Harbell J, Mattis AN, Wolfe AR, Benet LZ, et al. Mouse liver repopulation with hepatocytes generated from human fibroblasts. Nature 2014;508(7494):93–7.

[126] Yovchev MI, Xue Y, Shafritz DA, Locker J, Oertel M. Repopulation of the fibrotic/cirrhotic rat liver by transplanted hepatic stem/progenitor cells and mature hepatocytes. Hepatology 2014;59:284–95.

I'M GOING ON VACATION!
BUT DON'T WORRY—I ASKED A FRIEND TO FILL-IN FOR ME WHILE I'M GONE...
© Mark Weinstein

Index

N

O

P

Edwards Brothers Malloy
Ann Arbor MI. USA
January 4, 2016